China-Africa Science, Technology and Innovation Collaboration

"If you want a deep understanding of the dynamics of Africa-China relations, this is the book for you! China and Africa have a relationship that dates to the ancient history. From China's support to Africa's liberation movements to current policies aimed at liberating the continent from poverty and underdevelopment; this relationship has deepened over the years. However, there is a dearth of rigorous scholarship on recent developments and the changing dynamics of China-Africa relations especially in the domain of Science, Technology and Innovation Collaboration. This book brings together some of the best researchers from China and Africa dealing with the topics of mutual interest and their sustainable development. The book does not shy away from addressing the complex, yet thorny issues that shape China-Africa relations viz; higher education, agriculture and food security, climate change and environment, renewable energy, space technology applications, health and manufacturing. The multidisciplinary and pragmatic approach of the book is a feather in its cup, making it easy to read and appealing to all stakeholders.

As the final decade of action to achieve the Sustainable Development Goals (SDGs) gains traction, and the African Union's Vision 2063 (The Africa We Want) draws closer; the book provides evidence-based knowledge to inform policy, whilst the policy interventions provide an avenue for forging partnerships between stakeholders from Africa and China to address the issues of common interest.

This book on China-Africa Science, Technology and Innovation Collaboration is a must read for policy makers, researchers, practitioners, as well as other key stakeholders in international development and South-South cooperation."

—Prof. Dr Felix Kwabena Donkor, *University of Education-Winneba, Ghana*

"This book is not necessarily about persuading the pessimists and sceptics of the real interest of China in Africa. The China-Africa Science, Technology and Innovation Collaboration book is a pacesetter. The book among other issues has convincingly brings to the table, that the relationships between China and Africa goes beyond the rhetoric of neo-colonialism or 'expansionism'. Having summarily gone through this book, which I completely endorse, I will say *'res ipsa loquitor'* (the facts speak for itself). No colonialist or imperialist nor slave master ever negotiates with the colonies about the terms and conditions of colonialism or slavery and slave trade. Just some pointers: China claimed to have dealt with the enigma of absolute poverty and achieved SDG1, whereas Africa is still wallowing in abject poverty. The current Chinese African development pathways should be critically evaluated to discover the appropriate inclusive ways to eradicate poverty, inequality and put in place policy to achieve all 17 sustainable development goals for both parties of Africa and China. This book has demonstrated how such objective is realistically possible and in the context of mutual benefit!"

—Prof. Adewale Olutola, *Advocate of Law and Associate Professor of Science Policy*

"China is the largest developing country in the world, while Africa is the continent with the largest number of developing countries. China has always offered respect, appreciation, and support for the sustainable development of Africa. Technology and innovation is the core for development, China actively strengthens communication and coordination with Africa in terms of technology transfer and technological innovation. For decades, China and African countries have set up high-level joint laboratories, the China-Africa Joint Research Centre, an innovation cooperation centre and others. This wonderful book provides the state of art of China-Africa's collaborations in technology and innovation and will help to design a sound policy for the development of the indigenous capabilities of both sides."

"China-Africa Science, Technology and Innovation Collaboration provides comprehensive information and in-depth analysis on the China-African collaboration in the areas of higher education, science and technology, agriculture, food security, environmental management, telecommunication, space, digital technologies, finance, renewable energy, health care and manufacturing. The broad coverage of the topics represents the editors and authors' valuable effort to provide an updated account of the important collaboration. It is a must-read book for everyone interested in African-China relationship. "

Mammo Muchie · Angathevar Baskaran ·
Mingfeng Tang
Editors

China-Africa Science, Technology and Innovation Collaboration

 Springer

Editors
Mammo Muchie
DSI/NRF SARChI Research Chair on
Science, Technology and Innovation
Studies
Tshwane University of Technology
Pretoria, South Africa

Angathevar Baskaran
Department of Political Science, Public
Administration and Development Studies,
Faculty of Business and Economics
UM North-South Research Centre
(UMNSRC), University of Malaya
Kuala Lumpur, Malaysia

Mingfeng Tang
Sino-French Innovation Research Centre
(SFIRC), Faculty of Business
Administration
Southwestern University of Finance
and Economics
Chengdu, China

ISBN 978-981-97-4575-3   ISBN 978-981-97-4576-0 (eBook)
https://doi.org/10.1007/978-981-97-4576-0

This Springer imprint is published by the registered company Springer Nature Singapore Pte Ltd.
The registered company address is: 152 Beach Road, #21-01/04 Gateway East, Singapore 189721, Singapore

If disposing of this product, please recycle the paper.

This book is dedicated for all the children of Africa and China to serve and inspire by being the doers and makers of the shared mutually beneficial future for Africa-China enduring relations. Let all the children unite and become the exemplary model for all humanity to live with peace, happiness, love, sustainable development, and innovative knowledge to free the world from climate change mitigation, demographic explosion, poverty, inequality, and unemployment forever.

Foreword

Cooperation Between Africa and China in Science, Technology and Innovation: A Unique Strategic Partnership

The People's Republic of China and African countries enjoy long-standing and comprehensive cooperation in science, technology, and innovation. The Chinese Ministry of Science and Technology (MOST) over the years concertedly and strategically invested in supporting this cooperation, with African partner governments reciprocating with investments of their own, critically confirming the values of co-ownership and shared responsibility, which underpin the collaboration. Indeed, the Africa-China science diplomacy partnership can rightfully be regarded not only as an outstanding success story, but also as a best practice example of the cooperation required to respond to the major societal challenges of the twenty-first century.

Science, technology and innovation orientated cooperation between Africa and China comprises a rich and diverse portfolio of programmes and instruments. These include funding for joint research projects, student exchange, and researcher mobility programmes, as well institutional collaboration such as the creation of joint research centres and joint virtual laboratories. In recent years, exciting new programmes have also been launched to support innovation partnerships, including collaboration between entrepreneurs, supported by soft-landing and science park initiatives.

Africa and China's joining forces to generate and apply new knowledge is making decisive contributions in responding to global challenges such as adapting to and mitigating the impacts of climate change, ensuring food security and preparing our response to new pandemics. Africa is also benefitting significantly from sharing in China's huge experience and expertise in harnessing innovation as an instrument to eradicate poverty. Cooperation further includes an exciting and growing focus on frontier science, as evidenced by African participation in Chinese space programmes, for example related to the global navigation satellite systems, and astronomy partnerships, such as those enabled by the Square Kilometre Array (SKA) global radio telescope project.

This extensive contact and cooperation in science, technology and innovation between Africa and China also plays an invaluable role in strengthening political, economic, social, and development partnerships between the two partners. This is evidenced by the important enabling role science diplomacy is playing to support and deepen cooperation as part of the Forum for China—Africa Cooperation (FOCAC) and the Belt and Road Initiative. It is collaboration which reinforces China and Africa's shared commitment to multilateralism, global solidarity, and inclusive and sustainable international development.

The paramount strategic importance of this cooperation, thus, fully merits the critical assessment of the drivers and opportunities for, challenges associated with, as well as the impact of Africa-China partnerships, which Prof. Muchie and his esteemed group of scholars undertook, in presenting this unique publication. In delivering an invaluable knowledge resource for all those interested in Africa-China, Prof. Muchie and his colleagues deserve our gratitude and appreciation. With Africa-China cooperation in science, technology, and innovation (STI), continuing to rapidly expand and deepen, as science, technology and innovation is called upon to play an increasing important role in responding to the challenges of the twenty-first century, this publication will certainly demonstrate its worth as an authoritative reference.

Pretoria, South Africa

Daan Du Toit
Deputy Director-General
International Cooperation
and Resources at South African
Department of Science and Innovation

Preface

Mammo Muchie, Angathevar Baskaran and Mingfeng Tang (editors) with the very inspirational support of the Peoples Republic of China's embassy in South Africa worked together to successfully organise the China-Africa Science, Technology, Innovation Collaboration conference, which eventually resulted in publication of this book.

The principle of the China-Africa relations is founded to build enduring bridges where tangible, explicit, clear and measurable mutual benefit is the anchor and bedrock for full and sustainable collaboration. The principle of mutual benefit is how China and Africa relations is conceptually framed, founded, and anchored. The gains and losses in the relationship are fully transparent, open, tangible, explicit, measurable and known.

What China has been doing is assisting Africans to engage in an economic and political model relations to make Africa fully independent. The Africa-China social-economic bridge building has generated a new pathway to build the win-win mutually beneficial relationship with a shared future as a priority. China has the opportunity to lead with a win-win model relationship with Africa and all in the Global South. Together China and Africa should develop the unique role model where no one loses but all as partners gain. China has the opportunity with her bridge building African relations to demonstrate with concrete and reliable evidence to give real lessons to the former colonial powers that still keep making Africa a loser.

The China-Africa relation is currently being discussed and various positive and negative claims are spread globally. The decision to focus on the science, technology and innovation collaboration between China and Africa is motivated by the fact that the primary relation of China with Africa is driven principally in building the infrastructure that contributes to economic development in Africa. All the contributions in this book demonstrate in all spheres the China-Africa relation is driven by the principle of mutual benefit.

Africa must also learn from the great success China has achieved. There is a lesson Africa must learn how China managed to deal with and respond to external powers. What has been truly extraordinary today is that China has gone through the long history of the difficult journey and now finally has attained a development status that is recognised and appreciated even by the competing powers. Millions of Chinese have now come out of poverty. Unemployment is decreasing and inequality will decrease in the course of time by managing it with appropriate policy. What Africa should learn is how China achieved this status and managed how to deal with and respond to the global economy that has marginalised Africa to remain in this time of digital knowledge economy at the raw material, agriculture, and mineral stage of development.

The best gift China can give to Africa is to share frankly and honestly how the Chinese managed the difficult journey and achieved such a globally renowned success, especially building science, technology, and innovation capabilities. The other important lesson is to generate ideas for all to learn by discovering the actual relations to draw the model for building real African unity, liberty, and independence without external challenges. Africa and China share a legacy of external intervention and patterns of development influenced by external interests. China has demonstrated an alternative development pathway addressing uneven development with a dual circulation model for prioritizing domestic economic circulation and continuously welcoming international trade and investment. Growing economic, technological and scientific collaboration between Africa and China offers the opportunity to co-develop mechanisms to provide the skill base to move from discovery to invention, innovation, and implementation to the benefit of both.

At the third Summit of the Forum on China-Africa Cooperation in Beijing in 2018, China has pledged US$60 billion in grants and loans for infrastructure projects, medical programmes, clean-energy initiatives, and other projects in Africa. As part of the plan, China will train African scientists to improve African science in fields from agriculture and climate change to quantum physics and artificial intelligence. China will offer 50,000 scholarships for African students including scientists, to study in China, and will provide short-term training opportunities for another 50,000 people to travel to seminars and workshops. The action plan also offers scholarships for postgraduate training in China and at African institutions, such as the Sino-Africa Joint Research Centre at the Jomo Kenyatta University of Agriculture and Technology in Juja, Kenya. China has made commitment to build the Africa Centres for Disease Control and Prevention (Africa CDC), a Continent-wide Public Health Agency to safeguard Africa's health.

Other areas of collaboration include (i) Development of the China-Africa Joint Research Center, for conducting scientific research and training professionals with a focus on ecological preservation, bio-diversity protection, agriculture and food security, and water environment treatment; (ii) Setting up intermediary institutions (Technology Transfer Offices (TTO)) linking research and production, and cooperation on training, public awareness, and the system and practice of intellectual property rights examination and registration; (iii) The China-Africa Green Envoys Program to strengthen Africa's human capacity for environmental management, pollution prevention and control, and green development; (iv) Providing meteorological satellite data, products and necessary technical support such as remote-sensing application equipment, education and training to better equip African countries for disaster prevention and mitigation as well as climate change response; and (v) Strengthening climate change adaption capabilities through providing assistance in kind and capacity-building training.

On August 24, 2021, the China-Africa Partnership Plan on Digital Innovation initiative was proposed. Under the initiative China will share digital technologies with Africa to promote digital infrastructure connectivity, including fiber optic backbone networks, cross-border connectivity and new-generation mobile communication networks, and help expand Internet access in Africa's remote areas. It will also help adoption of new technologies such as cloud computing, artificial intelligence, the Internet of Things, and mobile payment. Further, the initiative proposes to strengthen e-commerce cooperation with Africa to bring more quality African products into the Chinese market. Africa can learn from the Chinese extraordinary successful journey to eradicate poverty and apply digital technology to deal with and respond effectively with the pandemic, climate change and demographic explosion with values of supporting all as one humanity in one world community.

The book we have produced enriches all stakeholders with scientific research, and clearly shows that the China-Africa relations is fully driven by the principle of mutual benefits and not self-interest to build a shared future for all without any exclusion by applying full inclusion.

Pretoria, South Africa   Mammo Muchie
Kuala Lumpur, Malaysia   Angathevar Baskaran
Chengdu, China   Mingfeng Tang

Acknowledgements

It is with great honour that we appreciate the People's Republic of China Embassy in South Africa and the South African Ministry of Science and Innovation for their full support for the China-Africa Science, Technology and Innovation Collaboration Book to be published. We appreciate the publisher Springer and Nature for their patience as the review process took more time than we expected.

We are also grateful for the support of Sino-French Innovation Research Center (SFIRC) and the Faculty of Business Administration in the Southwestern University of Finance and Economics, Chengdu, China; DSI/NRF SARChI, Tshwane Univetsity of Technology, Pretoria, South Africa; the University of Malaya North-South Research Centre (UMNSRC) and the Faculty of Business and Economics, University of Malaya, Malaysia; and the Association for South-South Cooperation in Innovation Systems Transformation (ASSIST).

Above all, we would like to thank all the authors who contributed the chapters, they deserve sincere appreciation for their hard work and patience.

Contents

About the Editors

Prof. Mammo Muchie Completed his undergraduate degree in Columbia University, New York, USA and received his postgraduate M.Phil. and D.Phil. in Science, Technology and Innovation for Development (STI&D) from the University of Sussex, UK. Professor, and DST-NRF research chair in Innovation Studies at the Faculty of Engineering and the Built Environment, Tshwane University of Technology (TUT), South Africa. Fellow of the South African Academy of Sciences, the African Academy of Sciences, Fellow of the Ethiopian Academy of Science, and the African Science Institute. Adjunct Professor at Bahir Dar and University of Gondar, Ethiopia; Senior research associate fellow at the TMD Centre of Oxford University and the Africa Centre of Excellence in Data Science at the University of Rwanda. Chairman of the advisory board of African Talent Hub of the Community Interest Company to raise funds for making Africa the talent, innovation, entrepreneurship, creativity, and knowledge hub. Special distinguished advisor to the Africa Union's Student Council and a mentor for the African Entrepreneurship award. He has initiated the African Unity for Renaissance and Knowledge Exchange. He is a founding scientific advisor to the African Solar network, founding chair of the Network of Ethiopian scholars. He is also a co-founding member of the Nano Technology Institute in TUT. He is a scientific and academic advisor to the local e-Governance research that involved ten African countries on ICT4D. He is the founder as the co-Chief Editor of the African Journal on Science, Technology, Innovation and Development (AJSTID). Since 1985, he has produced over

495 publications, including books, chapters in books, and articles in internationally accredited journals and entries in institutional publications. He has received many awards including outstanding contribution to Science, Engineering, Technology (SET) and innovation by NTSF in South Africa. He has taught at the University of Economics in Prague, part-time senior lecturer in Cambridge University, UK., Visiting Professor in Jawaharlal Nehru University in India and Tonji and Shanghai University in China, Honorary Professor Jiaxing University in China, Assistant Professor in Amsterdam University and visiting Professor in Carleton College, USA and Professor at Aalborg University in Denmark.

Prof. Angathevar Baskaran Head, University of Malaya North-South Research Centre (UMNSRC) and Professor (Innovation and Entrepreneurship), Department of Political Science, Public Administration, and Development Studies, Faculty of Business and Economics, University of Malaya. He is also a Senior Research Associate, SHARChI (Innovation Studies), Tshwane University of Technology, Pretoria, South Africa. He holds Ph.D. (Science & Technology Policy Studies, Sussex Univ., UK), M.Sc. (Financial Management, Middlesex Univ., UK), MPhil (International Studies, JNU, India), M.A. (History, Punjab Univ., India), and M.A. (Political Science, Madurai Univ., India). He was a Senior Lecturer at the Middlesex University Business School, London (from 1999 to 2014). His research interests include Economics of innovation; National Innovation Systems; Innovation and Entrepreneurship, Sustainable Development and Environmental Management with the main focus on multidisciplinary themes related to emerging and developing economies. He has produced over 140 publications, including 15 books and 40 journal articles. He has undertaken research projects for UNESCO, OECD and European Commission, and Economic Research Institute for ASEAN and East Asia (ERIA). He is the Co-Editor-in-Chief, *African Journal of Science, Technology, Innovation and Development (AJSTID)*; Member, Editorial Board in: *International Journal of Technological Learning, Innovation and Development, Asian Journal of Technology Innovation*, and *Innovation and Development*.

Prof. Mingfeng Tang Assistant Dean in charge of International Affairs Office, Founder and Director of the Sino-French Innovation Research Center (SFIRC) at the Faculty of Business Administration, the Southwestern University of Finance and Economics (SWUFE), Chengdu, China; Associated Researcher at Bureau d'Economie Théorique et Appliquée, Strasbourg, France; and Visiting Professor in the University of Leeds, UK. Her main research fields include technology business incubators, green innovation, innovation systems, entrepreneurial behaviour, etc. She has published 57 peer-reviewed academic papers in Chinese and international journals, and another 41 papers were accepted by domestic and international academic conferences. She published 4 book chapters and 2 English monographs. She presided over and participated in 38 domestic and foreign research projects financed by NSFC, CSC, OECD, EU, DAAD, and others. She also won the 2023 Excellent English Taught Course Award and the 2022 Excellent Hybrid Course Award of Sichuan Province, the 2023 Excellent Teacher Award, the 2020 Excellent Researcher Award, the 2018 Excellent Undergraduate Thesis Supervisor Award, the 2010 Outstanding Scientific Research Achievement Award, and other awards in the Southwestern University of Finance and Economics, the Best Paper Award on The Thirteenth Annual Global Information Technology Management Association GITMA World Conference 2012 and The Fifth International Technology Management Forum 2007 at Zhejiang University.

Chapter 1
Introduction: China–Africa Science, Technology, Innovation Collaboration for Their Win–Win Enduring Shared Future Relations

Mammo Muchie, Angathevar Baskaran, and Mingfeng Tang

The research title for the conference and the book selected was the China–Africa Science, Technology, Innovation Collaboration in order to discover in a specific domain the existing China–Africa relationship by undertaking evidence and data analytic research on the following thematic areas: China–Africa collaboration in higher education, China–Africa research collaboration and training, China–Africa collaboration in agriculture, food security and environmental management, China–Africa collaboration in telecommunications and space, Evolution of China–Africa collaboration in science, technology and innovation, China–Africa collaboration in digital technologies, China–Africa collaboration in finance and in renewable energy, China–Africa collaboration in health sector and China–Africa collaboration in manufacturing. China has been working with African states by undertaking several collaborative work in healthcare, poverty reduction and agriculture, trade promotion, investment, digital innovation, green development, capacity building, people-to-people exchanges, and peace and security (Muchie and Patra, 2020).

M. Muchie (✉)
DSI/NRF SARChI Research Chair on Science, Technology and Innovation Studies, Tshwane University of Technology, Pretoria, South Africa
e-mail: muchiem@tut.ac.za

A. Baskaran
Department of Political Science, Public Administration and Development Studies, Faculty of Business and Economics & UM North-South Research Centre (UMNSRC), University of Malaya, Kuala Lumpur, Malaysia
e-mail: baskaran@um.edu.my

M. Tang
Sino-French Innovation Research Centre (SFIRC), Faculty of Business Administration, Southwestern University of Finance and Economics, Chengdu, China
e-mail: tang@swufe.edu.cn

© The Author(s) 2024
M. Muchie et al. (eds.), *China-Africa Science, Technology and Innovation Collaboration*, https://doi.org/10.1007/978-981-97-4576-0_1

1

The science, technology and innovation collaboration between China and Africa can promote what China is achieving from the 17 SDG Goals before 2030 (Wang and Zhang, 2020). China has been recognized for achieving success of zero poverty. Africa can draw lessons from the Chinese sustainable development efforts to fulfil all the 17 SDG goals (Xie et al., 2021). The Chinese and African current development pathways must be critically evaluated to discover the appropriate inclusive ways to eradicate poverty, unemployment, conflicts, and inequality. The China–Africa mutually agreed partnership can generate the appropriate action and policy to implement the 17 SDGs. Clear indicators can be used to measure success and failure to find the necessary ways to achieve on time what is declared. China has not used the economic system that value and validate success with profit, market, and commercial gains. There is lessons to draw from the Chinese development experience for Africa endowed with rich resources but remaining poor with the opulence of Africa's wealth still exploited by those who used to colonise Africa.

It is acknowledged that China is likely to achieve SDGs before the deadline in 2030. There is real innovation success from China that Africa can learn from the specific Chinese economic development pathway. Africa has an economic system that is still copied from the capitalist economy from the colonial time. The China–Africa Science, Technology and innovation collaboration has to address in transforming the economic system in Africa by doing SWOT (Strengths, Weaknesses, Opportunities, and Threats) and PESTLE (Political, Economic, Social, Technological, Legal and Environmental Factors) analyses. It is a real challenge: how to transform the current economic system to include nature and human well-being to create an economic system that eliminates inequality, poverty, and unemployment to achieve both human and natural safety. A new system that synthesizes creativity and innovation, state with private, market with planning, and economics with politics is critical to make a real difference by evaluating the Chinese economic development pathway. A new economic success and failure validation criteria is necessary. There is a need to innovate economics from market to become social market, from commerce to become social commerce and from narrow and exclusive economics to become innovative and inclusive social and sustainable, smart and green economics. Scientific and indigenous knowledge should be aligned and not do the inclusion of scientific knowledge and the exclusion of indigenous knowledge. The science, technology and innovation collaboration model is essential to make social innovation and social entrepreneurship to create a new production and sustainable development path for the knowledge, industrial, manufacturing, and service global value chain to achieve climate change mitigation and renewable energy and the achievements of all the 17 SDGs goals. How did China create an economic system not to rely on aid, loans, debts, and foreign outward and inward investments? What lessons can Africa learn from the Chinese development innovation experience not to remain dependent on aid, loans, and debts. Economic development should be anchored on how the well-being, health and livelihood of people and nature are promoted to create a system where no citizen begs, where no one starves and where no one is homeless, and all have opportunities to learn and work to make everyone live a decent life. China has its own economic system where the challenges of poverty, health and education are addressed as priority with its own social, economic and knowledge production value and supply chain with Chinese characteristics.

An economic system is necessary that does not exclude the well-being, services provision, environmental protection, virus eradication, human well-being by removing the current dominant success validation by merely narrow market, commerce and economic profit including the current dominating e-commerce expansion. Economic development should be anchored on how the well-being, health and livelihood of people and nature are promoted to create a system where no citizen begs, where no one starves and where no one is homeless, and all have opportunities to learn and work to make everyone live a decent life. Economic gain to make profit is not sufficient. There must be social well-being gain and environmental protection gains. China–Africa Science, Technology, and Innovation Collaboration with the China–African Partnership to Promote Sustainable Development to Build a China–Africa Community with a Shared Future can generate the pathway for Africa to produce the system to make its rich wealth eradicate entirely its massive poverty.

The contributions of the book starts with three chapters in Part I (Chapters 2, 3 and 4) on the China–Africa Collaboration in higher education by articulating how significantly the research collaborations between African countries and the People's Republic of China in the last half-century after the People's Republic of China's reforms and open policy in 1978 have grown in the (1) Minerals sector, which includes the fields of geosciences, mining, mineral processing and metallurgy; (2) Transport and related fields, focusing on the establishment of transport universities in Africa; (3) Agricultural sector, specifically on the development of dwarf maize, high-yield crops, soil remediation and water-saving; and also in different collaborative programs. Chapter 3 addresses the primary form of the China–Africa higher education cooperation towards the cultivation of scientific and technological talents for Africa. The China–Africa higher education institutes need to further enrich cooperation forms and deepen cooperation areas based on mutual benefits and win–win results. The authors recommend sorting out the history of China–Africa higher education cooperation development, current challenges and opportunities, suggestions, and prospects for future development with strong and enduring higher education cooperation to complement each other's advantages and jointly train African scientific and technological innovation talents for enhancing Africa's talent pool, improving its innovation capacity, and breaking through development difficulties. Chapter 4 has selected one of the leading Chinese universities-Beijing Foreign Studies University as a case study and uses the theory of synergistic to explore how the collaboration of China with Africa started with infrastructure synergistic in 1976 by building the Tanzam Railway with the cooperation of China with Tanzania and Zambia. China has continued to support Africa with infrastructure since it launched the "Belt and Road" initiative in 2013, which has been widely supported by many African countries. Education, as the foundation of national prosperity and people's well-being plays a fundamental and leading role in China–Africa joint construction of the "Belt and Road" Initiative. The relationship of China with Africa is synergistic as the inputs generate outputs that are more than the sum of the parts by working together with cooperation. The mechanism of interaction and coupling between China–Africa universities, governments, international organizations (e.g. United Nations, UNESCO) to demonstrate cooperative actions, achievements, and future cooperation outlook of Beijing Foreign Studies University from the perspective of synergistic have been done.

Chapters 5, 6, 7 and 8 form Part II on the Africa–China research collaboration and training. Chapter 5 identifies that China's involvement in Africa has dramatically increased during the last few decades by using research collaboration data of African scientific publications from 1971 to 2019 in the Web of Science database. The findings reveal that China is gaining an increasingly important position in Africa's international research collaboration. China is now establishing contacts with African researchers through scholars from other countries. Surprisingly, publications involving Chinese researchers typically attract more citations. It demonstrates how scientific research collaboration differs from economic cooperation by nature. China–Africa scientific research collaboration is more about inclusive and win–win cooperation than an exclusive and zero-sum game.

Chapter 6 focuses on the need to undertake comprehensive, analytical, reflective, descriptive, and critical research on the China–Africa relationship in Science, Technology, and Innovation (STI) by including spheres in agriculture, manufacture, and services. Technology transfer in the history of China–Africa relations has not been systematically researched. The evidence-based technology transfer research must be used to generate policy learning both to China and Africa to develop mutually beneficial relationship that can serve as a model for the rest of the world.

Chaper 7 reviews the history and development of China–Africa educational cooperation and presents the talent education practice of Sino-Africa Joint Research Center, Chinese Academy of Sciences. It introduces the status of China Africa education in the new period, reveals the opportunities and challenges faced by current higher education of African students in China. It is suggested to establish an alumni mechanism to track African students' status after graduation, and give full play to talents who have the experience of studying in China for future cooperation in various fields between China and Africa. More efforts from China–Africa should be made to carry out inter-school exchange of students for joint training, international cooperative teaching, expert visits, and encouraging African youths to come to China for further study and training, continuously serving the needs of social and economic development of China and African countries.

Chapter 8 addresses the global mineral resource development problems such mining depth, declining resource quality, tightening environmental constraints, and increasing security challenges. The development and utilization technology of mineral resources is in urgent need of innovation. As leading mining countries, starting from the joint research project (JRP) to the establishment of the China-South Africa Joint Research Centre for Exploitation and Utilization of Mineral Resources (JRC), China and South Africa have cooperated in depth for over 10 years and made fruitful achievements. The fields of cooperation involve mining, automatic control of mineral processing, comprehensive utilization of tailings, waste catalyst recovery and lithium-ion battery materials.

There are three chapters (9, 10 and 11) in Part III on China–Africa collaboration in agriculture, food security and environmental management. Chapter 9 explores the collaboration between China and Africa in the field of agriculture and food security. China and Africa have a long-standing partnership in various sectors, and agriculture and food security have emerged as key area of collaboration in recent years. Through various initiatives and programs, China and Africa have been working together to address the challenges of food insecurity in Africa, which is a significant impediment to the continent's economic growth and development. The challenges and opportunities associated with collaboration are addressed. One of the main challenges is ensuring that the collaboration is based on reciprocal benefit, respect, and understanding, and that it does not lead to the exploitation of African resources or undermine local agriculture systems. The collaboration between China and Africa in agriculture and food security presents significant opportunities for addressing food insecurity and promoting sustainable development in Africa.

Chapter 10 reflects on the China–Africa formal contact over half a century ago. Understanding the compound factors in the interaction and contextualizing them with food security is an interesting subject that preoccupied academia and practitioners to build the China–Africa sustainable development collaboration. China's advancement in agricultural science, technology, and innovation makes it one of the competitive African partners. Reimagining China's presence in Africa within the member states' inimitable and multiple biographical narrative models can undercut the complexity of China–Africa food security collaboration.

Chapter 11 focuses on how environmental analyses and management systems are becoming the central themes in socio-economic development frameworks and plans of basically all countries, including China and African countries. These systems are tied to the efficacy of natural sustainable development targets as well as livelihood support and human health protection schemes of countries. Largescale, transboundary environmental issues are now commonly associated with energy, transportation, and agricultural sectors of national economies. Several factors that determine the environmental sustainability of any country frequently lie outside the geography and jurisdictional boundaries of the country. Pollution and associated risk drivers may have their origins in one country but some of their effects could precipitate in/on other countries. In this interconnected world, greenhouse gas emissions in China can imperil human health directly and indirectly in some African countries. Conversely, bush burning in Sub-Saharan Africa can affect ecological systems in China through some cascading environmental effects.

Part 4 is on China–Africa collaborations in telecommunications and space, which consists of three chapters (12, 13 and 14). Chapter 12 is on how the relations between Africa and China can deepen with the possibilities offered by new technologies for bridging the huge geographical distance between these two territories. Efforts are discussed to expand people-to-people exchange with educational scholarships, research collaboration, public dialogues with the use of the media space through Xinhua, CCTV and the CGTN-Africa. How the major broadcasting technology and technology transfer and cooperation are contributing to the nature of the

Africa–China relations for the mutual benefit is explored by using a case study of the Nairobi-based CGTN-Africa.

Chapter 13 on China-South Africa collaboration in astronomy shares common aspirations for scientific and technological development and for the growth of human capacity resources. Both as BRICS members are doing strong bilateral relations by developing the research in astronomy and astrophysics. Astronomers and astrophysicists from the two countries started their scientific collaboration in 2016 and shared a lot of resources in astronomy research, co-educated students and postdocs and fostering future generations of scientific talents and strengthened the scientific and technological collaboration.

Chapter 14 is on Technology transfer and technological spill overs from Chinese tech giant Huawei in Algeria. China, as an emerging economy with an exceptional development path, has shown a strong support to South-South cooperation and strong commitment to build radically different relations between nation-states. It translated by an increase in investments in high value-added sectors to boost cooperation in science and technology and its willingness to share with African countries its wealthy experience to help bridging the digital divide. Driven by its tech giants, China's Digital Silk Road (DSR) aims to promote connectivity by bringing advanced digital infrastructure to BRI (Belt and Road Initiative) countries, such as fibre optic cables, data centres, 5G networks, e-commerce platforms, and smart cities. In North Africa, the digital sector plays an increasingly important role in Sino-North African relations. With its strategic location, connecting Asia, Africa, and Europe, North Africa holds a central position in China's BRI and Algeria with its size and economic potential, managed to attract the giant tech Huawei which built a manufacturing plant in the country, the only African country to have such a plant. The capacity building component of China in DSR is promising mostly through the provision of training courses and technical assistance, with the aim of transferring technology mainly through knowledge spill over. The chapter draws from nascent literature on the topic and also from the first-hand experience by one of the authors who worked in the Chinese corporation Huawei Telecommunication in the North African region.

Part 5 is on evolution of China–Africa collaborations in science, technology, and innovation. There are five chapters (15, 16, 17, 18 and 19). Chapter 15 stresses that China–Africa relations go back to ancient times when scientific and technological exchanges already emerged, and are being greatly driven by cooperation in science, technology, and innovation (STI). There are generally three stages of the evolution of China–Africa cooperation in STI: "mutual exchanges" focused cooperation from the Han Dynasty to the founding of the People's Republic of China (PRC); "direct aid" focused cooperation from the founding of the PRC to the late twentieth century; and "capacity building" oriented cooperation since the beginning of the twenty-first century. In the new era, two versions of the China–Africa Science and Technology Partnership Program (CASTEP), namely, the "CASTEP 1.0" and the "CASTEP 2.0", are implemented and advanced, targeting at national capacity building to realize transformation and development, which have made certain achievements. China–Africa cooperation in STI sees new opportunities and challenges brought by profound changes unseen in a century, in the context of evolving international political and economic landscape, and the ongoing new round of technological revolution and industrial transformation. In response to this, capacity building-orientation is still the future direction of China–Africa cooperation in STI. Both sides need to further strengthen the strategic significance of China–Africa cooperation in STI and the consensus on development, expand China–Africa cooperation in the fields of new technologies, and diversify and improve mechanisms of cooperation.

Chapter 16 focuses on the lessons to be learned by African countries and the benefits Africa will enjoy by cooperating and collaborating with China's STI initiatives based on investigations on the STI challenges in African countries. Research is done on the mandate of how the China's Science, Technology, and Innovation can fit with the efforts to advance the use of knowledge for the sustainable development of Africa. How can the China STI initiatives play to generate action within the African continent is undertaken with reviews of some of the key African STI initiatives. This helps to assess how the China STI initiatives can contribute to and promote already existing structures and programs within Africa. The coherence and expected synergies between the China STI objectives and the ongoing African initiatives are investigated. The China STI initiatives are found to be fitting well to enhance the utilization of knowledge for the development of African countries through synergistic cooperation and collaborations.

Chapter 17 provides an overview of China–Africa collaborations. China is a growing developing Asian country that established trade organisations with Africa both specific bilateral and multilateral relations under Forum on China–Africa Cooperation (FOCAC), AFRODAD, Partnership for Growth and Development (PGD), Economic & Trade Cooperation Forum, and Joint Economic and Trade Committee (JETC). The research explored the overall China–Africa economic collaborations covering both China's contributions to Africa, and Africa's contributions to China. China has built massive infrastructures like roads, buildings, and railways for African countries including providing loans, and skills transfer. Africa–China relations are progressive and mutually beneficial. Africa–China collaboration must continue in a mutually beneficial sustainable manner. The political variable should be included in the new international economic models showing how politics will affect the number of products to be traded in bilateral trade and how politics plays a role in ascertaining which products should be traded and providing reasons for such decisions, especially considering Africa–China relations.

Chapter 18 addresses the evolution and prospects of China–Africa cooperation in Science, Technology, and Innovation (STI). Cooperation in science and technology offers an important way in which countries can establish mutually beneficial, win–win partnerships. STI has been a consistent and important component of China–Africa cooperation and is playing an increasingly prominent role in South-South cooperation and the development of China–Africa relations. As China–Africa relations continue to develop, China–Africa STI cooperation has been flourishing and borne fruit, providing strong impetus to both sides' economic and social development, and there will be even more space for improvement in the future. The progress, mechanisms and effectiveness of China–Africa STI cooperation are growing strong.

Chapter 19 is on China–Africa multifaceted collaboration highlighting China's presence in Africa with economic agreements signed with several countries and the cooperation in aid and development in infrastructure, health, education, and humanitarian assistance. China has pledged not to interfere in political affairs but has tried to establish relations of mutual benefit. China desires a friendly and multipolar international climate to conduct what it defines as its peaceful rise. For this reason, it has fostered the rhetoric of foreign relations based on trust and mutual benefit, equality and respect between countries and cooperation. Located within this framework of action is its policy of foreign aid for development towards Africa. There is a lack of political conditionality for granting Chinese development aid which demonstrates an alternative model of cooperation for development. Due to its staunch defence of national sovereignty, Chinese foreign aid is extremely attractive to African countries, which is still highly susceptible to foreign intervention due to the colonial heritage threat to Africa.

Part 6 is on China–Africa Collaboration on digital Technologies, which incudes two chapters (20 and 21). Chapter 20 addresses the present and future challenges of the China–Africa on-going science, technology, and innovation collaboration by exploring how China collaborates with Africa in digital technologies and helps Africa to solve digital technology-based problems. The examples from the digital Chinese companies like Huawei and Alibaba are used to analyse how these Chinese companies help Africa to understand the challenges and future of China–Africa collaboration in digital technologies.

Chapter 21 explores the development of Africa's economy and trade and China's investment in Africa by taking Africa's Business to Consumer e-commerce to sort out the development of e-commerce in typical African countries in terms of information technology, economic development, logistics facilities and e-commerce-related laws and regulations. Chinese enterprises invest in Africa mainly in the infrastructure, logistics industry, e-commerce platforms and third-party payment platforms.

Part 7 is on China–Africa collaboration in Finance and in Renewable energy (Chapters 22, 23 and 24). Chapter 22 examines the Sino-Ghanaian collaboration to aid Ghana in the development of its cryptocurrency future by exploring the opportunity offered to develop partnership through digital transformation in win–win cooperation. China is an epitome and a major propagator of digital transformation as a developing country and has high potential to help develop Ghana in terms of modern-day financial technologies. By drawing a nexus between cryptocurrency/blockchain, digital transformation and financial inclusion, an important empirical insight is provided into an interesting focal point on Sino-African collaboration in advancing cryptocurrency development in a new emerging digital economy. Chapter 23 explores the innovative technology of the smart transport and smart cities for sustainable infrastructure to support resilient economic system in China and Africa. Smart transport and smart cities are playing an important role in the world. Smart transport and smart cities depend heavily on good infrastructure information management systems. Smart transport and smart cities have been considered as the key drivers for supporting the national and local economic development of China over the last decade. Through collective efforts made by the central government, local governments and private innovative companies' huge investments, the great achievements on smart transport and smart cities have been secured. Significant lessons from the case of smart transport and cities as determinant of infrastructure development to support local economic development from Shenzhen and Southern China is regarded as significant as the Southern African Development Community (SADC) and Africa at large are still at infant stage in the desire to establish smart transport and cities. Much can be learned from more advanced countries like China, to excel in infrastructure development without making serious mistakes in the development of appropriate smart transport and cities technologies for Africa. Chapter 24 is on China's green energy investment in Kenya, Ethiopia, and Tanzania, as these countries deal with multiple challenges in the transition to a modern and sustainable energy system. They are faced with severe shortage of energy supply challenges. China's support on infrastructure development has been useful in alleviating the energy challenge in Africa. The Chinese development finance has been however split across several other sectors such as transport infrastructure, manufacturing, construction, agriculture, and tourism industry. The Chinese aid support is necessary to assist African countries to create a modern and sustainable energy system.

Part 8 is on China–Africa collaboration in the health sector, which consists of three chapters (25, 26 and 27). Chapter 25 shows that under the framework of the Belt and Road Initiative (BRI) as well as the Forum on China–Africa Cooperation (FOCAC), the China–Africa community has a shared future in a wide range of fields including medicine to create full health for both Chinese and Africans. China has been providing medical assistance to promote the African region's health systems. Potential challenges are identified to improve the China–Africa collaboration in the health sector.

Chapter 26 is on a critical evaluation of health diplomacy as an altruistic underpinning of China–Africa Relations. The focus is on China's altruistic overtures created on African public health issues. Work on health diplomacy has contributed to new interpretations of China–Africa relations by proffering alternative perspectives beyond the traditional conceptions of hegemonic power interests in Africa. These alternatives demonstrate the developmental trajectory pursued by a developing country, China, towards a developing African region. Arguably, China–Africa health relations are instrumental to Beijing's 'novel' internationalisation agenda, demonstrated by its concentration in countries with no strategic interests. In the broadest sense, China's health policy in Africa is encapsulated in the 1964 'Eight Principles for Economic Aid and Technical Assistance to Other Countries and the 2006 'China's African Policy' that promulgated the bedrock Forum on China–Africa Cooperation (FOCAC).

Chapter 27 is on Public Perception of Traditional Chinese Medicine in Africa. The research focused on how Traditional Chinese Medicine (TCM) has been used to prevent and cure the people affected by the Covid-19 pandemic and how this medicine service has been understood in Africa. The extent to which TCM has been acknowledged in Africa and the effort to internationalise TCM for pharmaceutical services have been explored.

Part 9 is on China–Africa collaboration in Manufacturing, which includes four chapters (28, 29, 30 and 31). Chapter 28 is on a framework for productive collaboration by riding the dragon that explores the prospect for effective STI collaboration between Africa and China by examining the comparative and competitive advantages. It identifies the obstacles and opportunities for effective China-Africa engagement with the current stage of globalisation. It evaluates the context of China's recent engagement with Africa, sets out the challenges presented by postcolonial relationships across the continent and identifies the material, institutional and human resources with which African countries can contribute to the challenges and opportunities presented by the threat of climate change and the potential of a Fourth Industrial Revolution.

Chapter 29 is on the role of China and Western Countries in Africa's economic globalisation. The focus is on "Who is behind the recent economic renaissance of Africa?". The growth model and econometric panel data analysis have been used from two categories of mineral and agricultural exporting African countries to and from China and the West. Although both China and the West have positively contributed to African GDP growth, China's FDI is more significant than OECD-FDI and the effect is stronger for the mineral exporters. Also, the share of OECD's export and imports are higher than China but declining sharply. However, OECD's aid has more significant effect than the Chinese for both categories. Both OECD and China are important to fasten economic growth. However, the Chinese influence is rising sharply and displacing the West. Without them, still there would be growth, but it is much slower. Therefore, African countries need to design strategies to seize the opportunities available from both China and the Western countries.

Chapter 30 is on whether the China–Africa economic partnership is in the right pathway. Although China was never a colonizing power, its economic relations with Africa can be traced to antiquity. However, meaningful economic partnership between China and Africa is a recent phenomenon. After a decade and more trade engagement with China many African countries have been claiming that there is economic growth momentum induced in their respective economies. Despite the progress achieved in the past decades, cooperation between China and African also faces numerous challenges ahead. Key problems associated to China–African partnership are neither fully addressed nor satisfactorily contextualized in the current debate regarding China and Africa. There is optimism about the rises of China and India, and other south-south partners of Africa hoping that such a partnership offers a great chance to Africa in claiming twenty-first century. Here it should be underlined that if China follows the same old western principles towards Africa, the new economic partnership with China may end up with usual win-lose scenario. However, such an extreme scenario may not happen in the twenty-first century when Africa has its own prominent continental institution like African Union, The United Nations Economic Commission for Africa, The New Partnership for Africa's Development (NEPAD) and other local and international independent academic and research institutions which can alert the continent of foreseeable risks.

Chapter 31 is on the impact of Chinese foreign direct investment on the productivity growth in the Ethiopian manufacturing sector. The research provides a new empirical model on the relationship between Chinese foreign direct investment (FDI) and the total factor productivity growth using panel data for a subset of 24 groups of the Ethiopian manufacturing sector between 2011 and 2016. The outcome indicates both positive and negative effects from Chinese and other countries' FDI through the growth of value addition to domestic firms. Especially, the impact of Chinese investments on TFP is positive and statistically significant regardless of the considered technology level classification. As for the other countries' investments, the impact is negative and statistically significant. The FDI impact on TFP is transmitted to local firms through demonstration and competition effects at a moderate level. But the negative impact of other countries' investment in TFP is in the scope of medium and high-level technology. The technology gap is a critical factor among the other determinants of FDI.

Finally, the research myriad contributions articulated in all the chapters on the 'Science, Technology and Innovation Collaboration' book demonstrate with clarity and concrete evidence the China–Africa relationship is bounded with real practice of the China–Africa mutual respect, understanding and benefits with core shared values of integrity, honesty, dependability, helpfulness, impartiality, courteousness, and fairness as both are travelling on the journey of the human life saviour community with a shared future.

There is a lot Africa can learn from the growing China technology capacity as China is very willing to share its technology and infrastructure without interference. This book only has focussed on the science, technology, and innovation side, but more research needs to be done on all spheres of the China–Africa relations. China–Africa trade is said to have started in 220BC. And the long history of China–Africa must be explored by including all dimensions of the China–Africa relationship. The mutual relation that has long historical background should continue with the principle of win–win relations. This book has fully highlighted with data evidence the mutual relations in higher education, agriculture and food security, climate change and environment, renewable energy, space technology applications, health, and manufacturing.

References

Muchie, M., & Patra, S. K. (2020). China–Africa science and technology collaboration: Evidence from collaborative research papers and patents. *Journal of Chinese Economic and Business Studies, 18*(1), 1–27.

Wang, X., & Zhang, X. (2020). *Towards 2030: China's poverty alleviation and global poverty governance.* Springler Link [part of the book series: International Research on Poverty Reduction (IRPR)].

Xie, H., Wen, J., & Choi, Y. (2021). How the SDGs are implemented in China: Comparative study based on the perspective of policy instruments. *Journal of Cleaner Production, 291*, 12593.

Part I
China-Africa Collaboration in Higher Education

Part I

Chapter 2
Africa–China Higher Education and Research Collaboration in Minerals, Transport and Agriculture

Thibedi Ramontja, Baojin Zhao, Yuchen Wang, Yinsuo Jia, and Junping Ren

2.1 Introduction

Minerals, transportation and agriculture play an important role in the socio-economic development of Africa and the People's Republic of China (China). Accordingly, the governments of African countries and China have initiated various partnerships through higher education and research in the fields of minerals, transport and agriculture in recent years. These partnerships, in the form of various collaboration programmes, gained traction and momentum following China's introduction of economic restructuring and open policy in 1978. The collaboration programmes now include the exchange of students, scholars and researchers, as well as the undertaking of research programmes.

T. Ramontja
School of Mining Engineering, University of the Witwatersrand, Johannesburg, South Africa

B. Zhao (✉)
Department of Environmental Sciences, University of South Africa, 28 Pioneer Ave, Florida Park, Roodepoort 1709, South Africa
e-mail: tzhaob@unisa.ac.za

School of Geoscience and Technology, Overseas Expertise Center for Deep Marine Shale Gas Efficient Development Innovation (111Center), Southwest Petroleum University, Chengdu, China

Y. Wang
South Africa-China Transport Cooperation Centre (SACTCC), University of Venda, Thohoyandou, South Africa

China-Africa Transport Strategy Research Institute (CATSRI), Chang'an University, Xi'an, China

Y. Jia
Richard Corporation of Science and Technology, Zhangjiakou, Hebei, China

J. Ren
Tianjin Center, China Geological Survey, Tianjin 300170, China

© The Author(s) 2024
M. Muchie et al. (eds.), *China-Africa Science, Technology and Innovation Collaboration*, https://doi.org/10.1007/978-981-97-4576-0_2

Students from China have been studying at higher learning institutions in countries such as South Africa, Nigeria and Ghana (Li, 2019). Similarly, an increasing number of students and scholars from African countries are being sent for short training courses and degree studies in China (Li, 2018). The former has mainly taken place through training programmes hosted by higher learning and research institutions sponsored by the Ministry of Commerce of China and China Scholarship Council (CSC), while the latter is hosted by various Chinese universities and supported by the Ministry of Education of China.

Whilst Muchie and Patra (2019) published their results about the collaborative achievements of the Africa–China Science and Technology (S&T) in terms of publications and patents, this chapter reviews and presents key collaborative programmes, based on the authors' experience, in four categories: (1) Minerals sector, including the fields of geosciences, mining, mineral processing and metallurgy; (2) Transport and related fields, focusing on the initiative of the establishment of four transport universities in Africa; and (3) Agricultural sector, specifically on the development of dwarf maize, high-yield crops, soil remediation and water-saving; and (4) Other collaborative programmes including BRICS' joint research programme.

Minerals play an essential role in the progressive advancement and industrialisation of society. With increasing world population growth, more minerals of various usages will be required to ensure prosperity for humanity. Therefore, the partnership between Africa and China in the minerals sector is an important one that requires continued reinforcement and innovation. In this regard, African countries and China have embarked on significant collaboration projects in terms of research and capacity building, specifically in geosciences, mining, mineral processing and metallurgy. This chapter presents selected key collaborative programmes between African countries and China in these fields.

2.2 Africa–China Higher Education and Research Collaboration in Geosciences

Whilst geosciences have a variety of fields, the most prominent collaborative programmes between Africa and China, which are discussed in this section, are in higher education, geoscience mapping and geo-parks.

2.2.1 Higher Education Collaboration in Geosciences

The collaboration between African countries and China in respect of higher education commenced in 1956 through the exchange of students between China and Egypt (Li, 2018). In recent years, the collaboration has been incorporated into the Forum on China Africa Cooperation (FOCAC) programmes, a head-of-states forum established in 2006. In line with the principles of the FOCAC and the promotion of higher education between African countries and China, a 20 + 20 Cooperation Programme was conceived at the Fourth Ministerial Conference of the China–Africa Forum held in Sharm El Sheikh, Egypt, in November 2009. The plan is a partnership programme between 20 Chinese and 20 African tertiary institutions in 17 countries (Gu, 2017). As part of the 20 + 20 Cooperation Programme, the Chinese government established a Confucius Institute (CI) at the University of Namibia to advance capacity-building programmes, including teaching the Chinese language.

Additionally, the China University of Geosciences, Beijing (CUGB), initiated one of the most prominent educational cooperation programmes. The programme actively recruits African students to further their studies in geosciences at CUGB. In 2012, about 30 African students were enrolled with CUGB, of whom five were staff members of the University of Namibia who registered for postgraduate courses with the support of the Chinese and the Namibian governments (China-Africa, 2012). In 2006, the CUGB and the University of Namibia (UNAM) signed an MoU to advance cooperation in geosciences, natural resources and the environment (ChinaAfrica, 2012).

2.2.2 Collaboration in Geoscience Mapping and Related Research

One of the earliest collaborations between African countries and China in geosciences was in geochemical mapping, which was in the form of a four-week capacity-building course known as "Geochemical Mapping and Environmental Geochemical Survey for African Countries". The course, offered to African geoscientists, was coordinated by the Chinese Ministry of Commerce and the China Geological Survey (CGS-China). At its inauguration in 2004, at least 13 African countries sent their geoscientists to attend the course. The course outline, as described by Wang (2012), included the following: (1) Regional and national-scale geochemical mapping, which has been successfully used for the exploration of ore deposits in China; (2) Geochemical data management and map generation; and (3) Environmental geochemical surveys and applications.

The Chinese government has also initiated and signed several collaboration agreements with some African countries, focusing on geoscience mapping and mineral development. For example, agreements in the form of Memorandums of Understanding (MoUs) were signed in 2016 between the Nigerian Geological Survey Agency (NGSA) and the CGS-China, as well as between the NGSA and the Shandong Provincial Bureau of Geology and Mineral Resources of China (SPBGMRC). As captured in their MoU, the NGSA and CGS-China cooperative agreement entails cooperation in research and development in respect of geology, geological mapping, regional mineralisation and geochemical mapping of Nigeria (Nigerian Investment Promotion Commission, 2016). The MoU between the NGSA and SPBGMRC is more focused on, among other things, cooperation in respect of detailed mineral exploration, geosciences, and evaluation and compilation of reports on the occurrence of gold and base metals (Nigerian Investment Promotion Commission, 2016). The CGS-China has also signed an MoU with the Council for Geoscience of South Africa (CGS-South Africa), which outlines specific cooperation in geoscientific research.

In 2013, the government of China signed an MoU with the government of Kenya, which allowed the former to advance a grant of $68 million for a geological survey programme and assessment of the mineral potential in Kenya (Bariyo, 2013). In Zambia, the CGS-China entered into a five-year agreement with the government of Zambia in 2016 to undertake geoscientific mapping, including geological, geophysical and geochemical surveys (Mulikelela, 2016). With regard to Sudan, following the signing of an MoU between the Chinese and Sudanese governments in 2010, the CGS-China embarked on a geochemical survey programme to enhance available geoscientific information (Ma, 2020). The geoscientific programme in Sudan formed part of the China-Arab States Countries Cooperation in the geological survey programme that includes Morocco and Saudi Arabia and has resulted in the CGS-China undertaking geological and geochemical surveys at 1:50,000 to 1:250,000 scales (China Geological Survey, 2017).

2.2.3 Africa–China Geoscience Research Centers

To coordinate the collaboration in geoscientific field mapping, research, exchange and training in Africa and implement the objectives of the Forum on China–Africa Cooperation-Dakar Action Plan (2022–2024), two major research Centres were formed under the CGS-China: The China-Southern African Geoscience Cooperation Centre r in Tianjin and the China-Northern African Geoscience Cooperation Center in Wuhan, Hubei Province, which are responsible for coordinating the activities in the southern and northern African regions, respectively. The equator was used to constitute the two regions, with the northern research centrecomprising countries in the northern hemisphere and the southern research centrecomprising southern hemisphere countries. Based on information from the China-Southern African Geoscience Cooperation Centrer, its major activities can be summarised as the following:

1. Establishment of a collaborative relationship with the relevant Geological Survey of twelve countries in the Southern African Region and communication relationship with about ten individual Embassies of the Southern African Region in Beijing;
2. Completed or ongoing for about 30 geological mapping and research projects, financially sponsored by the CGS-China and Ministry of Commerce of China;
3. Establishment of supporting and communication channel for attending and supporting the annual events of the Mining Indaba in Cape Town, South Africa and the China Mining Conference and Exhibition in Tianjin;

4. Sponsor bursaries to support students or technicians from the countries in the Southern African Region, such as nominating or recommending them to study for postgraduate degrees at China University of Geosciences, Wuhan or Chinese Academy of Geological Sciences;
5. Support researchers to apply for funding from the National Research Foundation of individual countries, jointly worked and shared all data as a team;
6. Recent ongoing and completed projects, for example, including the evaluation of copper and cobalt resources in Zambia, tantalum resources in Rwanda; and
7. Technical support for local mining development, promoting local social and economic development to improve people's living standards.

2.2.4 Collaborations in Geo-parks Development

Whilst geo-parks are not often established for mineral development and mining, they can be integrated into the mining environment, especially if they are linked with the education of people about the association of rock formations and minerals. They, therefore, play a vital role in ensuring that people, especially young ones, understand the natural environment holistically. China is one of the leading countries in establishing and managing geo-parks and has played an essential role in transferring and disseminating knowledge on geo-parks in Africa through the Global Geo-parks Network. This organisation provides a platform for promoting and sharing information on geo-parks at its various forums. One such forum was a workshop held in 2007 at the CGS-South Africa, where a presentation titled "China's Taishan Geo-park and Management of Geo-parks" was presented to share the experiences of establishing and managing geo-parks (Global Geo-parks Network, 2007). The collaboration in geo-parks has the potential to be strengthened by the recent initiative known as the "UNESCO-Africa–China Forum on World Heritage Capacity Building and Cooperation", which seeks to enhance collaboration between Africa and China in the sustainable management of World Heritage sites (UNESCO).

2.2.5 Africa–China Research Collaboration in Mining, Mineral Processing and Metallurgy

Unlike cooperation in geosciences, which has seen the African and Chinese governments implementing significant programmes, cooperation in mining, mineral processing and metallurgy is still in its infancy. Nonetheless, some notable projects have been initiated, such as the 2020 cooperation research agreement between the University of the Witwatersrand in South Africa and the China University of Mining and Technology. The agreement, signed in 2020 at a ceremony in Xuzhou, China, formed a framework for establishing the Joint International Research Laboratory of China–Africa Mining Geospatial Informatics (Benton, 2020). According to Benton

(2020), the collaborative programme aims to research a technology that can locate underground mineworkers. It builds on the initial collaboration that commenced in 2013 and focused on, among other things, underground communication systems, risk modelling and prediction of harm (Moore, 2016).

On 23 March 2022, an important advancement was made in mining research collaboration between China and Africa when a webinar titled "Rock-bursts Webinar: China and South Africa" was held. The webinar organised by the Southern African Institute of Mining and Metallurgy and the China University of Mining Technology provided an opportunity for mining engineers from both countries to share knowledge regarding the latest research areas in rock-burst (Southern African Institute of Mining & Metallurgy, 2022).

Regarding mineral processing and metallurgy, in 2018, China and South Africa launched the China node of the Joint Research Centre for the Development and Utilisation of Mineral Resources. The first node is located at the Beijing General Research Institute of Mining and Metallurgy (BGRIMM). The second node of the collaborative programme is planned to be established at Mintek, a mineral processing research institute based in South Africa (Department of Science and Innovation, 2018).

Another important project involves collaboration between the BGRIMM, the University of Limpopo and Mintek in material science modelling and mineral processing reagents. The partnership can potentially improve the configuration of battery solutions for solar energy storage in Africa, thereby addressing energy poverty (Ngoepe, 2020). Additionally, the Council for Scientific Industrial Research (CSIR) of South Africa and Gold Yard (a consortium of Chinese institutions) signed an MoU to collaborate in the fields of alternative and renewable energy (CSIR, 2018).

2.3 Africa–China Higher Education Collaboration in Transport

2.3.1 Overview

Transport plays a vital role in the socio-economic development of a country, region or continent. During the last 40 years, China has invested considerable funds in the transport infrastructure and made prominent social and economic developmental achievements. In 2000, Africa and China co-established the Forum of China–Africa Cooperation (FOCAC) in Beijing, China. Subsequently, FOCAC has been held every three years in Africa or China. It was successfully held in Eastern, Northern, Southern and Western Africa in 2003, 2009, 2015 and 2021, respectively. Africa and China have put considerable effort into higher education in the transport sector, which links to significant human resource improvement and skills development. For highly efficient management, capacity building and technology transfer play an important role. Based on the mutual understanding between Africa and China, both sides have cooperated on broad training at different levels. Available records show that China has assisted in the construction of over 6000 km of railway, 6000 km of high-level roads, and dozens of ports and airports in Africa since 2000. Currently, Africa has joined the Belt and Road Initiative (BRI) with China, and more transport infrastructure and logistics projects are expected to be developed in line with this initiative.

2.3.2 FOCAC Johannesburg Action Plan

From 4 to 5 December 2015, Chinese President Xi co-hosted the Johannesburg Summit of the FOCAC. At the opening ceremony (Fig. 2.1), an announcement was made about the co-establishment of Africa–China transport universities as listed in Sect. 3.3 of the FOCAC Johannesburg Action Plan (2016–2018) (FOCAC, 2015). The transport universities' objective is to facilitate high-level capacity building in respect of infrastructure connectivity and economic integration in Africa among other things.

The Johannesburg Action Plan on higher educational partnership has since promoted the Africa–China mutually beneficial infrastructure cooperation through infrastructure planning, design, construction, operation and maintenance. The partnership, which is aimed at enhancing Africa's sustainable development, has also supported Chinese enterprises to actively participate in the construction of African railways, highways, regional aviation, ports, electricity, telecommunications and other infrastructure.

Underdeveloped infrastructure is one of the bottlenecks hindering the sustainable development of Africa. Accordingly, the FOCAC has created an avenue for

Fig. 2.1 Opening ceremony of the Johannesburg summit of the FOCAC

Fig. 2.2 Hon. Minister Mr. Chibuike Rotimi Amaechi (A: Right 7) of Federal Transport of Nigeria attending the 2017 World Transport Convention

encouraging Chinese businesses and financial institutions to invest in Africa through various means, such as Public–Private Partnerships (PPPs) and Build-Operate-Transfer (BOT) programmes. It also supports the development of African flagship projects, notably the Programme for Infrastructure Development in Africa and the Presidential Infrastructure Championing Initiative. After the FOCAC Johannesburg Action Plan, four major transport infrastructure projects on higher education have been established and are discussed in the following sections.

2.3.3   Nigerian Africa–China Transport University

In June 2017, the Minister of Federal Transport of Nigeria, together with several executives, visited China and attended the first Belt and Road and African Infrastructure Forum hosted by the China Railway Construction Corporation (CRCC) in collaboration with Chang'an University and the Southern Africa–China Transport Cooperation Center (SACTCC) at the first World Transport Convention (WTC) held in Beijing, China (Figs. 2.2 and 2.3). At the forum, the Nigerian government held discussions with the CRCC to explore cooperation in infrastructure development.

The China–Africa Transport University Programme forms part of the Nigerian Modernised IKA Railway Project signed by China Civil Engineering Construction Co., Ltd. (CCECC) and the Nigerian Ministry of Transport. The total planned investment is about 50 million US dollars, including building and operating a new campus for five years. The collaboration will ensure that the project contract is fulfilled and effectively implemented and that high-level capacity building is guaranteed. At the same time, the partnership will advance Chinese-foreign educational cooperation and people-to-people exchanges, push the construction of a community with a shared future for humanity, and promote Chang'an University's "Higher Education Internationalisation" strategy. Chang'an University has cooperated with the CCECC under the CRCC to work jointly on the China–Africa Transport University in Nigeria.

Additionally, the School of International Education of Chang'an University held a video conference to exchange views on the design, construction and progress in terms of the implementation of the China–Africa Transport University. The preparation plan was presented at the meeting, outlining the university's operation model, management structure, professional construction and teaching organisation. The China–Africa Transport University is one of the capacity-building projects in the "Nine Projects" of Africa–China cooperation, which is of great significance to promoting the "Belt and Road" initiative, Chinese enterprises going global, and the "Double First-Class" construction of the Chinese universities. The first phase of the construction project is expected to be completed ahead of schedule by June 2023.

2.3.4   China–Africa Transport University for Portuguese–Speaking Countries

In April 2016, the Southern Africa–China Transport Cooperation Center (SACTCC) and the representatives of CRCC discussed with the Ministry of Transport and Communications (MTC) of Mozambique to co-establish a China–Africa University initiative in the country under the FOCAC Johannesburg Action Plan. Several developments have since taken place in respect of this initiative, and in July 2018, the representative of the MTC attended the 2nd China–Africa Infrastructure Forum at the World Transport Convention (WTC) in Beijing, China. At the forum, the SACTCC presented the framework of the China–Africa Transport University in Portuguese-speaking African countries (Fig. 2.4). The SACTCC, Mozambique and five other Portuguese-language countries are expected to take the process forward.

Regarding the implementation of the framework, among the six Portuguese-speaking countries in Africa, Mozambique, which currently has two transport-oriented higher education institutions that tend to cooperate with China, is to be prioritised for cooperation. One is the Advanced Transportation Institute (postgraduate), while the other is a comprehensive university in northern Mozambique. The SACTCC conducted a physical inspection of the two universities to investigate which institution would be the most appropriate for cooperation.

Fig. 2.3 H.E. Nigerian President making an announcement on the ground-breaking ceremony (top left), The Chinese ambassador (top right), CRCC undertook the construction (bottom left), and H.E. Nigerian President inspecting the blueprint of the campus (bottom right)

2.3.5 Africa–China Transport Strategy Research Institute in South Africa and China

On 15 July 2016, the China–Africa Transport Strategy Research Institute was opened at Chang'an University in China (Fig. 2.5) based on the MoU signed between the University of Pretoria and Chang'an University. The two universities subsequently organised the Africa–China session at the WTC in June 2017.

From 2017 to 2019 and 2021 to 2022, the Southern African Transport Conference (SATC) organised a student essay competition programme at Chang'an University every year, with winners receiving financial support to attend the SATC held in Pretoria, South Africa. The China–Africa Transport Strategy Research Institute has extended the working MoU for the collaboration from 2022 to 2027.

2.3.6 The Silk Road Energy Alliance of Industry, Education and Research

The Silk Road Energy Alliance of Industry, Education and Research was hosted at Xi'an Shiyou University with participants from African universities and non-profit

Fig. 2.4 SACTCC's Prof. Alex Visser presented the proposal of the China–Africa Transport University for Portuguese-Speaking Countries at WTC2018

organisations (Fig. 2.6). Some of the organisations that were represented included the University of South Africa, the University of Johannesburg and the Southern Africa–China Science, Engineering & Technology and Education Association (SETEA) and the SACTCC. The Alliance was established in Xi'an, Shaanxi Province in China asthe starting point of the ancient Silk Road.

Fig. 2.5   Unveiling the plague of the China–Africa Transport Strategy Research Institute (CATSRI) at Chang'an University, Xi'an, China

2.4 Africa–China Collaboration in Agriculture

2.4.1 Overview

Since the establishment of diplomatic relations between African and China countries and the creation of the Science and Technology Forum between China and Africa, several agricultural research projects have been initiated. For instance, the Ministries of Education, Science and Technology and Agriculture of China have introduced and utilised animal and plant germplasm resources in 53 countries in Africa, including the introduction and utilisation of germplasm resources such as crops, grasses, fruit trees, vegetables, flowers, and medicinal plants. In particular, the introduction of wheat and maize germplasm resources has played a significant role in China's and Africa's economic and social development.

A significant number of cooperative research programmes have also been initiated in soil improvement, plant protection, biological fertiliser, water-saving irrigation, agricultural machinery, and economy and development. The programmes have improved agricultural science and technology in research, increased grain output, and promoted the development of agriculture and animal husbandry in Africa and China. They further enhanced the relationship between African and Chinese people. Table 2.1 presents the list of germplasm resources, including maize, wheat, fruit, flower and medicinal plants in African countries. A total of four projects were carried out in respect of agricultural science and technology cooperation with 16 countries from Africa and are discussed below.

2.4.2 The Hebei Academy of Agricultural and Forestry Sciences Cooperation Project

The Hebei Academy of Agricultural and Forestry Sciences has established an international science and technology cooperation project sponsored by the Ministry of Science and Technology in China (MOST). The project titled "Cooperative Research on Exchange, Exploration and Innovation of Agricultural Germplasm between China and South Africa" commenced in October 2006 to December 2009 and was led by Profs. Y. Jia from China and G. Fraser from the University of Fort Hare (Fig. 2.7). The project, which the Chinese government-funded to the tune of 2,6 million Yuan, involved the following seven aspects of cooperative research:

1. South African fruit trees, including nectarines and apples, were introduced for the study of their adaptability and quality, and further addition of two plum varieties and two rootstocks;
2. Three high-quality protein maize (QPM) population materials were introduced;
3. Introduction of 1–2 characteristic flower varieties;

Table 2.1 The list of plant germplasm from the countries in Africa as of 2022

Country	Maize	Wheat	Fruit	Flower	Medicinal plants	Total
South Africa	436	67	24	25	34	586
Lesotho	38					38
Mozambique	56					56
Egypt	79	36			15	130
Namibia	60					60
Botswana	46					46
Ethiopia	82					82
Zimbabwe	285		14		8	307
Zambia	22					22
Uganda	48				4	52
Kenya	297					297
Malawi	16					16
Ghana	16				15	31
Angola	13					13
Nigeria	14					14
Burkina Faso	24					24
Grand total	1532	103	38	25	76	1774

Fig. 2.6 Establishment of the Silk Road Energy Alliance Commences

4. Research on wheat breeding by combining molecular markers assisting conventional breeding by use of the breeding system of sterilised male dwarf wheat, gradually building near isogenic introgression lines for the efficient exploration of favourable hidden genes, cultivating new drought-resistant wheat lines, which has realised the mutual exchange of advanced technologies between the two sides;

5. Research and application of Optimised Cultivation Technology (OCT) of water-saving wheat, mainly including water utilisation efficiency and high yield through simplified water-saving irrigation technology and efficient fertiliser utilisation,

while the organisations involved, including the Hebei Academy of Agricultural and Forestry Sciences and the University of Fort Hare for the past 20 years;
6. Remediation technology of degraded soils to be developed in China and South Africa in terms of ecological remediation, soil testing, micro-ecological regulation, and model fertilisation as the core technology;
7. Research on optimisation of sustainable development of small-scale agricultural economy—a typical survey conducted by selecting farmers with different incomes (high-, middle- and low-income households), identification of the main modes of increasing the income of different farmers to be evaluated according to the theory of agricultural economics, and finally, determination of the best ways and measures for the sustainable development of general and small-sized farmers in both South Africa and China.

2.4.3 Collaboration on Drought- and Disease-Resistant Maize Hybrids Between China and Kenya

This collaboration consists of the Hebei Academy of Agricultural and Forestry Sciences, Hubei Seed Group Co., Ltd., and Crop Institute of Chinese Academy of Agricultural Sciences, China, led by Profs. Y. Jia, and Rongo University (RUC) and the Agricultural Research Institute (KARI), Kenya, led by Gudu (Fig. 2.8). The project, which had total funding of 2.8 million Yuan, commenced in 2015, ended in 2018 and was sponsored by the MOST and Hebei Province.

A number of visits and exchange programmes were carried out in both Kenya and China. From 19 to 28 March 2016, a delegation of four Chinese researchers visited Uganda and Kenya to investigate joint agricultural research programmes, including drought and disease-resistant maize hybrids undertaken by the MOST for developing countries. In addition, the effect of Ryma on Maize Lethal Necrosis (MLN) disease in East Africa was investigated. H.E. Minister Kibirige Sebunya from Agriculture, Animal Husbandry and Fisheries of Uganda received the delegation and discussed progress relating to the cooperative research programmes (Figs. 2.9, 2.10, 2.11).

2.4.4 Identification of QTL Loci Related to Wheat Yield and Drought Resistance and Innovation

The project is entitled "Identification of QTL loci related to wheat yield and drought resistance and innovation of germplasm resources in China and Egypt" and was sponsored by the MOST. Profs. Y. Jia and M. He led the project with researchers from the Shijiazhuang Academy of Agricultural and Forestry Sciences, Hebei Province, China (Fig. 2.12).

Fig. 2.7 Group photos showing the collaboration with institutions in South Africa, including the University of Fort Hare, Agriculture Research Council, University of the Witwatersrand

2.4.5 Cooperative Research on Spices, Millet and Herb

Profs. Y. Jia from China and F. Dakora of Tshwane University of Science and Technology, South Africa (Fig. 2.13) carried out the project.

2.5 Africa–China Other Education and Research Collaboration Programmes

There are some significant programmes on a relatively larger scale of collaboration between Africa and China. These are worthwhile mentioning: (1) Confucius Institute (CI) Programme at the Chinese universities paired with African universities; (2) Bilateral research programme between South Africa and China; and (3) BRICS research programme.

Fig. 2.8 Joint collaboration on drought- and disease-resistant maize hybrids between China and Kenya

Fig. 2.9 Group photo showing the collaboration with the institutions from Kenya and Uganda

2.5.1 The Confucius Institute Programme

According to the Center for Language Education and Cooperation, the Ministry of Education, China, the mandate for all CIs globally is to:

- Develop and facilitate the teaching of the Chinese language and culture;
- Promote educational and cultural exchange and cooperation between China and other international communities; and

Fig. 2.10 Group photos showing the collaboration with the institutions from Zimbabwe and Egypt

Fig. 2.11 Group photos showing collaboration with the institutions from Zambia and Ethiopia

- Adhere to the principles of mutual respect, friendly negotiations and mutual benefit.

About 50 Chinese universities paired with African universities are co-hosting 55 CIs around the African continent. The main functions include teaching the Chinese language and culture, herbal medicine, and business and trade courses. These CIs

Fig. 2.12 Visit to the grass farm of the Alexandria University of Egypt

Fig. 2.13 Group photos showing the collaboration with the institutions from Ghana and South Africa

are located at the campuses of certain African universities and bridge the gaps in languages, cultures and communication between Africa and China.

The CIs in Africa have outstanding characteristics such as rapid growth, large-scale and good school-running system. While providing high-quality Chinese teaching, CIs have increasingly become an essential platform for Africa–China educational cooperation and people-to-people and cultural exchanges. However, the CIs urgently need expansion and improvement concerning faculty structure, school-running direction, cross cultural communications, expanding the resource dependence system, enhancing the teaching ability of vocational education, brand competitiveness and comprehensive strength.

As far as the mineral sector is concerned and as stated in Sect. 2.1 of this chapter, it is worthwhile mentioning that the CI in Namibia provides an opportunity for building strong partnerships between the China University of Geoscience Beijing, one of the Chinese universities specialising in minerals, and the University of Namibia. Although the CI focuses on the general mandate of language teaching and cultural exchange, it is believed that collaborations will increase in specialised fields, such as the mineral sector.

2.5.2 Africa–China Other Research Programmes Including BRICS

2.5.2.1 China and South Africa Collaboration

The science and technology cooperation agreement between South Africa and China was signed during the first session of the joint commission meeting on scientific and technological cooperation held in Pretoria in March 1999. Based on the agreement, a Bi-National Commission (BNC) was created to champion bilateral relations in seven sectoral committees, including science and technology. The purpose of the bilateral programme is to support the research exchange programme between South Africa and China. In this regard, funding will be made available for visits and exchanges of researchers conducting joint research in the identified research areas. At its inception in 2001, the programme initially supported ten collaborative research projects and eventually 15 in 2020, with each project cycle lasting two years. The National Research Foundation (NRF) in South Africa fulfils the role of an administrative organisation, while the MOST directly manages the programme's processes in China. Given the current financial situation, the funding scales from the MOST are generally larger than those of South Africa.

The joint research projects are generally divided into research areas, prioritised by the two countries. For example, the designated areas in the 2020 call for proposals included: (1) Smart Manufacturing; (2) Educational Science and Technology (ICT for Education); (3) Renewable Energy; and (4) Transportation Technology .

2.5.2.2 The BRICS Research Programme

The BRICS Science Technology Innovation Framework Programme was established to support excellent research in priority areas, best addressed by a multinational approach consisting of at least three of the BRICS countries. The programme facilitates cooperation among researchers and institutions in a consortium. For example, the 5th coordinated call for the 2021 BRICS multilateral projects was issued in 2020 (NRF, 2020) and supported research areas that included: (1) Transient astronomical events and deep survey science; (2) Antimicrobial resistance: technologies for diagnosis and treatment; (3) Simulation and big data analytics for advanced precision medicine and public healthcare; (4) High Performance Computing (HPC) and Big Data for sustainable development: Solving large-scale ecological, climate and pollution problems; (5) Innovation and entrepreneurship on photonic, nano-photonics and metamaterials for addressing bio-medicine, agriculture, food industry and energy harvesting issues; (6) Materials science and nanotechnology for addressing environmental climate change; agricultural, food and energy issues; (7) Renewable energy, including smart grid integration; (8) Ocean and polar science and technology; (9) Water treatment technology; and (10) Research in aeronautics and aerospace.

2.6   Concluding Remarks

2.6.1   Minerals Sector

The principles and objectives that mainly inform the current collaborative programmes between African countries and China are based on FOCAC, which was established in 2002. In the minerals sector, cooperation in research between African countries and China has evolved over many years and has extended across the entire minerals value chain. There are now collaborative programmes in geosciences, mining, mineral processing and metallurgy, and the mineral sector's capacity building.

With regard to the geosciences, the CGS-China and universities play a pivotal role. They are engaged in collaborative programmes that can be subdivided into two categories, namely geoscience mapping and capacity building. The geoscience mapping programmes primarily aim to assist African countries in developing their geoscience infrastructure, specifically regarding geological, geophysical and geochemical mapping. These programmes are vital in attracting investors to the minerals and mining sector. Interestingly, such support and financial assistance were in the past mainly provided through grants and loans from the European Union and the World Bank. Many geoscientific cooperative programmes between African countries and China reveal that vast areas in African countries need to be geo-scientifically mapped.

However, the mining, mineral processing and metallurgy collaborative programmes are far fewer than the geoscience programmes, and more partnerships need to be developed. Increased collaborative research projects and programmes are needed to create opportunities for knowledge exchange in these areas of the mineral industry. Nonetheless, some agreements and programmes involve universities such as the University of Limpopo and the University of the Witwatersrand. Additionally, in terms of the plans of the agreement between the governments of China and South Africa that established the Joint Research Centre for the Development and Utilisation of Mineral Resources, the second research node centre will be created at Mintek in

South Africa. Such a development could significantly advance mineral processing on the African continent. The establishment of two African Geological Centres by the CGS-China has provided a good opportunity for capacity building, research and technology transfer and teamwork in China and Africa.

2.6.2   Transportation Sector

Based on the Johannesburg Summit of the FOCAC, the co-establishment of Africa–China transport universities has the potential to develop skills needed in developing the African continent and China. These transport university establishments include (1) Nigerian China–Africa Transport University; (2) China–Africa Transport University for Portuguese–Speaking Countries; (3) China–Africa Transport Strategy Research Institute in South Africa and China; and (4) The Silk Road Energy Alliance of Industry, Education and Research.

2.6.3   Agricultural Sector

Through Africa–China agricultural cooperation, a total of 1774 African crop germplasm resources were introduced, among which there were 1532 maize resources which, through domestication and improvement in China, have played a significant role in breeding dwarf, dense, stress-resistant, high-yield inbred and new cultivars.

The Africa–China agricultural cooperation has advanced in terms of research quality in agricultural science and technology for China and Africa, strengthening the Chinese and African scientists' relationship and laying a foundation for future sustainable collaboration. Continued joint projects by researchers from African countries and Chinese universities and institutions have the potential to provide long-term win–win mechanisms that can advance scientific and technological knowledge. We believe that the collaboration has already produced tangible results, such as capacity building and fruitful agricultural products. To date, the various programmes have trained at least eight doctoral and master's students, two engineering and technical personnel, and delivered 256 papers, six innovation patents and 16 awards in China.

2.6.4 Other Collaborative Programmes Including BRICS' Joint Research Programme

There are several important programmes for bilateral research and multiple-country research involving African countries and China in minerals, transportation, and agriculture. Some projects are long-term, while others are often periodic and play an essential role in higher education and training in Africa and China. Some of these projects are: (1) CI Programme at African universities, (2) The bilateral research programme between South Africa and China, and (3) BRICS Research Programme.

2.6.5 Outlook

Africa is the host to a variety of the most important minerals, such as cobalt, copper, and platinum group metals, which play a critical role in the prosperity of human beings and in addressing global warming. This mineral endowment in Africa and advanced technological expertise in China provide an opportunity for a symbiotic relationship that has the potential to create sustainable benefits for the people of these jurisdictions. These benefits can be realised through continued and strengthened collaboration between Africa and China across the entire mining value chain, specifically in higher education and research. While most higher education and research collaboration programmes in minerals are in geosciences, it is foreseen that as geoscience programmes assist in discovering new minerals deposits, Africa–China collaborations in mining, mineral processing and metallurgy will significantly increase. It is also anticipated that with growing concerns about protecting the environment and society, future collaborations in the field of minerals will include research and capacity building in sustainable development, focusing on safeguarding the environment and addressing mining community issues. In addition, future partnerships will most likely involve collaborative research in the exploration, mining and beneficiation of technology minerals such as cobalt, platinum group metals exploring and lithium, which are critical in the decarbonisation of the world.

Transport infrastructure in Africa needs major improvement to cope with its economic development. The logistics costs in land and air transports between African counties have been unbalanced compared with China, Europe and northern America. Most African countries have understood the importance of increasing investment in transportation to strengthen their social-economic advancement. Following the directions of the FOCAC, Africa and China are exploring a better way to work together, and higher education and professional training is expected to play an important role in this regard. The Chinese government has offered over 5 thousand scholarships to Africa each year. Simultaneously, the Chinese government assists African countries in co-establishing transport universities and colleges in line with the 2015 FOCAC Johannesburg Summit. To date, several achievements in this field include the development of high-speed information technology (IT) and online higher education and training programmes. The achievements represent significant advancements in respect of the 4th Industrial Revolution, which is changing the world.

Regarding agriculture, Africa has a total population of 1.34 billion, relative to China (1.426 billion), which accounts for about 35% of the world's 8 billion. However, Africa is significantly less populated, with a land area of about 40 million km^2 and has vast fertile land in many countries, relative to the densely populated China, with a land area of 9.6 million km^2 (United Nations. The uncultivable arble land in Africa is over 60%. There is an opportunity for Africa to build sustainable food security. At present both Africa and China are large importers of food, and the former often experience challenges concerning food security and nutrition amongst its population (FAO, IFAD, UNICEF, WFP and WHO, 2022). The reliance on imports is not sustainable, and the two jurisdictions should continue with the current collaborative capacity-building and research programmes in agriculture to improve the food security. In addition, the increasing adverse effect of climate change necessitates increasing efforts to embark on innovative agricultural research programmes. Accordingly, Africa and China need and should work together to help each other and the world through cooperation, knowledge exchange and transformation of science and technologies in agriculture. Such partnerships can potentially improve food security and peace in Africa, China, and the world.

References

Bariyo, A. (2013). *China to finance Kenya's geological survey*. Fastmarkets.

Benton, D. (2020). *Research Boom: University of the Witwatersrand Collaborate with China University of Mining & Technology*.

China Geological Survey. (2017). *China-Arab States Cooperation on Geological Survey*. China Geological Survey, Ministry of Land and Resources, P. R. China.

ChinaAfrica. (2012). *China University of Geosciences in Beijing*. ChinaAfrica.

Council for Science and Industrial Research. (2018). *CSIR and Gold Yard Partner to Scale Up Industrial Development in Africa*. Council for Science and Industrial Research.

Department of Science and Innovation. (2018). *South Africa and China Celebrate Milestone with New Mineral Resources Joint Research Centre*.

FAO, IFAD, UNICEF, WFP and WHO (2022). The State of Food Security and Nutrition in the World 2022. Repurposing food and agricultural policies to make healthy diets more affordable. Rome, FAO. https://doi.org/10.4060/cc0639en

Forum on China-Africa Cooperation. (2015). *The Forum on China-Africa Cooperation Johannesburg Action Plan (2016–2018)*.

Global Geoparks Network. (2007). *Workshop on Geoparks, Geological & Mining Cons & Tourism Dev in South Africa-Global Network of National Geoparks*. Global Geoparks Network.

Gu, M. (2017). *The Sino-African Higher Educational Exchange: How Big Is It and Will it Continue?* World Education News plus. Reviews.

Li, A. (2018). *African Students in China: Research, Reality, and Reflection*.

Li, A. (2019). *The social and economic history of the Chinese overseas in Africa*. Jiangshu People's Publisher (Volume I, II, III), History of the foreign Chinese and Chinese in Africa (pp. 15–16).

Ma, X. (2020). *Deepen Mining Cooperation to Benefit the People of Sudan*. Embassy of the Peoples Republic of China in Sudan.

Moore, P. (2016). *China-Africa Mining Geospatial Informatics Research Lab Established—International Mining*. International Mining.

Muchie, M., & Swapan Kumar Patra, S. K. (2019). China–Africa science and technology collaboration: evidence from collaborative research papers and patents. *Journal of Chinese Economic and Business Studies*. https://doi.org/10.1080/14765284.2019.1647004

Mulikelela, M. (2016). *Zambia, China Ink Deal for Geological Mapping*. AllAfrica.Com.

National Research Foundation. (2020b). *South Africa/China Joint Research Programme 12th Call for 2021 Joint Research Project Proposals (JRP)*.

National Research Foundation. (2020a). *BRICS STI Framework Programme 5th coordinated call for BRICS multilateral projects 2021*.

Ngoepe, P. (2020). *Computational Modelling and Experimental Science for Sustainable Energy Storage, Mineral Processing and Alloy Development*. Synchrotron Techniques for African Research and Technology.

Nigerian Investment Promotion Commission. (2016). *Nigeria and China Sign Agreement on Mining Cooperation*. Nigerian Investment Promotion Commission.

Southern African Institute of Mining and Metallurgy. (2022). Collaborative Rockbursts Webinar—China and South Africa. https://www.saimm.co.za/saimm-events/saimm-webinars/collaborative-rockbursts-webinar-china-and-south-africa

Wang, X. (2012). Current status of applied geochemistry research in China. *Association of Applied Geochemist, 2012*, 1.

Chapter 3
The Path of China–Africa Higher Education Cooperation in Training Talents for Africa

Mingfeng Tang, Jue Wang, Yajie Yang, and Angathevar Baskaran

3.1 Introduction

At the Beijing Summit of the Forum on China–Africa Cooperation in 2018, China and Africa agreed to build a closer community with a shared future, and will deepen exchanges and cooperation in the fields of politics, economy, culture, education, science and technology, ecology and others. General Secretary Xi Jinping believes that youth are the hope of China–Africa relations. In the eight China–Africa Action Initiatives, many measures focus on cultivating young people and are committed to providing them with better space for development (Chinese Government Network, 2018).

In the age of global integration, the focus of communication between countries gradually spreads from material exchange to knowledge interaction and personnel communication (Zhang & Liu, 2022). The endogenous growth theory emerging in the 1980s introduced knowledge and human capital into the growth model, theoretically explaining the important role of specialized knowledge and human capital accumulation in economic growth (Liu & Fan, 2019). Since then, the theory of

M. Tang (✉) · J. Wang · Y. Yang
Sino-French Innovation Research Center (SFIRC), Faculty of Business Administration, Southwestern University of Finance and Economics, Chengdu, China
e-mail: tang@swufe.edu.cn

J. Wang
e-mail: juewang8236@swufe.edu.cn

A. Baskaran
Department of Political Science, Public Administration and Development Studies, Faculty of Business and Economics, University of Malaya North-South Research Centre (UMNSRC), University of Malaya & Senior Research Associate, SARChI (Science, Technology and Innovation Studies), Tshwane University of Technology, Pretoria, South Africa
e-mail: baskaran@um.edu.my

© The Author(s) 2024
M. Muchie et al. (eds.), *China-Africa Science, Technology and Innovation Collaboration*, https://doi.org/10.1007/978-981-97-4576-0_3

endogenous growth has been used to widely explain the relationship between education and economic growth, believing that technological innovation, research and development and population change are the most fundamental driving sources of economic and social structural changes and endogenous economic growth (Aghion et al., 2015).The key factor to promote international economic development lies in the training of high-quality talents with international horizon (Fang, 2020). The talent is the strategic resource for national economic construction, scientific and technological development, and cultural progress (Wang et al., 2020). Higher education institutions provide students with specific, professional and complex knowledge and technologies and prepare them to become responsible, innovative and high-performing society builders. Higher education institutions are the centre of knowledge generation.

Driven by the increasingly fierce international competition and the internationalization process of higher education, the whole world has higher requirements for the training of higher education talents. The development of the country will also be more and more closely dependent on scientific and technological progress and innovative talents to support. Therefore, the exchanges and cooperation between China and Africa in the field of higher education can help to promote mutual learning among civilizations. Through China–Africa higher education cooperation, it is of great significance to help African countries improve their education level and jointly train African scientific and technological innovation talents, so as to strengthen Africa's talent pool, improve its innovation capacity and break its development bottlenecks. At the same time, it will open up a broad space for China's international thinking and related theories, and contribute to the sustainability of China–Africa development and South-South cooperation.

Education (e.g., affordable vocational training, quality higher education) is one of the most powerful and proven vehicles for eradicating poverty, transforming lives and achieving breakthroughs in all the UN Sustainable Development Goals (United Nations, 2015; World Education Forum, 2015).

The establishment of FOCAC is an institutionalized mechanism of China's public diplomacy with Africa and an exclusive platform for multilateral exchanges between China and Africa. It plays an extraordinary role in China–Africa relations. It creates a platform to articulate common interests between China and Africa (Lemmy, 2020). Among them, education and high-level education play an important role in promoting the sustainability of China–Africa relations. The university will be used as a platform to link vocational training, formal higher education, academic research and think tanks on both sides, and pass on China's development experience to Africa. These different cooperative projects are a signal that the relationship between China and Africa has gone beyond a purely commercial level of determination (Caruso, 2020).

China–Africa higher education cooperation is of great practical significance. First, educational exchanges and cooperation between China and Africa are the shared aspirations of the two peoples, which is also the key to the enduring development of exchanges and cooperation. The China–Africa education community has reached broad consensus among the people of both sides. Also, the model of sustainable educational exchanges and cooperation will help enhance the friendship between the African and Chinese people and further deepen bilateral relations. Educational exchanges and cooperation need to be win–win and even multi-win beneficial relations. The win–win relation cannot only boost the development of African education and help the development and progress of the African continent, but also it can raise the level of internationalization of Chinese education. It can become an important way to promote people-to-people exchanges between China and Africa and enhance China's soft power. China–Africa educational exchanges and cooperation also can reflect the real need for cultural exchanges and mutual learning between the two sides. Especially since the beginning of the twenty-first century, China and Africa have become important forces promoting the progress of civilizations and building a harmonious world on a global scale. Africa is endowed with rich natural resources and broad market prospects, but it is also constrained by its lagging economic development with shortage of high-end talents, weak education and low level of science and technology. China respects Africa's sovereignty and supports African unity by attaching importance to Africa's propositions. China–Africa education cooperation embodies the vision of inclusiveness, equality, mutual learning and common development among different civilizations. The China–Africa educational exchanges and cooperation have helped Africa to achieve sustainable talent development. The achievements China has made since its reform and opening up have caught the world's attention. History has proved that Africa needs China for its development, especially in the field of education. For more than half a century, China has helped African countries train much-needed talents in various fields by building schools, training human resources and providing scholarships.

China–Africa higher education cooperation will play an important role in providing high-end talents by strengthening education diplomacy and realizing policy coordination. China's innovative education will become an important force driving the rapid, steady and innovative development of the Chinese and African economies.

3.2   Development History of China–Africa Higher Education Cooperation

China–Africa higher education cooperation has a long history. From the 1920s to the present, China–Africa higher education exchanges and cooperation have explored a development path from scratch, from sporadic exchanges to institutionalized cooperation, and from a single form to various forms (Li, 2021). Educational exchanges began in the early years of the Republic of China. As early as 1922, China sent students to study in Al-Azhar University in Egypt before the founding of the People's Republic of China at the time when China and Africa had little cooperation in higher education.

The early 1950s to the late 1970s were the beginning stage of China–Africa higher education cooperation. At that time, the main form of higher education exchanges between China and Africa was China's large-scale free assistance of books, teaching instruments, school supplies and other educational materials to Africa. China and

Africa also exchanged international students, teachers and education delegations. In the 1950s, both China and African countries were faced with the arduous task of developing the economy and improving people's livelihood. It became an urgent need for both sides to strengthen exchanges and cooperation in the field of education and help each other cultivate talents needed for national development. In the 1960s, more African countries gained independence and established diplomatic relations with China successively. China–Africa friendly relations developed rapidly. The establishment of diplomatic relations laid a solid foundation for China–Africa higher education cooperation and exchange (Li, 2021). In order to further promote the cooperation in higher education between China and Africa, China has provided government scholarships for African international students in China, which has resulted in a significant increase in the number and nationality of African international students (Wang, 2019). By the end of 1962, about 118 African students were attending Chinese institutions of higher education (Chkaif et al., 2022). By the end of 1966, more than 190 African students from 14 countries had studied in China. China has also sent teachers to more African countries to teach Chinese, mathematics, physics, chemistry and other courses, which has promoted the development of subjects in African universities and middle schools.

From the 1970s to the end of the twentieth century, due to the restoration of China's legal seat in the United Nations and the implementation of reform and opening up, China–Africa higher education exchanges and cooperation also developed rapidly and gradually became systematic. During this period, China's own education system began to be established and gradually improved. At the same time, international exchanges and cooperation in education was highly valued by the government (He, 2007). In addition to the government-led exchange of academic visits and exchange of international students in the aid stage, there were also forms of self-funded students, laboratory assistance, cooperative scientific research projects, professional research classes, and inter-university cooperation between Chinese and African universities (Wang, 2020). In order to help African countries to cultivate more high-level of management and scientific research personnel, China sent teachers to Africa to teach the frontier disciplines and natural science with the basic education courses and humanities extension that started the China–Africa education relation The level of qualification was also gradually upgraded from undergraduate to postgraduate with the increase enrolment ratio of doctoral students and master's students. Another important development is the beginning of scientific research cooperation between Chinese and African universities. African students in addition to continue to receive and send Chinese teachers to Africa, at the request of African countries, China began to selectively help colleges and universities in the new curriculum development in parts of Africa and set up all kinds of laboratories, scientific research, discipline construction and human resources training, for the economic development of African countries to cultivate more professional talent (He, 2007). In the 1980s, the number of African students sent to China increased to 2,245, and it increased to 5,569 in the 1990s, and the source country of international students increased to more than 50 (Compiling Group for China Africa Educational Exchange & Cooperation, 2005). With the 1990s, Beijing University, Nanjing University, Tsinghua

University, Southeast University, Beijing University of aeronautics and astronautics and other 40 universities have been identified as national pilot unit to study abroad. The universities directly sent aid to African teachers' autonomy and the number of African students to China also increased significantly. In the 1990s, China sent about 350 teachers and students to African countries (CRI Online, 2019).

In the twenty-first century, with the help of multilateral cooperation and consultation mechanisms such as the Forum on China–Africa Cooperation (FOCAC), China and Africa have closer ties. China–Africa exchanges and cooperation in higher education are gradually developing in various fields and levels. The Forum on China–Africa Cooperation, first held in 2000, has become an important starting point for China–Africa higher education exchanges and diversified cooperation. The African Human Resources Development Fund has been set up to help African countries train various talents and select outstanding domestic universities to participate in assistance projects. The areas of assistance include education, health care, training of administrative personnel and training of vocational and technical personnel. In addition to sending students to each other, under the guidance of the Addis Ababa framework signed in 2003, the cooperation between China–Africa has gradually diversified, such as holding Confucius Institutes and short-term vocational and technical training courses. Cooperation between universities has also gradually increased. In 2009, the Forum on China–Africa Cooperation (FOCAC) adopted the Sharm el-Sheikh Action Plan, proposing that China and Africa select 20 universities each to establish a new model of "one-to-one" inter-university cooperation, known as the China–Africa University 20 + 20 Cooperation Plan. The cooperation and exchanges between Chinese and African universities based on their respective strengths and characteristics have opened a new era in China–Africa higher education collaboration. After nearly 20 years of development, China–Africa higher education has basically taken shape in the field of China–Africa higher education. It is guided by the introduction of policies, supported by the establishment of institutions, driven by government and universities creating fund (Wang, 2020).

3.3 The Current Development Status of China–Africa Education Cooperation

According to the Blue Book on the African New Economy (2019 edition) jointly released by Good Hope Watch and Cloud Time Capital, Africa's education level is generally backward, and relatively few countries can popularize primary education. In 2017, the continent's literacy rate was only 64%, well below China's 95% and the world's 86%. In addition, more than 34 million children are out of school, the inequality in basic education, and the mismatch between the supply and demand for vocational education are all huge problems in Africa's education sector.

Since the reform and opening up, China's education has entered a period of comprehensive development. Compulsory education has been continuously improved, higher education has been gradually strengthened, and the national education level has been continuously improved. In 2018, the retention rate of 9-year compulsory education reached 94.2%; 28.31 million general college students increased by 32 times over 1978; and the average years of education for the population aged 15 years and above increased from 5.3 years in 1982 to 9.6 years. Since the 18th National Congress of the Communist Party of China, China has made new historic progress in education, with its overall development level ranking among the highest in the world, and a modern vocational education system has been initially established. In 2018, the gross enrolment rate of higher education has reached 48.1%, higher than the average level of middle and high-income countries; the secondary vocational education schools reached 10,340 (The Ministry of Education of China, 2019). Therefore, China has the ability and conditions to cooperate and help Africa in the field of education to cultivate talent reserve forces for the revitalization of the African countries economic development to enhance its capacity for independent development.

From the perspective of the current status of China–Africa higher education cooperation, over the past 70 years and more, China–Africa educational exchanges and cooperation have always been based on the people of both sides by highlighting the people-oriented characteristics focused on communication and understanding to achieve the purpose of mutual benefit. China's support for Africa's education, without strings attached, has made a historic transition from "giving people fish" to "teaching people how to fish". It has not only helped Africa make social and economic progress, but also gradually won the trust and friendship of the African people. Secondly, China attaches great importance to the practical effects of educational exchanges and cooperation with Africa. In response to the underdeveloped economic conditions and relatively scarce human resources on the African continent, in recent years, in addition to increasing educational assistance and the number of overseas students, China has also focused in helping African countries to train supporting talents in light of the actual conditions of African countries economic development. For example, vocational education and human resources training are taking up an increasing proportion of bilateral exchanges and cooperation. Meanwhile, targeted cooperation projects between universities have also realized the "precise delivery" of African personnel training.

At present, China–Africa cooperation in vocational education and language education has achieved a lot. Vocational education has trained many urgently needed talents for the economic transformation and development of African countries (Liu, 2018). Vocational education is for many African countries the strategic priorities for the development of education. The main form of China–Africa vocational education cooperation is to include and use vocational education workshop by sending from China vocational education teachers. This is to develop with African countries to vocational education students by setting up the Luban workshop in African countries with China's vocational education at colleges and universities, etc. (Li, 2021).

Over the past decades, China has trained more than 300,000 practical talents for Africa through various short-term professional training programs, covering 17 fields related to national economic and social development, including agriculture, forestry, environmental protection, public administration, transportation and medical care. For example, through various training programs jointly conducted with Cote d 'Ivoire and more than 10 African countries, relevant Chinese companies have trained more than 3,000 professionals for these countries. These people, in turn, pass on the skills they have learned to those around them, creating a "spillover effect" from the training, creating jobs for local people and enabling more people to benefit from Chinese knowledge, skills and experience. In 2017, the South African government began to fund and send local students to China's vocational education institutions to carry out 1-year study and internship programs. By 2021, more than 1,200 South African students have come to China for study and internship through such programs, with more than 20 Chinese universities participating in them. The cooperation projects are mainly related to information, construction and mechanical technologies in line with South Africa's technology development strategy.

In terms of language education, Confucius Institutes are important institutions for African students to learn Chinese language. Currently, there are nearly 200,000 registered students and more than 1 million students have been trained in Africa (Li, 2022). By 2019, China had set up 61 Confucius Institutes and 48 Confucius classrooms in 46 countries in Africa, which have enrolled more than 15,000 students in the past 14 years (Confucius Institute website, 2020). South Africa has the largest number of Confucius Institutes and Confucius classrooms in Africa. As of August 2019, South Africa had six Confucius Institutes and three Confucius classrooms, and the Chinese language has been included in its national education system. In addition, 45 local primary and middle schools have opened Chinese courses (People's Daily Online, 2019). This model can not only strengthen language education in Africa, but also help African young people, especially college students to understand Chinese culture and lay a foundation for them to study in China. It can also play a synergistic role in promoting the work of non-Chinese enterprises.

In terms of African students coming to study in China, 492,185 international students from 196 regions were studying in China in 2018, according to statistics from China's Ministry of Education. Among them, 59.95% are from Asia, while Africa accounts for the second largest proportion (16.6%) (see Table 3.1). China has become the largest destination country for overseas students in Africa (Ministry of Education, 2019). As can be seen from Fig. 3.1, the number of overseas students in all continents is on the rise, but Africa is rising at the fastest rate.

The number of African students in China has grown rapidly in recent years as China has implemented many preferential policies for helping African students to study in China. From 1999 to 2018, the number of African students studying in China increased from 1384 to 81,562, an 80-fold increase with an average annual growth rate of 24 percent. It is worth mentioning that before 2003, the annual growth rate of African students in China was less than 10%. After 2003, the number of African students in China has achieved rapid growth, and the annual growth rate from 2006 to 2010 has exceeded 30%, with the highest growth rate as high as 58% in 2007. Ten years later, despite the slowdown in the growth rate of the total number of students in China, the annual growth rate of African students in China has maintained around

Table 3.1 The number and percentage of overseas students studying in China from all continents

Region	Number	Percent
Asia	295,043	59.95
Africa	81,562	16.57
Europe	73,618	14.96
America	35,733	7.26
Oceania	6229	1.27

Statistics of Studying in China in 2018—Government Portal of Ministry of Education, PRC (moe.gov.cn)

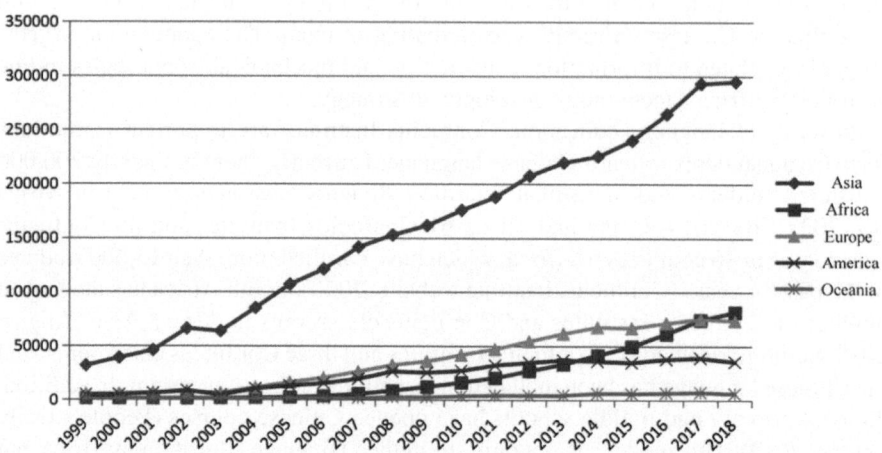

Fig. 3.1 Line chart of the number of Chinese students from 1999 to 2018. Department of International Cooperation and Exchange of the Ministry of Education: Brief Statistics of Overseas Students from 1999 to 2018

20%, maintaining a high growth trend. With this rapid growth, the number of African students surpassed the Americas in 2014 and Europe in 2018 to become the second largest source of students in China's overseas education after Asia (Li & Huang, 2021).

In addition, there is Open Distance Learning (ODL), such as MOOCS. However, there is currently no information on whether there are any MOOCs designed by Chinese and African scholars. ODL for university faculty professional development is an area that Chinese and African universities can further explore to enhance cooperation, facilitate knowledge transfer and exchange of ideas, and improve teaching and learning practices. Professionals from Chinese universities can co-design some online courses with counterparts from African universities to promote professional development and enrich students' learning experiences (Zhu & Chikwa, 2021).

According to the Journal named as West Asia and African, over the past half a century, the educational exchange between China and Africa has been continuously innovated and developed in form and content, from the initial exchange of international students to the current multi-level, multi-field and multi-form educational exchanges and cooperation. It has strengthened people connectivity both in China and Africa, built up knowledge/technology competences of African people, enhanced traditional friendship between China and Africa, promoted further connection of the world's two largest markets, further shared human, capital, technology, knowledge, raw materials and other resources for pursuing sustainable development and they are jointly building a China–Africa Community with a Shared Future.

3.4   Challenges and Suggestions for China–Africa Higher Education Cooperation

In view of the unique development history and practical needs of China–Africa higher education cooperation, in the context of globalization, China and Africa are faced with many problems in the new era of educational exchanges and cooperation, such as relatively single forms and subjects of higher education cooperation, unbalanced structure of educational cooperation, and misunderstanding caused by international public opinion. At present, China–Africa higher education cooperation and exchanges mainly take the form of African students studying in China. There are insufficient exchanges and interactions between ordinary students in Chinese and African universities. On the other hand, due to the inherent differences between China and Africa in language, culture, religious belief, lifestyle and other aspects, China and Africa have a certain deviation in their understanding of each other. Some Chinese students still think that Africa is poor and backward, and are reluctant to study and exchange in Africa. They believe that China's educational cooperation with Africa is purely unilateral aid. At the same time, influenced by Western public opinion, some Africans lack an objective, comprehensive and correct understanding of China and China–Africa educational cooperation, believing that China–Africa higher education cooperation belongs to China's "cultural colonization" and "cultural expansion". Therefore, improving cooperation and exchange mechanisms, innovating cooperation areas and forms, and encouraging two-way interaction between Chinese and African higher education are the key points for deepening the development of China–Africa higher education cooperation in the new era.

First, we should improve the policy design, cooperation and exchange mechanisms of China–Africa higher education cooperation, and jointly build cooperation platforms of various forms. China–Africa higher education cooperation is a strategic measure to serve the overall situation of China–Africa relations. The governments of both sides should improve the relevant top-level policy designs, and actively strengthen the exchanges and contacts between the education departments and universities of the two sides, to build platforms for the in-depth cooperation between Chinese and African Institutions of Higher Education. In addition to the Forum on China–Africa Cooperation established in 2000 and the 20 + 20

Cooperation Program jointly proposed in 2009, China–Africa can build inter-school cooperation platform between colleges and universities, such as China–Africa Joint Training Centre in university, China–African universities of science and technology innovation and entrepreneurial platform, China–Africa online teaching and scientific research platform, etc. It also calls for more aspiring young people to join the platform and provide them with a better training environment to show their talents, which will promote the deeper development of China–Africa higher education cooperation. At the same time, Chinese and African universities can jointly hold talent exchange activities. Affected by the epidemic, international conferences can be held online, with multi-themes, multi-levels and all-round aspects, to conduct personalized training for talents from different fields and disciplines.

Second, we should innovate the areas and forms of cooperation. China–Africa cooperation involves many aspects, each of which requires technical support and human resources. Higher education should give full play to its own advantages in scientific research and personnel training, broaden the field of cooperation and deepen the degree of cooperation. China and Africa should continue to explore new types of cooperation.

In addition to joint training, China and Africa can carry out cooperation in the form of joint scientific research, joint curriculum development and industry-university-research linkage. Through the cooperation of the government, industry, universities and research institutes, China and Africa can cultivate interdisciplinary, research-oriented, skilled and applied talents in an innovative way. Focusing on key areas that restrict African countries' economic and social development and enhance national competitiveness, China and Africa can continue to cultivate talents for scientific and technological innovation, in order to encourage universities and enterprises to build industry-university-research cooperation bases, and to enhance their capacity for independent scientific and technological research and development. Enterprises familiar with the industries of African countries are encouraged to participate in the training of talents in universities and colleges, in order to help enhance the majors and related courses of electronic information, social sciences, biological sciences, international economics and trade, etc., so as to train high-end talents in short supply for African countries or regions.

At the same time, we should pay attention to the training of interdisciplinary talents. With the integration of Internet technology and finance, culture and other industries, talents in the new era must have multi-disciplinary knowledge to meet the needs of social development. China–African universities can cooperation with each other to train engineering background students with management knowledge and social sciences background students with some nature sciences knowledge. They can also offer interdisciplinary double degree programs to students. That means students originally majoring in computer sciences can also major in business administration. If they can satisfy the graduation criteria of these two majors, students can get two degrees. One is computer sciences, the other is business administration.

Third, we should broaden the international perspective, encourage two-way inter-action between China–Africa higher education, give full play to the advantages of both sides, and jointly cultivate more scientific and innovative talents. China–Africa education cooperation has never been a one-way export from China, but a two-way interaction featuring exchanges and mutual learning among civilizations. Both China and Africa have their own advantages. The advantages of China, on the one hand, has strong capacity in building educational infrastructure. On the other hand, China's science, technology, educational experience and development model are relatively advanced and mature. In particular, in the field of basic education, such as literacy education, 9-year compulsory education and mathematical knowledge teaching, China has accumulated many experiences and methods for African countries to learn from. Africa's advantage is that its population continues to grow rapidly, with a young population and a growing demand for educational resources. Young Africans are eager to learn new knowledge and receive higher education. The African Union and many African countries support local higher education institutions to cooperation with oversea high quality universities for acquiring new knowledge and technology and cultivating local talents.

In order to ensure the quality of China–Africa higher education cooperation, win–win collaboration should be achieved on the basis of considering the demand of both sides. Therefore, it is important to maximize consensus and focus on areas of mutual concern, such as scientific and technological innovation. China and Africa can strengthen cooperation in scientific research fields through building substantive scientific research institutions. Chinese universities are superior in terms of research conditions and capacity, while African universities are superior in terms of local culture, geography and cultural knowledge. The two sides can leverage their respective strengths to jointly conduct research on Africa-related issues. China and Africa should make good use of the subjectivity, initiative and enthusiasm of universities to promote cultural and people-to-people exchanges by improving the education quality of international students and building Confucius institutes, so as to build a China–Africa Academic Community.

3.5 Future Prospects

In the future, there is still great room for the development of exchanges and cooperation in higher education between China and Africa. Especially in the post-pandemic era, the exchanges and cooperation will be more frequent, more comprehensive and can continue to strengthen.

With the promotion of FOCAC, China–Africa educational exchanges will continue to innovate and develop in both form and content. Compared with the past way of directly delivering scientific research results, the future educational exchanges and cooperation will promote the overall economic development of Africa, promote more social participation, cultivate innovative talents, and help Africa to conduct independent research and development of scientific and technological products.

Acknowledgements We would like to thank the Sichuan Provincial Philosophy and Social Science Planning Office and Wang Kuancheng Education Foundation for funding this research project under Grants (Nos. SC21ZD010 and 48).

References

Aghion, P., Akcigit, U., & Howitt, P. (2015). The schumpeterian growth paradigm annual review of economics. *Annual Reviews, 7*(1), 557–575.

Caruso, D. (2020). China soft power and cultural diplomacy: The educational engagement in Africa. «*Cambio: Rivista sulle trasformazioni sociali*», *10*(19), 47–58. https://doi.org/10.13128/cambio-8510

Chkaif, B., Marwan, S., Xu, M. H., & Hanane, T. (2022). African students' mobility to China: An ecological systematic perspective. *Trames Journal of the Humanities and Social Sciences, 26*, 185–206.

Compiling Group for China Africa Educational Exchange and Cooperation. (2005). *Educational exchange and cooperation between China and African countries*. Peking University Press.

CRI Online. (2019). China and UNESCO signed an agreement to support higher vocational education in Africa. http://www.sohu.com/a/347940060_115239

Fang, S. X. (2020). Research on the reform of the talent training mode of international education in the new era: Comment on "Research on the Talent Training Mode of Transnational Higher Education." *Chinese Journal of Education, 8*, 113.

Forum on China–Africa Cooperation (FOCAC). (2015). The forum on China–Africa cooperation Johannesburg Action Plan (2016–2018). September 11. https://www.fmprc.gov.cn/mfa_eng/zxxx_662805/t1323159.shtml

Forum on China–Africa Cooperation (FOCAC). (2018). Kenya roots for partnership with China to enhance skills upgrade for youth. *FOCAC Archives*, August 20. www.fmprc.gov.cn/zflt/eng/zxxx/t1546847.htm

He, W. P. (2007). A summary analysis of China–Africa educational exchanges and cooperation: Development phases and challenges. *West Asia and Africa, 3*, 13–18.

Hu, J. H. (2009). Reform of talent training system in colleges and universities under the vision of a powerful country in higher education. *Higher Education Research, 30*(10), 1–5.

Ji. P. (2013). *Educational exchanges and cooperation in the Central African historical process research and development*. Zhejiang Normal University, Master dissertation.

Lemmy, N. M. (2020). Forum on China–Africa cooperation: An assessment of modes of diplomatic engagements, its achievements, and aspirations. *Africa Journal of International Studies, 1*(1), 28–37.

Liu, W., & Fan, X. (2019). Internal logic and practice path of modern economic growth theory. *Journal of Peking University (philosophy and Social Sciences), 56*(3), 35–53.

Li, Y. J. (2021). A multiple-perspective analysis of China–Africa educational cooperation and exchange. *Journal of China–Africa Studies, 2*(4), 76–98.

Li, F. J. (2022). Discussion on the high-quality development of Confucius Institute in Africa under the new situation. *Comparative Study of Cultural Innovation, 1*, 162–165.

Li, B., & Huang, W. J. (2021). A study of the evolutionary trends of African students in China and relevant countermeasures based on 1999–2018 data. *Journal of Yunan Normal University (teaching and Studying Chinese as a Foreign Language Edition), 19*(6), 82–89.

United Nations. (2015). *Sustainable development goals: 17 goals for transforming the world*. https://www.un.org/zh/node/180631

Wang, Y. J., Wu, Y. N., & Lu, L. Z. (2020). Thoughts and outlook of China–Africa higher-education cooperation in the new era. *Journal of World Education, 489*(9), 29–37.

Wang, B. (2019). On the stability and long-term cooperation between Chinese and African higher education in the context of the belt and road. *Journal of Qiqihar University (physics and Sciences), 2019*, 168–170.

World Education Forum. (2015). *Final report*. https://unesdoc.unesco.org/ark:/48223/pf0000 243724

Zhang, W., & Liu, B. C. (2022). Soft power of education in the era of global competition: Connotation, challenges and response. *Tsinghua Journal of Education, 43*(1), 87–95.

Zhu, X., & Chikwa, G. (2021). An exploration of China–Africa cooperation in higher education: Opportunities and challenges in open distance learning. *Open Praxis, 13*(1), 7–19.

Websites

http://www.hanban.org/confuciousinstitutes/node_10961.htm
http://www.moe.gov.cn/jyb_xwfb/gzdt_gzdt/s5987/201904/t20190412_377692.html.
http://www.moe.gov.cn/jyb_xwfb/s5147/201907/t20190703_388746.html?authkey=boxdr3
http://www.moe.gov.cn/s78/A20/gjs_left/moe_850/tnull_1204.html
http://www.moe.gov.cn/s78/A20/s3117/moe_854/tnull_48799.html

Chapter 4
China–Africa Higher Education Collaboration from the Perspective of Synergistic: The Case of Beijing Foreign Studies University

Dan Yang, Mingfeng Tang, Yakun Zhang, Yifan Mei, and Mammo Muchie

4.1 Introduction

Since 2013 Chinese President Xi Jinping proposed *the Silk Road Economic Belt* and *21st-Century Maritime Silk Road (the Belt and Road)* cooperation initiative, it has been widely responded by the international community, especially African countries along the Belt and Road zones (Liao, 2020). In 2015, the African Union issued *the 2063 Agenda* which demonstrated more support will be given for in-depth China–Africa cooperation in various fields (Xinhua, 2021). Reaching the high-level consensus of peaceful development and upholding the development concept of extensive consultation, joint contribution and shared benefits, China and Africa make common unremitting efforts to build a unified and stable, prosperous and strong,

D. Yang
Beijing Foreign Studies University, Beijing, China
e-mail: yangdan@bfsu.edu.cn

M. Tang (✉) · Y. Zhang · Y. Mei
Sino-French Innovation Research Center (SFIRC), Faculty of Business Administration, Southwestern University of Finance and Economics, Chengdu, China
e-mail: tang@swufe.edu.cn

Y. Zhang
e-mail: karenzyk@163.com

Y. Mei
e-mail: meiyifan@smail.swufe.edu.cn

M. Muchie
DSI/NRF SARChI Research Chair on Science, Technology and Innovation Studies, Tshwane University of Technology, Pretoria, South Africa
e-mail: muchiem@tut.ac.za

© The Author(s) 2024
M. Muchie et al. (eds.), *China-Africa Science, Technology and Innovation Collaboration*, https://doi.org/10.1007/978-981-97-4576-0_4

peaceful and tranquil Africa with five major cooperation priorities: policy coordination, facilities connectivity, unimpeded trade, financial integration and people-to-people bond. President Xi proposed the vision of "jointly building a China–Africa Community with a Shared Future" at the Beijing Summit of the Forum on China–Africa Cooperation (FOCAC) in 2018, which further elevated China–Africa cooperation to an unprecedented level.

In China–Africa multi-disciplinary cooperation, higher education is an important element of the "people–to-people bond" in the construction of the "Belt and Road Initiative" and can play a key role in cultivating high-end talents, disseminating Silk Road culture, strengthening educational cooperation, and achieving policy convergence. Furthermore, it is also helpful to cement the public opinion foundation for deepening and solidifying China–Africa cooperation (MOST 2019). If we view China–Africa higher education cooperation as a subsystem of China–Africa Community with a Shared Future, talent training, scientific research, and social services are the components of the subsystem and a vital driving force to the rapid, healthy, and innovative development of the economy and society in China and Africa.

Africa is a continent with the world's youngest population. According to the data released by the African Union in 2021, the continent is home to more than 400 million people between the ages of 15 and 35, representing 40 percent of its population. By 2030, this segment of the youth is expected to represent 75 percent of Africa's population. This signals the continent has a competitive advantage over most developing and developed nations, as its young population makes the continent attractive to investors (Lefifi et al., 2021). Yet, young labour force without skills and competences is unable to promote the productivity of Africa. Therefore, improving the higher education system is crucial to achieve demographic dividend and economic growth of the African continent. Being the largest developing country in the world, China can provide African countries with various experiences in replicating the development model that enabled East-Asian countries to capture demographic dividends during the period covering the early 1960s to the 1990s (Groth et al., 2019). While seeking to advance its own development, China tries to offer any assistance possible to Africa without setting any political conditions, and to benefit African people through developmental advances (King, 2010).

Overall, China–Africa higher education cooperation has shifted from one-way aid to Africa, which began in the 1950s, to the current bilateral in-depth synergistic exchanges and cooperation. In addition to the government-led exchange visits of scholars and exchange of international students during the aid phase, new forms of higher education cooperation like laboratory assistance, cooperative research projects and inter-university cooperation between Chinese and African universities have also emerged, and they include around four modes: vocational education, language education, study in China, and cooperative schooling (Teng et al., 2016). In 2009 *Sharm el-Sheikh Action Plan* of FOCAC, China and Africa jointly proposed the "20 + 20 Cooperation Program of China–Africa Universities", which effectively promotes the deep-level cooperation and exchange between China and Africa. So far, 52 of the 54 African countries have signed the Belt and Road cooperation agreements with China. According to the data of Chinese Ministry of Education, in the past 20 years, the number of students from African countries coming to China has increased from less than 2000 in 2003 to more than 80,000 in 2018, cultivating groups of specialized and compound talents for African countries and contributing to the development of Africa in various fields. China has gradually become the favourite destination country for African students and the world's top 4 popular country for international students. In the future, how to further improve the level of China–Africa higher education exchanges and cooperation, and provide strong think-tank support for China–Africa win–win cooperation under the framework of *the Belt and Road* Initiative will become the focus of research.

Language inter-exchange is a prerequisite for people-to-people bond and a foundation for building *China–Africa Community with a Shared Future.* At present, nine of the top ten language universities in China have opened African institutes or African Studies Centres. Beijing Foreign Studies University (BFSU), an excellent representative among Chinese foreign language universities, has relied on its language discipline positioning, conducted a 60-year research on African languages and accumulated abundant experiences in African cooperation; 18 of the 20 African languages offered by BFSU are the unique national first-tier disciplines in China. Undoubtedly, BFSU is a typical example for studying China–Africa higher education cooperation.

As a sub-system of China–Africa cooperation system, higher education cooperation is an inseparable component and it is a multi-factor system, which requires the effective cooperation synergy of China–Africa universities, governments, enterprises and international organizations. Professor Hermann Haken first proposed the term "Synergistics" and synergistic approach to the science. He suggests to employ the concept of synergistic to understand various nonlinear complex systems and discoveries (Yakimtsov, 2018). Adopting the synergy theory (Haken, 2004) and case study method, this paper selects Beijing Foreign Studies University as an example and uses triangle sources (e.g. first- hand data, published documents and website) to explore its unique role and the mechanism of interaction and coupling among diverse elements in the subsystem of China–Africa higher education collaboration for analysing the specific measures and results achieved in the collaboration process. Based on the research results, we will find out challenges and future opportunities for China–Africa higher education cooperation, and then provide some implications for governments and universities in China and Africa.

4.2   Overview of China–Africa Education Cooperation

Inspiring by the theory of synergetic, we view China–Africa higher education cooperation as a subsystem under the China–Africa cooperation framework, and the elements in the subsystem such as government, universities and non-government organizations (NGO) have different positioning and unique roles, jointly taking an effect of interactive coupling on the China–Africa cooperation.

The goals and guidelines for this subsystem are set by the Chinese and African governments under their synergistic collaboration. There are two interwoven political factors of China's engagement with Africa. One derives from the friendship of long-term South–South cooperation; the other draws on China's role as a key member of the global community with its growing commitments to development assistance provision to other developing and under-developing economies (King, 2010). The *Sharm el-Sheikh Action Plan* adopted by the FOCAC in 2009 calls for a more sustainable, comprehensive, full-time and regular partnership between China and Africa in higher education through the *China–Africa University 20 + 20 Cooperation Program*. In July 2016, the Chinese Ministry of Education launched *the Education Initiative to Promote the Construction of the Belt and Road*. Being the flagship initiative of the *Silk Road Assistance Program*, it announced that the Belt and Road Education Community started to carry out in the countries along the route, which opens a new chapter of China–Africa governments' education cooperation (Chinese Government Website, 2016). In addition, *Opinions on Accelerating and Expanding the Opening of Education in the New Era* in June 2020 exhibits that the Chinese government will also cooperate with international and multilateral organizations, such as UNESCO, for ensuring the quality in China–Africa education cooperation field (MOE, 2020).

Universities are the most active element in this subsystem, but its effective operation cannot be separated from the lubricating role of the government. For example, Chinese government founded government scholarships to subsidize African students studying in Chinese universities or conducting scientific research. According to the Chinese Ministry of Education, the total number of African students among international students coming to China in 2018 was 81,562, accounting for 16.57% of the total incoming foreign students. China has become the largest destination country for African students (Wang, Wu, & Lu, 2020). The synergy of education resources between Chinese and African universities has entered a development stage since 2000, and gradually changed to a direction of multi-disciplinary and multi-level cooperation. In recent years, China has established new models including the *20 + 20 Cooperation Program between Chinese and African Universities* and Confucius Institutes. In order to achieve win–win cooperation, Chinese universities have sent delegations to visit Africa; they have provided assistance based on the requests of African parties, and offered hospitality and help for African delegations to China after signing cooperation agreements with their African partner institutions. In the process of cooperation, African universities have enhanced their capability for discipline construction, competence building and international academic exchanges. The cooperation has also helped to train plenty of specialized talents for Africa, improved the ability of African universities in teaching, researching and contributing to their local societies. For the Chinese side, the cooperation assists Chinese universities in exploring more training programs for Chinese students and more research programs with African researchers. The two-way mobility of students and staff between China and Africa strengthens mutual understanding, cooperation and people-to-people bond. Both sides have achieved remarkable success because of cooperation synergy.

In addition, the synergy between NGOs and universities also plays an important role in this subsystem (Nordtveit, 2011). The key task in the initial stage of the implementation of the "Belt and Road Initiative" is to train a group of high-quality talents who act as ambassadors between China–Africa communication and infrastructure construction through the synergy of education and industry, enterprises and other NGOs have adopted the science-industry linkage approach to offer offshore education programs in Africa through the Luban Workshops, making full use of the capital, management and social operation advantages to meet the synergistic need between education and industry in Africa (Wu et al., 2018).

4.3 Case Study: Beijing Foreign Studies University

The term "synergistic" was first proposed in 1983 and later used to explore the whole effect resulted from the interactions among the sub-elements of a system, to obtain effects that cannot be achieved by individual elements alone.

"Synergistic" originates from the Greek word—synergeia. It means cooperation, assistance, participation. It has been applied to the fields of management and pedagogy. Based on this perspective, Revyakina et al. (2021) consider the process of teaching foreign languages as a holistic dynamic system capable of self-organization and self-development, and the synergistic interpretation of the psychological and pedagogical teaching approaches based on digital technologies contributes to the personal growth of students and the development of a holistic dynamic system. Golin et al. (2022) proposed that in the context of Education 4.0 and the raging COVID-19 epidemic, various digital teaching tools need to be used synergistically. Jiang (2016) effectively explains the mutually supportive relationship between Confucius Institutes and university internationalization. Wu et al. (2018) explored Three Triple Helix Model of synergistic innovation by way of systematically analysing the positioning and function of government, industry, universities, research institutions, users and capital sectors.

China–Africa higher education cooperation involves the participation and cooperation of universities, governments and various international organizations. Synergistic is highly needed for the diverse participants to achieve expected goals. Thus, this paper adopts the theory of synergistic to analyse a typical case—Beijing Foreign Studies University, an excellent practitioner of China–African higher education cooperation. Based on triangle sources like first-hand data from BFSU, published papers and documents as well as relevant websites, we present what actions BFSU has taken in response to the China–Africa higher education cooperation initiatives, what cooperation results BFSU has achieved, and explore the interactive coupling mechanisms between BFSU and other diverse elements in this subsystem.

4.3.1 Profile of Beijing Foreign Studies University

BFSU adheres to the core strategy of "talent training, academic studies and global exchanges", practices the action guidelines of "global language, global culture, and global governance", and is committed to cultivating interdisciplinary talents with "patriotism, international perspective and professionalism". Thanks to its outstanding reputation in the academic field, BFSU has won the title "Cradle of Diplomats" in China. In recent years, it has further reformed talent cultivation model, expanded language training programs amounted to 101 foreign languages and established 52 research bases at national, provincial and ministerial levels. As a result, it has full coverage of the languages of countries with which China has diplomatic relations and gradually becomes the world's leading research platform in terms of nation-specific studies. Particularly, after more than 60 years of development, BFSU has become an important force for the common prosperity and progress of China–Africa education. In 2019, to further implement the vision of the *China–Africa Community with a Shared Future*, BFSU founded the School of African Studies to deepen research on African languages and cultures and to improve the China–Africa governance system.

4.3.2 Specific Measures

In response to the "Jointly building the Belt and Road Education Initiative" promoted by the Chinese government, BFSU has deeply understood the vital role of education in the *Five-Pronged Approach* (*policy coordination, infrastructure connectivity, unimpeded trade, financial integration, people-to-people bond*). It formulated a unique mechanism called "starting from language and spreading culture" to promote China–Africa higher education cooperation. Specific actions and their positioning in China–Africa educational cooperation are shown in the Table 4.1.

Table 4.1 Specific measures and positioning of China–Africa Education Cooperation at Beijing Foreign Studies University

Measures	Content	Position
Fundamental measures	China–African language construction	Prerequisite and guarantee
Supportive measures	China–Africa transnational teacher promotion program	Key and focus
	China–Africa joint talent training program	
	China–Africa research base building program	
Leading measures	First-class China–Africa governance platform construction program	Refinement and optimization

4.3.2.1 Fundamental Measures

Language intercommunication is an essential prerequisite for education cooperation. BFSU takes full use of the advantage of its language discipline to offer African language courses and does its best to promote Chinese culture in Africa and African culture in China.

African Languages Promotion in China

Firstly, it is necessary to produce synergy advantages between BFSU and African universities to increase the quantity and quality of teaching materials on African languages. Students in BFSU can learn 20 African languages, such as Swahili language, Hausa language, Zulu language, Amharic, Malagasy language, Somali language, Yoruba language, Afrikaans, Tswana language, Ndebele, Comorian languages, Creole language, Shona language, Tigrania, Lundi, Kinyarwanda, Chewa, Sesudo, Sango language and Tamazigat. BFSU has successfully achieved full language coverage within African countries along the "Belt and Road" zones by signing cooperation memorandums with first-class African universities. Secondly, BFSU has optimized the structure of its existing African language construction bases. For instance, BFSU takes different actions to handling teaching materials issue. For programs awarded national or Beijing's first-class undergraduate programs, BFSU persists in "reinforcing advantages" to explore the depth and breadth of teaching materials. For some African languages with a certain construction foundation but lack of strength, BFSU determines to "remedy shortcomings", focusing on the professional and core teaching materials at the early stage. For those languages which are not available in BFSU, BFSU takes actions to "fill in the blanks" and pays attention to the construction of various introductory series of educational textbooks. Thirdly, via educational information technology, BFSU builds China–African textbook databases, teaching resource Centres and online courses in order to attain efficient resource integration, remove barriers and advocate a hybrid teaching method among all African studies in the world (Yang, 2022).

Chinese Language Promotion in Africa

Firstly, BFSU organizes a series of in-depth dialogues with African embassies in China to keep up to date with the current needs of Africa. Based on the feedback, BFSU has strengthened the training of local Chinese teachers and held academic conferences related to Chinese cultures at Confucius Institutes in Africa, which indirectly facilitates the transmission of the Chinese language and culture. Secondly, BFSU encourages its staff to cooperate with African scholars to edit and translate a series of Chinese books into African languages, including the "*Key Concepts in Chinese Thought and Culture Project*" the "*Belt and Road Culture and Education Series*" and the "*Multilingual speaking of Chinese culture*". Thirdly, to reduce the language and cultural exchange barriers, BFSU sets up the Chinese-African and African-Chinese Dictionary project with highly qualified teachers from all over the world, which certainly promotes the in-depth mutual transmission of Chinese and African languages.

From the above two-way language promotion in China and Africa, BFSU does facilitate mutual understanding and strengthen people-to-people bond between China and Africa.

4.3.2.2   Supportive Measures

BFSU has adopted three supportive initiatives including the China–Africa Transnational Teaching Staff Promotion Program, the China–Africa Joint Talent Training Program, and the China–Africa Research Base Building Program, to truly accelerate cultural exchange based on language intercommunication. The construction of African language teaching staff is the key to the cultivation of African language talents, the main force for fostering talents, and the basis for the "101 Project—Africa Block" (Yang, 2022). 101 project here means 101 languages, which African languages account for over 50%.

There Has Been Continuous Deepening of the Collaborative Training Mechanism

Firstly, BFSU selects local teachers to study in African universities, and gradually expands the quantity and proportion of staff studying in African countries. Secondly, it recruits African teachers to enlarge the provision of African languages and enhance China–Africa cooperation in teaching and research.

In order to cultivate multidisciplinary talents to contribute to China–Africa cooperation, BFSU actively explores the joint cultivation methods between Chinese and African universities. On the one hand, it provides a better practice environment for local students to learn African languages, and on the other hand, it develops suitable training programs for African students in China.

BFSU Continues to Implement the "Cultivation of African Language Talents" Programme

At the undergraduate stage, freshmen and sophomores usually study some fundamental courses in China, like African language reading, listening, audio-visual speaking and writing, while juniors and seniors learn more practical skills through part-time job in translation, internship in Africa, exchange in Africa. On the other hand, BFSU is committed to interdisciplinary academic training. In addition to studying the language of the target country, students are encouraged to take minor courses in diplomacy, economics, law, and other courses. The purpose is to train students to conduct country-specific and interdisciplinary research based on their language advantage and better serve China–Africa cooperation.

BFSU is Further Strengthening Synergies with African Partner Universities that Have Signed Agreements

Firstly, in response to the needs of Africa, BFSU discusses cooperation forms and focus with African partners and adjusts the cooperation majors available for African students so as to train customized talents for Africa. Secondly, BFSU strives to deepen China–Africa cooperation strength, smooth the lecturing and teaching channels for China–African experts, and increase the number of incoming and outgoing students from/to African countries.

BFSU is Continuously Reinforcing Its Ties with African Governments, Embassies of African Countries and International Organizations for Better Formulating Joint China–Africa Talent Training Programs

On the one hand, BFSU makes great efforts to gradually remove the policy bottle-necks in China–Africa education exchange such as mutual recognition of credits and joint award of degrees among cooperative universities. On the other hand, BFSU devotes to buffer the cultural barriers in this cooperation and ameliorates mutual understanding about the needs of both sides.

4.3.2.3   The China–Africa Research Base Building Programme

BFSU established the African Language Studies and Culture Research Centre to build up the world's first-class African education resource platform. BFSU fully uses its advantages in African language resources and organizes numerous China–Africa related international conferences. Besides, BFSU invites scholars in China–Africa related fields to give a series of lectures and make interviews for broadening horizons of students and staff in scientific research cooperation. Furthermore, BFSU further deepens the collaboration with African-related academic institutions in Africa and other continents. For example, BFSU cooperates with the top African research centres in Europe for building an academic exchange platform with integrated resources, shared results, and complementary advantages through the tripartite collaboration among China, Africa, and Europe. BFSU has managed to achieve a synergistic effect on staff and student exchanges, joint Africa-related research, mutual lending of digital library collections, teaching resources and information sharing among the three continents. All these measures are in favour of China–Africa education development.

Finally, BFSU is committed to improving the diversity of Africa-related research. African language, literature, society, and culture are viewed as the main research objects in BFSU. BFSU places an emphasis on African native language research, African language teaching and acquisition, African official language policy and planning, African traditional social and cultural studies, and China–Africa intercultural communication and other topics, to better promote people-to-people contact, and meet the needs of in-depth cooperation between China and Africa.

4.3.2.4 Leading Measures

In addition to the above two types of initiatives, BFSU has taken measures in the following aspects to ensure high-quality education cooperation between BFSU and African partners along the Belt and Road zones under a positive and friendly environment, and truly achieve closer People-to-People Ties.

BFSU further expands high-level bilateral and multilateral consultation on People-to-People Exchange with African Belt and Road countries to better refine cooperation consensus, identify focus and direction (Chinese Government Website, 2016). It increases the signature of exchange and cooperation agreements with top-ranking African universities that were not cooperated before. The respective needs and advantages of signature universities were taken into consideration in the agreements. For those universities with which BFSU has had relevant cooperation experience, it continues to further improve cooperation projects and expand cooperation fields under the premise of current cooperation, to provide a solid foundation for China–African language construction and cultural exchanges.

Secondly, BFSU actively joins international cooperation platforms and develops their potential. It works closely with UNESCO, embassies of African countries in Beijing, national think tanks to better advance the programs related to educational and cultural exchange like teacher training, introduction of teaching resources and dictionaries, and oversea internships for students. The overseas internships for students between BFSU and African universities fully depend on the existing projects such as the "20 + 20 Cooperation Program between Chinese and African Universities".

Finally, BFSU is endeavouring to build a first-class China–Africa governance platform. In collaboration with African universities and African substantive departments, BFSU has provided specialized language services for major events, international conferences and exchanges, in purpose of creating a China–Africa language service platform. Meanwhile, BFSU has used educational information technology to break through the blockages and barriers in enrolment, training, employment, research, services, and supervision. Working together with African partners, BFSU has never stopped facilitating the modernization of China–Africa governance capability and system by the high level of global language governance competence (Yang, 2022).

4.3.3   Achievements

After more than 60 years of development, BFSU has achieved remarkable results in African languages research and talents cultivation. Specifically, it has successfully signed exchange and cooperation agreements with 18 leading universities and institutions in 10 African countries, and added 20 African languages to its exiting language training system. It is worth mentioning that 7 of them have been set up within the past 5 years. The African Language and Culture Research Centre in BSFU is under orderly construction and BFSU further deepens the ties with African partners. This signifies that BFSU truly practises the consensus of China–Africa on "*One Belt, One Road*" and "*Agenda 2063*". Therefore, this paper demonstrates the achievements of BFSU cooperation with African partners in accordance with the goals and vision of the *Education Initiative*.

4.3.3.1   Co-development of Education Levels in China and Africa

More frequent faculty mobility. At present, the School of African Studies at BFSU has 21 faculty. 71.4% of them are Ph.D. holders and the teacher-student ratio attains 3.33:1. African teachers account for 33.3% in the School, coming from Ethiopia, Nigeria, Rwanda, South Africa, Somalia, Madagascar and other African countries. In the past 5 years, BFSU has introduced a total of 83 long- and short-term foreign experts from 55 African countries, including 28 long-term full-time foreign teachers (more than 3 months) and 55 short-term foreign experts (visiting scholars, short-term lecturers, foreign consultants, etc.), which surely promotes the construction of more than 10 non-common African languages (e.g. Yoruba, Hausa and Malagasy) and the study of specific African countries. BFSU dispatched 2 faculty to Madagascar and 1 faculty to Ethiopia for a short-term visit. 66.6% of the full-time faculty in the School of African Studies have studied in African target countries for more than 1 year. Thanks to the excellent Chinese and African faculty, nowadays the School of African Studies at BFSU has become China's top institute in African language training.

Provide the most African languages training. 20 African languages are available at the School of African Studies at BFSU. The School owns the largest number of approved African languages and the largest non-common African language construction base in China. Swahili and Hausa have been selected in the national first-class undergraduate major construction list, whereas Zulu and Amharic have entered the Beijing first-class undergraduate major construction list. The School lays a solid foundation for cultural exchanges between China and countries along the "Belt and Road" in Africa. Ge'ez is an ancient language in Ethiopia. It is now being taught in 21 universities in different countries outside Africa. It might be considered to teach in the School in the future.

Achieve fruitful Africa-related research results. So far, the School has two second-tier disciplines: Afro-Asia Language and Literature; Afro-Asia Area

Studies. These two disciplines are authorized to recruit master students. Students can apply for doctoral programs in Oriental Nationalist Literary Studies and Francophone Area Studies. The School also has the Centre for African Languages and Cultures, devoting to study China–Africa languages, education, culture and relevant fields. The Centre has signed memorandums of cooperation with the Centre for African Studies at Leiden University in the Netherlands and the School of Oriental and African Studies at the University of London in the United Kingdom. In the past 5 years, 89 papers on African-related topics have been published; in the past 3 years, the team of the Centre for African Languages and Cultures has undertaken major research projects (e.g., key research program of National Social Science, the Youth Innovation Team Project of Beijing Foreign Studies University, and the key projects of the 13th 5-Year Plan of the National Language Commission). The Centre founded two journals named as "Asian-African Studies" and "African Language and Culture Studies", which provide an important basis for China–Africa people-to-people bond through fundamental research on African indigenous languages, societies, history and culture. In the past 2 years, 10 research reports have been adopted by relevant governmental departments. Other research outputs were: 5 textbooks, 1 edited book, 2 monographs, 1 translation, and more than 160 original tweets on academic public websites, achieving a total of more than 60,000 hits. These research achievements facilitate the in-depth cooperation and mutual learning in education between China and Africa (Chinese Government Website, 2016).

4.3.3.2 Remarkable Achievements in Chinese and African Talents Cultivation

According to the *Education Initiative to Promote the Construction of the Belt and Road*, BFSU has achieved remarkable achievements in cultivating a large number of talents urgently needed for the construction of "One Belt, One Road" and for supporting countries along the route to realize policy coordination, unimpeded trade, financial integration, people connectivity and facilities connectivity.

Great Success in China–Africa Joint Education

In the past 5 years, 54 scholars and students from 20 African countries came to BFSU for short-term exchange (within 1 year). They registered in 10 majors and are hosted by 5 colleges in BFSU. At the same time, 6 Chinese students from the School of African Studies went to Bayero University in Nigeria for a 4-month inter-college exchange. In terms of long-term exchange (1 year or more), there are 140 African students from 37 countries, enrolling in 9 colleges and 18 majors in BFSU. Meanwhile, 3 Chinese students from African College in BFSU studied in Ethiopia and another 5 students went to Tanzania for long-term study, majoring in local language and literature. In the past 2 years, 1 student from the School of Journalism got the government scholarship and went to Hargeisa University in Somaliland. In addition, BFSU has sent up to 30 students to the School of Oriental and African Studies of the University of London, one of the world's top African research institutions, for long-term study, and all of them have received full funding from the China Scholarship Council. The student mobility has largely improved the talent cultivation quality in terms of learning non-common languages and conducting specific African country studies.

Remarkable Contribution of BFSU Graduates to China–Africa Cooperation

Over the past 60 years, the School of African Studies has trained a large number of African language talents with solid basic language skills, broad knowledge and high competences for the common development of China and Africa. For instance, the number of Swahili language talents has reached 200. They are now mainly employed in national ministries, enterprises and institutions and universities at home and abroad, playing an active role in the friendly cooperation between China and Africa. Additionally, in the past 5 years, BFSU students have gone to African countries for exchanges and joined local teams for medical consultations and environmental education activities in local primary and secondary schools, effectively serving the local communities and meeting the needs of African people. African students in BFSU also actively participate in various social services, building a bridge between Chinese and African cultures. Finally, in the past 2 years, 100% graduates majoring in Hausa and Zulu at the School of African Studies got a slot in master programs. Some of them have gone to the University of Warwick, the University of Bristol and other top universities for further studies, and continued their studies in pedagogy, African culture, business trade and other professional fields. Being a multidisciplinary talent pool based on language and other disciplines like economy, politics and law, BFSU will continue to foster talents and help them grow to become the backbone of the "One Belt, One Road" construction in China and Africa in the future.

4.3.3.3 People Connectivity Strengthened Between China and Africa

Guided by the *Education Initiative to Promote the Construction of the Belt and Road*, BFSU has carried out a wider range, higher and deeper level of people-to-people cultural exchanges, and continuously promotes a closer, deepen mutual understanding of the people along the route.

Translate and Publish Books on China and Africa for Enhancing Mutual Understanding

In order to further deepen the African people's understanding of Chinese culture and to carry out the dissemination of Chinese excellent traditional culture research, BFSU translated and published *the series of Chinese Thought and Cultural Terminology Dissemination Project* and *Multilingual Chinese Culture series* in Hausa. BFSU has also compiled the volumes of most African countries in *the Senegalese Culture and Education Research*, which is helpful for the Chinese people to understand African history. The publication of the above books plays an important role in the exchange and dialogue between Chinese and African civilizations, and will have profound academic influence and good social benefits.

Participation in China–African Cultural Exchange Activities

The scholars from BFSU were invited to participate in the "*International Confucianism Forum—Rabat International Symposium*" co-organized by the Confucius Institute of Mohammed V University in Morocco, discussing themes such as the historical experience of different civilizations in Asia, Africa and Europe on the ancient Silk Road through exchanges, mutual learning and cooperation, as well as historical figures and stories who made outstanding contributions. In 2018, BFSU scholars went to Nigeria to participate in the 4th Nigerian Social Folklore Symposium to accumulate materials for further research. Meanwhile, BFSU and the University of London, a world-class African research base, jointly set up the British Confucius Institute. Both parties are committed to sharing resources and carry out in-depth research on African history and culture in Chinese.

Organize Diverse China–Africa Exchange Activities

In the past 5 years, BFSU takes full use of African language resources and the advantages of non-language majors to hold numerous international conferences related to Africa. The series of conferences include one domestic seminar on "Humanities Exchange and Cooperation under the Vision 2035 of China–Africa Relations", one China–Africa Law Forum, two international seminars on "African Languages and Cultures", one doctoral thesis on African studies, one national seminar on teaching African languages and national competition on African knowledge. In addition, 24 lectures were hosted by BFSU, including 12 lectures in the "BFSU Africa Lecture Hall" series, 3 lectures in the "Emerging Scholars" African Studies series, and 4 lectures in the GAFSU series co-organized with the International Affairs Office of BFSU and the Institute of African Studies of Zhejiang Normal University. Apart from these, BFSU received 8 delegations from African countries, including 2 from universities and 6 from embassies and consulates in China, with a total of 45 visitors. The specific results of cooperation are shown in the Table 4.2:

Table 4.2 Achievements of China–Africa Cooperation at Beijing Foreign Studies University. *Source* Data collected by the author from the web and internal documents of BFSU

Number of partner universities	10 countries, 18 cooperation agreements				
	South Africa 4, Nigeria 3, Ethiopia 2, Tanzania 2, Kenya 2				
	Madagascar 1, Rwanda 1, Comoros 1, Botswana 1 Burundi 1, UK 1, France 1, Netherlands 1				
Faculty members in Africa School of BFSU	Total faculty	21	Incoming long- and short-term visiting African experts	83	
	African faculty	7	Countries where incoming long- and short-term visiting African experts come from	55	
	Faculty with Ph.D. degrees	15	Outgoing BFSU faculty for short-term visits in Africa	3	
	Student enrolment	70 per year	BFSU faculty who have studied in target countries in Africa	8	
African Language and Cultural Studies	Available African languages	20	National-level first-class undergraduate major construction base	2	Beijing First-class Undergraduate Major Construction base	2
	Published academic papers (2017–present)	89	Textbook	5	Edited book	1
	Reports adopted by relevant national authorities (2021–present)	10	Monograph	2	Translation	1

(continued)

Table 4.2 (continued)

Number of partner universities	10 countries, 18 cooperation agreements				
	South Africa 4, Nigeria 3, Ethiopia 2, Tanzania 2, Kenya 2				
	Madagascar 1, Rwanda 1, Comoros 1, Botswana1 Burundi 1, UK 1, France 1, Netherlands 1				
China–African Cooperation in Education and Cultural Exchange	Creation of African-related journals	2		Related research projects	4
	Short-term exchange (last 5 years)		Students from Africa to China	54	Chinese Students to Africa	6
	Long-term exchange (last 5 years)		Students from Africa to China	140	Chinese Students to Africa	9
	China–African international conference	6	China–African lecture	24	Visits by delegations from African countries	45
	Visiting delegations from embassies and consulates in China	6	University delegation	2	National delegation	8
Talent Achievements	Undergraduate for master slots	100%	Main employment	Government ministries, enterprises and institutions and universities	

4.4 Conclusion: Future Opportunities, Challenges and Policy Recommendations

Following the advancement of the "China–Africa Community with a Shared Future", there will be huge opportunities for in-depth cooperation between China and Africa in various fields. At present, the number of Chinese universities cooperating with African universities is still limited. The quantity and quality of high-end talents they trained also remain insufficient for satisfying the expanding needs of China–Africa cooperation in various fields. Moreover, due to the different cultural attributes between Chinese and African people, many other factors should be taken into consideration besides language to promote people-to-people bonds (Liu et al., 2020).

From the synergistic perspective, the China–Africa cooperation system can achieve coordinated development only if the internal sub-systems are coordinated and coupled with each other. Otherwise, the internal friction of the entire system will increase, causing conflicts and discouragement. Therefore, through the platform of the Global Alliance of Foreign Language Universities, BFSU needs to further play a leading role in African language construction in the future. BFSU should share the experience, resources and achievements resulting from the existing cooperation with other foreign language universities to create a "high and deep level" model for China–Africa language cooperation. Moreover, it is important to conclude cooperation agreements which are based on the disciplinary characteristics of Chinese and African universities. In this way, both sides can shape a multi-disciplinary, multi-field related teaching and research force on China and Africa. Besides, the Chinese and African governments should identify cooperation focus, better synergy strategic planning and policy consultation, establish an ensured operation mechanism and improve the supervision and evaluation system (Chinese Government Website, 2016). Additionally, the intermediary role of UNESCO and embassies under the China–African cooperation framework, should be further brought into play. They can offer cooperation guidelines, strategic planning, smooth international cooperation paths, solve the problems of resources and policy bottlenecks in education cooperation and exchanges. Furthermore, the Chinese and African governments should continue to keep close contact, strengthen the overall planning and coordination among various participants involving in China–Africa higher education cooperation. It is necessary to coordinate between participants to ensure that universities and industries can take use of their respective advantages and exchange resources. The collaboration between universities and enterprises needs to be further deepened, such as technological innovation cooperation through joint R&D, outsourcing R&D, project funding support etc. The governments should give full play to the lubricating role of factors (e.g., funds and mobility of researchers, staff and students) and strengthen a solid industry-university-science-government linkage.

The wide application of information technology has provided unprecedented prospects for resource integration and academic exchanges between Chinese and African universities. AI, IoT and 5G have great potential to boost teaching and learning, and to inspire and support internal synergies in new generation educational ecosystems (Muraszkiewicz, 2019). Especially since the outbreak of the COVID-19 epidemic, the application of online platforms has made China–Africa cooperation effectively and efficiently regardless of time and space constraints. However, in order to promote the breadth and depth of exchange and cooperation, how to use online platforms to expand new ways of cooperation between Chinese and African universities deserves further consideration. Therefore, the Chinese and African governments should play a leading role in formulating relevant policies and regulations and improve cyberspace supervision capability to ensure the cyberspace security of China–Africa achievements. Chinese and African universities should innovate new cooperation models, further share scientific and technological innovation achievements, advance teaching methods and enrich teaching contents (Yang, 2022). BFSU has set up an excellent example for using online platforms to further expand the relevant cooperation with world-class African research bases such as Britain and France. China and Africa should speed up the online transformation of Confucius Institutes. The number of Confucius Institutes in Africa nowadays remains small, and the distribution is uneven. Their focus on Chinese teaching without other functions hampers the growth of these institutes. Therefore, it is of great importance to open online courses and conduct cultural exchange activities in local languages. Through this way, people-to-people bond and cultural exchanges between China and Africa can be enhanced. As for international organizations like UNESCO and embassies of African countries, they are expected to create diversified communication subjects (Wang & Wang, 2020). For example, they need to guide Chinese and African business leaders, scholars, and international students to participate in international subject communication. At the same time, do not forget to make full use of the publishing industry and other communication channels to forge a stronger bond among people (Sun, 2014).

Acknowledgements We would like to thank the Sichuan Provincial Philosophy and Social Science Planning Office and Wang Kuancheng Education Foundation for funding this research project under Grants (Nos. SC21ZD010 and 48).

References

Chinese Government. *Notice of the ministry of education on printing and distributing the "Educational Action to Jointly Build the Belt and Road" [EB/OL].* http://www.gov.cn/gongbao/content/2017/content_5181096.htm. 2016-7-23.

Goldin, T., Rauch, E., & Pacher, C. (2022). Woschank M. Reference architecture for an integrated and synergetic use of digital tools in education 4.0. *Procedia Computer Science, 20,* 407–417. https://doi.org/10.1016/j.procs.2022.01.239

Groth, H., May, J. F., & Turbat, V. (2019). Policies needed to capture a demographic dividend in Sub-Saharan Africa. *Canadian Studies in Population, 46*(1), 61–72. https://doi.org/10.1007/s42650-019-00005-8

Haken, H. (2004). Synergetics introduction and advanced topics. *Physics and Astronomy Online Library, 2004,* 758 blz.

Jiang, J. B. (2016). Research on the coordinated development of confucius institutes and universities in the internationalization of confucius institutes and universities from the perspective of synergy theory. *Education Review, 09*, 8–11.

King, K. (2010). China's cooperation in education and training with Kenya: A different mode? *International Journal of Educational Development, 30*(5), 488–496. https://doi.org/10.1016/j.ijedudev.2010.03.014

Lefifi, T., & Kiala, C. (2021). Untapping FOCAC higher education scholarships for Africa's human capital development: Lessons from haigui. *China International Strategy Review.* https://doi.org/10.1007/s42533-021-00074-y

Liao, G. X. (2020). Research on China–Africa education cooperation under the belt and road initiative. *University of South China.* https://doi.org/10.27234/d.cnki.gnhuu.2020.000292

Liu, H. W., & Lin, C. (2020). People-to-people exchanges promote China–Africa cooperation to be stable and far-reaching. *West Asia and Africa, 2*, 22–32.

MOE. (2020). Opinions of the Ministry of Education of the People's Republic of China, Ministry of Education and other eight departments on accelerating and expanding the opening up of education in the new era[N]. *People's Daily,* 23 June 2020.

MOST. (2019). B*eijing forum: The special forum on "Scientific and Technological Innovation and Transformation of Scientific and Technological Achievements of Central African Universities"*, Organised by the Ministry of Science and Technology Development, Ministry of International Cooperation [EB/OL]. https://news.pku.edu.cn/xwzh/0aed0ff6ad024cbeb6daaa39d61e6e84.htm. 2019-11-08.

Muraszkiewicz, M. (2019). The synergetic impact of AI, IoT, and 5G on information literacy and education. *Zagadnienia Informacji Naukowej—Studia Informacyjne.* https://doi.org/10.36702/zin.451

Nordtveit, B. H. (2011). An emerging donor in education and development: A case study of China in Cameroon. *International Journal of Educational Development, 31*(2), 99–108. https://doi.org/10.1016/j.ijedudev.2010.01.004

Revyakina, N., & Sakharova, E. (2021). Psychological and pedagogical support of the educational process: synergetic approach. *E3S Web of Conferences. EDP Sciences.* https://doi.org/10.1051/e3sconf/202127312124

Sun, X. M. (2014). Challenges and strategies: China's national image communication under the threshold of African local newspapers and periodicals. *Modern Communication (journal of Communication University of China), 36*(02), 53–56.

Teng, J., Li, X. X., & Chen, L. (2016). China: The historical foundation and future challenges of an emerging education donor: A critical literature analysis based on "China–Africa educational cooperation and exchange. *Journal of Beijing Normal University (social Sciences Edition), 1*, 17–30.

Wang, H., & Wang, L. J. (2020). The role, challenges and coping strategies of university think tanks in international communication: A case study of the Institute of African Studies. *Zhejiang Normal University, African Studies, 16*(01), 166-175-204–205.

Wang, Y. J., Wu, Y. N., & Lu, L. Z. (2020). Reflections and prospects of China–Africa higher education cooperation in the new era. *World Education Information, 33*(09), 29–37.

Wu, W. H., Chen, G. X., & Zhang, A. M. (2018). The three-three-helix model and mechanism of multi-subject collaborative innovation. *China Science and Technology Forum, 5*, 1–10. https://doi.org/10.13580/j.cnki.fstc.2018.05.001

Xinhua, Former Vice-Chairman of au Commission: The Belt and Road Initiative is in line with the development direction of THE AU's Agenda 2063[EB/OL]. http://news.youth.cn/gj/202112/t20211220_13359831.htm. 2021-12-20.

Yakimtsov, V. V. (2018). History and development of Haken's synergetics. *Scientific Bulletin of UNFU, 28*(9), 119–125. https://doi.org/10.15421/40280923

Yang, D. (2022). Strengthen global strategic planning and implementation to build a world-class foreign Chinese university. *China's Higher Education, 5*, 16–18.

Part II
China-Africa Research Collaboration and Training

Part II

Chapter 5
Exploring China's Emerging Role in Africa's International Research Collaboration

Ruoyan Zhu and Yin Li

5.1 Introduction

Over the past decade, Chinese influence in African countries has significantly increased (Grimm, 2014). Even though the epidemic poses significant barriers to globalization, China–Africa cooperation has grown significantly and made significant progress.[1] Nevertheless, given the asymmetrical nature of relations and Africa's changing attitude toward topics like humanitarian intervention, China–Africa cooperation has been running into impediments over time (Alden, 2005). The international society, principally some Western countries, is suspicious of China's engagement and its impact in Africa. In fact, China–Africa cooperation has evolved over time and taken many different forms apart from economic ties. International research collaboration, as an important form of bilateral and multilateral cooperation, has received a lot of attention from academics.

Co-authorship statistics have been widely used since the 1990s to measure international research collaboration (IRC). Chinese researchers' participation in international scientific collaborations has also gained popularity in contemporary scientometric literature (Pouris & Ho, 2014). Researchers are investigating the effects, modes, dynamics, and motives of collaboration in a continental research system, which is in an embryonic stage and in different stages of development from country

[1] Ministry of Commerce of the People's Republic of China, press conference of the MOFCOM (January 14, 2021), available at: http://www.mofcom.gov.cn/xwfbh/20210114.shtml, accessed on February 13, 2023.

R. Zhu (✉)
School of Government, Peking University, Beijing, China
e-mail: zhuruoyan@stu.pku.edu.cn

Y. Li
School of International Relations and Public Affairs, Fudan University, Shanghai, China
e-mail: yinli@fudan.edu.cn

to country (Wang et al., 2013; Guns & Liu, 2010; Zhang & Guo, 1997). China has been strengthening its international ties with cross-border collaborators, particularly with Africa, thanks to a significant increase in high-quality research output over the past few decades. Research partnerships between China and Africa follow misgivings over the Chinese approach to collaboration with Africa, assessing patterns of progress by the China–Africa scientific research collaboration cannot be ignored (Eduan & Jiang, 2019).

The need to survey and follow up on the China–Africa scientific research collaboration issue calls for statistical indicators sensitive enough to reveal the structure and changes of collaborative networks (Melin & Persson, 1996). In order to fill the research gap in China–Africa scientific research collaboration, this study uses the available data from research publications of African countries to examine how the research collaboration between African countries and other international countries outside the region, especially the collaboration with China, is evolving and how China's position in the larger landscape is changing, especially highlighting the dynamics of China's emerging role in Africa's international research collaboration for a more comprehensive understanding of the China–Africa relationship.

The rest of the chapter is organized as follows: Sects. 5.2 and 5.3 summarize the basic situations and review the current literature of Africa's IRC and China–Africa scientific research collaboration. Section 5.4 presents the data and methodologies used in this study, and Sect. 5.5 is the most important part as it presents the main findings, the detailed dynamics of collaboration between China and Africa, including country collaboration patterns, multinational collaboration patterns, field dynamics, and institutional collaboration patterns, followed by the conclusion in Sect. 5.6 and discussion in Sect. 5.6. Lastly, Sect. 5.6.1 discusses limitations.

5.2   International Research Collaboration in Africa

As a subdomain of the broader collaboration research and policy landscape, international research collaboration (IRC), which first attracted scholars' attention at least half a century ago, is now increasingly important as an emerging area of innovation studies (Chen et al., 2019). International collaborative behaviour among scientists is investigated by examining international co-authorship patterns for a number of scientific fields or a certain region or country (Kim, 2006). Scientific co-authorship by African or Chinese researchers has also become a fashionable topic in the recent scientometric literature (Pouris & Ho, 2014). Researchers are investigating the effects, modes, dynamics, and motives of collaboration in a continental research system that is in an embryonic stage and at different stages of development from country to country (Pouris & Ho, 2014). When talking about IRC in regard to Africa, it has been noted that scientific knowledge authored by African scientists has essentially been developed in an international context (Vieira, 2022). Foreign researchers contributed a high percentage of total scientific production (as measured by peer-reviewed literature) in several African countries (Megnigbeto, 2013; Pouris & Ho, 2014).

The existing studies on Africa's IRC could be categorized into the following three categories: The first type of research is to study the patterns of research collaboration in a certain African country or region, which also occupies the main body of current related research. For example, Boshoff (2009) examined aspects of both neo-colonial ties and neo-colonial science in research papers produced by Central African countries. Owusu-Nimo and Boshoff (2017) investigated patterns of research collaboration in Ghana to study why Ghanaian-affiliated researchers collaborate with others both within and outside of Ghana. Mêgnigbêto (2013) focused on the case of West Africa, studying their intra-regional collaboration, intra-African collaboration, and collaboration with the world in scientific publishing. Sooryamoorthy (2009) investigated the collaboration patterns of South African researchers, and Onyancha and Ochlla (2007) used co-authorship to measure country-wise collaborations in HIV/AIDS research in Kenya and South Africa from 1980 to 2005.

The second type of research focuses solely on studying the internal collaboration in the African region. For example, Boshoff (2010) identified South–South collaboration in research, specifically collaboration among the 15 countries of the Southern African Development Community (SADC) as well as between the SADC and the rest of Africa. Onyancha and Maluleka (2011) found that knowledge production through collaborative research among sub-Saharan African countries is minimal. Schubert and Sooryamoorthy (2010) showed that "a theory of scientific collaboration building on the notions of marginality and centre-periphery can explain many facets of South African-German collaboration," where South Africa is a semi-peripheral region, a center for the periphery, and a periphery for the centre.

The third type of research not only studies the internal interactions of African countries but also looks into the collaborative relationship between Africa and the world outside the region. For example, Adams's study presents a complex picture of diverse research collaboration links, internationally and within Africa, using data from 2000 to 2012 (Adams et al., 2014). Their study suggested that the collaboration pattern for countries in Africa is far from universal. Instead, it exhibits layers of internal clusters and external links that are explained not by monotypic global influences but by regional geography and, perhaps even more strongly, by history, culture, and language (Adams et al., 2014). Toivanen and Ponomariov (2011) draws on a bibliometric analysis of co-authorship of African research publications from 2005 to 2009 and proposes an empirically derived grouping of the African research community into three distinct research regions: southern and eastern, western, and northern. They presents an empirical analysis for the three African research regions and found that Africa's internal research collaboration suffers from structural weaknesses and uneven integration (Toivanen & Ponomariov, 2011).

Given the importance of IRC in Africa and its challenges and benefits, it is believed that IRC in Africa has attracted the attention of the scientific community on several aspects. However, with the gradual improvement of China's status in the international community and the deepening of China–Africa scientific research collaboration, the level of research on the topic of China–Africa scientific research collaboration has previously been noted to be relatively not as high as possible in the literature.

5.3   China–Africa Scientific Research Collaboration

In the existing studies on China's international research collaboration, much attention has been paid to the bibliometrics analysis of Chinese publication records recently. Researchers have made attempts and published a series of papers on the performance of China's international collaborations through quantitative analysis and evaluation of scientific publications (Melin & Persson, 1996). For example, Tang and Shapira (2011) examined the patterns of China's regional and China–US collaboration in nanotechnology. Zhou and his colleagues (2013) investigated China–UK collaboration in food and agriculture. He (2009) focused on the international scientific collaboration of China with the G7. Zhou and Lv (2015) paid attention to China–Germany collaboration in physics. Niu (2014) investigated the China–Australia collaboration. However, although there is a lot of research on China's international research collaboration or the international research collaboration between China and other international countries, there is not much literature on the international research collaboration between China and the African region (Wang et al., 2013).

Research studies on China–Africa scientific research collaboration are rarely observed in the literature of the bibliometrics analysis of Chinese publication records. Even though a few studies made attempts to summarize the status of China–Africa scientific research collaboration, most of them lacked comprehensiveness and did not study China's role and position in the whole landscape of Africa's international research collaboration by comparing China–Africa scientific research collaboration with Africa's international research collaboration with other international collaborators.

Lee (2018), for example, summarized the China–Africa flourishing friendship in science and technology through pure descriptions rather than rigorous quantitative research. Eduan and Jiang summarized the pattern of China–Africa scientific research collaborations from 2006 to 2016 from a bibliometric analysis perspective (Eduan & Jiang, 2019). Using InCites research data, their study reveals that the partnership is growing progressively in absolute terms while maintaining a high relative growth rate. A few of the African countries are more engaged than the others, and fortunately for Africa, the partnership involves the physical sciences, where the continent is greatly lacking and high impact is being registered (Eduan & Jiang, 2019). Muchie and Patra (2020) used co-authored scholarly journal publications and joint patents to map the S&T collaboration between China and African countries. They found that China–Africa S&T collaborations have evolved after China engaged actively with African countries, and the China–Africa collaboration is increasing and is likely to grow in the coming years as China is emerging as a leading global research hub in the world (Muchie & Patra, 2020).

To sum up, earlier empirical studies either focused on China–Africa scientific research collaboration for a short period or explained the partnership without examining Africa's connections to other international collaborators. They did not identify the long-term development pattern of the research collaboration between China and Africa, and were also unable to place China within the broad framework of African international research collaboration. Given the importance of China–Africa scientific research collaboration and the research gap in the literature, it is critical to conduct a detailed and rigorous study on China's emerging role in Africa's IRC to identify strengths, weaknesses, and opportunities for international research activities.

5.4 Data and Methodology

By mentioning international research collaboration (IRC), this study refers to co-authorship between authors from different countries. Africa's IRC includes publications produced in collaboration by African countries (Table 5.1) with countries outside the African region. Data consisting of cases of international research collaboration in Africa were obtained through a bibliometric analysis (Table 5.2). Bibliometric analysis is now a common statistical procedure involving written publications such as books or articles used for assessment of research performance in many studies (Eito-Brun & Rodríguez, 2016; Sooryamoorthy, 2014; Zhou et al., 2013). There is general consensus that the observed growth in multiple-authorship, or so-called co-authorship, is evidence of an increase in collaboration.

The often-used databases for extracting the data of co-authorships include PubMed, Web of Science (WoS), ScoPus, and so on. To facilitate studies, many Web-based tools also now exist to allow easy comparisons of research productivity and impact. Such tools include InCites (using the WoS) and SciVal (using Scopus), as well as software to analyze individual citation profiles using Google Scholar (Publish or Perish, released in 2007). AlRyalat's comparative analysis (2019) of these databases found that PubMed has a sophisticated keyword optimization service (i.e., Medical Subject Heading, or MeSH), while both Scopus and Web of Science provide search analysis tools that can produce representative figures. Archambault et al. (2009) show

Table 5.1 List of African countries. *Source* Ministry of Foreign Affairs, People's Republic of China

	Sample
Country	Algeria, Ethiopia, Angola, Benin, Botswana, Burkina Faso, Burundi, Equatorial Guinea, Egypt, Togo, Eritrea, Cape Verde, Gambia, Congo Brazzaville, Democratic Republic of Congo, Chad, Central African, Djibouti, Guinea, Guinea-Bissau, Ghana, Gabon, Zimbabwe, Cameroon, Comoros, Cote d'Ivoire, Kenya, Lesotho, Liberia, Libya, Rwanda, Madagascar, Malawi, Mali, Mauritius, Mauritania, Morocco, Mozambique, Namibia, South Sudan, South Africa, Niger, Nigeria, Sierra Leone, Senegal, Seychelles, Sao Tome and Principe, Eswatini (Swaziland), Sudan, Somalia, Tanzania, Tunisia, Uganda, Zambia

Table 5.2 Search strategy

	Search Strategy
Database	Web of Science (WOS) database
Keywords	*CU = Algeria OR CU = Ethiopia OR CU = Angola OR CU = Benin OR CU = Botswana OR CU = Burkina Faso OR CU = Burundi OR CU = Equat Guinea OR CU = Egypt OR CU = Togo OR CU = Eritrea OR CU = Cape Verde OR CU = Gambia OR CU = Rep Congo OR CU = DEM REP CONGO OR CU = Chad OR CU = Cent Afric Rep OR CU = Djibouti OR CU = Guinea OR CU = Guinea Bissau OR CU = Ghana OR CU = Gabon OR CU = Zimbabwe OR CU = Cameroon OR CU = Comoros OR CU = Cote d'Ivoire OR CU = Kenya OR CU = Lesotho OR CU = Liberia OR CU = Libya OR CU = Rwanda OR CU = Madagascar OR CU = Malawi OR CU = Mali OR CU = Mauritius OR CU = Mauritania OR CU = Morocco OR CU = Mozambique OR CU = Namibia OR CU = South Sudan OR CU = South Africa OR CU = Niger OR CU = Nigeria OR CU = Sierra Leone OR CU = Senegal OR CU = Seychelles OR CU = Sao Tome and Principe OR CU = Swaziland OR CU = Sudan OR CU = Somalia OR CU = Tanzania OR CU = Tunisia OR CU = Uganda OR CU = Zambia*
Time span	1971–2019
Source category	All documents from Science Citation Index Expanded (SCI-Expanded), Social Sciences Citation Index (SSCI) and Arts & Humanities Citation Index (A&HCI)
Academic field	All disciplines
Document type	All documents regardless of type (e.g., article, meeting abstract, proceedings paper, review, editorial material, book review, letter, note, etc.) were processed

that indicators of scientific production and citations at the country level are stable and largely independent of the database.

The dataset utilized in this study contains Africa's worldwide IRC records and was downloaded from the Science Citation Index Expanded of Thomson Reuters Web of Science (WoS). The benefits of using the WoS database over other databases are as follows: first, it covers a broader range of data because WoS covers or relates to many disciplines and fields of study; second, it offers more complex and focused search options, allowing the researcher to filter and refine queries, identifying international collaboration records; and third, it allows the researcher to extract detailed information about a group of publications from a single search retrieval. Muchie and Patra (2020) used the WoS database to conduct research on China–Africa scientific research collaboration, Nguyen et al. (2017) used it for international collaboration in scientific research in Vietnam, and Mêgnigbêto (2013) used it for analysis of international collaboration in scientific publishing in West Africa. These previous studies proved the reliability and popularity of the WoS database.

This study extracts relevant records using the advanced WoS searching technique with the Boolean words "and" and "or", with the time period of the publications for empirical analysis in the dataset being from 1971 to 2019. With this approach, we gather information from all publications pertaining to African authors published between 1971 and 2019. The benefits of this searching strategy are as follows: (1) the country name is a common expression in the WoS database, allowing to download the most comprehensive and accurate documents published by the country; and (2) the time span setting is most appropriate and comprehensive. On the one hand, China and Africa started working together on research projects in the 1970s. Africa also reached most other international collaborators in the 1970s. On the other hand, Africa did not have a great deal of scientific research output prior to the 1970s, so it would not significantly affect the results. (3) It may make the most of the database by including a variety of research outputs and achieving a high degree of comprehensiveness for international comparison.

We eliminate duplicates based on the WoS accession number (unique paper identifiers) since there are recurrent ones as a result of collaboration between different African countries. By grouping together non-African nations, we are able to further sort out Africa's IRC records with countries outside the region. This study defines an African IRC collaboration paper as one containing at least one international (non-African) author address in the organizational affiliations recorded in the publication (Tang & Shapira, 2011). Finally, a dataset encompassing 481,108 Africa's IRC records from 1971 to 2019 was obtained. All documents, regardless of type (such as articles, meeting abstracts, proceedings papers, reviews, editorial materials, book reviews, letters, and notes), were processed in this study.

We further identify records of Africa's IRC with each international collaborator. For example, a China–Africa collaborated publication is defined as one that includes at least one Chinese author and one African author. Here, China is delineated as mainland China and the two special administrative regions of Hong Kong and Macau (Tang & Shapira, 2011). We used VantagePoint (VP) text mining software to clean and analyse the data. VantagePoint text mining software is one of the most important and frequently used analytical tools for big data mining and analysis. It is a professional-grade desktop text mining application offering Analysts a broad suite of powerful refining, analysing, and reporting tools for scientific, technical, market and patent information, which is developed by the research team led by Professor Alan Port of Georgia Institute of Technology. This study uses VP to import the downloaded data records and helps analysis the data needed for the research. VP allows the researcher to directly import txt files downloaded from WoS and perform integrated analysis. It is always great at analysing text but also enhances statistical summaries that allow the researcher to better understand and visualize the numerical data. The techniques used in this study to clean and standardize bibliometric data for China are borrowed from previous literature (Tang & Walsh, 2011; Tang & Shapira, 2011). This resulted in 21,778 China–Africa collaborated records and 129,875, 95,728, 80,824, 44,707, 38,589 for US–Africa, France–Africa, UK–Africa, Germany–Africa and Saudi Arabia–Africa collaborated records respectively for the period 1971–2019.

5.5 Analytical Results

5.5.1 Growth of China–Africa Scientific Research Collaboration

From 1971 to 2019, the number of publications with co-authorship from China and Africa has increased dramatically. There is also an increase in China's proportion in Africa's IRC. In total, there are 21,778 records of co-authorship publications during our research period. The first record of research collaboration between China and Africa, published in 1977 for the reform and opening up of China, has aided in the development of the China–Africa relationship.[2] While moderate development is expected over the coming years, China and Africa have begun working together on basic research projects. Prior to 2000, fewer than fifty records were released annually.

With the overall development of the China–Africa relationship, China–Africa scientific research collaboration took off in the twenty-first century and began marching steadily forward toward prosperity in the recent decade. In 2019, there were 4756 China–Africa co-authored papers, ten times as many as there were 10 years prior. China's share of Africa's IRC is growing in lockstep with the growth in records. The percentage of China–Africa scientific research collaboration in Africa's IRC has climbed from zero to 9.17% between 1971 and 2019. In accordance, the proportion of publications co-authored by authors from China and Africa has increased in all Chinese publications, going from zero in 1971 to 0.59% in 2019 (Fig. 5.1).

5.5.2 Patterns and Dynamics of China–Africa Scientific Research Collaboration

The top ten African countries that took the lead in collaboration with China from 1971 to 2019 in terms of total co-author publications include South Africa (7416), Egypt (5779), Nigeria (1953), Morocco (1426), Kenya (1370), Ghana (1212), Algeria (873), Sudan (824), Tunisia (713), and Ethiopia (650) (see Table 5.3). Among which, South Africa and Egypt are the leading collaborators, taking up a large share of more than 60 percent of the total publications, with both records above 5000, while the other African countries all scored below 2000 records.

The distribution showed that the China–Africa scientific research collaboration had greater involvement from countries in the Maghreb: Egypt, Morocco, Algeria, and Tunisia. In terms of the top ten countries, Nigeria and Ghana were the leading nations in West Africa. Kenya was dominant in eastern Africa. From northeastern Africa, countries that participated less in the collaboration—Sudan and Ethiopia—fall in the last three among the top ten leading countries. For Southern Africa, only

---

[2] Li, A., Seventy Years of China–Africa Studies: Retrospect and Prospect (December 23, 2019), available at: https://opinion.huanqiu.com/article/9CaKrnKou4h, accessed on September 27, 2022.

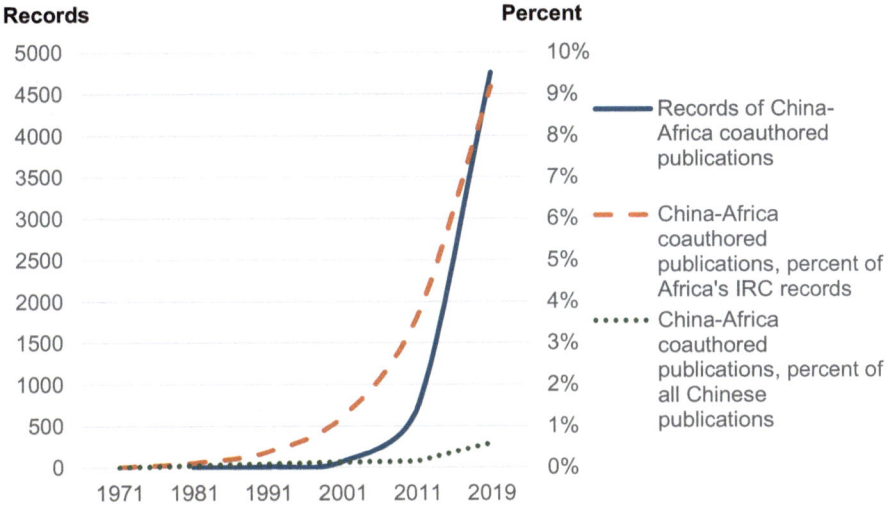

Fig. 5.1 Growth of China–Africa co-authored publications (records and percentages in Africa's IRC and all Chinese publications, 1971–2019). *Source* Web of Science database

Table 5.3 Top 10 African research collaboration partners with China (1971–2019). *Source* Web of Science database

Rank	Country (with China)	Records
1	South Africa	7416
2	Egypt	5779
3	Nigeria	1953
4	Morocco	1426
5	Kenya	1370
6	Ghana	1212
7	Algeria	873
8	Sudan	824
9	Tunisia	713
10	Ethiopia	650

South Africa took a prominent position, and the second Southern African country, Zambia, falls into the second group of the ten among the twenty leading countries (Fig. 5.2).

The results of social network analysis suggested that in the network of research collaboration between China and Africa, the degree of scientific research collaboration between China and different African countries is uneven. Some African countries play a key role in the entire collaboration network, which is not necessarily related to their weight in collaboration, as Egypt actually has a low degree of centrality compared to its heavy weight. These countries act as bridges in the whole network, including South Africa, Cameroon, Nigeria, Uganda, and Cote d'Ivoire (Fig. 5.3).

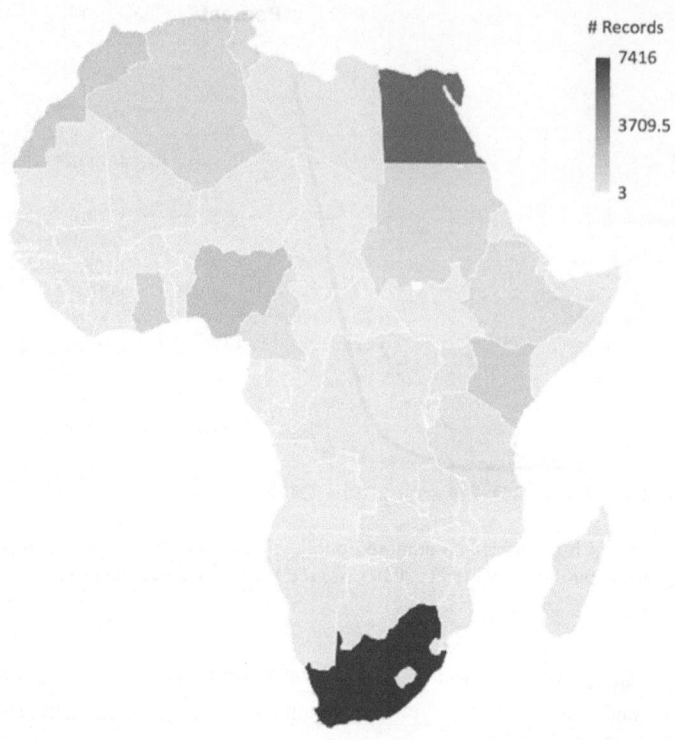

Fig. 5.2 Leading Participants of the China–Africa Scientific Research Collaboration (1971–2019). *Source* The authors' compilation of data on China–Africa scientific research collaboration obtained from the Web of Science database

There are 152 research areas of international research collaboration between African countries and international collaborators in total. The top-ranked research areas are mostly basic research areas, mainly in science disciplines and hardly in the humanities. Engineering had the highest number of collaborations, followed by Chemistry, Physics, Public, Environmental, & Occupational Health, Environmental Sciences & Ecology, Infectious Diseases, Science & Technology—Other Topics, Materials Science, Tropical Medicine, and Computer Science (Table 5.4).

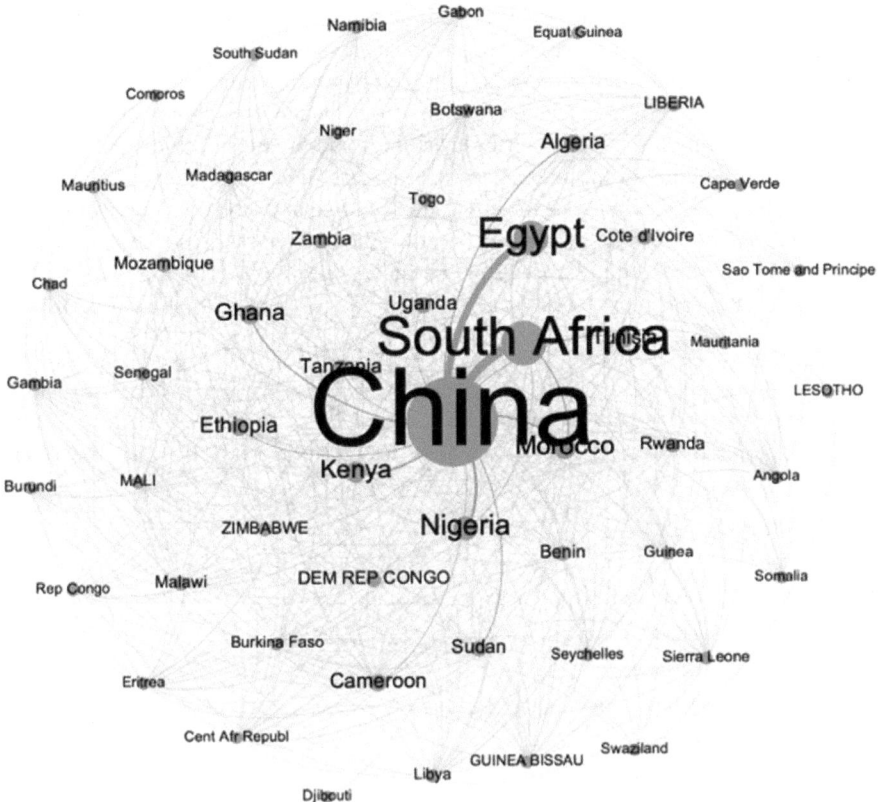

Fig. 5.3 SNA Analysis of the China–Africa Scientific Research Collaboration Network (1971–2019). *Source* The authors' compilation of data on China–Africa scientific research collaboration obtained from the Web of Science database[3]

---

[3] The nodes in the figure indicate the countries participating in the collaboration, and the connecting line between the nodes indicates that there is cooperation between the connected countries, and the thickness of the connecting line indicates the strength (weight) of the research collaboration. The size of the nodes indicates the weighted degree.

There are 149 research areas in the China–Africa scientific research collaboration. The top-ranked research areas are also basic research areas. Physics ranked first
with the highest collaboration records, followed by engineering, chemistry, environmental sciences & ecology, science & technology, materials science, astronomy &
astrophysics, computer science, agriculture, and mathematics (Table 5.5).

We further calculated the increase in China's percentage in every research area
from 1971 to 2019. The increase in China's percentage in the research area of physics
in Africa's IRC is 11.46%, the greatest among the top ten research areas in China–
Africa scientific research collaboration. There are some other research areas where
China is showing an increasingly greater percentage. Table 5.6 shows the top ten

Table 5.4 The top ten research areas of Africa's IRC and their percentages in Africa's IRC (1971–
2019). *Source* Web of Science database

Rank	Research areas	Percent of total (African's IRC) (%)
1	Engineering	9.56
2	Chemistry	8.21
3	Physics	6.64
4	Public, Environmental & Occupational Health	6.50
5	Environmental Sciences and Ecology	5.69
6	Infectious Diseases	5.56
7	Science and Technology—Other Topics	5.05
8	Materials Science	4.75
9	Tropical Medicine	4.44
10	Computer Science	4.02

Table 5.5 The top ten research areas in China–Africa scientific research collaboration and their
percentages (1971–2019). *Source* Web of Science database

Rank	Research areas	Percent of total (China–Africa co-authoring publications) (%)
1	Physics	16.80
2	Engineering	12.28
3	Chemistry	7.96
4	Environmental Sciences & Ecology	6.61
5	Science & Technology—Other Topics	6.30
6	Materials Science	5.61
7	Astronomy & Astrophysics	5.48
8	Computer Science	4.61
9	Agriculture	4.39
10	Mathematics	3.85

research areas with the highest growth percentage from 1971 to 2019. Those are mostly emerging fields and are correlated with China's economic ties with Africa.

Table 5.7 lists the top ten Chinese institutions involved in the China–Africa research collaboration. Chinese Academy Science takes the lead with a percentage of 28.28%, followed by universities including Nanjing University, Shandong University, University of Science and Technology of China, Peking University, Sun Yat Sen University, Hong Kong University, Tsing Hua University, the Chinese University of Hong Kong and the Hong Kong University of Science and Technology. It implies that the collaboration is mainly policy-driven and correlated with academic exchanges.

The physics departments at Chinese universities and African universities are the main institutions involved in the collaboration. Nanjing University has a long history in African studies, and the African Institute of Nanjing University, established in

Table 5.6 The top ten research areas with highest increase of China's percentage in Africa's IRC (1971–2019). *Source* Web of Science database

Rank	Research area	Increase of percentage (%)
1	Astronomy and astrophysics	14.09
2	Mycology	12.88
3	Instruments and instrumentation	11.61
4	Physics	11.46
5	Acoustics	9.28
6	Automation and control systems	8.88
7	Transportation	8.27
8	Energy and fuels	7.68
9	Allergy	7.64
10	Remote sensing	7.61

Table 5.7 The top ten Chinese institutions involve in Africa's IRC. *Source* Web of Science database

Rank	Institution	Percentage (%)
1	Chinese Academy Science (Branches Included)	28.28
2	Nanjing University	7.48
3	Shandong University	7.01
4	University of Science and Technology of China	6.47
5	Peking University	6.29
6	Sun Yat Sen University	4.57
7	Hong Kong University	4.52
8	Tsinghua University	4.35
9	The Chinese University of Hong Kong	3.87
10	The Hong Kong University of Science & Technology	3.49

Table 5.8 The top ten leading international collaborators in Africa's IRC (records and percentages of Africa's IRC records, 1971–2019). *Source* Web of Science database

Rank	Country	Collaborated papers with African authors (thousands)	Percent of Africa's IRC records (%)
1	USA	129.9	26.99
2	France	95.7	19.90
3	UK	80.8	16.80
4	Germany	44.7	9.29
5	Saudi Arabia	38.5	8.00
6	Canada	29.2	6.08
7	Australia	25.5	5.30
8	Italy	23.9	4.97
9	Netherlands	22.9	4.77
10	Spain	21.9	4.56
11	China	21.8	4.53

1964, is also the only domestic African research institution on geography in China.[4] Physics in the other universities also ranks at the top in terms of discipline in China. Besides, many Hong Kong universities are involved, which is also related to the historical origin of the British colony.

5.5.3 Comparison Among Top International Collaborators in Africa's IRC

From 1971 to 2019, the top ten collaborators with African countries were the United States, France, the UK, Germany, Saudi Arabia, Canada, Australia, Italy, the Netherlands, and Spain. Among the top ten collaborators, the USA, France, and the UK were responsible for over 63% of Africa's internationally collaborated publications. The United States is Africa's leading research collaboration partner, as over one-fourth of Africa's internationally co-authored publications in the 1971–2019 period involved at least one US-based researcher. China ranks 11th in terms of collaboration records with Africa but has the greatest collaboration records with Africa in Asia (Table 5.8).

The United States, France, the UK, Germany, Saudi Arabia, Canada, Australia, Italy, the Netherlands, and Spain are the top ten countries that have collaborated with African nations from 1971 to 2019. More than 63% of Africa's IRC involves collaborators from these top ten countries. Over one-fourth of Africa's IRC between

---

[4] Nanjing University., "African Development Studies" published by the Institute of African Studies of Nanjing University), available at: https://sgos.nju.edu.cn/83/fd/c8418a230397/page.htm, accessed on September 27, 2022.

1971 and 2019 involved at least one researcher stationed in the US, making it the continent's top partner for research collaboration.

Total internationally co-authored publications with African authors (1971–2019) = 481,108. Internationally collaborative publications may have co-authors from more than one country. (Since an internationally collaborative publication can involve more than two countries, the total of country percentage shares exceeds 100%).

The China–Africa scientific research collaboration began later, in the 1970s, than the top four international collaborators in Africa, and has grown in number in a similar manner to the Saudi Arabia–Africa scientific research collaboration (Fig. 5.4). Though China is now outmatched in numbers in the collaboration with Africa, the relative importance of China as Africa's international research collaborator has grown over the last two decades. In 2019, 9.17% of Africa's internationally co-authored publications involved at least one Chinese researcher, close to the proportion of Germany in that year. Saudi Arabia's share in the top countries has likewise dramatically expanded. Established countries like the United States and Europe experience several changing trends. From 1971 to 2019, the percentage of collaborated publications involving the United States fluctuated while remaining unchanged overall; France decreased after a rapid increase, the United Kingdom decreased continuously, and Germany increased gradually; in 2019, the proportions of collaborated publications involving these four countries are 26.43% and 15.11%, 16.8% and 9.55%, respectively (Fig. 5.5).

Though most of the internationally co-authored publications from Africa are bilateral—between two countries—which accounts for more than 66% of the total

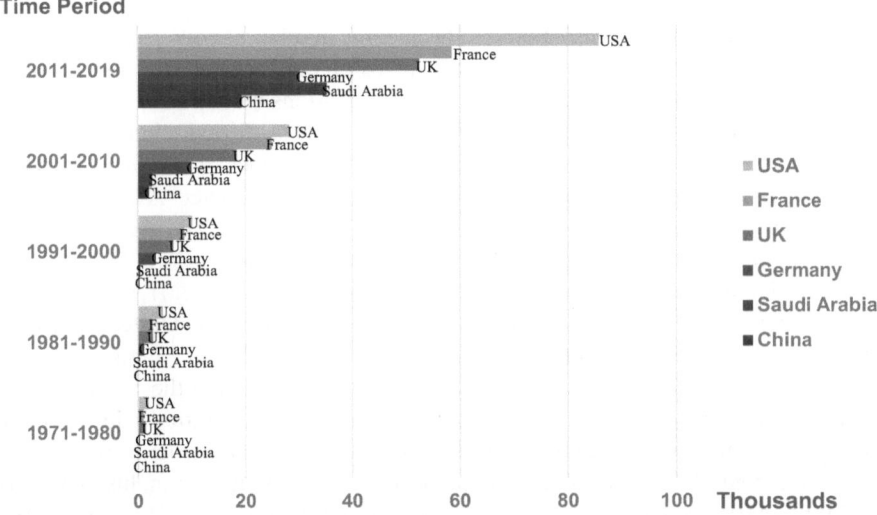

Fig. 5.4 Collaborating publications in 1971–2019 of the top five countries and China in Africa's IRC (1971–2019). *Source* Web of Science database

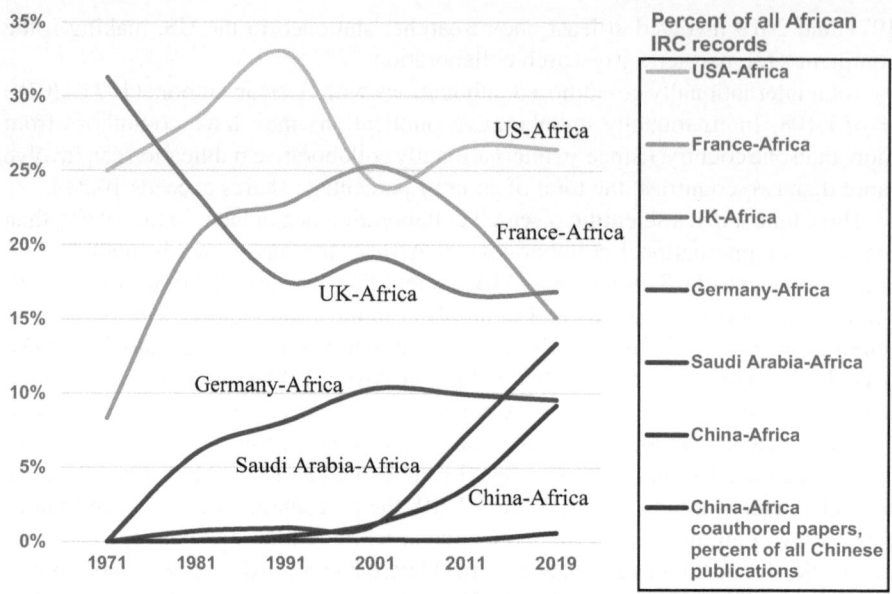

Fig. 5.5 Percentages of collaborating publications in 1971–2019 of the top five countries and China in Africa's IRC (1971–2019). *Source* Web of Science database

publications, Africa's international research collaboration also involves multinational collaboration. In our database, around 33% of records are collaborated on by researchers from at least three different countries, and more than 89% of publications involve at least three researchers. The average number of countries and authors participating in the China–Africa scientific research collaboration has substantially exceeded the top five countries' co-authored publications with Africa.

The average number of countries involved in the China–Africa scientific research collaboration is 8.76, which is more than double the number of countries involved in the US–Africa scientific research collaboration and the France–Africa scientific research collaboration, which are at 4.08 and 4.05, respectively. The average number of authors in the China–Africa scientific research collaboration is 246.87, five times that of the United States and four times that of France (48.91 and 62.01, respectively). The numbers for Britain and Germany are slightly higher than the United States and France, while also being far lower than China. Surprisingly, Saudi Arabia is not only much lower than China in these two figures but also lower than the other top four countries. It might imply that Saudi Arabia's research collaboration with Africa is mostly bilateral between Saudi Arabia and one African country (Fig. 5.6).

Involvement of multiple countries and more authors implies an inclusive pattern of China–Africa scientific research collaboration (Kaplinsky, 2013; Liu, 2019; Verkhovets & Karaoğuz, 2022). While it comes at a price, in the China–Africa scientific research collaboration, African countries have a lower percentage of first authors in the collaboration. The percentage of first authors addressed in Africa is

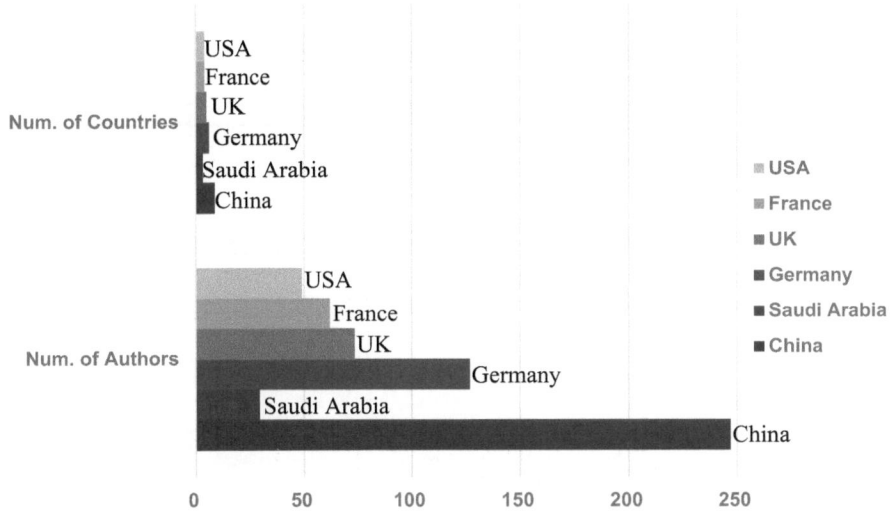

Fig. 5.6 Average number of countries and authors in Africa's IRC (with the top five countries and China, 1971–2019). *Source* Web of Science database

38.09%, and that of the top five countries co-authoring publications with Africa all exceeds that of China, which is 28.85%, 45.82%, 30.09%, 28.57%, and 36.83% for US–Africa, France–Africa, UK–Africa, Germany–Africa, and Saudi Arabia–Africa research collaborations, respectively.

In the China–Africa scientific research collaboration, only 12.59% of publications have African researchers as the first author. The proportion of Chinese researchers as the first authors accounts for 46.61%, also lower than that of the United States. While publications have researchers from other international (non-African and non-Chinese) countries as first author, they accounted for 40.8% of the total, much higher than the top five countries (Fig. 5.7). It implies that in China–Africa scientific research collaboration, other international countries (mostly American and European countries) play important bridging or leading roles (Table 5.9).

In the collaboration network, China is one of the leaders of one module with more international countries and fewer African participants. China has more collaborations with countries in BRICS, Asia, and Eastern Europe. The collaboration between China and Africa in scientific research is limited to a few African countries with better science and technology infrastructure. China exerts a smaller influence in the collaboration network. China is less likely to serve as a broker in the network and is less able to contact all other members of the network easily compared with several other major western economies (Fig. 5.8). The collaboration network between African states and international countries is mainly divided into three modules.

The first module is led by the United States, the United Kingdom, and Germany and covers most African countries. In the first module, the African countries that act as the main leaders in the collaboration are South Africa and Nigeria. The second

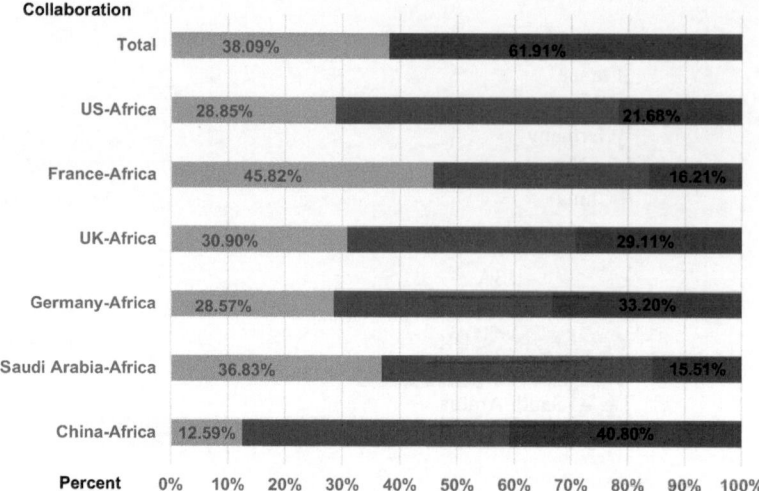

Fig. 5.7 Percentages of first author origins in Africa's IRC (with the top five countries and China, 1971–2019). *Source* Web of Science database

Table 5.9 The top ten international collaborators (non-African countries) in China–Africa scientific research collaboration (1971–2019). *Source* Web of Science database

Rank	Countries	Records (thousand)	Percentage (%)
1	USA	7.41	33.71
2	UK	5.38	24.50
3	Germany	4.61	20.95
4	France	4.49	20.43
5	Italy	4.00	18.22
6	Brazil	3.72	16.93
7	Australia	3.62	16.48
8	Spain	3.61	16.41
9	India	3.51	15.99
10	Switzerland	3.40	15.47

module is led by France and involves a small number of African countries. In the second module, the African countries that act as the main leaders in the collaboration are Egypt, Morocco, and Tunisia.

The third module mostly includes international countries like China, Spain, Italy, Brazil, and India as the main international participants, with even fewer African countries, i.e., Cape Verde. It is obvious that in the collaboration modules, China is the leader of the third module. However, the third module, where China is located, contains more international countries, including China, Spain, Italy, Brazil, and India, but with fewer African participants; only Cape Verde from Africa is in the third module. Besides, in the collaboration network, China is less likely to serve as a broker

Fig. 5.8 Social network analysis of Africa's IRC (1971–2019). *Note*: The nodes in the figure indicate the countries participating in the collaboration, and the connecting line between the nodes indicates that there is collaboration between the connected countries, and the thickness of the connecting line indicates the strength (weight) of the research collaboration. The size of the nodes in indicates the weighted degree and the colors/shapes of the nodes indicate different communities determined by the modularity algorithm and basically it shows which routers are more densely connected between each other than to the rest of the network. *Source* Web of Science database

with a lower betweenness centrality and is less able to contact all other members of the network easily with a lower closeness centrality compared with several powerful Western European and American countries, such as the US, UK, France, Spain, Canada, and so on (Figs. 5.8, 5.9).

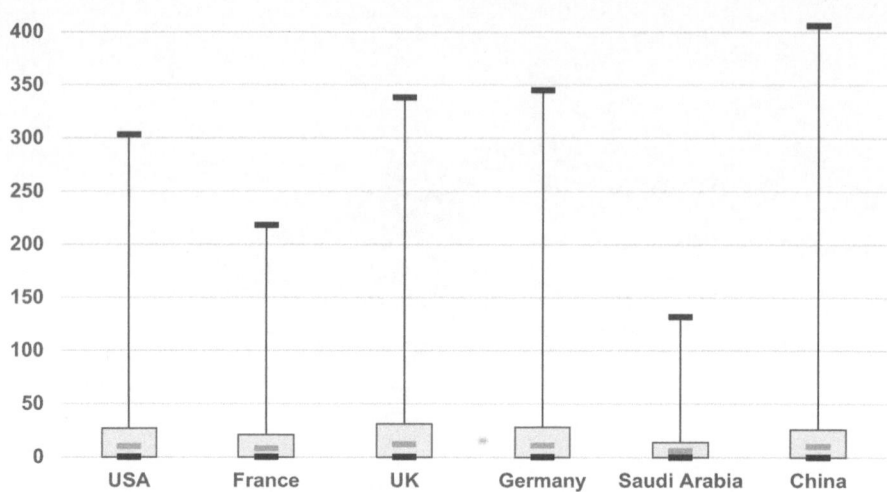

Fig. 5.9 Citation distribution box plot of publications by authorship type (the top five countries and Chinese co-authoring papers with Africa), publication output over 1971–2019. Lowest datum (within 1.5 * inter-quartile range of lower quartile) is overlapped by the 25-percentile due to number of zero-cited papers. Bold line across box plots is at the 50 percentile (median) point. Box top is at the 75 percentiles. Highest datum is within 1.5 * inter-quartile range of the upper quartile (Tang & Shapira, 2011). *Source* Web of Science database

5.6   Conclusion

This chapter analyses the emerging role of China in Africa's international research collaboration to highlight its dynamism and implications. This study offers an overall idea of Africa's IRC. It has been found that IRC has become increasingly important to Africa. We have also found that while China's research collaboration with Africa started later than that of other major international collaborators, China is gaining an increasingly important position in Africa's international research collaboration, especially with its rapid growth in the recent decade. In China and Africa, research collaboration has increased annually, starting with only one document in 1977 and increasing to more than 4500 in 2019. One of the important features of China–Africa international collaboration is that China's engagement in Africa's international research collaboration involves more international collaborators. It implies that China has not established stable bilateral research collaboration ties with African countries and is now connecting with African researchers via researchers from other international countries, primarily the United States and Europe, which play important bridging or leading roles in China–Africa scientific research collaboration.

However, we found that publications involving Chinese researchers generally attract more citations. Significantly, when a Chinese researcher is involved in the publication, there is an increase in citation quality compared with the publications involving the top five countries collaborating with Africa. In other words, China's engagement in the collaboration has increased the impact of the publications and brought more benefits to the co-authoring researchers. It further demonstrates how research collaboration differs in nature from international economic cooperation; when a Chinese researcher is involved in the publication, there is an increase in citation quality compared with the publications involving the top five countries collaborating with Africa. In other words, China's engagement in the collaboration has increased the impact of the publications and brought more benefits to the co-authoring researchers. It further demonstrates how research collaboration differs in nature from international economic cooperation. The research partnership between China and Africa is more about inclusiveness and win–win cooperation than exclusivity and zero-sum competition.

Lastly, we looked into the patterns and dynamics of research collaboration between China and Africa. The results reveal that the degree of cooperation between China and different African countries is uneven. The distribution showed that, except for South Africa, the China–Africa scientific research collaboration had greater involvement from countries in the Maghreb. The countries that have the most cooperation with China are South Africa and Egypt, which are possibly related to their own levels of science and technology development. In terms of research areas, as indicated by the statistics, physics is the most prominent field of study for the China–Africa scientific research collaboration. Aside from physics, China remains a modest presence in the most crucial research areas in Africa's IRC, despite an increase in records. Yet China is leading the way in collaboration with Africa in several emerging fields. The institutional patterns imply that the collaboration is mainly policy-driven. The primary organizations involved in the collaboration are the physics departments of Chinese and African universities.

The study offers some recommendations for developing and implementing S&T policies. China and Africa share common expectations and aspirations as they confront the enormous opportunity presented by a new round of scientific and technological revolutions. Scientific and technological cooperation and exchanges are a crucial component of the new China–Africa strategic partnership and will aid in fostering mutual benefit between the two sides. The good news is that China is currently the top Asian partner with Africa in terms of research collaboration after increasing its involvement over time. China still has a limited impact on the network of collaboration, nonetheless. China will continue to look for ways to enhance its research collaboration with Africa and aim to play a more significant role in the network of Africa's IRC. Physics is currently the principal area of research collaboration between China and Africa. It is essential that China and Africa expand their research into important areas and involve more institutions from diverse backgrounds in the collaboration.

In the meantime, we discovered that China has yet to establish long-term bilateral research collaboration ties with African countries. Instead, China is now connecting with African researchers through researchers from other international collaborators. While Saudi Arabia's experience suggests that China should not only actively participate in Africa's IRC, the Chinese government and individual researchers should also encourage African countries to participate more as the leaders of the research project in order to strengthen the African voice in the collaboration.

5.6.1 Limitations

The study has several limitations. Co-authorship measurement is by no means flaw-less. It is also acknowledged that multiple-authorship may not always accurately reflect research collaboration because not everyone listed as an author on a paper is accountable for the work and should not share the credit accorded to it (Katz & Martin, 1997). For instance, in an early case study to investigate collaboration, Hagstrom (1965) found evidence that some publications listed authors for solely social motives. More recently, the investigation of several instances of scientific fraud has revealed how common it has become to add coworkers as honorary co-authors. It has been clear from the examination of multiple cases of scientific fraud how widespread it has become to add coworkers as honorary co-authors (LaFollette, 1992). The complex nature of collaboration is perhaps not as readily amenable to assessment as previous authors have assumed. Bibliometric analysis of multiple-author papers can only be used as a partial indicator of collaborative activity.

Additionally, the Web of Science is internationally recognized as one of the leading standardized datasets for analysing scientific publications. However, there are coverage deficiencies in the Web of Science (Duque et al., 2005), particularly in the coverage of papers produced by African researchers in non-English language journals. Many African scientists still publish exclusively in domestic Chinese or African journals, most of which are not indexed in the Web of Science. This gives rise to potential biases. On the one hand, absolute counts of African paper outputs that rely on the Web of Science will underestimate total scientific productivity in Africa since papers in non-listed African language journals will not be included. However, China–Africa or US–Africa co-authored papers are less likely to be affected by this problem since such papers are apt to be published in English-language Web of Science journals (Tang & Shapira, 2011).

Moreover, researchers who only publish in domestic journals in their native African languages are less likely to collaborate with peers from other countries and may, on average, produce research of lower quality than African researchers who publish in journals listed in the Web of Science in English (Tang & Shapira, 2011). However, since all the records examined in this study were produced by worldwide collaborators in English, quality comparisons in this instance might not be prejudiced. Finally, our analysis focuses primarily on examining China's contribution to the IRC in Africa. The regional publications between African states and the international collaboration between Africa and other international countries are not investigated in depth due to the length of the paper, and at the same time, specific research areas and the mechanisms behind China–Africa scientific research collaboration are still underexplored and could all become future research directions.

References

Adams, J., Gurney, K., Hook, D., & Leydesdorff, L. (2014). International collaboration clusters in Africa. *Scientometrics, 98*(1), 547–556.

Alden, C. (2005). China in Africa. *Survival, 47*(3), 147–164.

AlRyalat, S. A. S., Malkawi, L. W., & Momani, S. M. (2019). Comparing bibliometric analysis using PubMed, Scopus, and Web of Science databases. *Journal of Visualized Experiments, 152*, e58494.

Archambault, É., Campbell, D., Gingras, Y., & Larivière, V. (2009). Comparing bibliometric statistics obtained from the Web of Science and Scopus. *Journal of the American Society for Information Science and Technology., 60*(7), 1320–1326.

Boshoff, N. (2009). Neo-colonialism and research collaboration in Central Africa. *Scientometrics, 81*(2), 413–434.

Boshoff, N. (2010). South-South research collaboration of countries in the Southern African Development Community (SADC). *Scientometrics, 84*(2), 481–503.

Chen, K., Zhang, Y., & Fu, X. (2019). International research collaboration: An emerging domain of innovation studies? *Research Policy., 48*(1), 149–168.

Duque, R. B., Ynalvez, M., Sooryamoorthy, R., Mbatia, P., Dzorgbo, D.-B.S., & Shrum, W. (2005). Collaboration paradox: Scientific productivity, the Internet, and problems of research in developing areas. *Social Studies of Science., 35*(5), 755–785.

Eduan, W., & Yuanqun, J. (2019). Patterns of the China–Africa research collaborations from 2006 to 2016: A bibliometric analysis. *Higher Education, 77*(6), 979–994.

Eito-Brun, R., & Ledesma Rodriguez, M. (2016). 50 years of space research in Europe: A bibliometric profile of the European Space Agency (ESA). *Scientometrics, 109*(1), 551–576.

Grimm, S. (2014). China–Africa cooperation: Promises, practice and prospects. *Journal of Contemporary China., 23*(90), 993–1011.

Guns, R., & Liu, Y. (2010). Scientometric research in China in context of international collaboration. *Geomatics and Information Science of Wuhan University, 35*(ICSUE 2010 special), 112–115.

Hagstrom, W. O. (1965). *The scientific community*. Basic books.

He, T. (2009). International scientific collaboration of China with the G7 countries. *Scientometrics, 80*(3), 571–582.

Kaplinsky, R. (2013). What contribution can China make to inclusive growth in sub-Saharan Africa? *Development and Change, 44*(6), 1295–1316.

Katz, J. S., & Martin, B. R. (1997). What is research collaboration? *Research Policy, 26*(1), 1–18.

Kim, K.-W. (2006). Measuring international research collaboration of peripheral countries: Taking the context into consideration. *Scientometrics, 66*(2), 231–240.

LaFollette, M. C. (1992). *Stealing into print: Fraud, plagiarism, and misconduct in scientific publishing*. London: Univ of California Press.

Liu, W. (2019). *The belt and road initiative: A pathway towards inclusive globalization*. Routledge.

Mêgnigbêto, E. (2013). International collaboration in scientific publishing: The case of West Africa (2001–2010). *Scientometrics, 96*(3), 761–783.

Melin, G., & Persson, O. (1996). Studying research collaboration using co-authorships. *Scientometrics, 36*(3), 363–377.

Muchie, M., & Patra, S. K. (2020). China–Africa science and technology collaboration: Evidence from collaborative research papers and patents. *Journal of Chinese Economic and Business Studies, 18*(1), 1–27.

Nguyen, T. V., Thao, P. H. L., & Ut, V. L. (2017). International collaboration in scientific research in Vietnam: An analysis of patterns and impact. *Scientometrics, 110*(2), 1035–1051.

Niu, X. S. (2014). International scientific collaboration between Australia and China: A mixed-methodology for investigating the social processes and its implications for national innovation systems. *Technological Forecasting and Social Change., 85*, 58–68.

Onyancha, O. B., & Maluleka, J. R. (2011). Knowledge production through collaborative research in sub-Saharan Africa: How much do countries contribute to each other's knowledge output and citation impact? *Scientometrics, 87*(2), 315–336.

Onyancha, O. B., & Ocholla, D. N. (2007). Country-wise collaborations in HIV/AIDS research in Kenya and South Africa, 1980–2005.

Owusu-Nimo, F., & Boshoff, N. (2017). Research collaboration in Ghana: Patterns, motives and roles. *Scientometrics, 110*(3), 1099–1121.

Pouris, A., & Ho, Y.-S. (2014). Research emphasis and collaboration in Africa. *Scientometrics, 98*(3), 2169–2184.

Schubert, T., & Sooryamoorthy, R. (2010). Can the centre–periphery model explain patterns of international scientific collaboration among threshold and industrialised countries? The case of South Africa and Germany. *Scientometrics, 83*(1), 181–203.

Sooryamoorthy, R. (2009). Do types of collaboration change citation? Collaboration and citation patterns of South African science publications. *Scientometrics, 81*(1), 177–193.

Sooryamoorthy, R. (2014). Publication productivity and collaboration of researchers in South Africa: New empirical evidence. *Scientometrics, 98*, 531–545.

Tang, L., & Shapira, P. (2011). China–US scientific collaboration in nanotechnology: Patterns and dynamics. *Scientometrics, 88*(1), 1–16.

Toivanen, H., & Ponomariov, B. (2011). African regional innovation systems: Bibliometric analysis of research collaboration patterns (2005–2009). *Scientometrics, 88*(2), 471–493.

Verkhovets, S., & Karaoğuz, H. E. (2022). Inclusive globalization or old wine in a new bottle? China-Led Globalization in Sub-Saharan Africa. *Globalizations, 19*(8), 1195–1210.

Wang, X., Xu, S., Wang, Z., Peng, L., & Wang, C. (2013). International scientific collaboration of China: Collaborating countries, institutions and individuals. *Scientometrics, 95*(3), 885–894.

Zhou, P., & Lv, X. (2015). Academic publishing and collaboration between China and Germany in physics. *Scientometrics, 105*(3), 1875–1887.

Zhou, P., Zhong, Y., & Yu, M. (2013). A bibliometric investigation on China–UK collaboration in food and agriculture. *Scientometrics, 97*(2), 267–285.

Chapter 6
China–Africa S&T Relationship: Selected Cases of Learning and Technology Transfer

Swapan Kumar Patra and Mammo Muchie

6.1 Introduction

China–Africa relationship is getting momentum since a couple of decades. With the beginning of the twenty-first century, China–Africa ties have been further strengthened in both the political and economic arena. Particularly, with the formation of the Forum on China–Africa Cooperation (FOCAC) in 2000, economic and trade cooperation has been further enhanced and revitalized (Forum on China–Africa Cooperation Beijing Action Plan 2019–2021). Both sides are very eagerly working towards a long term mutual, strategic, and sustainable partnership. China and Africa are working towards a real South-South cooperation for a win–win situation for both the sides.

Recent empirical evidence shows that now China is the largest trading partner of Africa. However, China–Africa economic and trade cooperation centred around mostly on bilateral trade and Chinese aid to Africa. According to Chinese State Councillor and Foreign Minister Wang Yi, the volume of trade between China and Africa was more than $200 billion in 2019. With this increasing trade, now China is the largest trading partner for Africa. The minister further stated that China's stock of direct investment in Africa has reached about $110 billion, and more than 3700 Chinese enterprises have operation in the African continent. Empirical evidence (Fig. 6.1) shows that the trade between China and Africa has been increasing over the year.

Even in the COVID-19 pandemic, China–Africa economic and trade cooperation is continuing. Presently, more than 1100 Chinese projects are operating in Africa.

S. K. Patra (✉)
Sidho-Kanho-Birsha University, Purulia, West Bengal, India 723104
e-mail: skpatra@gmail.com; skpatra@skbu.ac.in

M. Muchie
DSI/NRF SARChI Research Chair On Science, Technology and Innovation Studies, Tshwane University of Technology, Pretoria, South Africa
e-mail: muchiem@tut.ac.za

© The Author(s) 2024
M. Muchie et al. (eds.), *China-Africa Science, Technology and Innovation Collaboration*, https://doi.org/10.1007/978-981-97-4576-0_6

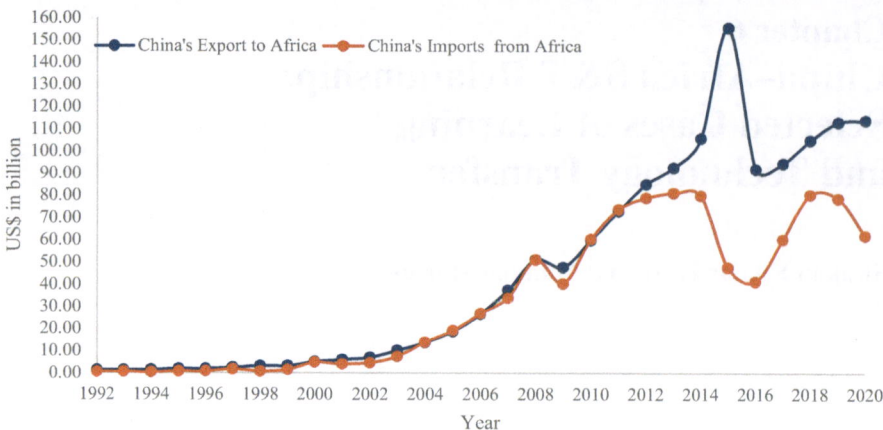

Fig. 6.1 China–Africa Trade. *Source* China Africa Research Initiative John Hopkins School of Advanced International Studies & UN Comtrade data

Moreover, during the first quarter of the year 2020, China's investment in Africa grew by 4.4 percent in comparison to the trade relation that took place in the year 2019. According to the statistics released by the Ministry of Commerce of China, the trade volume between China and Africa reached over $80 billion in the first half of the year 2020 (Yuanyuan, 2020). It can be observed that even during the COVID-19 time, the trade between China and Africa had not declined. It is assumed that the trade is likely to be increased further along with the growing China–Africa relationship.

Scholarly research on China–Africa relations have been carried out mostly in the fields of economics, politics, and diplomacy. However, China–Africa collaboration in Science and Technology (S&T) is very little explored. Hence, this study is an attempt to map the China–Africa cooperation through technology transfer and local learning in various sectors across the Global Value Chain (GVC) in the field of S&T.

It is observed that several technology transfers have happened from Chinese firms, and Chinese experts to various African countries. The cooperation has happened from low technology to high technology sectors in recent years. For example, Ethiopia has launched its second satellite into space in 2020 with the help of China. The ET-SMART-RSS nanosatellite is launched to provide earth observation services to China and other African countries. Ghana and Zambia have developed solar irrigation facilities for farmers across the country. These low-cost solar power facilities are the big relief for the farmers in the rural areas where there is no consistent electricity supply. Moreover, China is cooperating with many African countries with green technology transfer. For example, China is working with Ghana and Zambia on the China–Africa South-South cooperation on Renewable Energy Technology Transfer projects. The aim of this cooperation is to transfer renewable energy technology to many African countries and enhancing off-grid, community-based electrification. In agricultural sector, Chinese experts help to increase rice production in Mozambique and new irrigation techniques to increase crop yields.

6.2 Methods

This study mainly uses newspaper content analysis method. The English language newspapers for example Beijing Review (a national English news weekly) and China Daily were searched for cases of technological cooperations. After an extensive search, a few important cases are selected from these newspaper reports. Using selected case study in few sectors in African countries, this study will present the various cases of technology transfer from China and Africa and the benefit incurred in African economy. This study is based on the news published on English language in Chinese online newspapers. Hence there might be a language and Chinese bias. This is the limitation of this study. However, to the best of our knowledge, this type of study has not been carried out before, so, it will certainly have policy implication to understand the China–Africa relation in S&T.

6.3 Selected Cases of Technology Transfer from China to Africa

It is a well-established fact that African countries are lagging in terms of technology. With the growing influence of China in the continent, various types of collaborations happen between China and African countries. However, in case of China–Africa relations, technology transfer is one of the less studied subjects compared to the other aspects of China–Africa relations. In this context this section is an attempt to investigate the China–Africa technology transfer from the selected cases of technological cooperation in different sectors.

Technological transfer (TT) is the process by which the owner of the technology passes the technology 'know-how' of products, management, and sale along with its rights to the other parties through several ways. In this context Li (2016) has observed that along with the "Eight Principles for Economic Aid and Technical Assistance to other Countries in 1964" (Shixue, 2011),[1] TT also happens between China and Africa along with the other form of cooperations. Li (2016) found that China–Africa technology transfer happens in many ways. For example, TT occurs through technical assistance, knowledge transfer and knowledge sharing (Li, 2016).

---

[1] [The "Eight Principles" are set of guidelines given by the then Chinese Premier Zhou Enlai during his visit in Africa in 1963–1964. Those guidelines are: mutual benefit; no conditions attached; the no-interest or low-interest loans would not create a debt burden for the recipient country; to help the recipient nation develop its economy, not to create its dependence on China; to help the recipient country with project that needs less capital and quick returns; the aid in kind must be of high quality at the world market price; to ensure that the technology can be learned and mastered by the locals; the Chinese experts and technicians working for the aid recipient country are treated equally as the local ones with no extra benefits for them].

– Jiang Shixue (November 29, 2011). China's principles in foreign aid. Available at http://www.china.org.cn/opinion/2011-11/29/content_24030234.htm#:~:text=The%20% 22Eight%20Principles%22%20are%3A,project%20that%20needs%20less%20capital

6.4   Agriculture Technology

Agricultural technology transfer plays a significant role in transforming agricultural productivity in rural areas. Agricultural production is utmost important especially in current scenario where food demand exceeds the production capacity. TT facilitates the movement of soft and hard skills essential for improving farm production. Yet, the technical cooperation projects in Africa have been suffering from effectiveness and sustainability challenges while lacking responsiveness to local demand. Lessons from Japan and China's experiences in agricultural technology transfer projects in various African countries show that Japanese and Chinese agricultural technology has the potential to improve productivity and the livelihood of rural households in Africa. Further, the local governments have a proactive role to set policy environments and institutional frameworks that might encourage and support the agricultural technology transfer to benefit the rural farmers (Mgendi et al., 2019).

6.5   Chinese Influence in African Agricultural Sector

Beside infrastructure, agriculture has been identified as the priority area in the China–Africa strategic relationship. Agricultural knowledge and technology transfer were given stressed in the Forum on China–Africa Cooperation (FOCAC). China, along with Food and Agriculture Organization (FAO) of the United Nations, have deployed more than 1000 technical experts and technicians in Africa. The African continent benefited from the Trust Fund set up by China for the South-South Cooperation projects across the world (FAO, n.d.).

Many Chinese firms have acquired land in Africa for contract farming (Table 6.1). According to the estimate by China Africa Research Initiative of John Hopkins School of Advanced International Studies, Chinese firms have acquired 104.7 thousand hectares in Cameroon. Moreover, many Chinese agricultural experts are now engaged on the African continent for improving agricultural productivity. These experts could help African farmers to increase the crop yields, poverty reduction and sustainable development (Kaizhi, 2020a, 2020b).

There are many more examples of Chinese technological cooperation in various sectors. In this section some of the Chinese activities in African agriculture sector will be discussed.

6.5.1   China–Ethiopia Agricultural Cooperation

Several agricultural cooperation projects are playing key role in Sino-African relations. Among the many other agricultural cooperation, Ethiopia has joint agricultural projects with China. Both the countries have signed mutual agreements on

Table 6.1 Land Acquired by Chinese Companies in Africa, 1987–2016. *Source* China Africa Research Initiative John Hopkins School of Advanced International Studies

Sl. no.	Country	Land acquired (thousands of hectares)
1	Cameroon	104.7
2	Mozambique	31.2
3	Madagascar	30.5
4	Mali	26.2
5	Zimbabwe	14.5
6	Nigeria	14.0
7	Zambia	8.9
8	Tanzania	6.9
9	Benin	5.2
10	Guinea	2.4
11	Sierra Leone	1.8
12	Sudan	1.7
13	Togo	1.7
14	Côte dIvoire	1.6
15	DR Congo	0.7
16	Ghana	0.5
17	Uganda	0.5

the exchange of agricultural technology and specialists. An agreement was signed in the year 2000 where Chinese agricultural scientists and professionals impart training to the Ethiopian agriculture extension workers. Perhaps, this is one of the oldest and the most successful China–Africa agricultural education program. The project is known as "Sino-Ethiopian Agricultural Technical Vocational Education and Training" (ATVET) project. Under this project China has sent about 500 agricultural experts to Ethiopia. These experts trained Ethiopian agricultural extension workers at the ground level. With this program, Ethiopian agricultural personal trained for further development of vocational training and education within Ethiopia (CHINAFRICA, 2020).

6.5.2 China–Zambia Technology Transfer in Agricultural Sector

CAMCO Group of industries is China's leading company that is engaged in agricultural and engineering machinery. Its subsidiary, CAMCO Equipment (Zambia) Ltd., is playing very proactive role in the agriculture and construction sectors by supplying know how of modern firm equipments. The company is not only providing agriculture equipments but also providing after sell service of agricultural machinery to meet the local demand for advanced agricultural technology using both Chinese and Zambian technicians.

Moreover, the company generates employment for locals. According to the report, the company has employed about 680 local personals from Zambia. The company has also agriculture related operation base in other African countries including Kenya, Malawi, and Cameroon. Furthermore, the company has invested about $ 20 million all over Africa (Silimina, 2018).

6.5.3 China–Burundi Agricultural Technology Transfer

Burundi is an agrarian economy with more than 90 percent of its population depends on agriculture. The major bottleneck of agricultural development in the country is that, it is an underdeveloped country. Farmers still use old equipments and mode of production in agriculture is quite obsolete. The machinery used in agriculture is imported and the agricultural technology is mainly depend on imports of foreign technology. Further, most of the imported equipment, there is very little after sell services and sometimes it is difficult to repair or replace the damaged parts after breakdowns. There is no local manufacturing capability of agricultural equipment. This undeniably impaired local agricultural development in Burundi.

Many Chinese agricultural experts are now working in Burundi, where they provide technical assistance to the local people. Chinese experts helping local farmers and others agricultural extension workers to procure spare parts for related equipments they need. Also, they impart training to the local technicians to enhance their awareness of system maintenance.

6.5.4 China–Mozambique Agricultural Technology Transfer

According to an estimate, Mozambique has about 900,000 hectares of cultivable land. This arable land is mainly suitable for rice cultivation. Before the Chinese experts started their agricultural projects in Mozambique, the rice production was roughly 1.3 tons per hectare. Moreover, only about 300,000 hectares were under rice cultivation. Hence there were always the scarcity of rice production in Mozambique.

Chinese experts involved in Sino-Mozambican agricultural cooperation project are mainly doing research to improve agricultural productivity. Chinese experts opened rice research field in 'Mozambique Agricultural Research Institute.' Those experts teach and share modern rice cultivation techniques, and management methods. This project was materialized along with local agricultural experts. With their continuous support, the productivity of Chinese hybrid rice has increased many-fold. With proper advice and management from Chinese experts, even the yield of local rice variety has increased significantly. Moreover, the Chinese experts are advising local farmers for better yield for other crops also (Jing, 2020a, 2020b).

Chinese experts are also experimenting in Liberia on the productivity of domestic animals. It is reported that, because of the counselling of Chinese experts in terms of technical know-how, there is a growth in the yields of pigs using artificial inseminating techniques. In the West Africa, particularly in Nigeria, Chinese technical personnel were engaged for increasing in rice and fish production (Adoboe, 2017).

6.5.5   China Burkina Faso Agricultural Technology Transfer

Until very recently, Burkina Faso did not have diplomatic relationship with the People's Republic of China. It is one of the three African countries that was not a part of the Forum on China–Africa Cooperation (Khan, 2014). In 2018 China and Burkina Faso announced the resumption of their diplomatic relations.

With the establishment of diplomatic relations, the Burkinabe Government requested Chinese agricultural experts of Chinese Ministry of Agriculture and Rural Affairs to came to Burkina Faso. In Burkina Faso's rice-producing area, Chinese experts developed water and irrigation projects adapted to local agricultural conditions. Chinese agricultural expert team along with local experts identified several potential projects, including land reclamation and irrigation canal construction. Along with Chinese experts, many water conservations projects have been initiated and rice seed production has been achieved in some places (Jing, 2020a).

6.5.6   Rural Technology Transfer in Zimbabwe

Due to the lack of communication infrastructure travelling in and around rural Zimbabwe is quite difficult. Women, particularly the rural women suffer due to lack of road infrastructure. Rural women often walk long distances for buying things from the markets or fetch firewood and water.

In this context electric powered motorbikes imported from China to Zimbabwe are seen as a tremendous opportunity for rural women. Motorbikes can transform the lives of many rural women. With the new Chinese motorbikes, mobility has been increased and the motorbikes are becoming a major source of livelihood for many women in the remote areas. The motorbikes are widely used across China as sustainable and cost-effective transportation for small-scale farmers. This model of transportation has been replicated in Zimbabwean case. Many non-government organizations (NGOs) are talking the opportunity to help women in the villages. For example, an NGO called 'Mobility for Africa' are using Chinese motorbikes assembled in Harare. However, the motorbike parts are made in China and are imported to Zimbabwe. These bikes are used by the rural women to do their day-to-day activities. Thus, the mobility has increased, which has empowered women and linked them with many income generating projects.

The idea has been materialised using various partners both from China and Zimbabwe. Mobility for Africa in Zimbabwe has partnered with the Lab for Life-long Learning of Tsinghua University, Beijing, motorbike manufacturer China Hebei Dajiang, and other Zimbabwean partners, the Midlands State University (MSU), Zimbabwe Incubation Hub and Solar Shack. Tsinghua University research students and Midlands State University worked together to design the motorbike. This a good example of collaboration between universities in China and Africa. Research scholars from the universities have designed models suitable for Zimbabwean road conditions.

'Mobility for Africa' have donated motorbikes to the rural women. They also do local assembly using training from the Chinese technicians. In this way technology transfer is happening from the Chinese technicians to the local technicians. The technology of electric motorbikes and solar powered batteries are useful for Zimbabwean local condition (Masau, 2020).

6.5.7   Chinese Training Program in Zambia

In Zambia, a few Chinese companies are providing training to its local employees mostly through formal programs in agriculture and construction industries. This is done through mentoring and on-the-job training. For these reasons, Chinese investments in Zambia's key economic sectors have continued to change people's lives, especially the youth, with many direct and indirect jobs creation (Silimina, 2020).

6.5.8   Animal Husbandry in São Tomé and Príncipe

São Tomé and Príncipe is a lower middle income, small island nation in East Africa. Chinese expert helps São Tomé and Príncipe to develop its animal husbandry industry. In 2018, Chinese agricultural experts on animal husbandry came to São Tomé and Príncipe. Those experts did research on pig breeding, preparation of animal feed from agricultural stock and so on. Along with the local farmers, Chinese consultants developed breeding techniques for advanced pig production and fodder production suitable for the local climate condition (Kaizhi, 2020a). The technology imparted by the Chinese experts may help this country to elevate poverty level of common people.

6.5.9 Satellite Program in Ethiopia

China has played a significant and very proactive role to facilitate the development of high technology capability (for example space programs) in many African countries. China also helps in the process of technological capability building in satellite programs. The satellite technology uses many crucial technologies including remote sensing, communication, and navigation. These technologies are very valuable to address various immediate and pressing needs of African countries. For example, remote sensing satellites can provide scientific data useful to manage disasters, weather forecasting, monitor agricultural activities, facilitate urban planning, and development. Moreover, many African countries are now eager to actively participate in space program. Africa no longer want to be the passive users of satellites services from the developed countries. With this objective, many African countries are now creating or strengthening their satellite programs. For example, South Africa and Nigeria are already in the race. These two countries are now collaborating with other developing as well as developed countries such as India, China, Brazil, Argentina, and South Korea (Wood & Weigel, 2010).

Ethiopia is the latest candidate in this field. Ethiopia is focussing on developing their technological capability in satellite technology. Ethiopia's first remote sensing satellite, ETRSS-1 was developed by the China Academy of Space Technology (CAST). The collaboration happened between CAST China, and Ethiopian Space Science Technology Institute (ESSTI), Ethiopia. The cost of the project sheared between the two countries. This Satellite will provide data for Ethiopian scientists for the environment monitoring purpose. In this process Ethiopian scientists, trained on this project as part of the technology-transfer agreement between China and Ethiopia. The launch of ETRSS-1 is a milestone in Ethiopia's space program. Before Ethiopia, Kenya (1KUNS-PF launched in 2018) and Rwanda (RwaSat-1 launched in September 2019) are the countries that already have their satellite into the space (Ibeh, 2019).

Moreover, Ethiopia is also planning to launch its second satellite into space with Chinese help. The proposed satellite, named ET-SMART-RSS is jointly developed, assembled, and tested in China. The satellite is developed to provide earth observation services not only to Ethiopia but also to other African countries (Ibeh, 2019). According to the report (Ibeh, 2020) Ethiopia is also planning to launch a third satellite with Chinese help. The proposed communication satellite will be used for commercial telecommunications and broadcasting services.

ET-SMART-RSS is a remarkable achievement of Ethiopian space ambitions. Along with the development of this satellite, Ethiopia going to accomplish its long-term plan of developing indigenous capabilities in building satellite systems and space technologies. In the long run the technological learning will benefit Ethiopia and other African countries.

6.6   Concluding Remarks

This exploratory research on China–Africa relations with technology transfer reveals many interesting cases from different African countries. China has created the Beijing Action Plan (2019–2021) by allocating $50 Billion from the Government and $10 billion from various Chinese companies. This will cover all spheres of China–Africa collaboration including technology and knowledge transfer, training and learning. These issues have been significantly stressed in the growing China–Africa collaboration. China is not only transferring technology, but also engage in human resource development in Africa. Staff from China collaborate and undertake vocational training to the African students. Practical training and many workshops are conducted to provide vocational training to African workers. Experts from China has opened opportunities for staff training in Africa. China is sharing the development models and practice from her own journey for Africa to learn and draw lessons from the Chinese experience. Knowledge creation through technology transfer from China to Africa is a unique model that has been applied in addressing problems and discovering appropriate and relevant solutions. What is required from the African side is the encouragement and right motivation among the African human resources (staffs) to be the transferee to learn from the science, technology, innovation, knowledge transferees to become inventors, knowledge creators with curiosity, wonder and imagination.

The China–Africa economic relations have been grown since the 1950s. What is to be clarified that the understanding on both sides of Chinese and African challenges has not been fully stated. The principle of mutual benefit and win–win relations needs to be explained with tangible and measurable quality output results. Africa needs the science, technology, engineering, and innovation infrastructure. The experience and knowledge acquired through the Chinese transformation, performance and development on every sphere, Africa can learn and acquire the right and sustainable skills. China has to offer real vocational training in all the areas such as agriculture, manufacturing, commerce, railways, infrastructure, and services. The transfer and exchange of information is necessary but not sufficient. What is needed is real knowledge and experience acquisition and achievement by the African side from the Chinese knowledge and experience through the practical training.

We have observed in this chapter what China has done and contributed to different African countries. The area where we did the research and found very interesting evidence is on agriculture along with other sectors. There is a need to create an Africa–China agriculture training for Africans to learn how the Chinese have managed their agriculture value chain. China has learned from other countries and sharing the experience with the rest of Africa the strength and weakness China went through will be useful.

The Africa–China technology and knowledge transfer collaboration is critically significant. As China is growing rapidly using digital technology, machine learning, data science, artificial intelligence and Blockchain, there is a lot that Africa can gain in the spheres of production such as agriculture, manufacture, infrastructure, education, research, and vocational training. This exploratory research has opened the opportunity to undertake more original data evidenced discovery to make sure the real benefits and costs are known freed from all the media exposures. The China–Africa technology transfer relations can generate the cases to demonstrate the mutual benefits between them. There is a need to do more research to generate contributions and recommendations based on research evidence anchored data of the weaknesses and strengths to produce policy learning for both China and Africa.

References

Adoboe, J. L. (2017). Interview: Tech transfer key to China–Africa agricultural cooperation: FAO official, *Xinhua*, 16 October. http://www.xinhuanet.com/english/2017-10/16/c_136683831.htm

CHINAFRICA. (2020). *Agricultural cooperation projects play a key role in Sino-African relations* (Vol. 12, October 2020). http://www.chinafrica.cn/Homepage/202010/t20201016_800223822.html

FAO (no date). *FAO-China joint efforts through south-south cooperation*. https://www.fao.org/china/programmes-and-projects/success-stories/south-south-cooperation/es/

Ibeh, J. (2019). Ethiopia launches first satellite named ETRSS-1 from China, *Space in Africa*, 20 December. https://africanews.space/ethiopia-launches-first-satellite-named-etrss-1-from-china/

Ibeh, J. (2020). Updated: Ethiopia to launch a second satellite in October with the help of China, *Space in Africa 2020*, 26 August. https://africanews.space/ethiopia-to-launch-a-second-satellite-in-october-with-the-help-of-china/

Jing, L. (2020a). A Chinese agricultural expert in Burkina Faso shares new irrigation techniques to raise rice yields. *CHINAFRICA*, 14 September. http://www.chinafrica.cn/Homepage/202009/t20200914_800220668.html

Jing, L. (2020b). A Chinese expert helps increase rice production in Mozambique. *China Africa*, 11 August. http://www.chinafrica.cn/Homepage/202008/t20200811_800217275.html

Kaizhi, L. (2020a). Burundi's rice cultivation benefits from Chinese expertise. *CHINAFRICA*, Vol. 12, 08 December. http://www.chinafrica.cn/Homepage/202012/t20201208_800229358.html

Kaizhi, L. (2020b). Chinese expert helps São Tomé and Príncipe develop its animal husbandry industry. *CHINAFRICA*, Vol. 12, 12 November. http://www.chinafrica.cn/Homepage/202011/t20201112_800226786.html

Khan, M. G. (2014). The Chinese presence in burkina faso: A Sino-African cooperation from below. *Journal of Current Chinese Affairs, 43*(1), 71–101.

Li, A. (2016). Technology transfer in China–Africa relation: Myth or reality. *Transnational Corporations Review, 8*(3), 183–195. https://doi.org/10.1080/19186444.2016.1233718

Masau, P. (2020). Chinese motorbikes change lives of Zimbabwe's rural people, especially women. *CHINAFRICA*, 31 August. http://www.chinafrica.cn/Homepage/202008/t20200831_800219113.html

Mgendi, G., Shiping, M., & Xiang, C. (2019). A review of agricultural technology transfer in Africa: Lessons from Japan and China case projects in Tanzania and Kenya (Review). *Sustainability (switzerland)*. https://doi.org/10.3390/su11236598

Shixue, J. (2011). *China's principles in foreign aid*. 29 November. http://www.china.org.cn/opinion/2011-11/29/content_24030234.htm#:~:text=The%20%22Eight%20Principles%22%20are%3A,project%20that%20needs%20less%20capital

Silimina, D. (2018). Mechanizing Zambia's Agricultural Sector. *CHINAFRICA*, Vol. 10 November. http://www.chinafrica.cn/Homepage/201811/t20181101_800145887.html

Silimina, D. (2020). Chinese business technology transfer provides benefits for Zambia, *CHINAFRICA*, 31 August. http://www.chinafrica.cn/Homepage/202008/t20200831_800219 114.html

Wood, D., & Weigel, A. (2010). Building technological capability within satellite programs in developing countries. In *61st international astronautical congress 2010, IAC 2010* (pp. 1663–1676).

Yuanyuan, X. (2020). Twenty years on, FOCAC drives stronger China–Africa cooperation on multiple levels. *CHINAFRICA*, Vol.12 October. http://www.chinafrica.cn/Homepage/202009/t20200927_800222004.html

Chapter 7
Current Enrolment and Training Situation of African Students in China: Case Study of Sino-Africa Joint Research Centre

Xu Shao, Huaibo Liu, and Xiao Zhang

7.1 Introduction

The educational exchanges and cooperation between China and Africa are important aspects of China's educational internationalization process. It is also an important component of China's foreign relations, especially China–Africa diplomatic relations. The emergence and development of educational exchange between China and Africa depends on the development of China–Africa relations and the evolution of China's foreign policy. China–Africa education cooperation has a history of more than 70 years. With the development of China–Africa relations, education has gone through three stages of development, from the initial exchange of international students to the current multi-level, multi-disciplinary, and multi-form cooperation (He, 2006).

7.2 China Africa Education Cooperation: First Stage (Early 1950s to Late 1980s)

In the early 1950s, African National Liberation Movement took on the stage. When the People's Republic of China was founded, China's economic strength was very weak in the world. To improve the international status and promote economic development, Chinese government decided not to only support African National Liberation

X. Shao (✉) · H. Liu
Sino-Africa Joint Research Centre, Chinese Academy of Sciences, Wuhan, China
e-mail: shaoxu@wbgcas.cn

X. Zhang
China-Africa Innovation Cooperation Centre, Wuhan, China

© The Author(s) 2024
M. Muchie et al. (eds.), *China-Africa Science, Technology and Innovation Collaboration*, https://doi.org/10.1007/978-981-97-4576-0_7

Movement, but also carried out cooperation with Africa in the fields of economy, politics, culture and education. In 1956, 4 Egyptian students came to study in China, representing the start of China–Africa educational exchange and cooperation. Throughout the 1950s, a total of 24 African students from Egypt, Cameroon, Kenya, Uganda, and Malawi studied in China. With African National Liberation Movement came to an end, the establishment and development of relations between China and Africa thrive, so did the educational exchanges and cooperation between China and Africa. By the end of 1966, a total of 164 foreign students from 14 African countries had come to China to study.

After the beginning of the "Cultural Revolution" in the 1960s, China–Africa educational cooperation temporarily stopped, and it was resumed in early 1970s. In the 1970s, China established diplomatic relations with 25 African countries, and a total of 648 African students from 25 African countries studied in China.

By the end of the 1980s, 43 African countries had sent students to China, and a total of 2245 African students studied in China, majoring in a variety of natural sciences and humanities, such as agriculture, computer science, biology, applied chemistry, biochemistry, medicine, mechanical engineering, architecture, water conservancy engineering, food engineering, international relations, economics, and Chinese.

7.3 China Africa Education Cooperation: Second Stage (1990s)

During this period, China–Africa education cooperation developed rapidly. After the reform and opening-up in 1978, China's national strength has gradually increased, and has made outstanding achievements and development in education, science and other fields. In the 1990s, the number of African students studying in China reached 5569. Compared with western countries, the tuition and living expenses of studying in China are relatively lower. The number of self-funded African students in China has also increased rapidly. In 1989, the first two African self-funded students came to China, and its number increased to 1580 from 42 African countries in the following 10 years.

According to the actual situation of African students and the characteristics of China's higher education. Chinese government adjusted the academic level and training mode of African students studying in China, increased the proportion of Master's and Doctoral degree students, while reduced the undergraduate student's number. In 1999, a total number of 666 students from 38 African countries came to China to study for graduate study (Cheng et al., 2018).

7.4   China–Africa Education Cooperation: Third Stage (Since 2000)

The First Forum on China–Africa cooperation in 2000 was a new milestone in the history of China–Africa relations. The forum published two important documents, "Beijing Declaration of the Forum on China–Africa Cooperation" and "China–Africa cooperation program for economic and social development". From this time on vocational education gradually became an important part of China–Africa educational cooperation in the new era. With the support of the "African human resources development fund" established by the Chinese government, vocational education, Chinese teaching and other professional skills training projects sprouted up in China and African countries.

In 2006, The Chinese government published the China's Africa policy paper, which greatly promoted the development of China–Africa educational exchanges and cooperation. China–Africa education cooperation covers academic education, vocational education, and short-term workshops. In 2012, a total number of 27,052 African students received various academic and non-academic education in China. Since the Belt and Road Initiative brought up by H.E. President Xi Jinping in 2013, a total number of 41,677 African students came to China in 2014. In 2018, Forum on China–Africa Cooperation was successfully held in Beijing such that China–Africa cooperation became more closer. The number of African students ranked the second largest group of overseas students in China. A total number of 81,562 African students came to China to study in 2018 (data from the Department of International Cooperation and Exchange of the Ministry of Education).

China–Africa education cooperation has profound connotation and far-reaching significance. It is not only the cooperation between China and African countries in talent training and human resource development, but also the embodiment of different cultural exchanges and integration. Education and talent cultivation are the basis and guarantee for the rapid development of a country and the overall progress of society.

The educational exchanges between China and Africa are conducive to the stability and development of African countries. Promoting educational exchanges and cooperation between China and Africa is not only conducive to the comprehensive development and consolidation of China–Africa relations, but also serves the common interests of the Chinese and African people who seek common development in the era of economic globalization (Zhang et al., 2004).

7.5   Overview of Talent Cultivation of Sino-Africa Joint Research Centre, CAS

Sino-Africa Joint Research Centre (SAJOREC) is an open and inclusive platform for joint scientific research and talent cultivation. It's the first overseas large-scale comprehensive science and education institution built by the Chinese government based on the exchange of official letters between Chinese and Kenyan Governments in 2013, and it's also the first batch of overseas science and education centres of the Chinese Academy of Sciences.

SAJOREC carries out scientific and technological cooperation and talent cultivation in biodiversity protection, geographic science and remote sensing, ecological environment monitoring, microbiology, modern agricultural demonstration, etc., strives to improve the scientific and technological level of African countries such as Kenya, and becomes a platform for scientific research and talent training between China and Africa, as well as a bridge for win–win cooperation and friendship promotion.

With the participation of 16 CAS-affiliated institutions and 15 academic organizations from 8 African countries, SAJOREC recruits 30 African master's students every year with full scholarships through the English Teaching Master's program opened by the University of the Chinese Academy of Sciences. The majors include botany, zoology, genetics, microbiology, ecology and other multidisciplinary fields. Students with excellent performance can take part in the PhD evaluation examination in the second and third academic year, and strive for scholarships. At present, 84 excellent Master's students have passed the examination and transferred to the doctoral study.

Since 2013, SAJOREC has recruited 320 African students, including 233 Master's students and 87 doctoral students. The students came from Kenya, Ethiopia, Tanzania, Egypt, Nigeria, Rwanda, Zimbabwe, Madagascar, the Democratic Republic of the Congo, Burundi, Uganda, Ghana, Senegal and Benin.

During their stay in China, African students not only enjoy the study of professional knowledge, but also can deeply understand and learn Chinese culture, participate in SAJOREC's projects, and lay the foundation for future cooperation between China and Africa in various fields. At the same time, SAJOREC actively organizes various cultural and sports activities to continuously enhance the exchanges and friendship between Chinese and African students.

7.6 SAJOREC's New Mode of Talent Cultivation

7.6.1 Enrolment Activities Advance with Times

Since its establishment in 2013, SAJOREC mainly recruited African students through the recommendation of its African partners. With its development, SAJOREC has gradually established its own official website and WeChat account. In addition to the recommendation of partners, the enrolment information also published through SAJOREC's official information platform. Besides, SAJOREC international students' enrolment brochures were made and introduced to African partners and visitors of SAJOREC headquarters in Kenya. In addition, SAJOREC recruit students through the project and attracts them to China for further study. In 2020, affected

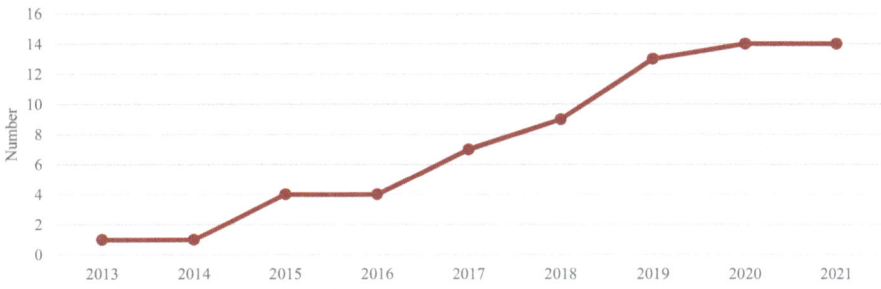

Fig. 7.1 Number of countries of students' origin from 2013 to 2021

by the global COVID-19, international cooperation was greatly limited. SAJOREC carried out online publicity activities to recruit international students, and let African students fully understand SAJOREC through online information session, so as to achieve the purpose of expanding the source of students. Through 9 years of continuous efforts, the number of countries of origin of students in SAJOREC has increased from one country in 2013 to fourteen countries in 2021. (Fig. 7.1), and the quality of students has also been significantly improved.

7.6.2   Comprehensive Talent Cultivation Mode

With the participation of several CAS-affiliated institutions, SAJOREC divided its various joint research programs into five areas, i.e., Biodiversity and Conservation, Geographic Science and Remote Sensing, Microbiology and Epidemic Disease Control, Water Resource Management, and Modern Agriculture Demonstration, all of which have established collaborative relationship with African partners. SAJOREC headquarters has six high-end professional laboratories, as well as fully functional scientific research greenhouses and herbarium. Moreover, SAJOREC has jointly established the African natural medicinal botanical garden with Masai Mara University, the modern agriculture demonstration zone with Jomo Kenyatta University of Agriculture and Technology, and set up regional offices and joint laboratories in Ethiopia and Madagascar. A joint research station on East African great lakes and urban ecology was established with Tanzanian Fisheries Research Institute. A joint laboratory of East African natural resources and environment research centre was established with University of Lay Adventists of Kigali (UNILAK), Rwanda. At the same time, it integrates the advantageous resources of the participating units in China, providing good conditions for students from different countries and different majors to carry out scientific research practice.

Through the "two-stage" teaching and "dual-tutorial" training mode, African students not only obtain a solid theoretical knowledge foundation in botany, ecology and other professional fields, but also cultivate the ability to think independently and engage in scientific research.

The full-time academic training for African graduate students studying in China is mainly the same as that of Chinese students, which is not only aimed at solving the needs of African countries. SAJOREC innovatively implements the dual-tutorial mode, namely, students are trained by the Chinese tutors together with the African tutors who carry out cooperative projects in Africa. In this way, students can learn a wealth of knowledge and advanced Chinese technology in a more specific path, and can apply what they have learned to solve local practical problems in Africa.

The China–Africa Centre has always adhered to the concept of individualized education, combining students' interests and mentors' projects with practical problems in Africa. According to the development needs of African countries, different training types are adopted, such as academic type, professional technology type, application type and so on. SAJOREC also actively encourages African students to participate in domestic and international academic exchange activities and professional training, broaden their scientific research ideas, keep abreast of the latest scientific research trends, and train African students' ability to independently engage in scientific research. Through 3–5 years of postgraduate training, graduate students can take important positions in universities, research institutions, farms and other units or departments of the country, and be committed to solving local agricultural development, ecological environmental protection and other livelihood issues, and promote the improvement of higher education level, so as to cultivate more talents, forming a virtuous cycle.

During the training process, the tutor will designate a special person to be responsible for the overseas students, including learning, daily life and mental health. At the same time, institute unites the graduate association in dealing with various problems and affairs after coming to China. In addition, SAJOREC encourages African students to elect their own presidents and class monitors of the International Student Association, who will assist in the daily affairs of studying and deal with international student conflicts and disputes.

Apart from teaching and research practice, SAJOREC also focuses on cultural communication and humanistic care, so that African students can fully integrate into China and become a team that knows China better, which can promote the rapid development of China–Africa bilateral cooperation in the coming days.

7.6.3 Remarkable Tutorial System

Domestic postgraduate training largely depends on tutor responsibility system. College tutors can recruit more than one student every year, while SAJOREC limits the tutor training indicators, so that the number of students trained and guided by each tutor will not be very large, which greatly ensures the training quality of students. As a national strategic scientific and technological force, the tutor from Chinese Academy of Sciences has undertaken various sophisticated scientific projects, and students can participate in tutor's projects and engage in scientific research during their study. The Chinese Academy of Sciences has always adhered to the tradition of educating and insisted on cultivating innovative and entrepreneurial talents in high-level scientific research practice. Graduate students have become the new force of national scientific and technological innovation and already participated in some of the major achievements and outputs of the Chinese Academy of Sciences.

7.6.4   Obvious Scholarship Setting Advantage

Most African students come to China for degree study based on full scholarships. Therefore, more and more universities have introduced different scholarship policies to attract foreign students to apply. Compared with the international student scholarship settings of colleges and universities in key domestic cities such as Beijing, Shanghai, Hangzhou, Guangzhou and Wuhan. SAJOREC's scholarship setting has obvious advantages in attracting African students.

In addition to the state funded part, tutors also match part of the living expenses to ensure that the master's degree is not less than 3000 CNY per month and the doctor's degree is not less than 3500 CNY per month. Especially excellent doctoral scholarships can reach 7000 CNY per month, which make SAJOREC's African student's grant much higher than that of ordinary colleges and universities.

Secondly, as the highest academic institution of natural science, the highest consulting institution of science and technology, and the comprehensive research and development centre of natural science and high technology in China, Chinese Academy of Sciences has 11 branches, more than 100 scientific research institutions, 3 universities, with more than 130 national key laboratories and engineering centres, 68 national field observation and research stations, and 20 national science and technology resource sharing service platforms. It is a collection of scientific research institutions, academic departments, national strategic scientific and technological force integrated educational institutions. In terms of international student enrolment, the participating units of SAJOREC can recruit students according to their own disciplinary advantages and needs and give recommendations and consideration to students from key cooperation fields and cooperation institutions in the scholarship evaluation, while universities and colleges are mainly guided by national policies.

Thirdly, for the sake of its own development needs, SAJOREC has set up scholarships supporting African youth who plays an important role in the development of SAJOREC and in their country. Namely, SAJOREC will subsidize them to study in China in accordance with the scholarship standards of the University of Chinese Academy of Sciences.

7.6.5 Knows Better About Africa

At present, colleges and universities train African students and other international students in a unified manner. Therefore, they lack experts, scholars and managers who study Africa, and lack in-depth understanding of the objects and target countries. However, located in Nairobi Kenya, SAJOREC not only has established joint offices and laboratories in many countries in Africa, but also has in-depth communication and connection with African universities and research institutions. SAJOREC has sent staff to work in Africa since its establishment in 2013. As time goes by, SAJOREC's staff can provide first-hand information for the education and training of African students in terms of cultural customs, social development and livelihood demands of African countries.

7.6.6 Adhere to the Concept of Win–Win Cooperation

In addition to cooperating with African universities and research institutions, SAJOREC also works closely with international organizations such as the United Nations Environment Programme, The World Academy of Sciences, International Livestock Research Institute, and Alliance of International Science Organizations in the Belt and Road Region. At the same time, SAJOREC actively contacts enterprises who undertake practical projects in Africa and gives full play to the role of these institutions and enterprises as practical internships and employment bases during the education and training of African students studying in China, providing more opportunities and possibilities for the African students' career development.

7.7 Training Achievements and Successful Cases

Relying on the University of Chinese Academy of Sciences, excellent teachers and perfect facilities, SAJOREC has made remarkable achievements in education. A number of excellent international students took on the stage. One of them won the "outstanding overseas student scholarship" by the Ministry of education, and 38 won the title of "Excellent international student/graduate" of the University of Chinese Academy of Sciences.

7.7.1 Case One

Veronicah Mutele Ngumbau (Kenyan) came to SAJOREC for Master study in 2014. Under the guidance of her tutor, Veronicah carried out research on plant richness, phylogenetic diversity of coastal forests in East Africa and a comprehensive list of plants in the coastal areas of Kenya. With the interest on plant taxonomy, Veronicah passed the evaluation examination in 2016 and entered the doctoral study. In July of the same year, Veronicah, as the representative, participated in the eighth Chinese ethnobotany academic seminar and the seventh Asia Pacific ethnobotany forum and made a report. Her excellent performance stood out from more than 100 young scholars and won the excellent paper award. In October, she joined the National Museums of Kenya as an assistant researcher. Through hard study and research, Veronicah published 5 first-author papers in journals such as "Phytotalxa", "Phytokens" during study, and acquired the doctoral degree in 2020, winning the title of "Excellent international graduate" of the University of Chinese Academy of Sciences in 2020. From March 2021 on, Veronicah serves as a senior researcher and the director of the education group at the National Museums of Kenya, continuing to carry out research on the plant diversity in East Africa and the compilation of Flora of Kenya with her tutor and SAJOREC.

7.7.2 Case Two

Through the platform of SAJOREC, Raphael Ohuru Nyaruaba (Kenyan) noticed that the Chinese Academy of Sciences is among the highest level scientific research institutions in China and even the world's top scientific research institutions. In 2017, Raphael was successfully admitted as a Master's student by SAJOREC after careful preparation and the recommendation of his tutor. When he first came to China, Raphael only had the basic professional knowledge and experimental technology of Microbiology. With the patient guidance of his tutor, professional courses studying and the practice of experimental skills, Raphael became more professional for microbial scientific research, and more aware of the importance of biosafety for human beings. Raphael's research direction is molecular diagnosis of biosafety grade III respiratory tuberculosis plasma and acute respiratory syndrome type 2 coronavirus.

After 3 years of hard work, he achieved excellent academic results and published a number of research papers as well. During the study, Raphael gained not only advanced scientific and technological knowledge, but also the rigorous and persistent academic attitude and selfless dedication spirit of Chinese scientists. During the outbreak of covid-19 in Wuhan in early 2020, Raphael participated in the fight against the epidemic. He joined the team of identifying the type of pathogen, formulating a detection plan, and making efforts to prevent the spread of novel coronavirus. In September 2020, Raphael was awarded "Excellent International Graduate" by the University of Chinese Academy of Sciences. At present, he has begun his doctoral study and research. Raphael hopes to set up his own research team in the future and go back to Kenya on sharing advanced scientific knowledge to the young generation of Africa and work together with Wuhan Institute of Virology and SAJOREC to make his own contribution to the research and prevention and control of severe infectious diseases in Africa.

7.7.3 Case Three

Under the guidance of his tutor, Samwel Maina Njuguna (Kenyan) investigated the concentration and distribution of potentially toxic elements (heavy metals and trace elements) in fresh water in Kenya. The research focused on analysing the risks to people's health, wastewater recovery and related risks, and discussed the potential application of plants in water environment restoration. Three related papers were published in Journal of Environmental Management, Process Safety and Environmental Protection and Environmental Monitoring and Assessment. After graduation with a doctoral degree in 2019, Samwel engaged in research work in the university in Kenya, maintained good communication with SAJOREC and continued to participate in the cooperative research of SAJOREC, providing personnel support for SAJOREC on water resources monitoring and environmental protection in Kenya.

7.7.4 Case Four

In 2018, Samuel Paul Kagame (Kenyan) came to SAJOREC for his Master study in Botany. Under the guidance of his tutor, he carried out research on the phylogeny, biogeography and diversity of Alpine ferns, alpine dominant lobelia, and savanna bluegrass in East Africa. The relevant results were published in the Journal of Phytokeys. He was not only studied very hard, but also actively helped his classmates and participated in academic and cultural activities, awarded the "Excellent Graduate" of host institute. In 2021, he received offers from Ghent University (Belgium) and the Center for Excellence in Molecular Plant Science, Chinese Academy of Sciences for doctoral degree study. In order to broaden horizons and gain more professional experience, Samuel chose to work at the John Innes Centre (JIC) in England to accumulate relevant experience and continued to study for a doctorate at Kew Gardens, the Royal Botanical Garden, in March 2022. Now Samuel has become bridge of the tripartite research cooperation project and will contribute to the future development of biodiversity conservation of SAJOREC.

7.8 Opportunities and Challenges of Current Education of African Students

7.8.1 Language Barriers of African Students

Although basic mandarin courses are offered to African students in the first year, their Chinese level is still relatively weak. Many problems appear during the training and management of African students. For example, the documents and education system issued by the school are all in Chinese, therefore, it needs to spend a lot of time and manpower to assist African students to complete all training processes and help them understand the key contents of documents. The existence of language barriers naturally increases the barriers for African students to get on well with Chinese students or Chinese culture, which in turn affects African students' rapid adaptation to living environment and study in China.

Language is the tool for communication. In addition to offering professional courses taught in English, and in order to fundamentally eliminate the barrier, the curriculum should increase and extend the proportion and classes of Mandarin courses to each semester and give incentives to students who pass the certain level of Hanyu Shuiping Kaoshi (HSK), such as exemption from basic Mandarin courses. By improving the Chinese language level of African students and gradually increasing their sense of identity and belonging to Chinese society and culture, it can not only effectively improve the cultural adaptation of African students, but also become an important means of China–Africa cultural exchange and integration, and promote win–win cooperation and social development.

7.8.2 Different Education System Between China and Africa

Many countries in Africa have a colonial history by Western countries. Therefore, their education system has been affected deeply. French speaking countries have completely followed the French education system. Their academic education and training requirements are quite different from those in China, and students may face problems such as weak professional foundation and inability to adapt after admission. Besides, the academic and scientific research capabilities of African students fluctuate drastically, their scientific research basis and the graduate students' cultivation mode in colleges and universities lay a big gap.

With the development of social economy, the internationalization of China's education mode also has a long way to go. Nowadays, the application review system is gradually replacing the traditional paper examination. For students with different cultural and educational backgrounds, their assessment and comprehensive evaluation indicators should be adjusted accordingly. On the other hand, we should gradually improve the construction of the curriculum system, and set up learning courses at different levels according to the basis and educational characteristics of African students (Zheng et al., 2017). At the same time, students can choose professional courses according to their own interests and levels, so as to lay a solid professional foundation for subsequent scientific research and practice activities.

7.8.3 Lack of Continuous Attention of African Students

In order to promote the cooperation with African countries, it is an important step to establish a good partnership with African governments, research institutions and universities through recruiting African students. At the same time, the experience of studying in China can help African countries to know China comprehensively. However, many colleges and universities pay little attention and care to the African students after their graduation, such as whether their studies are useful, how they will develop in the future, and whether they will become the "spokesperson" of China–Africa friendship. Recruiting African students has only become a "task", losing the original intention of talent education and training.

Most African students gradually lose contacts with their alma mater, which deviates from the original intention of talent training. At the same time, a large number of African students took the job which has little or nothing to do with what they have learned. Due to highly competitive pressure in their own countries, some students are unemployed for a long time.

Therefore, the recruitment and training of African students should be changed from "quantity" to "quality". It is suggested to establish an alumni mechanism to track African students' status after graduation, and give full play to talents who have the experience of studying in China for future cooperation in various fields between China and Africa.

7.9   Conclusion

The recruitment of African students is an important part of the educational exchange and cooperation between China and Africa, which is of great significance to the internationalization of Chinese education and has a more lasting and far-reaching influence on the social development of African countries. Although the education of African students still faces many problems and challenges, the training of African students has achieved many results. The effect of talent cultivation is very significant. In the future, the government should increase investment, encourage, and support educational cooperation and exchanges between Chinese universities and African institutions, so as to complement each other's strengths and learn from each other. We should also share more efforts to carry out inter school exchange of students for joint training, international cooperative teaching, expert visits, and encourage African youths to come to China for further study and training, continuously serving the needs of social and economic development of China and African countries.

References

Cheng, W. H., Dong, W. C., & Liu, X. G. (2018). Problems and countermeasures of African graduate education in China. *Academic Degrees and Graduate Education, 8*, 54–58.

He, W. P. (2006). Overview of China–Africa educational exchanges and cooperation—Development stages and future challenges. *China–africa Human Resources Development and Cooperation Forum, 2006*, 1.

Zhang, X. P., Xue, Y. Q., Qiang, Y. P., & Luo, J. B. (2004). Educational exchanges and cooperation between China and African countries. *West Asia and Africa, 3*, 24–28.

Zheng, S., Zuo, X., Yu, H. L., Dan, Y. Y., Zhu, X., & Gao, S. (2017). Explore the training mode of African graduate education in China. *Science and Technology Vision, 10*, 23–24.

Chapter 8
From Joint Research Project (JRP) to Joint Research Center (JRC): A 10-Years Cooperation in Efficient Utilization of Mineral Resources

Yangge Zhu, Long Han, Da Zhang, Dan Zhang, Xiyu Gao, Zhenguo Song, Chongjun Liu, Guoqiang Wang, Wei Xiong, Qingchao Zhao, Xingrong Zhang, Jun Wang, Yafei Liu, Yanbin Chen, Tao Song, and Bangsheng Zhang

8.1 Introduction

9th Oct. 2009, North Stradbroke Island Australia, a sunny and windy day, a group of professors from China, South Africa, Sweden, the UK, India, and Australia, gathered in the famous James Cook Lookout Point for lunch after visiting the CRL (Consolidated Rutile Limited) at the island. They are Council members of IMPC, the International Mineral Processing Congress, coming all the way for the Council meeting in Brisbane. Two of them are Prof. Cyril O'Connor from University of Cape Town (UCT), South Africa and Prof. Han Long from Beijing General Institute of Mining and Metallurgy (BGRIMM), China. Enjoying the nice lunch coffee and beautiful ocean view, they chatted about a possible joint research project between

Y. Zhu · L. Han (✉) · D. Zhang · D. Zhang · X. Gao · Z. Song · C. Liu · G. Wang · W. Xiong · Q. Zhao · X. Zhang · J. Wang · Y. Liu · Y. Chen · T. Song · B. Zhang
China-South Africa Joint Research Center for Exploitation and Utilization of Mineral Resources, BGRIMM Technology Group, Beijing, China
e-mail: hanlong@bgrimm.com

Y. Zhu · L. Han · D. Zhang · D. Zhang · X. Gao · Z. Song · C. Liu · G. Wang · W. Xiong · Q. Zhao · X. Zhang · J. Wang · Y. Liu · Y. Chen · B. Zhang
China-South Africa BRI Joint Laboratory On Sustainable Exploration and Utilization of Mineral Resources, BGRIMM Technology Group, Beijing, China

J. Wang · Y. Liu · Y. Chen
Beijing Easpring Material Technology Co., Ltd, Beijing, China

B. Zhang
Jiangsu BGRIMM Metal Recycling Science & Technology Co., Ltd, Xuzhou, China

the two institutions. At that moment, none of them realized that their conversation in front of the Pacific Ocean would initiate a special journey across the ocean lasting for more than 10 years.

14th Sept. 2018, Johannesburg South Africa, same sunny and windy day, the launching ceremony of the China–South Africa Joint Research Center for Exploitation and Utilization of Mineral Resources (JRC) was held in Mintek. Mr. Xu Nanping, the vice Minister of the Chinese Ministry of Science and Technology (MOST), and Dr. Phil Mjwara, Director General of DST were present the ceremony and officially unveiled the JRC. One month later, on 15th Oct 2018, a similar ceremony was held in BGRIMM Beijing, Ms. Mmamoloko Kubayi, the Minister of South African Department of Science and Technology (DST), and Mr. Zhang Jianguo, the vice Minister of MOST unveiled once again, for the official establishment of the JRC, which is the outcome of a special journey from Joint Research Project (JRP) to JRC, an almost 10 years' cooperation practice between China and South Africa.

In the journey from JRP to JRC, thanks to continuous support from both Chinese and South African governments, three phases of joint research program were successfully conducted with 10 Chinese and South African institutes and universities engaged and 15 joint research projects carried out. The research programs aimed to address those grand challenges global mining industry is facing, including deep mining, smart mining, mine safety, evacuation automation, low grade complex ore utilization, mine waste disposal, various environment and community constraints as well as digital transformation of traditional industry, with specific projects focusing on safety monitoring technology and instrumentations, high efficient grinding methodology, electro-chemical flotation technology, flotation reagent modelling, and tailings utilization etc., that not only reflects the technological trends of mining industry, but also reflects the common interests and complementarity of Chinese and South African research strengths in these areas.

In the journey from JRP to JRC, all the parties involved input great efforts and enthusiasm into this international cooperation. Numerous meetings, workshops, and talent exchanges were implemented. The research teams come from universities and institutes, consisting of renowned professors, senior researchers, young lecturers, as well as engineers and students, who learned from each other, shared experiences and fresh ideas, overcome various technical difficulties, geographical distance, language barriers and trans-culture challenges in particular, successfully achieved all project objectives to an extent of high satisfaction from both governments and third-party assessment.

In the journey from JRP to JRC, a number of outcomes and achievements have been accomplished. More than 10 high-quality joint papers were published in first-class international journals or conferences, one of the papers received the Young Author Award in 28th IMPC in 2016. More than 20 students completed higher education and got their Master's or Doctoral degrees by participating the joint research projects. Based on the joint research outcomes, dozens of patents were obtained, and the microseismic monitoring instrumentations and novel flotation reagent developed were commercially applied in the mining and processing operations.

The pragmatic cooperation has been greatly supported and highly affirmed by the governments of China and South Africa, especially South African Department of Science and Technology (DST) and Ministry of Science and Technology of the People's Republic of China (MOST). In July 2018, the BRICS Summit was held in JBG, South Africa. In the associated event of "China–South Africa Scientists High Level Dialogue", the representatives from BGRIMM and University of Limpopo (UL) were invited to deliver a speech. The joint research cooperation outcomes were presented in the "China–South Africa Cooperation Achievements Exhibition". In April 2019, the "Belt and Road" International Cooperation Forum was held in Beijing, Prof. Han Long on behalf of the JRC made a presentation to introduce the cooperation practice. A few months later, the JRC was entitled by MOST with the China–South Africa BRI Joint Laboratory on Sustainable Mineral Resources Development and Utilization. In 2021, Prof. Phuti Ngoepe from UL won the Chinese Government Friendship Awards, the highest commend for the outstanding achievement of the international cooperation practice.

"From JRP to JRC" is one of the Best Practices on international research cooperation, furthermore it builds a bridge between China and South Africa, connecting knowledge, understanding, friendship, and most importantly, the people.

8.2 Research: From Mineral Processing to Mining and Material

With the JRC platform, scientists from China and South Africa have carried out three phases of cooperation projects so far. The cooperative research fields of China and South Africa have gradually expanded from the initial mineral processing to mining, and further developed to hot topics such as metallurgy, battery materials and automation.

8.2.1 Intelligent Mining

In-depth cooperation between BGRIMM and CSIR have been carried out in the areas of efficient, safe, green and intelligent mining, with a focus on the intrinsically safe design of deep mines, automatic ground pressure monitoring and early warning, high precision 3D laser scanning and evaluation, and common key technologies. A BSN high-precision microseismic monitoring system and a vehicle-based BLSS 3D laser scanning system have been successfully applied in South African and Chinese mines.

Major mining nations around the world, including Australia, South Africa, Canada, America, and China, have expanded underground mining into deeper levels than 1000 m to keep up with the expanding mineral consumption. Deep mining takes place in a severe environment with high in situ stresses, high compressive strength, rheological characteristics, and mining disturbance. Massive effort and expansion are put into the excavation support system in order to guarantee the long-term stability of mine infrastructure, the effectiveness of mine excavation, and to safeguard the safety of the workforce. While deep mining techniques and support systems are important and essential, the most efficient and cost-effective design must be developed by keeping a close eye on the progress of the mine, the performance of the support, the redistributed stress field, and the changes in energy. Microseismic monitoring system, which can capture the emitted elastic wave when rock masses subject to critical differential stress fracture, has been widely applied in the monitoring and analysis of mining-induced seismicity in deep mines.

To ensure safely mining production, the BSN high-precision microseismic monitoring system is utilized to analyse the macroscopic stress distribution and capture the microscopic development of rock fracture in the underground mines. It can also be adopted for early warning and real-time evaluation of roof stability and optimizing support and mining strategies based on microseismic data. With the help of the microseismic monitoring system, the mining production is able to maximize benefits under the premise of safe operation. The BSN high-precision microseismic monitoring system attained key technologies such as high-sensitivity microseismic sensor arrays, long-distance enhancement of microseismic signals, 3D dynamic inversion of rock mass velocity field, high-precision location of microseismic events, and source parameter calculation (Zhang et al., 2022a). An intelligent safety monitoring and early warning cloud service platform (Fig. 8.1) is built (Zhang et al., 2020a) based on the case library and model library. The compact service of the BSN microseismic monitoring system has been successfully applied in Sanshandao Gold Mine in Shandong, and has been applied in Wulan Lead–Zinc Mine in Mongolia, Shanhu Tungsten Mine in Guangxi.

In underground deep mining, the stability and safety of infrastructure is significant to workforce's insurance and the sustainability of mining production. The stability analysis and management of underground mines require accurate models of the ore-pass, goaf, stope and tunnels. 3D laser scanning technology, which has strong advantages of anti-interference ability, high detection accuracy, simple and fast operation,

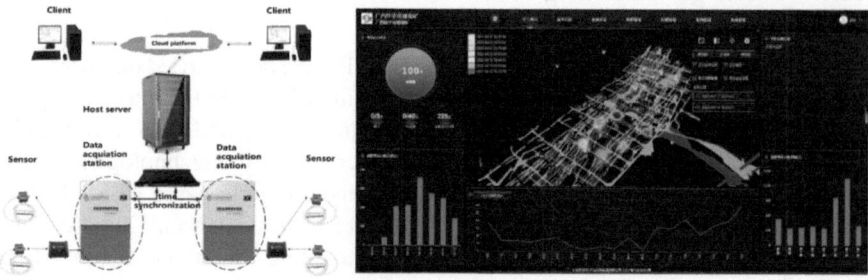

Fig. 8.1 Composition of BSN high-precision microseismic monitoring system (left) and the Cloud service platform (right)

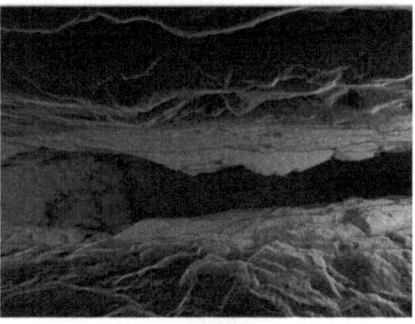

Fig. 8.2 The vehicle-based BLSS 3D laser scanning system (left) and the 3D point cloud of underground mine (right)

and better visualization, is an efficient method to construct the stress and displacement distribution cloud and to image the boundary of the ore-pass, goaf, stope, and tunnels. The scanning point cloud can also benefit to assess the production and efficiency of the mining and investigate hidden dangers. A vehicle-based BLSS 3D laser scanning system developed by BGRIMM provides a fast and high-precision 3D spatial shape measurement solution.

The vehicle-based BLSS 3D laser scanning system (Fig. 8.2, left) (Zhang, 2014) is mainly used to independently explore and efficiently obtain the 3D point cloud of the underground environment (Fig. 8.2, right) without GNSS, so as to realize 3D space modeling and provide spatial data for mine digitization. In order to optimize the follow-up recovery, the loss dilution rate is simultaneously assessed by several observations while drift detection and error correction are fine processed. The vehicle-based 3D laser scanning technology and instrument are suitable for complex underground spaces and harsh environments, which can realize efficient measurement of 3D space shapes in underground mines (Chen et al., 2013). In order to ensure the working safety, the stability of the rock mass is also examined and evaluated by scanning the rock displacements, joints and fissures. The BLSS system can be designated in underground high-performance 3D laser scanning, intelligent identification of rock mass structure based on high-density point cloud, extraction of large deformation based on multi-phase comparison, and high-precision 3D laser scanning measurement for deep mining. The system has been successfully applied in Shizhuyuan Tungsten Mine in Hunan, Sanshandao Gold Mine in Shandong. It has been planned to adopt in mines of VMR Group in South Africa.

8.2.2 Efficient Grinding

In the process of mineral processing, the choice of comminution process has an important influence on energy consumption, product characteristics and flotation

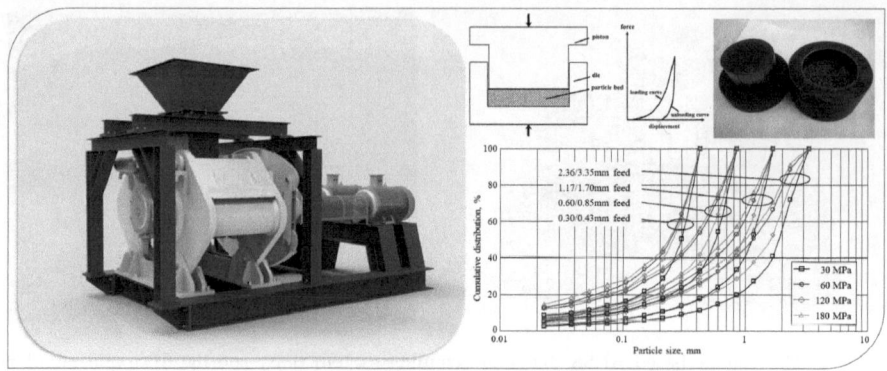

Fig. 8.3 Particle bed ballast test of HPGR

index. In the field of grinding, for the representative nonferrous sulfide ores, the influences of total energy consumption of crushing and grinding, the characteristics of selected products (particle size distribution, particle morphology, mineral dissociation degree, chemical properties of slurry solution) and flotation effect were investigated for the flowsheet option of "conventional crushing—ball milling", "conventional crushing—HPGR—ball milling" and "conventional crushing—HPGR". The equipment configuration of crushing and grinding operation is measured from the technical indexes of the whole process of crushing, grinding and separation (Liu & Sun, 2016). Among them, the selection and fragmentation function are used to establish the material model of particle bed ballast crushing. The characteristic parameters and energy distribution function of this model can predict the particle size distribution of particle bed ballast crushing products under the conditions of specific feeding size and specific energy consumption. The narrow particle bed ballast test mould and product particle size distribution was showed in Fig. 8.3. Compared with wet ball milling, the flotation feeding by dry crushing (particle bed ballast or dry ball milling) can effectively improve the floatability of copper minerals, especially facilitate the achievement of key achievements such as flotation recovery of coarse copper minerals, and provide a technical foundation for the popularization and application of HPGR technology in nonferrous sulfide ore dressing technology (Liu et al., 2018b).

8.2.3 Environmentally Friendly Flotation Reagents

Froth flotation is a physicochemical separation technique that exploits the difference in the surface wettability of mineral particles. The process strongly depends on the physicochemical surface properties of minerals, and is controlled by modifying these properties through addition of flotation reagents. In the field of flotation reagents, the collaboration between China and South Africa was mainly focused on the research and development of flotation reagents on sulfide ores and platinum group containing

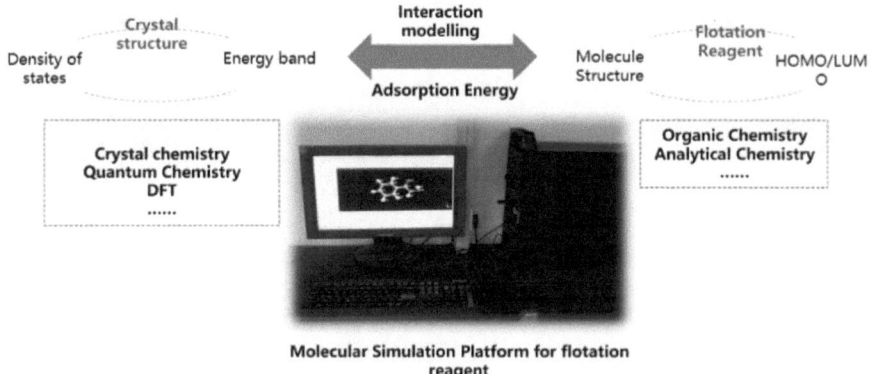

Fig. 8.4 Research and development process for flotation reagent

ores. The research was carried out by the molecule design of new and green flotation reagents under the computer aided molecular design as shown in Fig. 8.4, and then followed by the research of synthetic and flotation performance. 16 novel molecules were designed in total, and finally one of them was commercialized as the depressant applied in the separation of Mo/Cu sulfide ores named BK510 (Zhang et al., 2013). The following collaboration research was concentrated on the design and application of oxide ore flotation reagents with good fluidity under low-temperature. The flotation reagent design methodology and molecular fragment assembling technology were formed by the investigation of the migration behaviour of mineral lattice ions at the solid–liquid interface, and the interaction mechanism between minerals and reagents at the solid–liquid interface with the introduced of the theory of crystal and interface chemistry, solution chemistry, quantum mechanics. An example of the adsorption mechanism of flotation reagent on mineral was shown in Fig. 8.5. BK412 was finally exploited as the collector of cassiterite mineral, and positive feedback was received from the industrial application (Shang et al., 2016; Zhang et al., 2017a, 2017b). The current collaboration program is focused on the R&D of novel and green collectors for precious ores, and the project is completing as planned.

8.2.4 Electrochemistry of Flotation Process

The electrochemistry of the flotation of sulphide minerals has been studied for over half a century. And the electrochemical investigations of sulphide minerals/ thiol collectors have been widely proposed in literature (Chander, 2003; Javadi, 2015; Smith et al., 2012; Woods, 2003). But the investigations on other minerals/ collectors system were very limited before the joint adventure. In the field of electrochemistry of flotation process, the collaboration between China and South Africa was mainly focused on sulfide minerals/esters collector system, platinum group

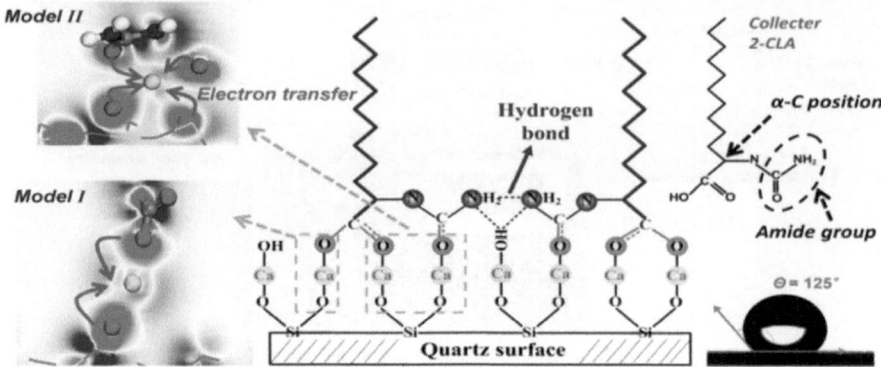

Fig. 8.5 Adsorption mechanism of flotation reagent on mineral

metal minerals, and oxide minerals to understand reagent-mineral interactions using electrochemical measurements and controlling electrochemical conditions in the flotation process. For sulphide minerals, the relationship between electrochemical properties and flotation behaviour in chalcopyrite-galena-Z200 flotation system and chalcopyrite-pyrite-xanthate flotation system was studied, and the optimal flotation parameters for chalcopyrite-galena separation were obtained. The pH-E_{Pt}-ε curves for different mineral-collector systems are shown in Fig. 8.6.

For platinum group metal minerals, an electrochemical study of rest potentials was conducted to determine the selectivity of thiol collectors towards selected platinum group minerals (PGM), viz. moncheite ($PtTe_2$), merenskyite ($PdTe_2$), cooperite (PtS) and vysotskite (PdS) along with the pure metals platinum and palladium. Mixed potential of different platinum minerals and platinum minerals was shown in Fig. 8.7. Results showed that all these minerals reacted with the collector in question and that Pt minerals were more anodic than Pd minerals. Dixanthogen was more likely to form on $PtTe_2$ than PtS but the reverse is evident for $PdTe_2$ and PdS. The results indicate that differences in floatability between sulphides and tellurides may be as a result of differences in the nature of the surface species of the collector formed on the minerals.

For oxide minerals, the study conducts the flotation electrochemistry investigation of by focusing on the role of two different electrochemistry parameters (viz. E_h and E_s) in the flotation system of Pb/Zn oxide minerals, viz cerussite and zinc oxide minerals. The flotation performances were investigated under different E_h, E_s levels and also different Na_2S dosage, as shown in Figs. 8.8 and 8.9. It was indicated that E_h and E_s played an important role in recovery of Pb/Zn oxide minerals. Optimum flotation recoveries occurred at a certain E_h and E_s range for both cerussite and zinc oxide minerals. Poor recoveries were obtained at potentials more positive or negative than the optimum range, which possibly due to mineral under-sulphidisation or over-sulphidisation.

The electrochemical regulation of flotation was carried out for sulfide-oxidized mixed copper ore, and the effect of sodium sulfide concentration on the potential and

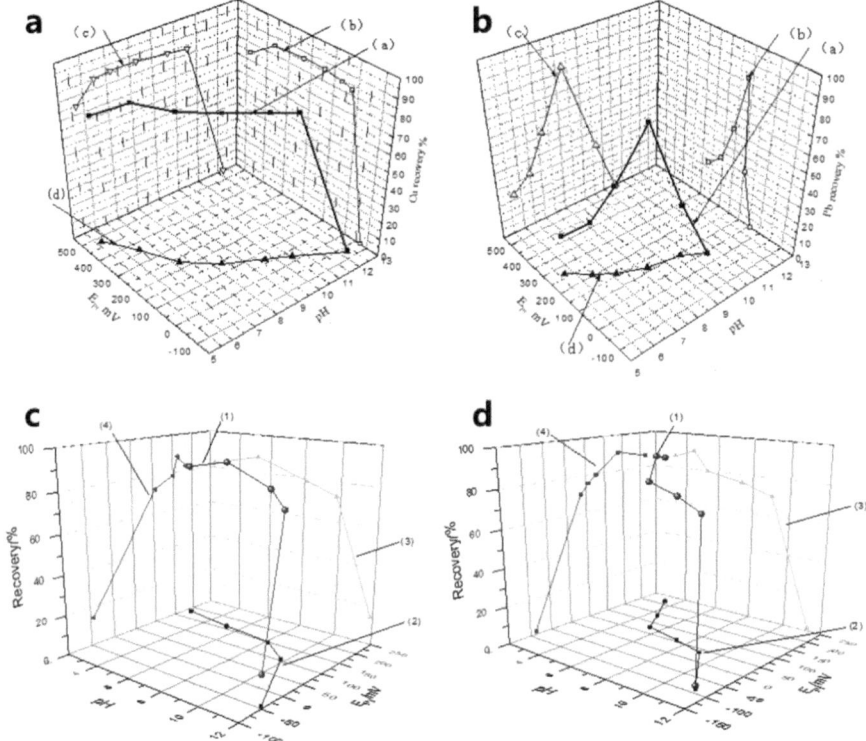

Fig. 8.6 The curves of pH-E$_{Pt}$-ε in chalcopyrite-Z200 system (**a**), galena-Z200 system (**b**), galena-KEX system (**c**), pyrite-KEX system (**d**)

floatability of chalcopyrite and malachite slurry was studied, as shown in Fig. 8.10. The results show that for chalcopyrite and malachite, sodium sulfide has very significant inhibitory and activation effects, respectively. The front-line orbital diagram of the interaction between hydrosulfide ion and malachite and chalcopyrite is shown in Fig. 8.11. The hydrosulfide ion could interact with both chalcopyrite and malachite, thereby playing an inhibitory or activating role. At the same time, the effect of hydrogen sulfide and chalcopyrite is stronger than that of malachite. Therefore, in the presence of chalcopyrite and malachite, sodium sulfide would inhibit the flotation of chalcopyrite. It is necessary to control the appropriate dosage and slurry surroundings.

8.2.5 *Automation on Mineral Processing*

Following the integration of industrial internet, big data, artificial intelligence, and other new generation information technology with the traditional process technology

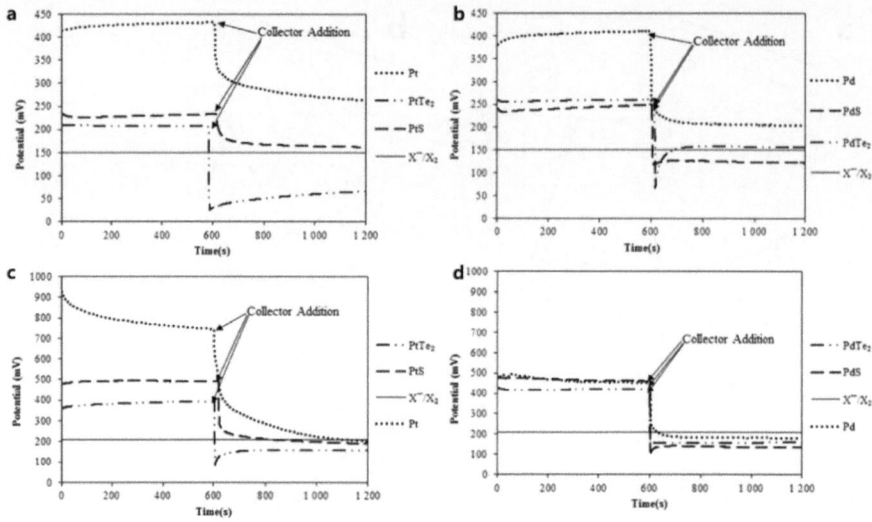

Fig. 8.7 Mixed Potential of Platinum Minerals (**a**) and Palladium Minerals (**b**) in the presence of Ethyl Xanthate (pH 9.2), Platinum Minerals (**c**) and Palladium Minerals (**d**) in the presence of Ethyl Xanthate (pH 4.2)

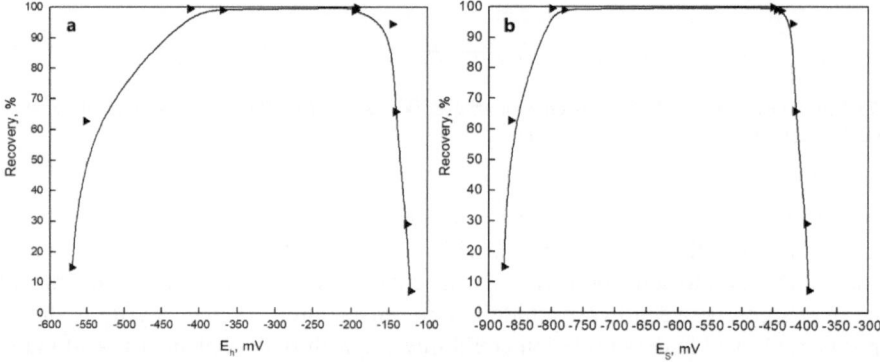

Fig. 8.8 Cerussite flotation recovery as a function of E_h value (**a**) and E_S value (**b**)

in the mining industry, the digital and intelligent upgrade of the mineral processing techniques has become the development trend. From 2016, BGRIMM has started researching and developing the optimal control algorithm and commercial software, specially used in mineral processing plant. Through the combination of digital twin technology and the characteristics of mineral processing industry, a digital simulation platform has been developed since 2018 (Song et al., 2022), which will enhance the application effect of equipment improvement, process optimization and intelligent control in the whole production cycle of a concentration plant.

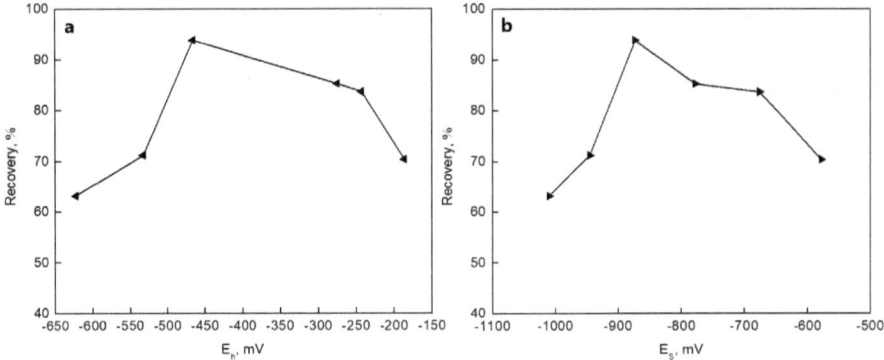

Fig. 8.9 Zinc oxide minerals flotation recovery as a function of E_h value (**a**) and E_S value (**b**)

Fig. 8.10 The floatability of chalcopyrite and malachite as a function of Na_2S concentrate (Butyl xanthate: 2.5×10^{-4} mol/L)

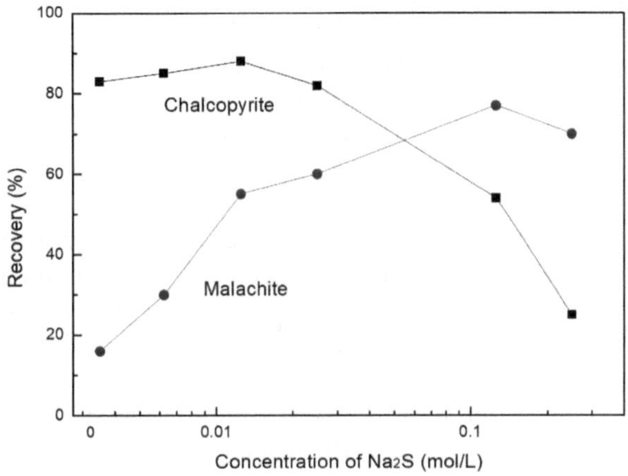

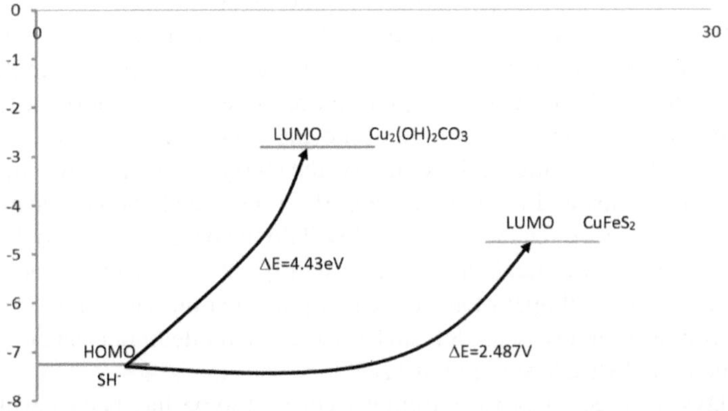

Fig. 8.11 Frontline orbital diagram of the interaction between hydrosulfide ion and malachite and chalcopyrite

The Centre for Minerals Research (CMR) in University of Cape Town has been involved in the development of flotation models and the use of these models for the design, integrated simulation and optimization of mineral processing concentrators since the mid-1980s. The CMR has spent the past 20 years working on developing software simulations of mineral processing concentrators as part of the AMIRA P9 project, including the contributions to the development of comminution, classification and flotation models within JKSimMet, JKSimFloat and more recently the state-of-the-art Integrated Extraction Simulator (IES), a cloud-based integrated multicomponent simulator which is able to simulate from the mine to the metal product.

In 2020, based on the cutting-edge research ability on fundamental mineral process modelling from CMR, and the advanced digital platform from BGRIMM, BGRIMM and CMR began doing research of on-line product quality (e.g., grade, recovery) analysis and control technology for flotation process. An entirely novel theory of froth flotation has been developed by M.C. Harris and D.A. Deglon from CMR. And the theory was used to define a relatively simple algorithm for steady state single cell simulation to predict the froth variations in copper and nickle concentrate production by BGRIMM research team. The objective of the proposed research is to use the new theory to specify a mechanistic model amenable for application on-line, as a digital twin of an industrial flotation circuit.

The model from CMR provides a gauge-symmetric solution of the evolution of the stress–strain relationship of the gas–fluid ensemble in a flotation cell that requires a minimal degree of parameterization: a characteristic length scale defined by the cell geometry, the ensemble viscosity and the steady-state interfacial surface tension gradient (Marangoni force). Consequently, the majority of the parameters required to run an online model arise from the representation of the flotation potential of the feed to the circuit, which comprises a discretized archetypal particle matrix that maps particle size and composition to flotation potential (interfacial interaction in the pulp zone) and froth transport potential (interfacial interaction in the froth zone in combination with hydraulic transport). The interaction of the flow field of the gas–fluid ensemble with the potential field of the particles in the feed conserves gauge-symmetry, and provides a complete space–time representation of the ergodic average composition and velocity field of the gas–fluid–solid ensemble in a cell.

Together with the flotation cell geometry, the slurry level, gas flow rate and froth height measurements, and the bubble size, life time, speed and colour information from online froth image sensors, the model will do the predictions for pulp zone, dry froth and wet froth. A critical gas flow rate will be calculated by online inputs and the efficiency of the cell operation can be recognized at the same time, and finally a flotation process digital twin model can be built. The model inputs and outputs were listed in the model structure in Fig. 8.12.

Nowadays, a flotation process digital twin prototype has been developed and tested with plant data from JINCHUAN concentrator in Gansu, China (Fig. 8.13, left). An intelligent control system has been put into practice in plant, including crushing intelligent control module, grinding digital twin module and unit flotation process digital twin module. With the progress of plant intelligent control in crushing,

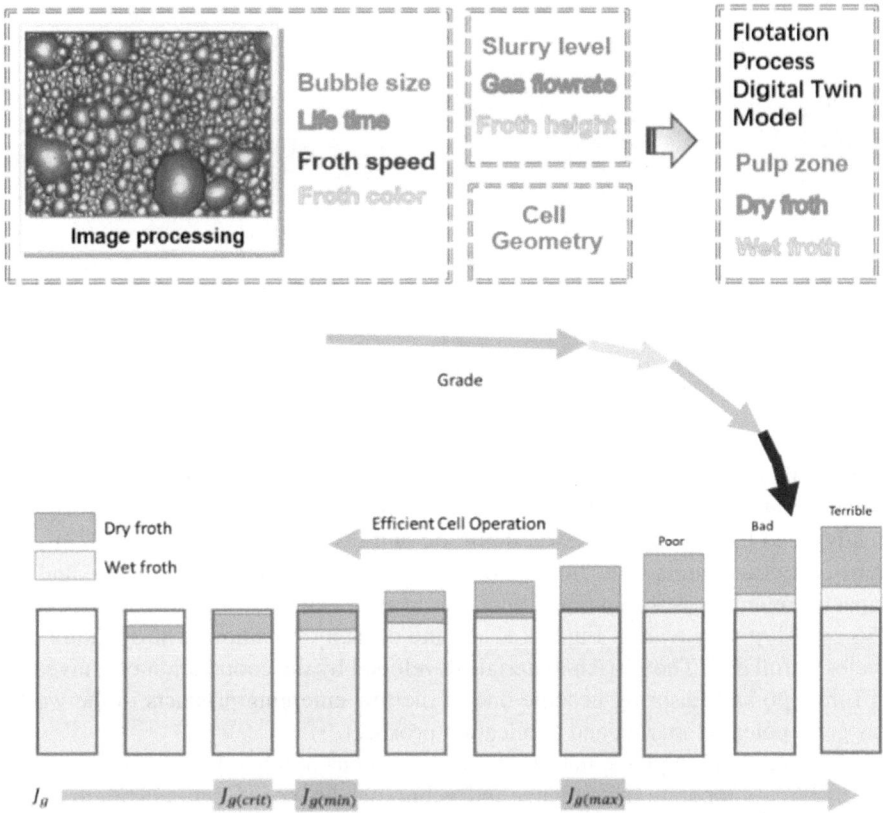

Fig. 8.12 Model function structure for flotation process digital twin

grinding, flotation and thickening process, the intelligent core technology of mineral processing and industrial practice have been carried out to provide users with a new perspective of production management, play a positive role in stabilizing the industrial process, improving production efficiency and realizing technological radiation. In the future, more application cases will help concentrators both in China and Africa fulfilling highly efficient and green mineral production (Fig. 8.13, right).

8.2.6   Advanced Lithium-Rich Materials

Since the project was established, the micro-structure of new Li-rich materials has been designed based on the theoretical simulation calculation results of the South African side, and technical problems such as low capacity, low press density, poor cycle life and large voltage decay of Li-rich materials have been solved successively by precursor preparation optimization, bulk doping, surface modification and

Fig. 8.13 Optimal process control technique applications in China mainland (left JINCHUAN concentrator, Gansu) and in Africa (right, COMMUS concentrator, Congo)

industrialization technology. The Li-rich materials with high specific capacity, good cycle life and high pellet density advantages have been developed, and were sent to advanced battery enterprises at domestic and overseas (Wang et al., 2022). The third-party test results show that the 0.1C capacity can reach 295.1 mAh/g, and the capacity retention is 94% after 100 cycles in coin cell. The 0.2C capacity reaches 266.2 mAh/g as shown in Fig. 8.14, and the capacity retention is 86.1% after 300 cycles in full cell. The Li-rich materials developed by the cooperation of University of Limpopo and Easpring become one of the few emerging products in the world, has great potential market and application prospect.

The successful implementation of this project has promoted efficient, innovative and practical cooperation on science and technology between China and South Africa under the Belt and Road Initiative, enhanced political mutual trust and developed a new strategic partnership between China and South Africa, and achieved win–win results. It has improved the layout of national scientific and technological innovation, improved the competitiveness of international market, and provided advanced materials for 350 Wh/kg Li-ion batteries by 2025.

8.2.7 Comprehensive Utilization of Solid Waste

Scientists from BGRIMM and MINTEK jointly conducted a research project on the preparation of fine-grained tailing-based permeable materials and simultaneous curing technology of hazardous components, using gold tailings and diamond tailings from South Africa as research objects.

In the preparation of permeable materials with tailings as the main raw material, a certain amount of aggregate is usually chosen to be added in order to enhance its permeability coefficient. The addition of aggregate gives the permeable material a richer connected pore structure (Liu et al., 2022). But the research on the preparation of high performance permeable materials from tailings in full volume has not been

Fig. 8.14 CV cycle and cycle performance of Li-rich materials with high specific capacity

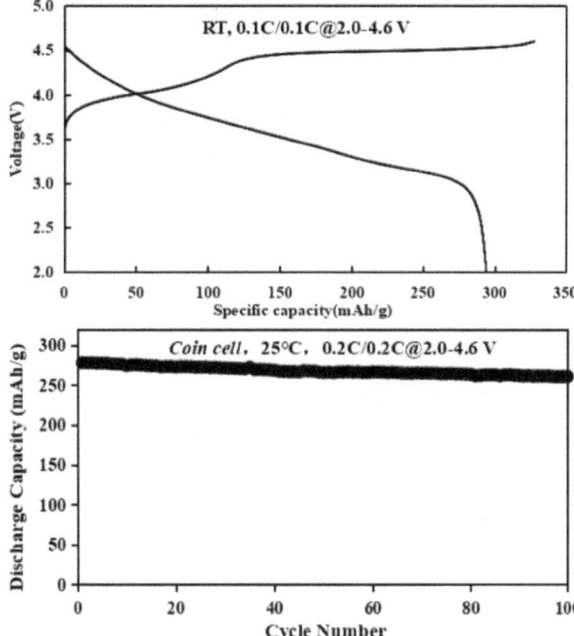

carried out in depth, the fundamental reason is that the tailings are fine and the strength and pore structure construction of permeable materials has been a contradictory issue.

The characteristics of tailings such as fine particle size, large differences, complex composition, containing unfavourable components, and poor activity of potential volcanic ash. Based on the mineralogical characteristics, the unfavourable component pretreatment and removal equipment is developed to realize the dry pretreatment and removal of unfavourable components in the tailings. In response to the problem of poor activity and wearability of the potential volcanic ash of tailings, an active activator was developed based on the crystal-chemical characteristics, and the regulation of the activity of the potential volcanic ash of tailings was realized through the compound action of chemical and mechanical means. The tailings used are very typical ridged materials, and in view of their poor plasticity and difficult forming, a new process of "two types and one sintering" is proposed, i.e., tailings pelletizing and forming, pellet billet forming and raw billet high-temperature firing. Combined with orthogonal experimental design, the construction and control mechanism of the high permeability connecting pore structure in the whole process of tailings granulation, coating, forming and sintering were investigated, and the problems of construction and strength balance of non-aggregate connecting pores of fine-grained tailings were solved. The process of connecting pore construction and strength balance of fine-grained tailings-based permeable material was showed in Fig. 8.15. The migration pattern and solidification mechanism of toxic components in tailings were identified, and lattice solidification and solid solution solidification of toxic components

Core-shell structure of the basic particles composing the water-permeable material

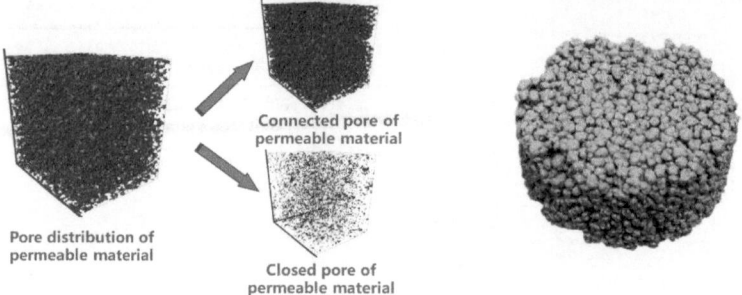

Connected pore of
permeable material

Pore distribution of
permeable material

Closed pore of
permeable material

Fig. 8.15 The process of connecting-pores construction-building and strength balance of fine-grade tailings-based permeable materials

in tailings-based permeable materials were achieved by means of physical phase reconstruction (Zhang et al., 2020b). Studied the characteristics of dissolution and dissociation of toxic components in multi-phase complex systems and developed a technology for the preparation of permeable materials from fine-grained tailings and the synergistic deep curing of toxic components (Song et al., 2021). Combined with the analysis of thermodynamic parameters and phase change characteristics of tailings and other raw materials, the thermodynamic parameters of phase reconstruction of tailings with a large blending ratio and the optimal material proportion were developed. The permeable material with excellent performance was prepared, and it was determined that the permeable material was compounded by gold tailings and diamond tailings according to 4:1, the sintering temperature range was 1135–1155 °C, the flexural strength grade of the permeable material reached more than Rf4.5, and the permeability Grade A.

8.2.8 Treatment of Waste Catalyst

In view of the low enrichment efficiency of spent automobile catalyst, China and South Africa have cooperated to carry out smelting technology research with iron and copper as collectors, study the physicochemical properties and phase equilibrium law of smelting slag phase under different slagging agents and additives. Through the exploration of multi factor condition test, the slag type with low melting point and low grade of platinum group metals in the slag is sought to reduce smelting energy consumption, reduce the corrosion of smelting slag and improve the enrichment efficiency of platinum group metals (Xie et al., 2020). Through the experimental optimization of smelting temperature, time and slag type, two high-efficiency capture technologies of pyrometallurgical smelting have been successfully developed-low temperature pyrometallurgical smelting key technology by iron capture and copper capture. The effect of temperature on the recovery of Pt, Pd and Rh was showed in Fig. 8.16. At 1400–1425 °C, the technical indicators of platinum and palladium capture rate are more than 99.7%, rhodium capture rate more than 97%, the platinum group metal grade in the slag was reduced to less than 10 g/t, reaching the industry-leading level. The appearance of iron alloy and smelting slag at optimum conditions was showed in Fig. 8.17. Due to the reduction of the capture temperature, the formation of ferrosilicon alloy in the smelting process is avoided. The silicon content in the platinum group metal concentrate is less than 0.5%, which significantly improves the leaching solubility of the alloy and reduces the subsequent treatment cost. Based on the new developed technology, an invention patent is applied: a method of pyrometallurgical smelting to recover platinum group metals from waste catalysts (Zhang et al., 2022b).

Through cooperative research, China and South Africa have successfully overcome the key technical problems of low enrichment rate of platinum group metals in the high-temperature smelting of spent automotive catalysts, revealed the binding morphology and distribution of collector and platinum group metal elements during smelting, proved the trapping mechanism of the collector on the target element and regulation mechanism of slag-alloy phase equilibrium in capture smelting process, improved the separability of slag phase and platinum group metal alloy phase, reduced platinum group metal residues in slag phase, improved the smelting and enrichment efficiency of platinum group metals. This technology is helpful to solve the existing technical bottleneck of pyrometallurgical recovery of platinum group metals from spent automotive catalysts and is expected to become an important development direction to realize the waste resource recovery strategy in China.

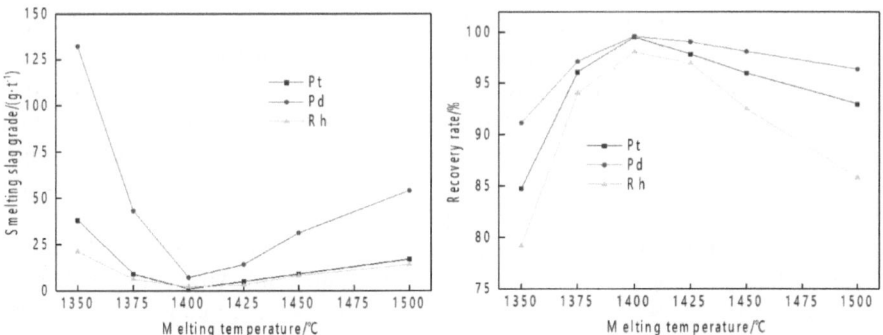

Fig. 8.16 Effect of melting temperature on the recovery of Pt, Pd and Rh

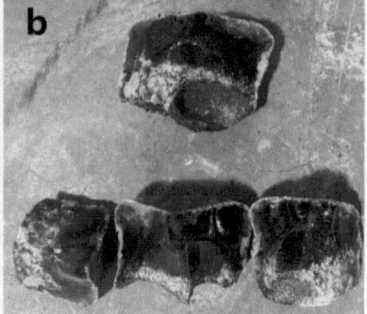

Fig. 8.17 a Iron alloy; **b** smelting slag

8.3   Personnel: From Communication to Exchange

During the cooperative research, the exchange between scientists of the two countries is the basis for realizing cooperative research and development, and it is also an important way to promote in-depth understanding between the two sides. For more than 10 years, the two sides have made much research progresses in efficient grinding, green reagents, and flotation electrochemistry through joint seminars and exchange of visiting scientists.

8.3.1   Seminars

Since 2012, during the implementation of the two cooperation projects, China and South Africa have established a stable academic exchange mechanism. Through regular workshops, the exchanges between China and South Africa at the level of project research have been strengthened. During the implementation period of the cooperation project, project workshops were held in Cape Town (2012), Beijing (2013), Limpopo (2014, Fig. 8.18), Beijing (2015), Quebec (2016), Johannesburg (2017), Limpopo (2018), and Beijing (2019), two parties carried out in-depth exchanges and cooperation in various research fields such as grinding technology, flotation technology, and lithium battery functional materials. Through academic reports, group discussions, experimental interactions, etc., strengthen the exchanges between researchers from both sides in terms of project research content, research methods, research results, etc. academic progress on both sides.

In order to further promote the cooperation between the two sides and strengthen the project cooperation research, since 2014, China and South Africa have exchanged researchers to carry out multi-level personnel exchanges and mutual visits to carry out cooperative research.

Fig. 8.18 Workshop held in South Africa (2014)

8.3.2 Exchanges of Scientists

In the field of mineral processing technology, BGRIMM has carried out in-depth cooperation and exchanges with University of Limpopo and the University of Cape Town. BGRIMM dispatched Dr. Zhang Xingrong and Dr. Liu Chongjun to visit University of Limpopo in 2016 and 2018 respectively, and achieved good cooperation results in mineral crystal simulation and flotation reagent molecular simulation. They jointly published 3 high-level academic papers in the field of collector design technology applied on flotation of copper sulfide minerals and pyrite. In 2014, 2016 and 2018, BGRIMM dispatched Dr. Song Zhenguo, Dr. Liu Jianyuan, and Dr. Zhang Xingrong to the University of Cape Town to carry out research on mineral processing technology and reagent microcalorimetry, and established a good foundation for cooperation with the University of Cape Town. The two parties jointly published 6 high-level academic papers on flotation technology, grinding environment and mineral flotation, and molecular design of flotation reagents. Relying on the good cooperation foundation in the early stage, Dr. Peace of University of Limpopo was supported by the national "International Outstanding Youth Program". Dr. Peace came to BGRIMM in 2019 to carry out scientific research work for 6 months, and conducted in-depth cooperative research on the molecular design and simulation of collectors for precious metal pyrite. After the funding period ended, he gave an academic report entitled "Design of new flotation collectors for pyrite minerals: Computational and experimental studies" in BGRIMM. After project cooperation, the two parties have formed two cooperative results, which were published in Minerals Engineering Journal and IMPC Conference. Relying on the joint research results, Dr. Zhang Xingrong won the "Young Author Award" at the 28th International

Fig. 8.19 Academic comminutions and visiting scientists

Mineral Processing Conference (IMPC) in 2018. The academic communications and talent exchanges are recorded in Fig. 8.19.

8.3.3   In-Depth Collaborative Research

In 2014, Dr. Song Zhenguo of BGRIMM completed a 1-year postdoctoral research at UCT. In his investigation, balls and rods consisting of different material are used to grind a BMS ore (Nkomati ore) to determine the effect of different rod/ball material on the slurry chemical properties (pH, Eh and DO) and flotation response of the ore.

In the study, grinding and flotation experiments using both balls and rods consisting of different material were performed to determine the effect of media shape and composition on slurry chemical properties (pH, Eh and DO) and flotation response for a base metal sulphide ore (Nkomati) (Corin et al., 2018; Song et al., 2016).

It was observed that in the case of using mild steel for the BMS is expected as this reactive media may have enhanced the inter-mineral galvanic interactions and promoted the reduction of the pulp oxygen. From the amount of EDTA extracted copper and nickel, more chalcopyrite and pentlandite were oxidized after grinding with chrome steel balls (BMCS) and stainless-steel rods (RMSS) than with forged steel balls (BMFS) and mild steel rods (RMMS), which was accordance with the mixed-potential theory.

As can be seen in Fig. 8.20, different grinding media had significant effect on slurry chemical properties and flotation performance of Nkomati ore. Mild steel rods

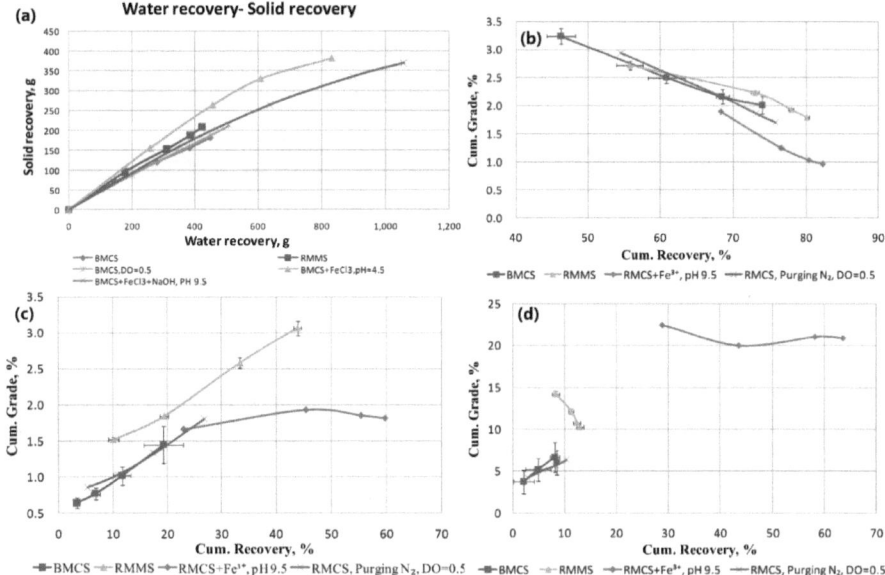

Fig. 8.20 **a** Solids-water recoveries **b** Cu grade-recovery **c** Ni grade-recovery **d** Pyrrhotite grade-recovery

(RMMS) produced the highest solid recovery, Ni and Po recovery, and a little higher Cu recovery, meanwhile the highest EDTA extracted Fe amounts and the lowest DO level. Chrome steel balls and stainless-steel rods produced the lowest metal recovery, EDTA extracted Fe amounts and the highest DO level. The addition of either Fe^{2+} or Fe^{3+} ions (using $FeCl_2$ or $FeCl_3$) significantly increased the solids and water recovery especially in the case of Fe^{2+}. There was clearly a greater mass pull when these ions were added. However, the flotation results had indicated an increase in grades and hence this increase in mass pull was selective towards the recovery of pentlandite and pyrrhotite.

In 2016 and 2018, Dr. Zhang Xingrong of BGRIMM completed a nearly 3-month study on mineral crystal simulation and calculation and reagent adsorption mechanism at the University of Limpopo and the University of Cape Town, respectively. In his research, mainly for chalcopyrite collectors and copper-lead separation inhibitors, related work such as mineral crystal simulation, chemical molecular design and simulation has been carried out. On the chalcopyrite capture simulation work, he completed the research on the performance of oxycarbonyl-thiocarbamate collectors on chalcopyrite by DFT method. The study shows that adsorption of the oxycarbonyl-thiocarbamate collectors preferred the Cu atom over Fe atoms, and that BBCTC and BECTC are better collectors for the selective flotation of chalcopyrite minerals, which indicated that a straight hydrocarbon chain gave stronger adsorption than a branched hydrocarbon chain, as shown in Fig. 8.21 (Mkhonto et al., 2021).

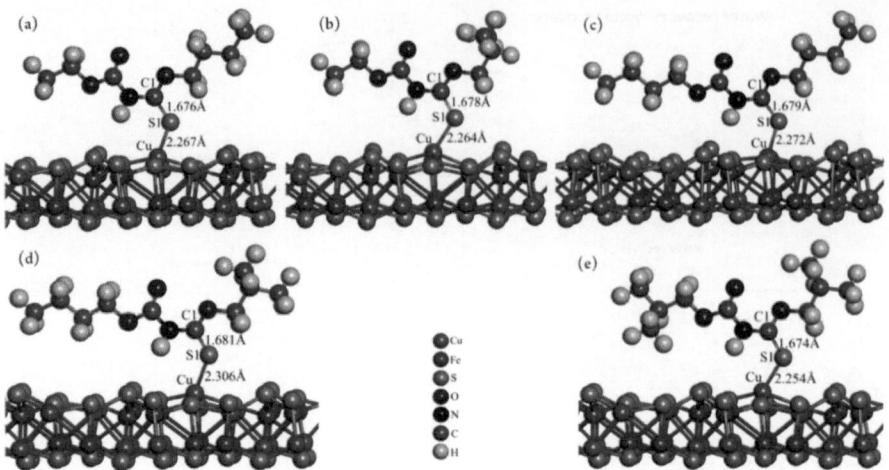

Fig. 8.21 Relaxed geometries of oxycarbonyl thiocarbamate collectors on the CuFeS2 (112) surface: **a** Fe–BECTC, **b** Fe–IBECTC, **c** Fe–BBCTC, **d** Fe–IBBCTC and **e** Fe–IBIBCTC

He also completed the research on adsorption mechanisms of thiocarbamate collectors in the separation of chalcopyrite from pyrite minerals from electronic structures of the collectors on the $CuFeS_2$ Rec-(112) surface and FeS_2 (100) surface (as displayed in Fig. 8.22) (Mkhonto et al., 2022), finding the thiocarbamate collector adsorption on chalcopyrite Cu sites decreased in the order ADEDTC > IPDETC > IPETC, while on pyrite Fe sites, the adsorption strength decreased in the order ADEDTC > IPETC > IPDETC. This confirms that a clear correlation between experiments and DFT predictions and therefore suggest that these thiocarbamate collectors may be useful in the floatation separation of chalcopyrite from pyrite minerals.

In the simulation work of copper and lead separation inhibitors, two new organic macromolecular inhibitors for copper and lead separation, PAM-ATU and DTC-PAA,

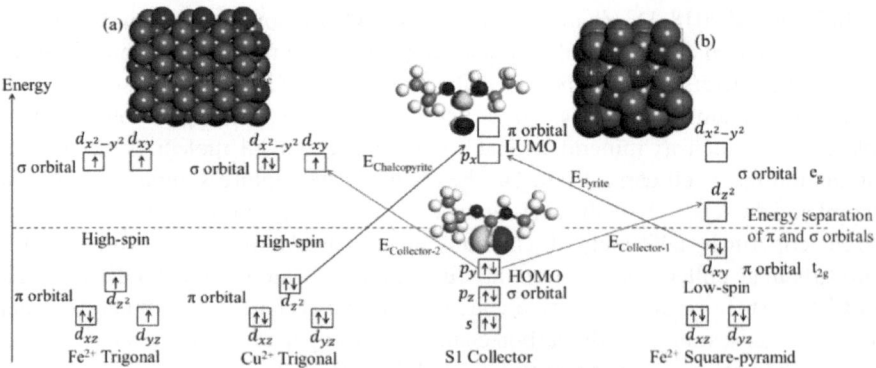

Fig. 8.22 Coordination model of collectors with $CuFeS_2$ Rec-(112) surface and FeS_2 (100) surface

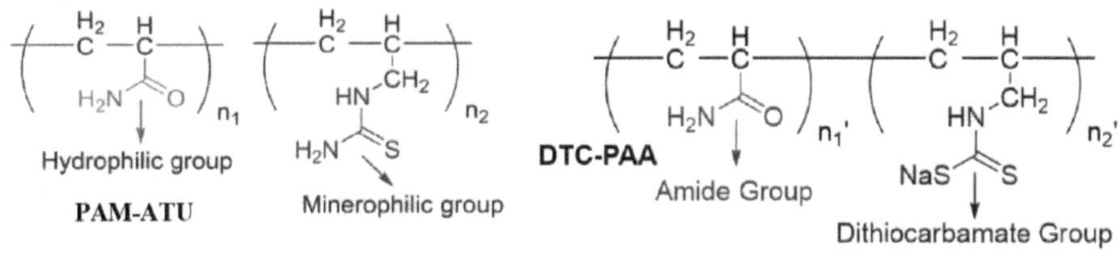

Fig. 8.23 The designed PAM-ATU and DTC-PAA depressants

were designed and prepared (Fig. 8.23) (Zhang et al., 2019, 2021). The study found that both PAM-ATU and DTC-PAA could be chemically adsorbed on the surface of galena, which is stronger than the effect of collectors on the surface of galena; while on the surface of chalcopyrite, the effect of collectors is stronger than that of inhibitors. Therefore, the separation of copper and lead is realized under the effect of competitive adsorption.

From July to August 2016, Dr. Liu Jianyuan of BGRIMM went to visiting research at the Center of Mineral Research, University of Cape Town (CMR/UCT), South Africa. During his visit as a visiting scholar, Dr. Liu studied the effects of lime addition method and different grinding conditions, such as dry and wet grinding, on the electrochemical properties of pulp and the flotation behaviour of copper ore.

Comparing the electrochemical parameters of the selected pulp obtained by the three grinding conditions, compared with "wet grinding-adding lime after grinding, the pH of the selected pulp obtained by "wet grinding-adding lime in the mill" is slightly lower, Eh Slightly increased (Lujiu ore) or almost unchanged (Dexing ore), DO is almost unchanged; the pH of the slurry prepared by "dry grinding—adding lime after grinding" is almost the same, Eh is slightly increased, DO almost unchanged (Liu et al., 2018b).

Grinding conditions have little effect on the relationship between concentrate rate and flotation time during the flotation of the two ores, and have a slight effect on the relationship between copper recovery and flotation time. When flotation of Dexing ore, there is little difference between dry grinding and wet grinding, but the copper recovery obtained by adding lime and wet grinding is slightly lower; In the flotation of Lujiu ore, the use of dry grinding or wet grinding with lime achieve higher copper recovery than adding lime after wet grinding (Fig. 8.24).

In February–March 2018, Dr. Liu Chongjun of BGRIMM went to University of Limpopo, South Africa for academic visit. During this time, the influence of hydrocarbon group structure on the properties of thiamine esters (Liu et al., 2018a), such as symmetry of frontline orbital (Fig. 8.25), highest occupied orbital and lowest non-occupied orbital energies, Mulliken atomic charges, total and partial density of states (Fig. 8.26), were discussed in depth with Prof. Phuti and Dr. Peace of Limpopo University. The results of quantum mechanical calculations were verified by pure mineral experiments, and the results showed the ability to donate electrons decreases in the order IPDETC > IPETC > ADEDTC, and the ability to accept electrons decreases in the order ADEDTC > IPETC > IPDETC. This implied that

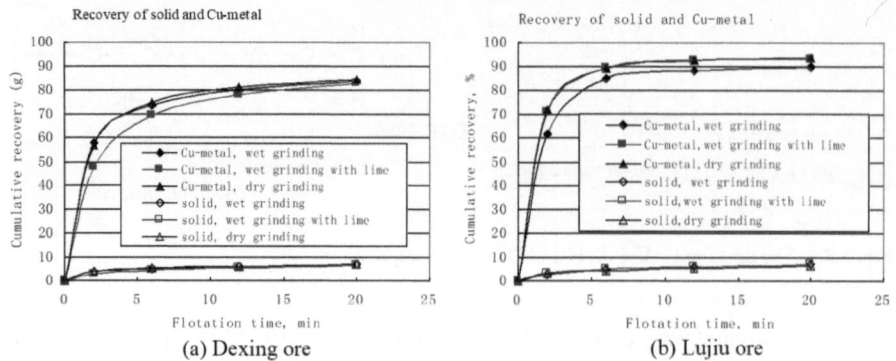

Fig. 8.24 The concentrate yield and copper recovery as a function of flotation time

ADEDTC, IPETC and IPDETC can react with the mineral surface through normal covalent bonds and backdonation covalent bonds. It can be seen that quantum mechanical calculations can provide theoretical support for the flotation collector design (Mkhonto et al., 2022).

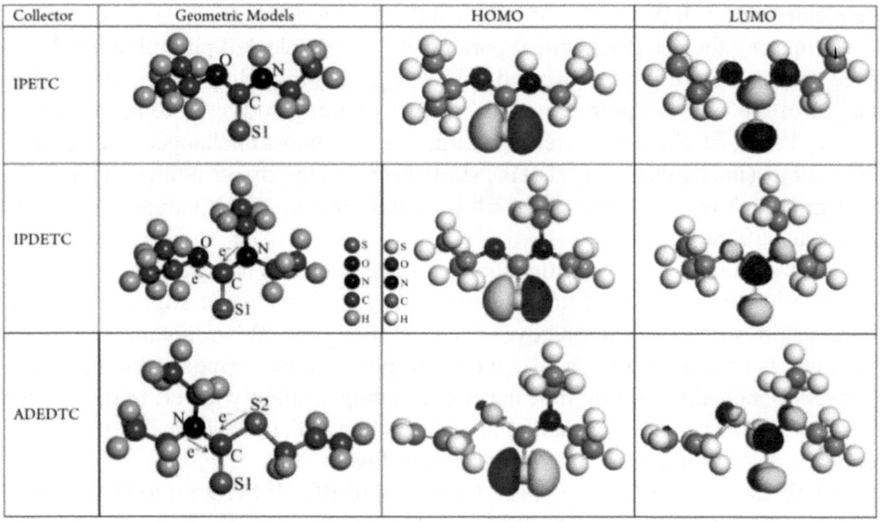

Collector	Geometric Models	HOMO	LUMO
IPETC			
IPDETC			
ADEDTC			

Fig. 8.25 Symmetry of Frontline Orbital of Thiamine Esters with Different Hydrocarbyl Structures

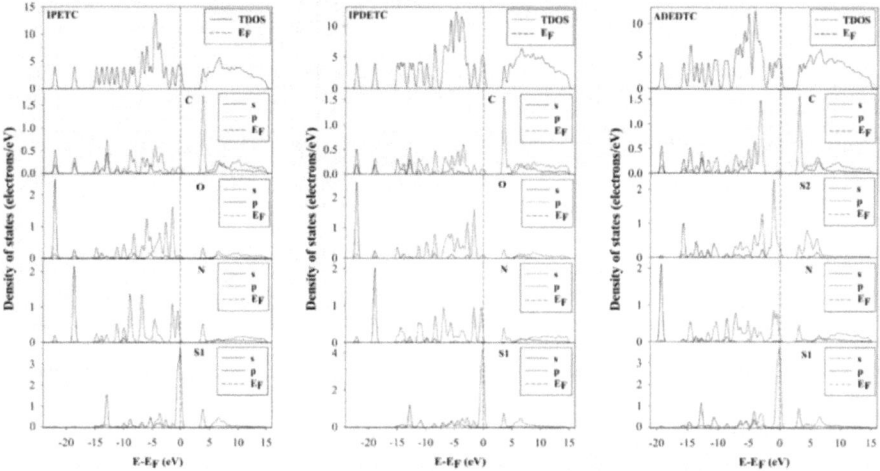

Fig. 8.26 Total and Partial density of states of Thiamine Esters with Different Hydrocarbyl Structures

8.4 Progress: From Achievements to Platform

The cooperative research has not only promoted the progress of theoretical research, but also achieved many application results, which have contributed to the technological progress of the industries of the two countries. Based on those fruitful cooperative research outcomes from JRP, the JRC has established.

8.4.1 Application of Outcomes

In the field of mining, BGRIMM cooperated with the CSIR and University of Witwatersrand to carry out joint research on the safety monitoring technology standards for deep mining (Fig. 8.27), ground pressure monitoring and environmental monitoring technology, and destress mining technology for deep mines. In the field of mineral processing, BGRIMM has cooperated with University of Cape Town, University of Limpopo, and MINTEK to carry out research on high-efficiency grinding, clean mineral processing reagents, flotation electrochemistry, etc., and has achieved a number of high-level research results, among which the new mineral processing reagent (Fig. 8.28) has been popularized and applied in mining enterprises in China and South Africa.

During the cooperation period, the two sides jointly held more than 10 special academic seminars, exchanged more than 20 visiting scientists, introduced 7 South African experts to work in China, jointly-educated 1 postdoctoral fellow, and published more than 10 academic papers in international authoritative journals such

Fig. 8.27 Localized real-time microseismic monitoring equipment and its appliance in Sanshandao Gold Mine

Fig. 8.28 Application of green and high-efficiency copper-molybdenum separation and flotation reagents in large-scale molybdenum mines in China

as Mineral Engineering. These cooperative researches have not only promoted the technological progress of China and South Africa in the field of mineral resources, but also established a long-term and stable scientific and technological partnership and a high-level cooperative innovation team, which has promoted cultural exchanges between the two countries.

8.4.2 Establishment of JRC

On the basis of the good cooperation between the two sides, the cooperation between China and South Africa has gradually extended the cooperation at the international joint research project (JRP) level to the cooperation on the platform of the international joint research center. In September 2018, relevant institutes of China and South Africa jointly promoted the establishment of the "China–South Africa Joint Research

Center for Exploitation and Utilization of Mineral Resources" (JRC), which further deepened the bilateral scientific and technological partnership. In April 2019, the center was successfully identified as one of the first national "Belt and Road Initiative" joint laboratories, and the scientific and technological cooperation between the two parties has been further elevated to a new level.

At present, all parties involved in China–South Africa cooperation have carried out deeper joint research relying on the JRC cooperation platform. The areas of cooperation between the two parties, on the basis of the initial cooperation on mining and mineral processing, have been expanded to the fields of automatic control of mining and metallurgy, comprehensive utilization of tailings, waste catalyst recovery and lithium-ion battery materials. The cooperation form between the two parties has been innovated by jointly carried out an international training program. On October 21st, 2021, hosted by the China–South Africa Joint Research Center for Exploitation and Utilization of Mineral Resources (JRC), and undertaken by the China–South Africa BRI Joint Laboratory on Sustainable Exploitation and Utilization of Mineral Resources, the "China–South Africa Modern Mineral Processing Technology (Phase 1) Online Training Program" supported by the Chinese Embassy in South Africa was successfully held. More than 100 trainees from 27 mining companies in 9 countries including China, South Africa, Zambia, DRC, Namibia and Algeria participated in the training (Fig. 8.29). Seven experts from BGRIMM, CSIR, the University of Cape Town and the University of Limpopo from the member units of the JRC made online technical lectures.

Fig. 8.29 Attenders of Webinar on Modern Mineral Processing Technology

8.5 Conclusion and Future Work

From JRP to JRC, BGRIMM and South African partners have carried out a series of fruitful cooperative research programs on the common challenges faced by not only China and South Africa, but also the global mineral resources development field, such as deeper mining levels, declining resource quality, severe security situation and tightening environmental constraints. The cooperative research involves a wide range of fields such as mining, mineral processing, automatic control of mineral processing, comprehensive utilization of tailings, waste catalyst recovery and lithium-ion battery materials. During the cooperation period, the two sides jointly held more than 10 special academic seminars, exchanged visiting scientists and jointly trained more than 20 postdoctoral fellows, and jointly published more than 10 academic papers in international authoritative journals such as Mineral Engineering. Relying on the cooperation achievements, Professor Phuti. Ngoepe of Limpopo University won the Chinese Government Friendship Award, and Dr. Zhang Xingrong, a young scientist of BGRIMM, won the Outstanding Young Author Award at the 28th International Mineral Processing Conference. At the same time, the two sides jointly launched the "Webinar on Modern Mineral Processing Technology" training, which improved the professional skills of the personnel of the mining enterprises of the two countries.

In the future, we will rely on the JRC and work with South African partners to actively practice the principle of "co-consultation, co-construction, and sharing", and actively align with the AU "2063 Agenda". We will continue to deepen the content of cooperation, expand the fields of cooperation, and work together on technological innovation, talent training and achievement transformation in the exploitation and utilization of mineral resources between China and South Africa. We will actively explore the development of third-party market services in the field of mineral resources in African countries, establish a collaborative innovation network and mechanism, and further enrich and innovate international scientific and technological cooperation models, to provide scientific and technological support for the sustainable exploitation and utilization of mineral resources in the "Belt and Road Initiative" countries, to promote international production capacity cooperation in the field of mineral resources, and to help the construction of the "Belt and Road Initiative" to deepen and solidify, and make new and greater contributions.

References

Chander, S. (2003). A brief review of pulp potentials in sulfide flotation. *International Journal of Mineral Processing, 72*, 141–150.

Chen, K., Zhang, D., & Zhang, Y. S. (2013). Development of a 3D laser scanning system for the cavity. *Proceedings of the SPIE, 2013*, 87691F.

Corin, K. C., Song, Z. G., Wiese, J. G., & O'Connor, C. T. (2018). Effect of using different grinding media on the flotation of a base metal sulphide ore. *Minerals Engineering, 126*, 24–27.

Javadi, A. (2015). *Sulphide Minerals: Surface Oxidation and Selectivity in Complex Sulphide Ore Flotation.* Ph.D. dissertation. Luleå tekniska universitet.

Liu, C., Zhu, Y., Wu, G., Liiu, L., Liu, H., & Phuti, E. N. (2018a). First principles study on electronic structure and lead activation mechanism on cassiterite surface. *Conservation and Utilization of Mineral Resources, 38*, 17–21.

Liu, J., Han, L., Corin, K. C., & O'Connor, C. T. (2018b). A study of the effect of grinding environment on the flotation of two copper sulphide ores. *Minerals Engineering, 122*, 339–345.

Liu, J., Shi, X., Zou, Q., Zhao, T., Zheng, J., Liu, T., Han, L., Ke, Y., & Wang, Q. (2022). Recycling of water quenched slag and silica sand tailing for the synthesis of an eco-friendly permeable material. *Construction and Building Materials, 357*, 129310.

Liu, J. Y., & Sun, W. (2016). Experimental study on comminution and mineral liberation of a copper ore by interparticle breakage. In *XXVIII international mineral processing congress*, Quebec, Canada.

Mkhonto, P. P., Zhang, X., Lu, L., Xiong, W., Zhu, Y., Han, L., & Ngoepe, P. E. (2022). Adsorption mechanisms and effects of thiocarbamate collectors in the separation of chalcopyrite from pyrite minerals: DFT and experimental studies. *Minerals Engineering, 176*, 107318.

Mkhonto, P. P., Zhang, X., Lu, L., Zhu, Y., Han, L., & Ngoepe, P. E. (2021). Unravelling the performance of oxycarbonyl-thiocarbamate collectors on chalcopyrite using first-principles calculations and micro-flotation recoveries. *Applied Surface Science, 563*, 150332.

Shang, Y., Xie, Y., Zhang, X., Zhu, Y., Han, L., Sun, C., & Zhang, Y. (2016). The first-principles study of quartz (100) surface adsorbed by hydroxyl calcium. In *XXVIII international mineral processing congress*, Quebec, Canada.

Smith, L., Davey, K., & Bruckard, W. (2012). The use of pulp potential control to separate copper and arsenic—An overview based on selected case studies. In *Proceedings of the XXVI international mineral processing congress (IMPC)*, New Delhi, India (pp. 24–28).

Song, T., Yang, J., Zou, G., Wang, J., & Zhou, J. (2022). Exploration on the application of digital twin technology in gold processing. *Gold, 43*(78–82), 90.

Song, W., Cao, J., Wang, Z., Geng, X., & Lu, J. (2021). Glass-ceramics microstructure formation mechanism for simultaneous solidification of chromium and nickel from disassembled waste battery and chromium slag. *Journal of Hazardous Materials, 403*, 123598.

Song, Z. G., Corin, K. C., Wiese, J. G., & O'Connor, C. T. (2016). Effect of using different grinding media on the flotation of a base metal sulphide and UG2 ore containing platinum group minerals. In *XXVIII international mineral processing congress*, Quebec, Canada.

Wang, J., Cui, Z., Wang, C., Liu, Y., & Chen, Y. (2022). Understanding Co roles towards developing Li-rich layered oxide cathodes for lithium-ion batteries. *Solid State Ionics, 379*, 115912.

Woods, R. (2003). Electrochemical potential controlling flotation. *International Journal of Mineral Processing, 72*, 151–162.

Xie, X., Qu, Z., Zhang, B., Liu, G., Zhang, F., & Zhang, B. (2020). Enrichment of platinum group metals from spent automobile exhaust purification catalysts. *China Resources Comprehensive Utilization, 38*, 105–109.

Zhang, D. (2014). Key technology study on 3D laser scanner for mine. In *International symposium on Test Automation and Instrumentation* (pp. 588–592).

Zhang, D., Ji, H., Ji, H., & Lou, Q. (2020a). A cloud platform-based micro-power microseismic monitoring IoT system for underground mine. In *2020 2nd international conference on industrial artificial intelligence (IAI)* (pp. 1–6).

Zhang, D., Zeng, Z., Shi, Y., Chang, Y., Dai, R., Ji, H., & Han, P. (2022a). An effective denoising method based on cumulative distribution function thresholding and its application in the microseismic signal of a metal mine with high sampling rate (6 kHz). *Frontiers in Earth Science, 10*, 1.

Zhang, J., Zhang, B., Zhang, F., Liu, G., Wang, F., & Xie, X. (2022b). Research status of pyro-recovery process of spent automotive exhaust purification catalysts. *China Resources Comprehensive Utilization, 40*, 97–103.

Zhang, R., Wei, X., Hao, Q., & Si, R. (2020b). Bioleaching of heavy metals from municipal solid waste incineration fly ash: availability of recoverable sulfur prills and form transformation of heavy metals. *Metals, 10*, 815.

Zhang, X., Qian, Z., Zheng, G., Zhu, Y., & Wu, W. (2017a). The design of a macromolecular depressant for galena based on DFT studies and its application. *Minerals Engineering, 112,* 50–56.

Zhang, X., Wei, M. A., Zheng, G., & Liu, L. (2013). Study on a new method of synthesis for thiocarbamate flotation reagents. *Modern Mining, 10*(21–23), 55.

Zhang, X., Xiong, W., Lu, L., Qian, Z., Zhu, Y., Mkhonto, P. P., Zheng, Y., Han, L., & Ngoepe, P. E. (2021). A novel synthetic polymer depressant for the flotation separation of chalcopyrite and galena and insights into its interfacial adsorption mechanism. *Separation and Purification Technology, 279,* 119658.

Zhang, X., Zhu, Y., Xie, Y., Shang, Y., & Zheng, G. (2017b). A novel macromolecular depressant for reverse flotation: Synthesis and depressing mechanism in the separation of hematite and quartz. *Separation and Purification Technology, 186,* 175–181.

Zhang, X. R., Zhu, Y. G., Zheng, G. B., Han, L., McFadzean, B., Qian, Z. B., Piao, Y. C., & O'Connor, C. (2019). An investigation into the selective separation and adsorption mechanism of a macromolecular depressant in the galena-chalcopyrite system. *Minerals Engineering, 134,* 291–299.

Part III
China-Africa Collaboration in Agriculture, Food Security and Environmental Management

Chapter 9
China–Africa Collaboration in Agriculture and Food Security: Prospects and Challenges

Tamika Kampini and Jessie Kalepa

9.1 Introduction

Food security remains a significant challenge in Africa, with millions of people facing chronic hunger and malnutrition. The continent's food systems are vulnerable to a range of factors, including climate change, conflict, and economic instability, which have undermined food production and distribution, leaving many people without access to adequate food (Food and Agriculture Organisation, 2015). Africa is a continent where more than half of the population is dependent on agriculture, specifically subsistence farming. However, most African countries are facing massive challenges concerning food security. For example, according to FAO (2019), the most recent estimates show that 281.6 million people on the continent, over one-fifth of the population, faced hunger in 2020, which is 46.3 million more than in 2019. The causes of food insecurity in Africa are due to various factors, some of which are natural while some are artificial depending on the circumstances and the countries affected.

China, as African Partner have been helping Africa to address the challenge of food security. China has been sharing its experience and expertise in agricultural technology, such as hybrid rice, irrigation, and pest control, with African countries (May, 2022). Berthelemy (2011) pointed out that Chinese agricultural experts have been working with their African counterparts to develop new farming techniques, introduce new crop varieties, and improve the efficiency of agricultural production. China has also been investing heavily in agriculture infrastructure projects in Africa, such as irrigation systems, storage facilities, and transportation networks (May, 2020). According to Gashu et al. (2019), these investments have helped to

T. Kampini (✉)
Governance and Regional Integration, Lilongwe, Malawi
e-mail: Kampinitamikah@gmail.com

J. Kalepa
Governance and development expert, Lilongwe, Malawi

© The Author(s) 2024
M. Muchie et al. (eds.), *China-Africa Science, Technology and Innovation Collaboration*, https://doi.org/10.1007/978-981-97-4576-0_9

increase agricultural productivity, reduce post-harvest losses, and improve access to markets for African farmers. Furthermore, China has been providing financial and technical support to African countries to enhance their food security. This includes providing loans and grants to support agricultural development, improving access to credit for smallholder farmers, and supporting research and development in the field of agriculture.

9.2   The Concept of Food Security

There are several definitions of food security, but this chapter will focus on two definitions which cover important elements in line with food security. We will start with a definition that relates food security to a secure world. In this definition, a food-secure world is described as one where all people have access to safe, nutritious, and affordable food that provides the foundation for active and healthy lives (FAO, 2006). Furthermore, according to the United Nations Systems Standing on Nutrition (2009), food security is defined as meaning that all people, always, have physical, social, and economic access to sufficient, safe, and nutritious food that meets their food preferences and dietary needs for an active and healthy life. The two definitions are closely linked and apply to the African context, as food insecurity has been a chronic problem for a decade. The region still remains at the bottom of the ranking of food insecurity (Global Food Index, 2022).

Food security is categorized into three aspects: acute, which refers to severe hunger and malnutrition to the point that lives are threatened immediately (e.g., famine); occasional, when food insecurity occurs due to a specific temporary circumstance; and chronic, which refers to the ability to meet food needs consistently or being permanently under threat (FAO, 2006). Following these two definitions and categorizations, most African countries have stagnated in food insecurity despite being agrarian societies and being land and labour-intensive.

Just to mention a few examples, Somalia has been plagued by conflict and drought for decades, which has led to widespread food insecurity. In 2020, the United Nations estimated that over 5 million people in Somalia were facing acute food insecurity (Integrated Food Security Phase Classification, 2022). South Sudan has also been grappling with civil conflict and displacement for many years, which has severely impacted its agriculture and food production. According to the World Food Programme (2022), over 7 million people in South Sudan are currently facing acute food insecurity. Nigeria is another example, where the growing population is struggling to keep up with food demand. In addition, climate change is having a significant impact on agriculture in the country, with droughts and floods becoming more frequent (FAO, 2015). The World Food Programme (2021a, 2021b) estimated that over 13 million people in Nigeria were facing acute food insecurity. Ethiopia has also been hit hard by drought and other climate-related shocks in recent years, which has led to crop failures and widespread food insecurity. According to the World Food Programme (2021a, 2021b), over 13 million people in Ethiopia needed food

assistance in 2020. Lastly, Zimbabwe has experienced economic and political instability in recent years, which has led to food insecurity. In addition, the country has
been hit by drought and other climate-related shocks, which have further impacted
food production. The World Bank (2022) estimated that over 5.5 million people in
Zimbabwe were facing acute food insecurity.

However, it must be pointed out at the outset that the countries of Africa are diverse
and not uniform in terms of area, population, and endowment of natural resources
such as cultivable land and water. Therefore, the agricultural potential of different
countries varies, and food security is under massive threat from population growth,
land scarcity due to the growing population, and worsening ecological degradation
because of climate change in most African countries. Furthermore, COVID-19 has
had a major effect on agriculture, worsening the status of food insecurity across
the continent in 2020 (African Centre for Strategic Studies, 2021). Border closures,
lockdowns, and lost jobs have reduced food access and food production. In addition,
food and livestock traders experienced a surge of spoiled goods due to prolonged
transit times.

9.3 Understanding the Landscape of the African Agriculture Sector

Agriculture is one of the essential sectors in Africa which is crucial to its food security
and poverty reduction. The development of the agricultural sector and the improvement of the food security situation will strongly influence other sectors of Africa's
economy (Kleeman, 2012). Africa has around 930 million hectares biophysically
suitable for agriculture (Vibeke, 2020). Further, the African Union recognizes that
African countries are labour and land-intensive, and supporting agricultural productivity is the way to achieve development and hence tackle the challenges of food security. For example, the Comprehensive African Agricultural Development Program
(CAADP) was introduced as the basic continental framework guiding the process
of agricultural development in Africa (African Union, 2021). CAADP focuses on
four 'pillars' as well as recognizing the importance of addressing a range of cross-
cutting issues and of integrating livestock, fisheries, and forestry into the agricultural
planning processes. Among the many priority issues identified by CAADP for transforming African agriculture. This framework is believed to unlock the agricultural
potential in Africa. Sub-Saharan Africa has the fastest-growing population in the
world and the most rapid rate of people moving from rural to urban areas. Most
African countries have serious land tenure issues and other challenges that have
stymied the development of agricultural projects (Kleeman, 2012). Further, in most
countries, due to the absence of adequate water sources, irrigation is not an option in
many areas; it is also expensive and comes with environmental downsides. Recognizing the challenges, Africa has depended on foreign intervention like that of China.
This is evidenced by different forums aimed at deepening China–Africa relationship,

for example, the Forum on China–Africa Cooperation, Dakar Action Plan (2022–2024), the 8th Ministerial Conference of the Forum on China–Africa Cooperation (FOCAC) was held in Dakar, Senegal China–Africa Business Council, 2021). The discussion points focused on the two sides giving full play to the (China–Africa agricultural cooperation mechanism, holding the second Forum on China–Africa Cooperation in Agriculture, and convene the first meeting of the China-AU Joint Committee on Agricultural Cooperation, to innovate cooperation methods, enrich the content of cooperation, and ensure the effective implementation of agricultural cooperation in fields of shared interest. In addition, they agreed on strengthening food security cooperation by giving full play to the China-AU coordination mechanism for Belt and Road cooperation and making good use of the South-South Coopera-tion Assistance Fund. Further, the two sides agreed on launching trials in such areas as the reduction of post-harvest grain loss in light of the AU's needs and China's strengths, to support African countries to improve their food supply capacity with existing production capabilities (Bucley, 2013).

Africa's growth in agricultural productivity is still low compared to other regions. This low productivity can be attributed to several factors such as the effects of climate change, low-scale investments in agricultural land, environmental degradation, and highly volatile world market prices, especially in this era of Covid 19. From the African perspective, the fight against hunger is the most important social challenge and there is a dire need to improve productivity specifically for smallholder farming and the more effective use of large farms to meet the dietary needs of the rural poor (Kleeman, 2012). Not only the creation of markets for agricultural products is crucial to economic growth. The mutual interest of China to promote agriculture and food security in Africa has become increasingly prominent. China is increasing engagement with African agriculture through aid, trade, and investments throughout the continent. Furthermore, the China–Africa Cooperation Vision 2035 clearly states that China–Africa agricultural cooperation will extend to the whole industrial chain in the next 15 years.

In addition, continued underinvestment by national governments in agricultural research, technology, and infrastructure further aggravates the food productivity decline in Africa. Africa has been trying all means to use technologies to boost food production hence culminating in the challenge of food insecurity (FAO, 2015). However, it is one of the regions in the world where a 'green revolution' was meant to revamp food production and countries are relying on rain-fed agriculture and manual labour due to a lack of new technologies.

Institutions like the World Bank and the Organization of Economic Cooperation and Development have been sending support to Africa as food aid and monetary aid to support the agriculture sector. New donors like China have also emerged in recent times which are supporting Africa in the agriculture sector to revamp food security. Africa has been labelled a hopeless continent, considering most countries are landlocked, populated, and with a lot of economic and socio challenges like food security. The African agriculture is not satisfying, food production is low and thousands of people are living below the poverty line. In 1943, Maslow developed a Hierarchy of Needs to explain the five levels every human must progress through to self-actualization, and food was highlighted as one of the basic need (Hopper, 2020). Unfortunately, in most African countries, these basic needs cannot be met without aid from donors. For a person to be a productive citizen, good health is a requirement and food is one of contributing factors to good health. A meal can be what's needed for a person to focus on obtaining higher needs, perhaps a meal is a foundation step to development for most African countries that are stagnated in poverty as healthy citizens are productive citizens.

In addition, the African Union has been trying to address the challenge of food security in Africa through Agriculture and Food Security Division. The division coordinates four continental initiatives to address the challenge; Comprehensive Africa Agriculture Development Programme (CAADP) Partnership for Aflatoxin Control in Africa Seed and Biotechnology Programme (ASBP), the Ecological Organic Agriculture Initiative, Geographic Indications Strategy for Africa (GISA) (AU, 2022) However, it is worrisome if Africa will meet the sustainable development goals, especially on Zero hunger and no poverty considering the massive challenges countries are facing concerning agriculture and food security. With this context in mind, it is no secret that Africa must find other means and seek foreign intervention to help culminate the challenge of food insecurity.

9.4 Sustainable Livelihood Theory and Food Security

Sustainable livelihood theory emphasizes the importance of maintaining and enhancing the well-being of people and communities through the sustainable use of natural resources, economic opportunities, social relationships, and political stability (Hussen & Nelson, 1998). Food security is a crucial component of sustainable livelihoods, as access to adequate and nutritious food is essential for maintaining health and well-being. Sustainable livelihood theory offers a framework for addressing challenges by focusing on the interconnections between social, economic, and environmental factors. Sustainable livelihoods theory (SLT) and food security are interconnected concepts that have gained increasing attention over the years. SLT emphasizes the importance of sustainable and resilient livelihood strategies that enable individuals and communities to cope with environmental, economic, and social shocks and stresses (Hussen & Nelson, 1998). As food security means availability, access, and utilization of safe and nutritious food at all time slot, recognizes that livelihood strategies are influenced by various factors, including natural resources, social networks, economic opportunities, and political institutions. The theory also highlights the importance of considering the different dimensions of well-being, including food security, when assessing the effectiveness of livelihood strategies.

Food security is one of the key indicators of sustainable livelihoods. To achieve sustainable livelihoods, communities must have access to adequate food and nutrition. This requires not only access to food but also access to the resources and knowledge necessary to produce, store, and prepare food. Sustainable livelihood strategies that promote food security may include activities such as agricultural diversification, agroforestry, and community-based natural resource management. Moreover, food security is also influenced by wider factors such as climate change, market volatility, and political instability. SLT emphasizes the importance of addressing these broader factors to promote sustainable livelihoods and food security. Sustainable livelihoods theory provides a useful framework for understanding the complex relationships between livelihood strategies and food security in Africa. In Africa, adopting a holistic approach that considers the different dimensions of well-being, SLT can help to inform policies and interventions that promote sustainable and resilient livelihoods and ensure food security for all.

9.5   Historical Background of China–Africa Collaboration and Current Priority Areas

China and Africa have a long history of cooperation in the field of agriculture. China's engagement with Africa in agriculture can be traced back to the 1950s and 1960s when China supported African countries in their efforts to achieve food self-sufficiency (WFP, 2021). During this period, China provided technical assistance, training, and equipment to African countries to improve agricultural production and productivity. In the 1970s and 1980s, China expanded its cooperation with African countries in agriculture, focusing on joint research and development of new crop varieties, livestock breeding, and aquaculture (Bräutigam & Xiaoyang, 2009). This cooperation led to the introduction of new high-yielding crop varieties, improved livestock breeds, and innovative farming techniques that helped boost agricultural productivity and food security in Africa. In the 1990s, China's engagement with Africa in agriculture shifted towards commercial interests. China began to invest in large-scale agricultural projects in Africa, including the production of cash crops such as cotton, tea, and rubber (Ndoricimpa et al., 2022). This investment led to the establishment of large-scale agricultural estates and the introduction of modern farming techniques and technologies in Africa.

In recent years, China has increased its investment in Africa's agriculture sector, focusing on food security, rural development, and poverty reduction. China has established several agricultural demonstration centres in Africa, where Chinese experts work with local farmers to improve agricultural practices, develop new crop varieties, and promote sustainable farming techniques Jiang and (Alden, 2019). The centres have been established in Kenya, Tanzania, Uganda, Zimbabwe, and Zambia. These centres have been successful in improving agricultural productivity and income for local farmers and promoting agricultural trade between China and African countries. Despite that these centres are meant to promote agriculture in Africa, critics have accused China of promoting its interests at the expense of African farmers and undermining local agricultural systems. However, China's engagement with

Africa in agriculture has helped improve agricultural productivity and food security in Africa and has provided opportunities for technology transfer and capacity building (Nalwimba & Mudimu, 2019).

China is a great example for Africa of a country that has managed to face the challenge of food insecurity despite the growing population without changing the stability of the agro landscape. Africa has should take advantage of this collaboration and learn from China to improve its agriculture sector and food security. For China to revamp production, the traditional biodiversity-friendly management of agricultural landscapes included various elaborate techniques such as the use of organic manure, traditional integrative farming approaches like dyke–pond and rice-fishery systems, crop rotations, and intercropping as well as the preservation of traditional agricultural landscapes including diverse natural and semi-natural elements (Huang, 2017). China completed 220 agricultural aid projects in Africa from 1960 to 2010 to enhance their partnership. In the 1950s, China mainly focused on the development of agricultural cooperatives that emphasized new techniques and better agricultural inputs, while also implementing large state farms (Global Times, 2023). With its vast experience in agricultural production and food security, China has much to offer to Africa.

Further, China has been supporting Africa in various aspects such as agricultural demonstration centres, sending agricultural experts and technicians, providing training courses, and participating in the Food and Agriculture Organization's programs for food security. However, the demonstration centres have faced criticism, with some viewing them as centres that boost production for export to China, rather than responding to the needs of the host country (Mgendi et al., 2019). Just to give an example, a study of centres in Mozambique and Benin revealed that the research and training offered did not align with the host country's needs, but rather served the strategy of the Chinese companies in charge of the centre (Sergo, 2014). Moreover, the centres did not work together with the national agricultural research efforts of the two countries. Meanwhile, the technology demonstration centre has become a significant aspect of China's agricultural aid program to Africa, along with sending experts. As of 2006, China established 15 agriculture centres and plans to set up another seven (Jiang, 2016).

Further, to boost food security, Chinese aid has focused on the transfer of technologies and training, and the establishment, and often operation, of demonstration farms. China has actively used these forms of assistance in forging or supporting strategic partnerships in Africa since at least the 1960s. In the 1960s and 1970s, China assisted in building over 80 farms in Africa, covering a territory of roughly 45,000 hectares (Jiang, 2016). Some 184 of these farms, such as the Mbarali farm in Tanzania, included long-term technical assistance and, at one stage, were responsible for roughly one-quarter of total domestic rice production, providing important support for national food security. Other important operations originating in this period included the Ubungo Farm Implements plant, also in Tanzania, which produced 85 percent of the country's handheld farm tools (Gebrekidan et al., 2020). Further, other farms include the Fano farm in Somalia, and the Chipembe farm in Uganda. The Chinese ministry of commerce has further claimed that over 1100

Chinese agricultural experts are currently stationed in Africa, maintaining at least 11 agricultural research stations and over 60 agricultural investment projects throughout the continent (Jiang, 2016).

9.6 Prospects of China–Africa Collaboration on Agriculture and Food Security

The collaboration between China and Africa on agriculture and food security presents several prospects that can transform the agricultural sector in Africa. Some of the prospects include:

9.6.1 Enhanced Agricultural Productivity

China has advanced technology and expertise in agriculture that can significantly improve agricultural productivity in Africa. The collaboration between China and Africa can lead to the transfer of knowledge and technology that can help African countries increase their agricultural productivity which as of current low. China launched projects to achieve breakthroughs in core agricultural technologies and key projects of seed and plant breeding, with breakthroughs in nurturing core seeds as well as agricultural mechanization in hilly areas. Africa can take advantage of these technologies. China has launched the 5-year plan (2021–2025) agricultural modernization plan and African agriculture experts can learn from this plan. The plan aims to strengthen the support of modern agricultural technology and material equipment, adhere to the self-reliance and self-improvement of agricultural science and technology, improve the stable support mechanism for basic research in the field of agricultural science and technology; establish several innovation base platforms, carry out in-depth scientific and technological support actions for rural revitalization, support colleges and universities to provide intellectual services for rural revitalization, build a modern rural industrial system, to rejuvenate the nation, the countryside must be rejuvenated (Pandaily, 2020). China Africa collaboration can also focus on investing in African Universities. China invests in research and technology through collaboration with universities, a good example is a collaboration between the Chinese Academy of Agricultural Sciences and Peking University collaboration to strengthen scientific and technological collaborative innovation, work together to build a national rural high-end think tank, and jointly explore a new model of deep integration of agricultural science and education, Africa might benefit through collaboration with these universities.

9.6.2   Food Security

The collaboration between China and Africa can help increase food production, reduce food waste, and improve food distribution systems. The partnership can also promote the development of a more diversified and sustainable agricultural sector in Africa. China and Africa have been working together to improve food security in Africa through various initiatives and collaborations. One of the key initiatives is the Forum on China–Africa Cooperation (FOCAC), which was established in 2000 to promote economic and trade cooperation between China and African countries. Under FOCAC, China has pledged to provide financial and technical assistance to African countries to improve their agricultural productivity and food security. China has provided grants, loans, and technical assistance to African countries to improve their agricultural infrastructure, increase food production, and enhance the capacity of African farmers to produce more food. Africa can take advantage of FOCAC to do several investments with the aim of improving its agriculture land scape and food security. Just like what is being implemented through, the Chinese Academy of Agricultural, China can also invest in agricultural research and development in Africa. For example, focusing on crop breeding, soil fertility management, and integrated pest management. In addition, China has provided training to African farmers and technicians in modern farming techniques, such as conservation agriculture, agroforestry, and precision farming.

9.6.3   Trade

The collaboration between China and Africa can promote trade in agricultural products between the two regions. China can export agricultural technology, expertise, and equipment to Africa while African countries can export their agricultural products to China. The prospect for agricultural trade between China and Africa is significant, with both sides showing an interest in expanding their agricultural cooperation. Africa has vast agricultural potential, with a significant portion of its land suitable for farming, and China has a large and growing population that requires food imports to meet its demand. China has been investing in Africa's agricultural sector, with a focus on promoting sustainable farming practices and improving productivity. Chinese companies have also been involved in building agricultural infrastructure such as irrigation systems, processing plants, and storage facilities in Africa. On the other hand, African countries have been exporting agricultural products to China, including coffee, tea, nuts, and fruits.

There is a growing demand for these products in China due to increasing interest in healthy diets and natural products which Africa can take advantage of. This can create a win–win situation for both regions. Africa's potential for industrial development and the benefits it brings will be delayed if it continues to export raw materials to China without any value addition. Moreover, if the current situation in Africa persists, benefiting only a select few whiles leaving the majority behind, the people of Africa will continue to struggle with poverty and food insecurity, as no direct benefits will be directed toward their immediate development needs.

9.6.4 Infrastructure Development

China's engagement, based on developing win–win outcomes within the framework of south-south cooperation, has provided very substantial resources for critically-needed infrastructure. China's assistance to Africa has changed from grants and soft loans to commercial loans at competitive rates for projects viewed as financially viable, although some raised concerns that Chinese involvement could impair debt sustainability (Mlambo, 2022). Chinese investment can contribute to improving African infrastructure considering that infrastructure development in most African countries, is still a challenge. China has been implementing infrastructure projects in some African countries. Some of notable Chinese projects include roads and bridges in the Democratic Republic of Congo (DRC), railways in Angola, and power stations in Zambia (Kuo, 2016). Further, in the rail sector, China's largest deals include the construction of mass transit systems in Nigeria and the construction of new lines linked to mining developments in Gabon and Mauritania (Institute of Developing Economies, 2023). The largest ICT project with Chinese involvement comprises the rollout of a national communications network in Ethiopia. Further, China has been involved in several infrastructure projects in Africa, including roads, railways, and ports. The development of infrastructure in Africa can enhance the transportation of agricultural products, improve access to markets, and reduce the cost of transportation. One of the key challenges in analysing China's infrastructure investment in Africa is that Chinese investments are fragmented across several government institutions, including the China Development Bank, China EXIM Bank, and the Ministry of Commerce (Schiere & Rugamba, 2011).

9.6.5 Employment Opportunities

The collaboration between China and Africa has potential to create employment opportunities in the agricultural sector. The partnership can promote the development of agro-processing industries, which can create jobs for millions of Africans. China–Africa relations have been growing steadily over the past few decades, with China becoming Africa's largest trading partner and investing heavily in various sectors in African countries, including infrastructure, mining, and agriculture. While the impact of China's engagement in Africa on employment is a subject of debate, there are both positive and negative aspects to consider.

On the positive side, China's investments in infrastructure projects, such as roads, railways, ports, and airports, have created employment opportunities for local workers. These projects have often required large numbers of workers, and Chinese firms have hired local workers to carry out the construction work though majority of workers are Chinese. Further, Chinese investments in manufacturing industries have also led to the creation of jobs for local workers. For example, the establishment of Chinese factories in Ethiopia has created jobs for thousands of Ethiopians (Aalen, 2021). On the negative side, Chinese firms have been criticized for bringing in their workers rather than hiring local workers for certain jobs. This practice has been particularly prevalent in the mining sector, where Chinese firms have been accused of importing Chinese workers to carry out jobs that could have been done by local workers. Additionally, Chinese firms have been criticized for paying lower wages than local companies, which has led to accusations of exploitation. Further, while Chinese investments in infrastructure and manufacturing have created job opportunities for local workers, there have also been instances of Chinese firms importing their workers and paying lower wages. Both Chinese and African governments need to work together to ensure that Chinese investments in Africa benefit local communities and promote sustainable development.

9.7 Challenges of China–Africa Collaboration on Agriculture and Food Security

Despite the prospects of collaboration between China and Africa on agriculture and food security, several challenges must be addressed. Some of the challenges include:

9.7.1 Language as a Barrier to Knowledge Transfer

As already highlighted in this chapter, technology transfer is an important aspect in deepening the African–China relationship. However, while China has advanced agricultural technology and expertise, transferring this knowledge to African countries is still a challenge. Language barriers, cultural differences, and institutional differences hinders effective knowledge transfer. For example, to facilitate technology transfer, China and Africa agreed on the establishment of zones. In 2006, the FOCAC agreed to establish 8 zones to facilitate technology transfer. Of the eight zones that have were approved so far, only one, located in Egypt, is operational. Another, in Zambia, is half-operational, and the others are still under construction or have yet to attract investors. Alves (2012) identifies some of the barriers to the success of the zones. These includes cultural and language which hinders transfer of technology.

9.7.2   Lack of Local Involvement

Chinese companies involved in agricultural products have been criticized because of not involving local communities in their projects, leading to a lack of community ownership and support. Some examples of projects which faced community back rash include; a project in Mozambique in 2018, where Chinese companies were granted a concession to establish a $3 billion agricultural project in Mozambique. However, the project was met with resistance from local communities who claimed that they were not consulted and that the project would destroy their farmland and water sources (Anesi & Fama, 2015). Further, in 2017, Chinese investors were granted a lease to establish a $2.3 billion agricultural project in Uganda. However, the project was met with resistance from local communities who claimed that they were not consulted and that the project would affect their livelihoods (Mayer & Burungi, 2019). Another good example is in 2019 where Chinese investors were granted a lease to establish a $2 billion agricultural project in Ghana. However, the project was met with resistance from local communities who claimed that they were not consulted and that the project would affect their access to land and water (Smith, 2019).

9.7.3   Environmental Impact of a Chinese Project in Africa

There have been several instances of Chinese agricultural projects facing backlash from African communities due to concerns about environmental and social impacts. Some notable cases include, in 2017, a Chinese company was accused of polluting the local environment through its large-scale farming operations in the Chongwe district of Zambia. Residents claimed that the company had polluted the local water supply and caused the death of livestock (Dynamic, 2020). Another one was in Mozambique in 2019, a Chinese-owned company was accused of forcibly evicting local communities from their land to make way for a large-scale agricultural project in the Nacala Corridor. The company was also accused of destroying forests and polluting water sources (Anesi & Fama, 2015). In 2020, a Chinese-owned company was accused of destroying local crops and grazing lands in the Matabeleland South province. The company had reportedly been granted a lease for over 30,000 hectares of land by the Zimbabwean government, but residents claimed that they had not been consulted or compensated for the loss of their land (Mishra, 2020). Lastly, in 2021, a Chinese company was accused of violating environmental regulations by dumping waste from its fish processing plant into the local river in Kisumu County (Yi, 2021). Residents and environmental activists staged protests, calling for the company to be held accountable for its actions. Chinese companies and African governments need to work together to ensure that agricultural investments are environmentally sustainable and socially responsible, taking into account the needs of local communities and the long-term impacts on the environment. Large-scale irrigation systems can lead to the depletion of water resources, which can have negative impacts on both agriculture and local communities. Further, they can also cause biodiversity loss through habitat destruction and the introduction of non-native species, and Increased use of fertilizers and pesticides can lead to soil and water pollution.

9.7.4 Quality and Safety of Imported Food

African countries have raised concerns about the quality and safety of food imports from China, which can undermine efforts to improve food security. There have been concerns about the safety and quality of Chinese food products, which has led to a decline in trust and demand for Chinese imported food in Africa. One of the main concerns is the use of chemicals in Chinese food production, which has led to several health concerns. For example, there have been reports of Chinese food products containing high levels of pesticides, heavy metals, and other harmful substances. In some cases, Chinese food products are contaminated with bacteria or other pathogens, which can cause foodborne illnesses. Another issue is the lack of transparency and regulation in the Chinese food industry. There have been reports of Chinese food companies falsifying labelling information, using substandard ingredients, and engaging in other fraudulent activities. This has led to a lack of trust in Chinese food products and a decrease in demand for them in Africa. To address these concerns, Chinese food companies and the Chinese government have taken steps to improve the safety and quality of their food products. For example, the Chinese government has implemented stricter food safety regulations and established a national food safety commission. Chinese food companies have also invested in quality control measures and are working to improve the transparency and traceability of their supply chains.

9.8 Conclusion

To conclude, the relationship between China and Africa in the context of agriculture and food security is complex and multifaceted. China has been investing heavily in agriculture in Africa, providing financial and technical assistance, as well as sharing its expertise in agricultural technology and infrastructure development. One of the important elements to consider in this collaboration is the transfer of technology. China has been providing African countries with modern agricultural technologies and techniques, such as high-yield seeds, irrigation systems, and farm machinery. This has helped to increase agricultural productivity in some areas of Africa, particularly in countries with favourable conditions for agriculture. In addition to technology transfer, China has also been investing in infrastructure development in Africa, particularly in the construction of roads, railways, and ports. This infrastructure development has helped to improve the transportation and distribution of agricultural products, enabling farmers to reach wider markets and reduce post-harvest losses.

China's investments in agriculture and food security in Africa have not been without criticism, however. Some have raised concerns about the environmental impact of large-scale agricultural projects, particularly in areas with fragile ecosystems. Others have raised concerns about the social and economic impact of these projects, including issues related to land acquisition, labour practices, and food sovereignty. Despite these concerns, the China–Africa relationship in agriculture and food security remains an important area of cooperation, with the potential to drive sustainable agricultural development and improve food security in Africa. Moving forward, it will be important for China and African governments to work together to ensure that these investments are socially, economically, and environmentally sustainable.

9.9 Recommendations

From the literature reviewed and discussed case studies, the study makes the following recommendations to deepen the China–Africa collaboration:

9.9.1 Sustainable Food Systems

China and African countries can work together to promote sustainable food systems, including reducing food loss and waste, promoting local food production, and supporting small-scale farmers. This would help to create more resilient and equitable food systems. Africa–China relations should focus on promoting agroecology, a system that emphasizes working with nature rather than against it. It involves using natural resources, such as water and soil, sustainably and reducing reliance on synthetic fertilizers and pesticides. Agroecology can improve soil fertility, increase crop yields, and enhance food security while reducing greenhouse gas emissions. Further, to achieve the food system the partnership can focus on encouraging small-scale farming as it is an important part of Africa's food system and supports rural communities. Policies that support small-scale farmers can help to improve their productivity, access to markets, and resilience to climate change. In addition, reducing food waste is another means of reducing significant problems in Africa, where up to 40% of food produced is lost or wasted (Jiang, 2016). Reducing food waste can improve food security and reduce greenhouse gas emissions. Finally Promoting sustainable fishing is another sustainable food systems management. Africa's fishing industry is vital for food security and livelihoods, but overfishing is a growing problem. Sustainable fishing practices, such as using selective fishing gear and implementing fishery management plans, can help to conserve fish populations and ensure a sustainable supply of fish.

9.9.2 Climate Change Adaptation

China and African countries can work together to address the impacts of climate change on agriculture, including droughts, floods, and soil erosion. This could involve the development of climate-resilient crop varieties, water management systems, and soil conservation techniques. Though Africa and China have been working on climate change on different projects such as China and Africa have been working together to address climate change through 2015, they jointly established the China–Africa Climate Change Partnership, which aims to promote cooperation in areas such as clean energy, sustainable development, and climate change adaptation. China has been providing financial assistance to African countries for climate change adaptation. In 2018, the Chinese government announced a $60 billion package of financial assistance for Africa, which includes funding for climate change adaptation (Yi, 2021). China has also been sharing its expertise and technology with African countries to help them adapt to the impacts of climate change. For example, China has provided training and technical support to African countries on renewable energy technologies, such as solar and wind power.

9.9.3 Green Investments

China can invest in green technologies and infrastructure in Africa, including renewable energy systems, sustainable agriculture, and waste management. This would promote sustainable development and reduce the environmental impact of food production. Africa should emphasize adapting Chinese agricultural technologies to increase food production the importance of science-based technology in transforming traditional agriculture and spurring growth, development, and transformation has already been discussed. It is also important to highlight the need for government intervention. First, China has extensive experience in green technology and renewable energy. By investing in green technology, China can help African countries to develop their renewable energy resources, which can be used to power agricultural production and improve food security. Second, China is a major investor in Africa and has already established close economic ties with many African countries. By investing in green technology and renewable energy, China can help to create sustainable agricultural practices that will benefit both Chinese investors and African communities. Third, green investment can help to address the environmental challenges that are currently facing many African countries. Climate change and environmental degradation are affecting agricultural productivity and food security in many parts of Africa. By investing in green technology and renewable energy, China can help African countries to mitigate the impact of these environmental challenges and develop sustainable agricultural practices. Fourth, green investment can help to create jobs and boost economic growth in African countries. By investing in renewable energy and green technology, China can help to create new opportunities for local communities, which can in turn help to reduce poverty and improve food security.

9.9.4 Local Involvement

One way to mitigate the negative impacts of the deepening China–Africa relationship on agriculture and food security is through local involvement. This means involving local communities, farmers, and stakeholders in the development and implementation of agricultural policies and projects. This can be achieved through Participatory Approaches where governments, investors, and other stakeholders should involve local communities in decision-making processes. This includes engaging them in consultations, consultations and seeking their feedback on proposed agricultural projects. Secondly, the capacity building where local communities, farmers, and stakeholders should be equipped with the necessary knowledge and skills to effectively participate in agricultural projects. This includes providing training on farming techniques, crop management, and marketing. Thirdly through knowledge sharing where China and Africa can work together to share their knowledge and expertise in agriculture. This can be done through joint research projects, training programs, and knowledge-sharing platforms. Thirdly, China and African governments should work together to ensure that agricultural projects are environmentally sustainable. This includes conducting environmental impact assessments before starting any project and adopting best practices for land management. Lastly, agricultural projects should be designed in a way that allows local communities to own and benefit from the land and resources. This can be done through joint ventures, cooperatives, and other ownership structures that allow local communities to have a stake in the project.

References

Aalen, A. (2021). *China's role in Africa's industrial growth—The case of Ethiopia [Lecture presentation]*. Chr. Michelsen Institute. https://www.cmi.no/publications/7805-chinas-role-in-africas-industrial-growth-the-case-of-ethiopia

African Centre for Strategic Studies. (2021). *Food insecurity crisis mounting in Africa*. https://africacenter.org/spotlight/food-insecurity-crisis-mounting-africa/

African Union. (2021). *The comprehensive african agriculture development programme*. https://au.int/en/articles/comprehensive-african-agricultural-development-programme

African Union. (2022). *Agriculture and food security*. https://au.int/en/directorates/agriculture-and-food-security

Alves, A. C. (2012). Chinese economic and trade co-operation zones in Africa: Facing the challenges. *Policy Briefing 51*, South African Institute of International Affairs.

Anesi, M., & Fama, R. (2015). China accused of Stealth land grab over Mozambique's great rice project. *The Ecologist*, November 30. https://theecologist.org/2013/nov/30/china-accused-ste alth-land-grab-over-mozambiques-great-rice-project

Berthelemy, J. C. (2011). China's engagement and aid effectiveness in Africa. *Working paper series 129*, African Development Bank Group.

Bräutigam, D. A., & Xiaoyang, T. (2009). China's engagement in African agriculture: "Down to the countryside." *The China Quarterly*, 199, 686–706. http://www.jstor.org/stable/27756497

Buckley, L. (2013). Narratives of China–Africa cooperation for agricultural development: New paradigms. *Working Paper 053*, Future Agricultures Consortium. http://www.future-agricultu res.org/publications/research-and-analysis/working-papers

China–Africa Business Council. (2021). Forum on China–Africa Cooperation Dakar Action Plan (2022–2024). http://en.cabc.org.cn/?c=policys&a=view&id=38

Dynamic B. (2020). China's environmental footprints: The Zambian example. *The China Story Yearbook 2016: Control*. Australian National University. https://www.thechinastory.org/yearbo oks/yearbook-2016/forum-diasporic-dilemmas/chinas-environmental-footprint-the-zambian- example/

FAO. (2006). Food security. *Policy Brief*. http://www.fao.org/publication

FAO. (2015). Climate change and food security: Risk and responses. http://www.fao.org/public ation

FAO. (2019). *Africa—Regional overview of food security and nutrition 2021: Statistics and trends*. http://www.fao.org/policy-support/tools-and-publications/resources-details/en/c/1470145/

Gashu, D., Demment, M., & Stoecker, B. (2019). Challenges and opportunities to the African agriculture and food systems. *African Journal of Food, Agriculture, Nutrition and Development, 19*(3), 14595–14603. https://doi.org/10.18697/ajfand.84.BLFB2000

Gebrekidan, B. H., Heckelei, T., & Rasch, S. (2020). Characterizing farmers and farming systems in Kilombero Valley floodplain, Tanzania. *Sustainability, 12*(17), 7114. https://doi.org/10.3390/ su12177114

Global Food Index. (2022). Exploring challenges and developing solutions for 113 countries. *The Economist Intelligence Unit*. http://impact.economist.com/sustainability/project/food-security- index/

Global Times. (2023). China makes progress in agricultural technology innovation: Official. *Global Times*, February 22. https://www.globaltimes.cn/page/202302/1284012.shtml

Hopper, E. (2020). Maslow's hierarchy of needs explained. *ThoughtCo.*, June 25. https://www.tho ughtco.com/maslows-hierarchy-of-needs-4582571

Huang, P. C. C. (2017). The three models of China's agricultural development: Strengths and weaknesses of the administrative, laissez faire, and co-op approaches. *Rural China, 14*, 488–527. https://doi.org/10.1163/2213

Hussein, K., & Nelson, J. (1998). Sustainable Livelihoods and Livelihood Diversification. *IDS Working Paper 69*. Brighton: IDS.

Institute of Developing Economies. (2023). China's Infrastructure Footprint in Africa–*China in Africa*. https://www.ide.go.jp/English/Data/Africa_file/Manualreport/cia_10.htm

Integrated Food Security Phase Classification. (2022). *Somalia acute food insecurity situation overview rural, urban and IDP: Food security outcomes*. http://www.ipcinfo.org/

Jiang, L. (2016). *Beyond ODA: Chinese way of development cooperation with Africa—The Case of Agriculture*. Doctoral thesis, London School of Economics and Political Science. http://eth eses.lse.ac.uk/3639/

Jiang, L., & Alden, C. (2019). *Chinese agricultural demonstration centres in Southern Africa: The New Business of Development*. https://www.researchgate.net/publication/334363274

Kleeman, L. (2012). *Sustainable agriculture and food security in Africa: An overview*. http://hdl. handle.net/10419/67346

Kuo, L. (2016). China is on a mission to modernize African farming—And find new markets for its own companies. *Quartz Africa*. https://qz.com/africa/788657/a-chinese-aid-project-for-rwa ndan-farmers-is-actually-more-of-a-gateway-for-chinese-businesses/

May, G. (2022). China Focuses on Food Security: What Xi Jinping's Latest Comments on Food Security Suggest About China's Priorities for 2022. *The Diplomat*, February 18. https://thedip lomat.com/2022/02/chinas-focus-on-food-security/

Mayer, J., & Burungi, J. (2019). Chinese Investment in Uganda, New Impetus for Sustainable Development. *Policy Brief*. ISBN:9781784316501

Mgendi, G., Shiping, M., & Xiang, C. (2019). A review of agricultural technology transfer in Africa: lessons from Japan and China case projects in Tanzania and Kenya. *Sustainability, 11*(23), 1–19. https://doi.org/10.3390/su11236598

Mishra, A. (2020). China's Green Promise in Africa: The Case of Zimbabwe's Sengwa Coal Power Plant. https://www.orfonline.org/expert-speak/china-green-promise-africa-case-zimbabwe-sen gwa-coal-power-plant-67578/

Mlambo, C. (2022). China in Africa: An examination of the impact of China's loans on growth in selected African states. *Economies, 10*, 154. https://doi.org/10.3390/economies10070154

Nalwimba, N., Qi, G., & Mudimu, G. T. (2019). Towards an understanding of Chinese agricultural technology demonstration centre(s) in Africa. *European Journal of Social Sciences Studies, 4*(2), 148–158. https://doi.org/10.5281/zenodo.2858007

Ndoricimpa, S., Xiaoyang, L., & Sangmeng, X. (2022). China's agricultural assistance efficiency to Africa: Two decades of forum for China–Africa cooperation creation. *Journal of Agriculture and Food Research, 9*, 100329. https://doi.org/10.1016/j.jafr.2022.100329

Pandaily. (2020). China aims to develop smart agriculture. https://pandaily.com/china-aims-to-dev elop-smart-agriculture/

Schiere, R., & Rugamba, A. (2011). Chinese infrastructure investment and regional integration in Africa. *Working Paper Series No. 135*. African Development Bank. http://www.afdb.org/

Sergio, C. (2014). *Chinese agricultural investment in Mozambique: The case of Wanbao rice farmers*. https://www.econstor.eu/handle/10419/248181

Smith, E. (2019, November 21). China's $2 billion deal with Ghana sparks fears over debt, influence and the environment. *World Economy News*. https://www.cnbc.com/2019/11/21/chinas-2-bil lion-ghana-deal-fears-over-debt-influence-environment.html

United Nations System Standing on Nutrition. (2009). A world free from hunger and all forms of malnutrition is attainable in this generation. http://www.unscn.org/

Vibeke, B., Henning, B., & Andre, F. (2020). Why agricultural production in sub-Saharan Africa remains low compared to the rest of the world—A historical perspective. *International Journal of Water Resources Development*. https://doi.org/10.1080/07900627.2020.1739512

WFP. (2021). *China supports WFP's lifesaving assistance to vulnerable people in four African countries*. https://www.wfp.org/news/china-supports-wfps-lifesaving-assistance-vulnerable-people-four-african-countries

World Food Programme. (2021). *What the world food programme is doing in Ethiopia*. https://www.wfp.org/countries/ethiopia

World Food Programme. (2021). *What the world food programme is doing in Nigeria*. https://www.wfp.org/countries/nigeria

World Bank. (2022). *Three challenges and opportunities of food security in Eastern and Southern Africa*. https://blogs.worldbank.org/

World Food Programme. (2022). *What the world food programme is doing to respond to the south sudan emergency*. https://www.wfp.org/emergencies/south-sudan-emergency

Yi, X. (2021). *Kenyan coal project shows why Chinese investors need to take environmental risks seriously*. China Dialogue. https://chinadialogue.net/en/energy/lamu-kenyan-coal-project-chinese-investors-take-environmental-risks-seriously/

Chapter 10
Contextualizing China–Africa Collaboration on Food Security: Biographical Narrative

Biraanu Gammachu⊙

10.1 Introduction

Historical evidence indicates, China and the African continent's interaction dates back to fifteenth century linked to the Chinese famous admiral who was assumed to have brought with himself a Giraffe painting placed in the Ming Imperial Tomb in Beijing during one of his expeditions to Eastern coast of Africa during the early period of the century (Stein & Uddhammar, 2021). In 1911, following the provisional reign of Sun Yat-sen as a leader of the Young Republic of China, it established contemporary contact with South Africa, and quite formally during the 1950s remarkably during the Bandung Conference of 1995 in Indonesia on peace, economic development, and decolonization as a principal party among the constituting assemblies from Asia, Latin America, and African countries 'third world' largely aiming at adding third force to the international order (Ibid., p. 3).

The modern China–Africa collaboration engagement, which was essentially ideological at the beginning, has slowly but successfully evolved into multisector collaboration for over five decades now. The end of the Cold War with the adoption of the Chinese grand strategy towards Africa probably marks a new beginning—in terms of approach and scope of the China–Africa cooperation. In recent times, agricultural production investments, energy, mining, and infrastructure development among others are China's increasingly favoured areas of interest in Africa (Taylor, 1998; Jakobsan, 2009).

This chapter contextualizes the phenomenon of food security in China–African collaboration through employing national biographical narrative as a framework of analysis. In the following sections it investigates China's external behaviour's foundation i.e., core political or socioeconomic justification, and the major features that define Africa's map in the global order. With this, the study briefly discusses

B. Gammachu (✉)
Peace and Security (Horn of Africa), Independent Research Consultant, Addis Ababa, Ethiopia

© The Author(s) 2024
M. Muchie et al. (eds.), *China-Africa Science, Technology and Innovation Collaboration*, https://doi.org/10.1007/978-981-97-4576-0_10

177

the two most prominent international relation theories—realism and liberalism, and characterizes the factors in framing national biographical narratives. Then it outlines policy recommendations and draws conclusion.

Why is it important to frame realism and liberalism as principal ideologue informing China–Africa relation? Presumably, China–Africa accord was mainly shaped by the deterministic international political perspectives—realism and liberalism, as in the base of conceptualization of state. The first as notably advanced by seminal works of Mearsheimer (1990) and Waltz (2014) that principally overlaps over the conception of the notion of a state as single legitimate political entity which is always engaged in power struggle, assumes global system as anarchic in nature, and 'remains power-seeking behaviour'. Realism mainly adopts and puts great emphasizes on character of the state, national interest, and power in international system. On the other hand, Liberalism as the leading theory of international relations accentuates harmonious relationship among states. Unlike realism, liberalism does not assume state as a single unitary actor, it rather acknowledges the significance of non-state actors in the realm beyond politics to economic and social life. It is the mainstay approach of our global order today i.e., the UN system which is equally known as the liberal peace order. According to the works of Kant (1795), Keohane (1984) and Paine (1776), liberal peace prescribes that states embrace liberal institutions and promote supposedly mutual cooperation, peaceful co-existence, and overcome war through liberal interstate diplomatic engagements. Transboundary trade and commerce, economic cooperation, and the exchange of liberal values are imperative to collective peace. Interdependence is far-reaching, and 'democracies won't make war' with one another (Keohane, 1984). Nevertheless, the global system often subjects the weak states to fit into the policy and development prescriptions through the Bretton Woods institution, World Trade Organization, and other global obligations (Jomo & Von Arnim, 2012).

In this regard, states in the African continent did not stand out from the crowd. However, China's presence in Africa in areas of agricultural investment, infrastructure development, and the mining sector is sound to provide an alternative development policy initiative partner with a relatively less stringent and more flexible engagement model. Nevertheless, more fine-tune needed into reimagining the constituting element to enable the African states to sustainably manage emerging food security challenges. Food security is defined as "when all people, at all times, have physical, social, and economic access to sufficient, safe, and nutritious food that meets their dietary needs and food preferences for an active and healthy life" (FAO, 2011).

According to AGRA (2017) and FAO (2021), food security is one of emerging challenges in domestic political agenda in African context and the same time another national frontier in areas of agricultural sector cooperation with China. The report revealed that 'the world has not been generally progressing either towards ensuring access to safe, nutritious and sufficient food for all people all year round or to eradicating all forms of malnutrition'. The report further indicated the number of people in the world affected by hunger continuously increased in 2020. It was virtually remaining unchanged from 2014 to 2019, and the prevalence of undernourishment climbed up from to around 9.9 percent in 2020, from 8.4 percent a year earlier

(FAO, 2021). The situation is much worse in developing countries and alarming in Sub-Saharan Africa.

Relevant literature shows China's presence as a legitimate entity in the African continent is principally guided by its state biographical narrative perceptions. The paper argues the China–Africa collaboration needs to explore broader and deeper factors beyond confinement to mirroring the global North's intervention in the continent and adopt multilayer state and non-state actor imperatives.

10.2   Methodology

This paper followed qualitative research design, a qualitative case study based on both primary and secondary data sources. Textual analysis of government policy papers, statements, plans, constitution, academic papers, books, speech acts, and media reports were used. Creswell (2018) cogently explained that research methods are the strategies, processes or techniques utilized in the collection of data or evidence for analysis in order to uncover new information or create a better understanding of a topic. Further, arguing the relevance of this design, Denscombe (2007) asserted that qualitative research design is preferred over quantitative design because it primarily strives to understand the meaning and experience of people. With this note, the paper employed qualitative discourse analysis in order to make an in-depth investigation of the topic. Qualitative case studies are a design of a scientific inquiry in use in many fields, especially evaluation, in which the researcher develops an in-depth analysis of a case, often a program, event, activity, process, phenomenon, or one or more individuals (Creswell, 2018).

The subject of this investigation is too broad. However, thematically the paper focuses on contextualizing China–Africa collaboration in food security from national biographical narrative perspective. The national biographical narrative is a framework of interpreting or seeking meaning to a given phenomenon from a state narrative perspective (Berenskoetter, 2014). The concept of "national biographical" is deliberately used to refer to the self-image and continuity of a state, a political unit in an international relation.

The study is essentially interested in characterizing the data than the quantities in effort to imply relatively the reasonable research puzzle. Nevertheless, the examination experienced data collection limitation and empirical evidence as there is dearth of literature on the topic. The data source constraint was both in terms of coverage and depth. However, this study contributes certain insights on why national biographical narrative is an important lens to better understand the China–Africa collaboration.

10.3 China–Africa Collaboration

The early 20th-century democratization endeavours and the contemporary political history of China especially the struggle led by Mao Zedong against imperialism, colonialism, and aspiration towards institutionalizing socialism relate it with much of the African contemporary state-building process. The African counterpart mostly some of the student-led movements during the Cold War period against European colonialism and the Imperial regime were parallel and aligned (Jakobson, 2009; Stein & Uddhammar, 2021; Taylor, 1998).

According to Taylor (1998) and Jakobson (2009) who wrote on China's foreign policy towards Africa, the China–Africa's multifaceted interaction is ever expanding—but still with predictable challenges—is encouragingly getting deep-rooted as many African states seem more comfortable with the relatively flexible Chinese approach—astoundingly less stringent and indifferent about much of African internal affairs. The constitutional provisions relating to a framework of its foreign engagement with developing countries including the African continent signpost that China took global mutual collaboration seriously. Expounding on China's self-definition and contrast as being-in-the-world and its continuity—framing national biographical narrative and how it operates in international system the preamble of the Constitution of the People's Republic of China (1982, p. 3) clearly states:

> "The future of China is closely bound up with the future of the world. China pursues an independent foreign policy, observes the five principles of mutual respect for sovereignty and territorial integrity, mutual nonaggression, mutual noninterference in internal affairs, equality and mutual benefit, and peaceful coexistence, keeps to a path of peaceful development, follows a mutually beneficial strategy of opening up, works to develop diplomatic relations and economic and cultural exchanges with other countries, and promotes the building of a human community with a shared future. China consistently opposes imperialism, hegemonic and colonialism, works to strengthen its solidarity with the people of all other countries, supports oppressed peoples and other developing countries in their just struggles to win and safeguard their independence and develop their economies, and strives to safeguard world peace and promote the cause of human progress".

Quite arguably, at the beginning of the end of the Cold War, China seems to embrace a substantial policy shift towards developing countries including the African conti-nents. The policy shift is essentially to accommodate broader economic relations, especially with the global south (Jakobson, 2009; Muchie & Patra, 2019; Stein & Uddhammar, 2021). During the periods from 1979 to 2008 China made unprece-dented steadfast economic progress that wasn't the case elsewhere in other developing countries (Todaro & Smith, 2011). The China–Africa collaboration takes different political, economic, and environmental forms—science, innovation, knowledge, and technology transfer, infrastructure development, foreign aid, agricultural investment etc. Todaro and Smith (2011) elaborated on various factors attributed to China's recent incredible leap in economic development. The economic relation is increas-ingly expanding as the leadership in China embrace the engagement to achieve the country's development goals. Relatively, within a few decades of commitment, according to Stein and Uddhammar (2021), China has emerged as Africa's biggest bilateral trading partner, Africa's biggest bilateral lender, as well as one of the biggest foreign investors in the continent. Chinese companies have entered almost all African markets. Now the African continent hosts over 1000 Chinese companies operating in African market which roughly attracted about one million Chinese citizens. In terms of trade, 'Chinese exports to Africa amounted to USD 113 billion in 2019, while imports from Africa reached USD 78 billion; the volumes have been steadily increasing for the past 16 years'.

China's agricultural investment in the African continent is another important indicator of the China–Africa collaboration. Among others, in recent times, food security is taking a significant stage as an important political agenda beyond the continent's boundary to influence China–Africa collaboration AGRA (2017). Assessing Progress towards the Millennium Development Goals Report (2013) indicated that the food security agenda—usually food stability, food availability, food accessibility, and consumption and use of food, goes to touch various wider factors contributing to food security governance. The food dynamic covers a broad range of factors that contribute to food security, from 'sufficient quantities and types of food to individuals' or households' incomes and sustained ability to purchase or produce food in sufficient quantities and types and to how it is stored, processed and consumed' (p. 101).

The International Institute of Tropical Agriculture (IITA) and China's Academy of Tropical Agricultural Sciences (CATAS) have begun a partnership to improve food security and create jobs for the growing youth population in Africa. The new partnership is part of China's efforts to deepen collaboration with Africa in the area of agricultural development with the cooperation of the International Institute of Tropical Agriculture (Ozor et al., 2013).

The following subsections of the paper quite briefly discuss realism vs liberalism, economic globalization, multiple biographical narratives, and food security as a factor in China–Africa collaboration in the context of China–Africa interaction. Sustainable China–Africa cooperation in areas of food security is better conceivable in the sense that it is considerate of the multiple actors and factors involved during this heightened globalization and global forces at play from the subunit perspective of the African continent.

10.3.1 Realism Versus Liberalism Dimension

Realism and liberalism—with of course certain defining departing points—but substantially remained the two most dominant features of international relation in operation (Grigsby, 2005; Waltz, 2014; Mearsheimer, 1990; Rousseau, 2017; Morgenthau, 1960). Any attempt to understand the dynamics of China–Africa collaboration in whatever area of engagements without looking into the underlying factors of these political ideologies arguably adds no value neither to the existing body of literature nor have no significant practice or policy implication.

Realism assumes the state as a territorial unit, which presents it as solid, clearly delineated, and closed-off ... the notion that defines the state as indivisible and autonomous entity was central to the Hobbesian conception of state projected to the Westphalian state (Morgenthau, 1960; Waltz, 2014). Realist further defines states as a legal entity that exercises sovereignty as 'supreme power over certain territory (Morgenthau, 1960, p.312). Liberalism is a term rooted in the Latin word liber, i.e., free. The classical works of seventeenth-century John Locke and the eighteenth-century literary works of Adam Smith are assumed a backstop of the theory of liberalism. Grigsby (2005, p. 93), in "Analyzing Politics", presented that liberalism over the century evolved from the classical—where an individual is assumed rational to make choices and more important than the state, promotes limited state intervention and economic freedom to economic inequality, to modern liberalism—where government intervention into individual and social life is justifiable, freedom is conceptualized as broad, expansive, and positive, and sceptical about economic inequality. Further, admitting interstate interdependence which is an essential element of social life and another significant dimension for the existence of states, 'liberals highlight political, commercial, and other democratically institutionalized linkages across state borders and their interaction with domestic structures' (Berenskoetter, 2014, p. 265).

With this note, these two prominent political ideologies discussed are at play in shaping the evolution of China–Africa interaction for over half a century now. In both dimensions of the realist and the liberal conception of state and its external behaviour they put no or little attention to important non-state actors and commonly oblige the national perspective or dominant narrative at unit, subsystem, or international system level. For instance, liberalists advance that democratic or republican government of the world inclined to be peaceful as their assumed democratic institution and tendency help them to negotiate the dispute amicably (Paine, 1776; Keohane, 1984). Kant did not totally reject the assumption of 'democratic peace'; however, he didn't submit to Paine's buoyancy that transition to democracy and peace would be quick and smooth. He rather envisioned a slow and rough transition towards democracy. At the centre of their prepositions, all people are assumed rationale, cooperative and transparent while engaging national interest. Critiquing the realist and liberal theories of international relation is not the concern of this paper, it only interested to imply the entities missed while these international political dimensions are the founding framework of operation. In international system, the power dynamics influence multiple important narratives at subunit and individual level to either be deliberately overlooked or sometimes considered value-empty (Ayoob, 1997; Bertrand, 2018; Waltz, 2014).

Therefore, we can argue, for sustainable and more impactful engagement on both sides the China–Africa collaboration in food security should take on a comprehensive and deeper conception of the underlying factors of the collaboration. The discourse of the relationship needs to be strategically reimagined from the subunit level in addition to unit-level in order to accommodate the emerging value-laden multiple realities in the African continent.

10.3.2 The Economic Globalization Dimension

One of the remarkable dimensions of globalization is economic trade, political and social being the remaining two; economic globalization is a key phenomenon that has been stimulating significant change in the world order often involving free international trade, investment, financial, capital, labour, science and technology, and specialization. It is mainly concerned with how integrated countries are in the global economy (Noman et al., 2012; Todaro & Smith, 2011; Begg et al., 2011). In the relations between countries in the current interdependent global system, it is understood that the economic interests of a country can hardly be pursued in isolation from other countries and economic entities (Ibonye, 2017). This global phenomenon has essentially affected China–Africa's collaboration in terms of its approach and content.

With certain alterations to state sovereignty, economic globalization brings with it free mobility of goods and services, and narrows the distance among borders, and motivates international commerce and trade. Nevertheless, because of several factors, developing countries including the African states are experiencing stagnation in the face of capacity and capability limitations to adjust to and manage the conditions and consequences of globalization (Todaro & Smith, 2011; Stein & Uddhammar, 2021; Noman et al., 2012). The process stimulated substantial policy shifts, especially among the global south.

In the last three decades, drastic change in Africa's development policy landscape is being observed. The continent began to practically experience globalization—international trade, economic liberalization, and privatization in the early 1980s. According to Jomo and Von Arnim (2012), however, liberalization measures meant to attract private investment have replaced state intervention and public investment. The experience is assumed to have benefited Africa from opening up to global economic forces, turned different, and resulted into substantial adverse consequences. 'Increase in capital flight limiting availability of financial resource for productive investments, premature foreign trade liberalization which undercut economic development, attracted Foreign Direct Investment (FDI) that is principally narrowed to mineral resources—mineral extraction, and China's significant raw material demand impaired African diversity production and export endeavour' (Jomo & Von Arnim, 2012, p. 499).

Further, Africa—especially due to external pressure and global trade conditionals, is at comparative disadvantage with agricultural trade. Subsequent change in Africa's export indicates no significant increase in activities in which African countries ostensibly had comparative advantage. 'High growth in China is likely linked most to increase in primary commodity price especially minerals then inducing strong response from Africa' (Jomo & Von Arnim, 2012, p. 513). The implications for sustainable development and food security have become apparent, and as food price begun to rise sharply it stimulated renewed attention to the food security issue. Therefore, the continent has experienced an impulsive and unfavourable trade environment for much of the post-independence period (pp. 510–511).

Jomo and van Arnim (2012, pp. 525–526) suggested, against external harmful and erroneous policy advice and conditionality, developing countries including states in Africa are encouraged to increase policy space for government to be able to pursue localised and manageable policies for meaningful development. External actors can mutually intervene in financial resources for the development. On the other side, Stein and Uddhammar (2021) discussed, Chinese foreign direct investment (FDI) in the African continent tends to have longer time space compared to the FDI from the West Countries—USA, Europe, Japan, or Canada. The durability is attributed to the fact that unlike the case with the Chinese which are usually state regulated, these are private, independent, and profit-maximizing companies. China–Africa collaboration is operational in the unfolding features of economic globalization. While welcoming the millennium, China seemed to be paying great attention to economic cooperation with the African states. In October 2000, China held its first high-level engagement known as The Forum on China–Africa Cooperation (FOCAC) in Beijing. The forum aimed at the promotion of political cooperation and creating a favourable condition for China–Africa business and trade. On the same meeting China committed to furnish special fund to support Chinese enterprise invest in African countries, send extra technical aid (medical), and initiate student exchange (Stein & Uddhammer, 2021).

Data sources reviewed clearly indicate, the global economic phenomenon has essentially affected China–Africa's collaboration in terms of its approach and content. While China made a steady economic growth (around 9% constant annual growth) since early 1980s for about three decades, the African continent has experienced impulsive and unfavourable trade environment for much of the post-independence period. Some scholars have argued that Chinese investments in Africa have had an adverse impact on local businesses and communities (Adisu et al., 2010; Marafa et al., 2019). Therefore, the ever-expanding China–Africa cooperation in many ways ought to draw the necessary attention to maintaining a win-win balance.

Globalization commonly tends to cost the weak actor. It significantly undermines local values, everyday economic activities, skills, and knowledge, and more importantly less considerate to multiple narratives at unit, subunit, and individual level. In comparative terms, the study therefore explores how the collaboration between China and Africa can make a real difference.

10.3.3 The Multiple Biographical Narrative Dimension

In the study of international relations, the contrasting frame or image of the self of being-in-the-world and the continuity is fundamentally determined by one's biographical narrative which takes, in this case, the state narrative, the government / regime narrative, or the political entity/party narrative. In the anarchic nature of the global system, the biographical narrative of a state actor or non-state actor at different level induces a security response that then constitutes its ontological security (Berenskoetter, 2014; Ringmar, 1996; Steel, 2005; Subotic, 2016).

The notion of ontological security is intricately explained by British sociologist Anthony Giddens in his book The Constitution of Society (1984) and reported by Mitzen (2006) is an important framework of analysis in discussing how one can explore the meaning of the self. For Rumelili (2015), ontological security is an idea that has been developed with the individual in mind. Further relevant scholarly work defines, the conception of ontological security as the 'confidence that most human beings have in the continuity of their self-identity and in the surrounding social and material environments of action'—an identification that is emotive, not always an overt cognitive experience.

Further, the concept of national security is ubiquitous because it lies at the heart of the condition of statehood and the landscape of the international system. In Buzan (1991) work, People, States and Fear; The National Security Problem in International Relations, the conception of national security has many referent objects, and that it should be engaged at the international system, the subsystem, and state level. The referent object—the core value that usually states want to protect, can be related to how and what constitutes the state as an individual self.

In his seminal work, Buzan (1991) outlines three principal constituent elements of the state, that is, its 'physical basis in territory and population; the idea of the state held in common by its population; and the institutional expression legitimized by the idea'. The crux of his arguments is the notion that states face security threats on all levels. Simply put, 'The idea of the state, its institutions, and even its territory can all be threatened as much by the manipulation of ideas as by the wielding of military power…', and the belief that it is a mistake to treat all states alike in terms of security problems (Johnston, 1991, pp. 150–153).

The biographical narrative is dynamic, and it is essentially about the meaning of power relations. Entities on hold of political, economic or social capital greatly influence the content and perspective of biographical narrative. Berenskoetter (2014), noted that with competing accounts of the past, present, and future a group of people commonly never experience a homogenous way of life and always hold potential for alternative. At the same time, there is demand not to let the multiple consideration reach a stage for deliberation in order to unsettle a stable sense of being in the world.

According to investigations of International Strategic Relations and China's National Security assessment on Africa's security situation the outbreak of pandemics and the pervasive act of terrorism are highlighted as serious non-traditional security dynamics and security governance challenges. Africa's contemporary security situations continuously attracted the keen attention of the international community (Xu et al., 2016, p. 263).

The analysis indicates the dynamics of biographical narrative substantially impacts the China–Africa food security collaboration. Much of the existing cooperation's operates within the liberal institutions and power-seeking dynamics of the global system. Hendriks (2005), Lotter (2014) discussed, food security much as it is a national agenda, it is equally an everyday reality at subunit and individual level.

10.3.4 Food Security as a Factor of China–Africa Collaboration

The earliest known connection between China and Africa dates to the fifteenth century (Stein & Uddhammar, 2021) and this relationship has evolved from a political to an economic focus, and in recent decades it has been further revived according to commerce, economic trade and agriculture. Why food security become an important agenda item in the relationship between China and the African continent is discussed in the following sections.

Food security literatures argue that the conception and practice of food security in the African continent is an emerging international relations element particularly resonating in the China–Africa collaboration spectrum (Chen et al., 2016; Grimm, 2014; Ibonye, 2017; Marafa et al., 2019; Mugwagwa et al., 2010). Unpacking the dynamics of food security as an agenda item in the China–Africa cooperation might take further deep and comprehensive investigation.

Marafa et al. (2019) in a jointly authored article titled, Upscaling Agriculture and Food Security in Africa in Pursuit of the SDGs, stated that China and Africa constitute more than a third of the world's population and China has become Africa's largest trading partner. Agriculture plays an important role in the Chinese economy and the African continent, contributing approximately 9% and 17% to GDP in 2015 respectively (World Bank, 2016). While China has successfully lifted hundreds of millions of farmers out of rural poverty to improve food security and people's livelihoods, many African countries are still grappling with food security challenges. The food security challenge in the African continent can be attributed to a number of factors— availability of proper market, agricultural inputs, leadership, climate change, socioeconomic and political landscape, knowledge and technology, etc. (Chen et al., 2016; Marafa et al., 2019; Mugwagwa et al., 2010).

Haggblade and Hazell (2010) argue that since the mid-twentieth century, agricultural performance in Africa has lagged other developing regions. It is evident that most agricultural productivity indicators show little improvement in case of the African continent. For example, while agricultural value added per worker nearly doubled for all countries in the world, from $1149 in 1990 to $2206 in 2015, that of the Sub-Saharan Africa increased from USD 774 in 1990 to USD 1240 in 2015 (World Bank, 2016). In China, the change was dramatic, going from $561 in 1990 to $1465 in 2015. The effect of those differentials in productiveness on food security is evident (Marafa et al., 2019). The improvement of the agricultural productivity in the case of China is arguably linked to refinement of policy towards agriculture and rural development, appropriate leadership, market, and advancement in agricultural input among others (Grimm, 2014; Adisu et al., 2010).

Yet agriculture is a major component of Africa's growth and development, as well as the livelihoods of its rural and urban populations. Although its contribution varies considerably, from 2% of value added to GDP in countries such as South Africa and Botswana to 60% in Sierra Leone, agriculture provides an average of 17% of value added to GDP for the continent (World Bank, 2016) and more, if further processing is considered. The sector compromises about 40% of the continent's export and adds 55% of job opportunities to the economy (Marafa et al., 2019). While China provides food for 20% of the world's population on 9% of the arable land, Africa cannot provide enough food for 15% of the world's population on a quarter of the arable land of the world. The success that China has achieved on agriculture and rural development is significant to Africa's food security that is on board in the equation of the China–Africa collaboration (Marafa et al., 2019).

Reviewed literature indicates, China is now becoming a major international investor in Africa, especially in the agricultural sector (Chen et al., 2016; Ibonye, 2017; Marafa et al., 2019; Mugwagwa et al., 2010). According to Marafa et al. (2019), China's attention to Africa's agriculture has begun in 1959 during which China supplied food aid to Guinea. The collaboration, overtime, went on from the generous food aid with less political implication to extensive state-owned farms involving agro-processing, irrigation, and agricultural technology transfers motioning global interest.

In China, the number of people living below the extreme poverty line has fallen from 67% in 1990 to 11% in 2010, with a rate reaching 5% in the final 2015 Millennium Development Goal Report (World Bank, 2016). On the other hand, the poverty rate in Sub-Saharan Africa has not changed since 1990 all the way through 2002, and more than 40% of the population is reported to be below the extreme poverty line in 2015 (Marafa et al., 2019). Development success in China has been achieved in part through large investments in agriculture and rural development and through the adoption of the Chinese Agricultural Model. The growing engagement between China and Africa over the past two decades could provide an opportunity for African countries to transfer aspects of the China model to improve agricultural productivity and subsequently food security. Since this commitment is not unilateral, ways will have to be found to align this relationship for the mutual benefit and to minimize the challenges and risks of exploitation (Chen et al., 2016; Marafa et al., 2019; Mugwagwa et al., 2010).

As China continues to increase agricultural investment in the African continent, it can also provide a market for African agricultural products. China is currently the second largest food consumer in the world. While it is currently meeting domestic demand, it has considered long-term strategies, as it continues to face dwindling local resources of arable land and irrigation water. Thus, Africa, with its abundant unused arable land, water and energy resources, could be part of China's long-term plan to meet its food security needs, if policies and agricultural action are introduced to facilitate production for both local consumption and export (Marafa et al., 2019; Mugwagwa et al., 2010).

Critical analysis into relevant documents do imply food security is an important factor that seeks proper attention from both parties—China and the African continent to further enhance the collaboration for truly mutual benefit and sustainability. However, China should be very careful while engaging the African contents on areas of food security as there is a serious state-building contestation (Ayoob, 1997; Bertrand, 2018) and concern for policy bias towards the rural people, quite comprehensively the "bottom billion" borrowing from Paul Collier (2007) in the region.

10.4 Conclusion and Policy Recommendations

In conclusion, the study noted many pieces of the literature on China–Africa inter-action within the context of global system essentially discuss the perspectives of China in Africa, not significantly the other way. Chinese presence in Africa is ever-increasing, and it has emerged from political focus to trade, agricultural investment, mining, and technology transfer. The African states seem to be more keen for collaboration with China because of its flexible and less ideological engagement, which is true to ensuring food security.

In Africa, economic globalization appears to have constrained the scope and ability for government policy intervention. It is as well apparent in international economic relations but also true to domestic policy environment where the Bretton Woods Institutions [World Bank, and IMF, the World Trade Organization (WTO)] and other international obligations have markedly transformed the space for national economic development policy initiatives. As such, the qualitative policy spaces and process in the African contexts are often for reasons of loopholes of globalization and internal power dynamics among others are not cognizant of the multiple realities at subunit and individual levels. In a similar manner, China's foreign policy and engagement towards collaboration with the African continent follow the state narrative orientation. However, China's relatively flexible engagement with the states in Africa is not a potential to be undermined.

The discussion on China–Africa relations includes the Chinese motivation, official justification and politico-economic drivers for interactions, as well as the practice and developmental impact of cooperation in African states and societies. All of these factors shape the perception and expectations of African actors. The relationship between Chinese perspective and African expectations is an interesting aspect in itself that merits a closer look in further research.

10.4.1 Policy Recommendations

China's multisector engagement including food security investment in Africa is informed by its national biographical narrative—it's a single legitimate state representing its core values that squarely lay at the centre of its external behaviour. Nevertheless, Africa is a region with multinational biographical narratives. Therefore, China's involvement in Africa needs to refine its foreign policy and reimagine its much-needed intervention in terms of Africa's multiple realities including the significant space of the non-state actors.

Africa's competing partners compared to China commonly move beyond passive non-interference principles and are involved in political, trade, and economic international relations to induce a strong response that is favourable to the global socioeconomic and political actors. China's sustainable interaction with African states depends on its propensity to mutually intervene in Africa's desired policy space to ensure its effort to adopt and implement sustainable development policy initiatives.

China's policy makers should be consciously sensitive while engaging the African countries in the area of food security as there is a serious state-building contestation and concern for policy bias towards the rural people, the "bottom billion".

References

Adisu, K., Sharkey, T., & Okoroafo, S. C. (2010). The impact of Chinese investment in Africa. *International Journal of Business and Management, 5*(9), 3.

AGRA. (2017). *Africa agriculture status report: The business of smallholder agriculture in sub-Saharan Africa* (Issue 5). Nairobi, Kenya: Alliance for a Green Revolution in Africa (AGRA).

Ayoob, M. (1997). Defining security: A subaltern realist perspective. In K. Krause & M. C. Williams (Eds.), *Critical security studies: Concepts and cases*. Minnesota Press.

Begg, D., Vernasca, G., Fischer, S., & Dornbusch, R. (2011). *Economics* (10th ed.). McGraw Hill.

Berenskoetter, F. (2014). Parameters of a national biography. *European Journal of International Relations, 20*(1), 262–288.

Bertrand, S. (2018). Can the subaltern securitize? Postcolonial perspectives on securitization theory and its critics. *European Journal of International Security, 3*(3), 281–299.

Buzan, B. (1991). *People, states and fear: An Agenda for security analysis in the post-cold War Era*. Weatsheaf.

Chen, W., Dollar, D., & Tang, H. (2016). Why is China investing in Africa? Evidence from the firm level. *The World Bank Economic Review*. https://doi.org/10.1093/wber/lhw049

Creswell, J. W., & Creswell, J. D. (2018). Mixed methods procedures. *Research Defign: Qualitative, Quantitative, and Mixed Methods Approaches*.

FAO. (2011). The State of Food Insecurity in the World 2011. https://www.fao.org/4/i2330e/i2330e.pdf

FAO. (2021). The State of Food and Agriculture 2021. Making agrifood systems more resilient to shocks and stresses. https://doi.org/10.4060/cb4476en

Grigsby, E. (2005). *Analyzing politics: An introduction to political science* (3rd ed.). CA.

Grimm, S. (2014). China–Africa cooperation: Promises, practice and prospects. *Journal of Contemporary China, 23*(90), 993–1011. https://doi.org/10.1080/10670564.2014.898886

Haggblade, S., Hazell, P., & Reardon, T. (2010). The rural non-farm economy: Prospects for growth and poverty reduction. *World development, 38*(10), 1429–1441.

Hendriks, S. L. (2005). The challenges facing empirical estimation of household food (in) security in South Africa. *Development Southern Africa, 22*(1), 103–123. https://doi.org/10.1080/037683 50500044651

Ibonye, V. (2017). China–Africa cooperation: Struggling commodities and the silver-lining in the innovation economy. *International Area Studies Review, 20*(2), 160–178. https://doi.org/10.1177/2233865916688845

Jakobson, L. (2009). China's diplomacy toward Africa: Drivers and constraints. *International Relations of the Asia-Pacific, 9*(3), 403–433. https://doi.org/10.1093/irap/lcp008

Johnston, R. J. (1991). A question of place: Exploring the practice of human geography. In *A question of place: exploring the practice of human geography*. Blackwell.

Jomo, K., & Von Arnim, R. (2012). Economic liberalization and constraints to development in sub-Saharan Africa. In A. Noman, K. Botchwey, H. Stein, & J. Stiglitz (Eds.), *Good growth and governance in Africa: Rethinking development strategies* (pp. 499–527). Oxford University Press.

Kant, I. (1795). Toward perpetual peace. In *Theories of federalism: A reader,* (pp. 87–99). New York: Palgrave Macmillan US.

Keohane, R. O. (1984). *After hegemony.* Princeton University Press.

Lotter, D. (2014). Facing food insecurity in Africa: Why, after 30 years of work in organic agriculture, I am promoting the use of synthetic fertilizers and herbicides in small-scale staple crop production. *Agriculture and Human Values, 32*(1), 111–118. https://doi.org/10.1007/s10460-014-9547-x

Marafa, L., May, J., & Tenebe, V. A. (2019). Upscaling agriculture and food security in Africa in Pursuit of the SDGs: What role does China play? *Africa and the Sustainable Development Goals.* https://doi.org/10.1007/978-3-030-14857-7_16

Mearsheimer, J. (1990). Back to the future: Instability in Europe after the Cold War. *International Security, 15*(4), 5–56.

Mitzen, J. (2006). Ontological security in world politics: State identity and the security dilemma. *European Journal of International Relations, 12*(3), 341–370.

Morgenthau, H. (1960). *Politics among nations.* Alfred Knopf.

Muchie, M., & Patra, S. K. (2019). China–Africa science and technology collaboration: Evidence from collaborative research papers and patents. *Journal of Chinese Economic and Business Studies.* https://doi.org/10.1080/14765284.2019.1647004

Mugwagwa, J., Wamae, W., & Outram, S. M. (2010). Agricultural innovation and food security in Sub-Saharan Africa: Tracing connections and missing links. *Journal of International Development, 22*(3), 283–288. https://doi.org/10.1002/jid.1688

Ozor, N., Umunnakwe, P. C., & Acheampong, E. (2013). Challenges of food security in Africa and the way forward. *Development, 56*(3), 404–411. https://doi.org/10.1057/dev.2014.10

Paine, T. (1776). Common sense: 1776. Infomotions, Incorporated.

Ringmar, E. (1996). On the ontological status of the state. *European Journal of International Relations, 2*(4), 439–466. https://doi.org/10.1177/1354066196002004002

Rumelili, B. (2015). Identity and desecuritisation: the pitfalls of conflating ontological and physical security. *Journal of international relations and development, 18,* 52–74.

Steele, B. J. (2005). Ontological security and the power of self-identity: British neutrality and the American civil war. *Review of International Studies, 31*(3), 519–540.

Stein, P., Uddhammar, E. (2021). China in Africa: The role of trade, investments, and loans amidst shifting geopolitical ambitions. *ORF occasional paper no. 327,* August 2021, Observer Research Foundation. India: New Delhi. ORF_OccasionalPaper_327_China-Africa.pdf (orfonline.org).

Subotic, J. (2016). Narrative, ontological security, and foreign policy change. *Foreign Policy Analysis, 12,* 610–627. https://doi.org/10.1111/fpa.12089

Taylor, I. (1998). China's foreign policy towards Africa in the 1990s. *The Journal of Modern African Studies, 36*(3), 443–460. https://doi.org/10.1017/s0022278x98002857

The Government and People of the Empire Ethiopia and The Government of the People's Republic of China, Joint Communique, 1970. 24 November 1970. Addis Ababa, Ethiopia. fmprc.gov.cn

The People's Republic of China, Constitution of the People's Republic of China, 1982. 4th December 1982. Beijing, China. The 1982 Constitution of the People's Republic of China (npc.gov.cn)

Todaro, P. M., & Smith, C. S. (2011). *Economic development* (11th ed.). Pearson.

Walker, C. T., & Rousseau, L. D. (2017). Liberalism: A theoretical and empirical assessment. In D. M. Cavelty & T. Balzacq (Eds.), *Routledge handbook of security studies* (pp. 52–64). Routledge.

Waltz, K. N. (2014). Realist thought and neorealist theory. In *The Realism Reader* (pp. 124–128). Routledge.

World Bank. 2016. *World Bank Annual Report 2016.* Washington, DC. https://doi.org/10.1596/978-1-4648-0852-4. License: Creative Commons Attribution–NonCommercial–NoDerivatives 3.0 IGO (CC BY-NC-ND 3.0 IGO).

Xu, X., Li, X., Qi, G., Tang, L., & Mukwereza, L. (2016). Science, technology, and the politics of knowledge: The case of China's agricultural technology demonstration centers in Africa. *World Development, 81*, 82–91. https://doi.org/10.1016/j.worlddev.2016.01.00

Chapter 11
Setting China–Africa Collaboration to Address Climate Change and Other Environmental Issues

Hilary I. Inyang, Thokozani Simelane, and Charles Hongoro

11.1 Introduction

China and Africa have targeted sustainable development to provide jobs and services to their teeming populations which continue to put pressures on the environment in both global regions. Industrialization has raised the standard of living of most of China's 1.4 billion people and that of the majority of Africans since the 1970s. According to Kan (2009), China's gross domestic product growth rate (GDP) averaged 10% during the period 1989–2009. However, infrastructure development has occurred with the opportunity cost of global climate change (with other regions as contributors as well) and environmental degradation. Environmental degradation in both China and Africa is not limited to the impacts of global climate change but manifests as pollution by industrial and domestic wastes, land degradation by unsustainable mining and agricultural practices, emissions of inadequately treated effluents, deforestation, surface and groundwater pollution, acid rain and a host of other derivatives of these environmental hazards, some of which can be exacerbated by climate change. There is serious pollution of land, water and air in China with air pollution being the most damaging category. As reported in the Proceedings of the United States National Academy of Sciences by Liang et al. (2020), between 1.5 million and 2.2 million people died prematurely from air pollution in China from 2000 to 2016. Air pollution also caused millions of deaths across Africa in 2019 according to the WHO Global Health Observatory. UNICEF (2019) reports that deaths due to outdoor air pollution in Africa increased from 164,000 in 1990 to 280,000 in 2017 with an annual GDP loss in excess of US$215 million.

The emission of greenhouse gases (carbon dioxide, methane, nitrous oxide, and fluoro-gases) is mostly responsible for observed increase in the average global surface

H. I. Inyang (✉) · T. Simelane · C. Hongoro
Africa Institute of South Africa (AISA), Human Sciences Research Council,
Pretoria, South Africa
e-mail: h.inyang26@gmail.com

© The Author(s) 2024
M. Muchie et al. (eds.), *China-Africa Science, Technology and Innovation
Collaboration*, https://doi.org/10.1007/978-981-97-4576-0_11

temperature to current levels that are roughly 1 °C on average above that of the pre-industrial (1850–1879) reference period. It should be emphasized that the 1 °C warming level is an average across the earth and is exceeded presently in some locations within the range of 1.5–2.0 °C. The warming stems from rapidly increasing concentrations of the greenhouse gases. They trap heat resulting in the warming of the earth and the atmosphere. The emitted greenhouse gases envelop the earth and allow passage of most of the incoming solar radiation to the earth's surface. Upon reflection of that radiation upward from the earth, the gases trap the infrared band of the radiation, resulting in heating of the gases and redirection of a fraction of the generated heat to the earth, hence global warming. The heat entrapment process by atmospheric processes has always occurred, even during the pre-industrial times but has intensified since the Industrial Revolution due to increased release of greenhouse gases into the atmosphere. In the IPCC (2014) report, it is indicated that carbon dioxide, methane and nitrous oxide have reached unprecedented concentrations in the atmosphere over the previous 800,000 years. Also, IPCC (2014) indicated that between 1750 and 2011, 2040 ± 310 GtCO$_2$ of carbon dioxide was cumulatively released into the atmosphere. About 40% of that quantity has remained in the atmosphere. The remainder has been removed through absorption by the oceans (with increased ocean acidification), absorption by soils and respiration by plants. As further indication of the increasing pace of climate change processes, IPCC (2014) indicated that about 50% of the CO$_2$ that was emitted between 1750 and 2011 was released within the 50-year period between 1936 and 2014.

More recently, the World Meteorological Organization (WMO) quoted by Babugura (2021), has determined that the highest average global temperatures on record occurred from 2014 to 2020. In 2020, this temperature level was 1.25 °C higher than the pre-industrial level. Although Africa has been in the same trajectory as China in terms of carbon-fuelled development, it contributes the least (4%) of the total carbon dioxide emissions regionally (Richie, 2018) to global GHG emissions. China has been developing control and management measures for associated environmental hazards and health risks at a faster pace. China and African countries have committed to sustainable development in their national socio-economic development plans and have signed international treaties, exemplified by the Paris Agreement in this regard. China has pledged to be carbon–neutral by 2060 while most African countries are developing plans to green their economies and address environmental pollution legacies. AfDB (2019) indicates that seven out of the ten countries that are most vulnerable to climate change are African countries. As in the case of water resources/security of Africa as discussed by UN Water (2000), Africa's relative disadvantage is driven by inadequate institutional and governance arrangements for domestic and transboundary challenges, inadequate deployment of policy and technical measures, and a large set of environmental stressors.

Herein, climate change mitigation is defined according to the perspectives of the African Development Bank (2018) as "reinforcement of the global response to the threat of climate change by keeping a global temperature rise well below 2 °C above pre-industrial levels, and thus pursuing efforts to limit the temperature increase even further to 1.5 °C". Adaptation is then defined by African Development Bank (2018) as "enhancement of adaptive capacity, strengthening of resilience and reduction of vulnerability to climate change with a view to contributing to sustainable development and ensuring an adequate adaptation response in the context of the temperature goal". The commonality of threats to the environment and life support systems in both China and Africa, especially in the post-COVID-19 era in which industrial activities are expected to accelerate, provides opportunities for collaboration between China and African countries en bloc or bilaterally on implementation of systems for environmental stewardship in general, and global climate change mitigation and adaptation in particular.

The instruments on which such collaborations can be formed, are regulatory development, technical support, regional monitoring systems, research and development, market incentives, enforcement, and knowledge diplomacy. In this analysis, approaches to using these mechanisms by both regions through collaboration, to address global climate change and environmental degradation are discussed. A critical advance in China–Africa technical cooperation as regards environmental management is the recent establishment of the China–Africa Environmental Cooperation Center in Nairobi, Kenya (Njeru, 2022) which is scheduled to commence operation in 2023 under the auspices of the United Nations Environmental Programme (UNEP), several African countries and Chinese agencies.

11.2 Scaling Green and Brown Environmental Challenges

The necessary infrastructure development and operation by both African countries and China derive from the very large populations that they have to support with social services. Environmental challenges that arise from operation of infrastructure can be referred to as "brown environmental problems". They include land pollution, water pollution and air pollution. Second and subsequent stage environmental problems that result from the aforementioned categories can be referred to as "green environmental problems". They include biodiversity loss, deforestation, and anomalous environmental radiation.

11.2.1 The Economic Development Imperatives of China and Africa

The need to support large populations in China and Africa to raise the standard of living (not to be confused with quality of life) is the driver of infrastructure development efforts of both regions to provide goods and services. Since the First Industrial Revolution, economic development has been carbon-fuelled with global climate change and direct pollution of the environment as the result. Using the examples of economic development circumstances of Asian countries depicted in Fig. 11.1, it is

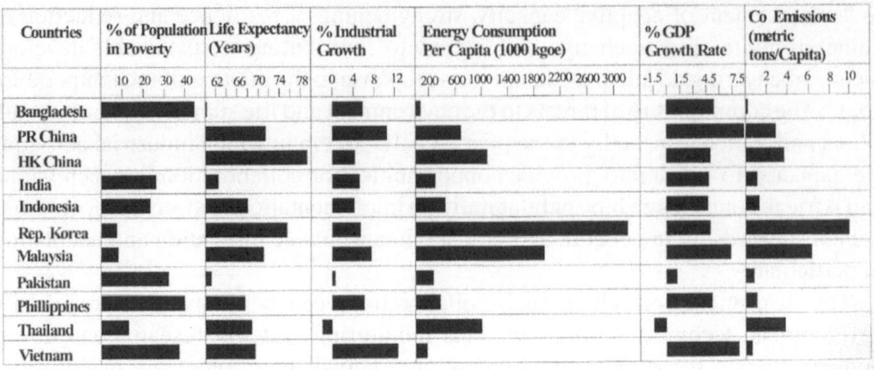

Countries	% of Population in Poverty	Life Expectancy (Years)	% Industrial Growth	Energy Consumption Per Capita (1000 kgoe)	% GDP Growth Rate	Co Emissions (metric tons/Capita)
	10 20 30 40	62 66 70 74 78	0 4 8 12	200 600 1000 1400 1800 2200 2600 3000	-1.5 1.5 4.5 7.5	2 4 6 8 10
Bangladesh						
PR China						
HK China						
India						
Indonesia						
Rep. Korea						
Malaysia						
Pakistan						
Phillippines						
Thailand						
Vietnam						

Fig. 11.1 Carbon emissions associated with economic development status of Asian countries using 2002 World Bank data

obvious that there has been positive correlation among carbon emissions, reduction in national levels of poverty and industrialization. IMF (2015) estimated that 18 million jobs per year need to be created in Sub-Saharan Africa up to 2035 to accommodate the multitude of youth who enter the labour market each year. China has a population of about 1.4 billion to cater for. Figure 11.2 is an illustration of the relationship among infrastructure development/operation, waste generation, environmental impacts, and human health impacts.

11.2.2 Climate Change and Its Cascading Impacts in China and Africa

It should be noted that not all anthropogenic environmental problems are attributable to global climate change. However, climate change is the phenomenon that has exhibited the largest scale and variability of environmental impacts. The impacts of global climate change also cascade and exacerbate other environmental stresses. Climate change impacts in China and Africa are illustrated in Fig. 11.3 and are briefly categorized and discussed below.

11.2.3 Ecological Impacts and Air Pollution

For the period 2009–2018, the partitioning of total carbon emissions to land, ocean and the atmosphere as determined by the Global Carbon Project (Friedlingstein et al., 2019), were 29%, 23% and 44%, respectively, the remainder being unaccountable. Along with many other categories of pollutants that are dumped into the environment in China and African countries, GHGs have significant ecological impacts. Kan

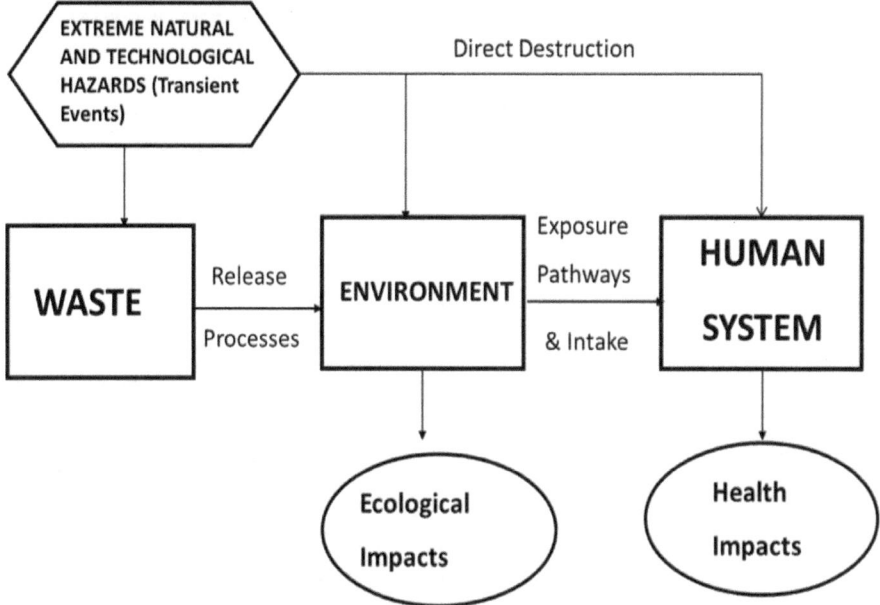

Fig. 11.2 Illustration of the relationship among infrastructure development/operation, waste generation, environmental impacts and human health impacts

(2009) notes that at its average GDP growth rate of 10% over the past two decades, air pollution sources in China have diversified from conventional coal combustion sources to a combination of coal combustion and motor vehicular emission sources. For millennia, China has practiced intensive agriculture with the result that a significant fraction of the country's 9.6 million square kilometres of land has been deforested. MEPC (2009) reported annual average concentration of respirable particles in China to be 89 $\mu g/m^3$ for PM_{10}, 48 $\mu g/m^3$ for sulphur dioxide, and 34 $\mu g/m^3$ for nitrogen oxides. These levels of emission of pollutants have been injurious to human health and the environment in China.

Ecological damage is observable throughout Africa as well. The most common large-scale ecological degradation that has been observed are loss of biodiversity, deforestation, desertification, loss of soil fertility, water resources pollution by erosion and sedimentation, and air pollution. Changes in hydrology and direct human activities combine to degrade ecological systems where there are inadequate management measures. The types of responses that can be expected as ecological damages due to stresses imposed by global climate change are summarized in Table 11.1. Ecological systems have complex interactions of stressors and reactions at various spatial scales. These interactions cascade and change with time and follow pathways that may not always be predictable. An example of the cascade of climate change impacts is initial desiccation of agricultural soil due to decrease in seasonal rainfall. Drought may occur and enable the generation and transport of huge quantities of dust

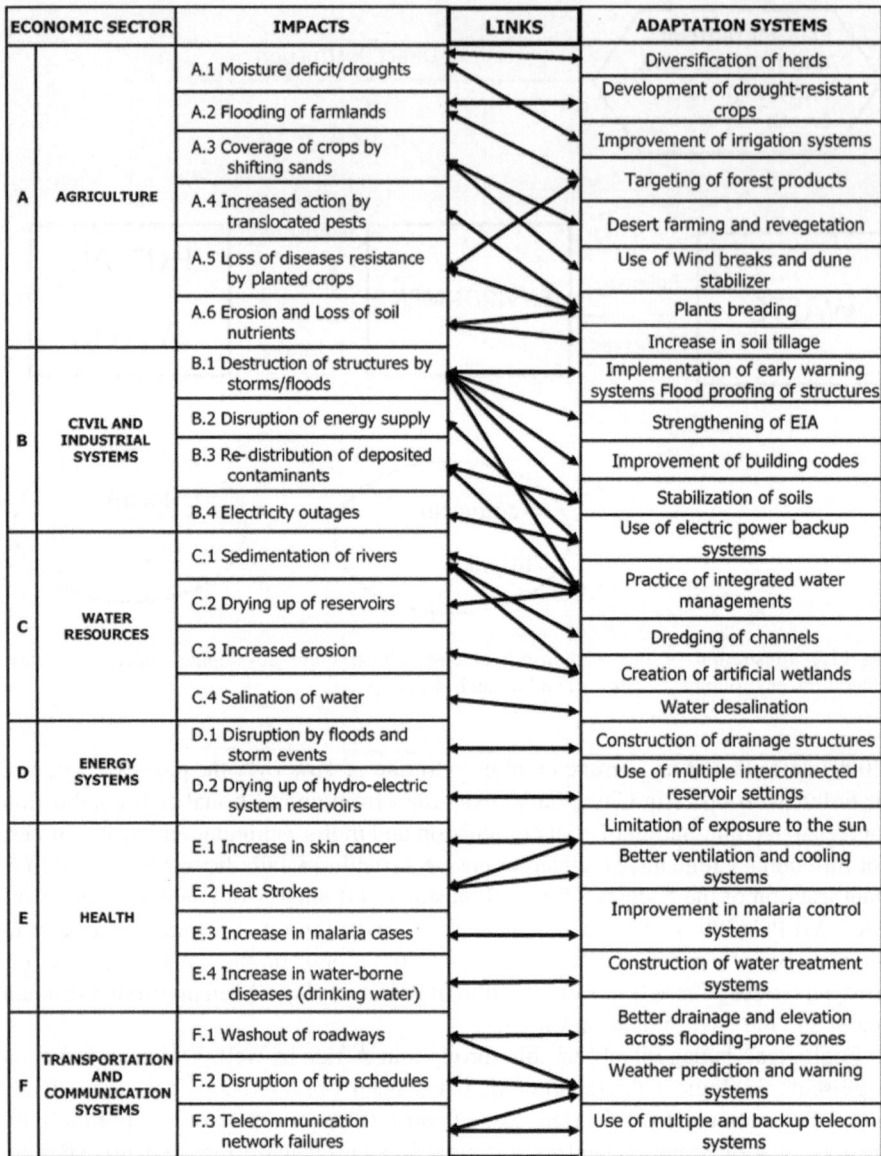

Fig. 11.3 Categorization of impacts of global climate change in China and Africa along with possible control measures

Table 11.1 Types of responses that can be expected as ecological impacts of global climate change in the environment

IMPACTS ON INDIVIDUAL ORGANISMS	Reduced growth and altered organ development
	Reduced reproduction and high mortality
IMPACTS ON FLORAL AND FAUNAL POPULATIONS	Reduced abundance and changes in spatial distribution
	Changed gene pool and age structure
IMPACTS ON ECOSYSTEM FUNCTIONS	Reduced organic decomposition
	Reduced ecosystem productivity
	Food web alterations/degradation
	Reduction in nutrient stocks
IMPACTS ON ECOSYSTEM STRUCTURE AND DYNAMICS	Switching of dominant species
	Community compositional changes
	Extinction of population
	Biodiversity loss

into ecological and urban systems. Many studies exemplified by Han et al. (2009), have assessed the emission of dust from road surfaces in various regions of China (PM_{10}).

These studies typically show that proximal ecological systems are negatively affected by the settling of dust on vegetation and water bodies as illustrated in Fig. 11.4. Dust settlement contributes to turbidity of water bodies and impairs photosynthesis upon covering plants. Expansion of desert areas in China's west as well as the Sahara and Kalahari deserts in northern/western regions, and southern region of Africa, respectively, is a contributor to the generation and travel of large quantities of aeolian dust. Lallanila (2013) observed that about 2.6 million km^2 of land in China was under desertification before 2013, and that Chinese deserts were expanding at the rate of 2469 km^2 per year. Ecological damage through the impacts of global climate change in China and Africa has negative implications on food security in both regions.

11.2.4 Hydrological Impacts

Indeed, the most commonly discussed impacts of global climate change are those that pertain to global, regional and local hydrological factors. The central meteorological parameters of air temperature, atmospheric humidity and wind circulation patterns are very significant with respect to the impacts of climate change on hydrological systems of a geographical region. Of course, terrain parameters such as elevation, slopes, geomedia type and cultural practices also play roles in vulnerabilities and resilience. Water security under climate change and other stresses is of great concern

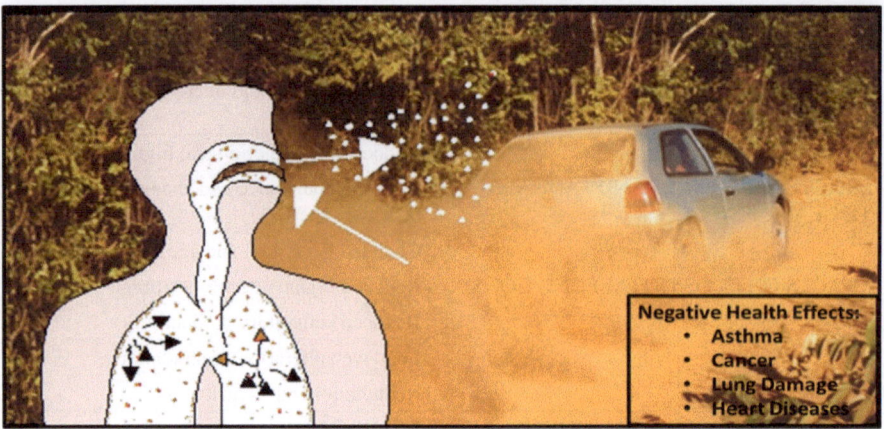

Fig. 11.4 An illustration of dust generation and its impacts on proximal ecological systems and human health

in China as well as in African countries. Water shortages as well as quality decay are significant challenges in both regions but there are great variabilities in water availability and seasonality across large expanses of both regions. These variabilities have been exacerbated by climate change. Droughts are common in Northern and Western China, the Horn of Africa, Sahelian and Southern Africa. Several parts of China and Sub-Saharan Africa have experienced devastating floods, cyclones and storms. Life support systems on the flanks of Chinese Yangtze and Yellow Rivers as well as those of African large rivers such as the Nile, Niger, Congo, and Zambezi are frequently threatened by floods that are increasing in frequency and severity due to global climate change (He et al., 2013).

The Tibetan glaciers that supply water to most of China's large rivers are melting and reducing in stock, with the possibility of exhaustion within the next few decades. Fallow land with increased precipitation has intensified gully erosion (see Fig. 11.5) and sheet erosion of terrain including agricultural land in both China and Africa, resulting in land degradation as well as excessive turbidity of surface water resources. Excessive evaporation due to higher temperatures has concentrated turbidity and other contaminants in lakes, streams, and rivers, rendering them unsuitable for use in industrial and domestic operations without treatment. In this regard, it should be noted that in spite of the coverage of the earth's surface by water more than terrain, only about 2.5% of surface water on earth (Oki & Kanae, 2006) is fresh water. While water scarcity is a problem in some regions of China and Africa, seasonality of available water or constraining of annual rainfall into a few weeks each year is increasingly causing damaging floods in the two regions. Coastal and river plains in both areas are particularly vulnerable to seasonal floods.

Fig. 11.5 Gully erosion sites in Uyo, Nigeria following many seasons of above-normal rainfall recorded by the first author

11.2.5   Occupational and Agricultural Impacts

China and Africa have population levels of 1.4 billion and about 1.0 billion people respectively. This places significant demand on food security in both regions. Across both regions, rice, wheat and other cereals, as well as tubers are staple food crops. However, climate change impacts on their cultivation through such factors as land fertility, erosion parameters and their physiographic adaptation to changing temperature and rainfall patterns. These factors translate to the livelihood and occupational engagements of many people in China and Africa which are still agrarian with respect to the engagement of residents. Changing temperature, rainfall and humidity patterns have displaced pests, leached soil nutrients, disrupted food chains and inundated farms with negative overall impacts in some communities. He et al. (2013) noted that economic losses from the 2009 drought in China reached about 150.9 billion RMB. In their studies, they found a Southeast-Northwest aligning pattern of agricultural drought risk in China. The highest risk areas are the eastern part of the Northeast Plain, Central part of Inner Mongolia, the Loess Plateau, North Xinjiang, north–south Yangtze Plain and the Yunnan-Guizhou Plateau. Babugura (2021) reports that in Southern Africa, the 2018–2019 drought reduced crop yields; caused loss of livestock and raised food prices. Unfortunately, that drought was preceded by heavy rains.

Thus, the two kinds of climate change-enhanced phenomena caused environmental damages, low agricultural productivity, and widespread hunger in the region.

11.2.6 Health Impacts

There are very profound interlinkages among climate, the environment and human health. However, disentanglement of cause-and-effect relationships between environmental changes induced by climate change and their impacts on human health is a very complex undertaking. This is the problem of attribution which can be addressed by qualitative and quantitative models that describe exposures and dose–response relationships. Often, the impacts are displaced in time away from the exposures such that many other extraneous factors are opportune to intercede, making their decoupling to verify the true effects of exposures very difficult. Extreme weather events such as environmental heat, flooding, and dust emission from dried out terrain have already endangered human health in China and Africa through direct stresses and cascading and interacting impacts. Extreme weather events as well as progressive change in the magnitudes of climate change parameters spatio-temporally, have caused redistribution of microbes and disease vectors. For the period from June 1997 to May 1998, Epstein (1999) plotted the locations of outbreaks of diseases such as cholera, encephalitis, malaria, hantavirus pulmonary syndrome, Rift valley fever, dengue fever and other respiratory diseases globally, following extreme weather events recorded in China and African countries.

Greater opportunities for drainage of waste constituents by flood water from contaminated sites and polluted facilities have enhanced the frequencies of ailments such as cholera, typhoid, and dysentery in rural areas of China and Africa. Desertification and droughts in both regions have implications on dust emissions, and human intake of respirable dust. Han et al. (2009) and Westerdahl et al. (2009) have discussed the health impacts of respirable dust emissions in China. Bae et al. (2006) developed an empirical dust emission equation for use in estimation of the quantity of dust that can be generated by an exposed ground surface with desiccation rate as one of the significant parameters. Climate change influences the desiccation rate of exposed ground through impacts on temperature and relative humidity. Inyang (2017a) has provided detailed analyses of the scenarios of typical health impacts as well as other consequences of global climate change with focus on Nigeria, an African country.

11.2.7 Infrastructure Impacts

One of the profound impacts of global climate change in China and Africa is sea level rise and the resulting inundation and impairment of coastal and hinterland facilities. Both regions have extensive coastlines flanked by storms and are susceptible to damage by those storms and inundation of coastal areas due to sea level rise.

Particularly, many large cities of coastal African countries are located on continental margins. Among them are Lagos, Rabat, Algiers, Tunis, Dakar, Conakry, Monrovia, Abidjan, Accra, Libreville, Cape Town, Durban, Dar es Salam, Mogadishu, Maputo and Asmara. Also, many large Chinese cities such as Dalian, Tianjin, Qingdao, Shanghai, Fuzhou, Shenzhen, and Xiamen are located on the coast. Their infrastructures have been built up at huge financial costs and cover civil industrial facilities, pipelines, communications networks, and port facilities.

The realistic modelling scenarios used by IPCC (2014) and designated as RCP2.6, RCP4.5, RCP6.0 and RCP8.5 to model prospective sea level between 2046 and 2065 indicate the range of 0.24–0.30 m for the mean and the likely nominal range of 0.17–0.38 m. The regime of expected sea level rise would cause disruptions of facilities and services in most of the cities if interventions are not made. In hinterland areas, China and many African countries depend on many reservoirs created using dams of various sizes, for irrigation and hydroelectric power generation. Figure 11.6 shows a small-scale reservoir for irrigation in Jiangsu Province of China. Changing patterns of precipitation affect water availability for cost-effective and efficient operation of dams and their functions in China and Africa.

11.2.8 Inter-Community Conflicts

Socio-political struggles that primarily derive from struggles over natural resources have become more common in China and Africa. Stronger enforcement of rules over use of natural resources has lessened the intensity of such struggles in China. However, in Africa where there are hundreds of ethnic groups and several countries with claims over natural resources, climate change-exacerbated conflicts have become more frequent. Drought and desert expansion have forced the migration of herders into regions that are traditionally occupied by pastoral farmers, thereby, causing conflicts in Sahelian Africa, East Africa and some parts of Southern Africa.

11.3 Other Environmental Pollution Problems in China and Africa

Beyond the direct and indirect impacts of global climate change, there are other environmental challenges in China and Africa. Most of these challenges are attributed to increase in population and the development and operation of industrial and civil support infrastructure. Essentially, there are brown environmental problems, examples of which are briefly described below:

Fig. 11.6 Counter-interacting drought with water impoundments to support agricultural irrigation in Jiangsu Province of China. **a** Top: The first author with a Chinese farmer in charge of irrigation in Xuzhou, Jiangsu Province, China. **b** Bottom: A completed water reservoir for use in irrigation during droughts in Xuzhou, Jiangsu Province, China

11.3.1 Solid Waste Generation

A combination of manufacturing strength and large population produces large quantities of solid waste in China. In a few African countries, particularly, South Africa, Egypt and Mauritius, manufacturing strength is moderately high and huge quantities of wastes are heaped and unutilized for other purposes. Coal mining in China and South Africa generates significant quantities of solid wastes. So do other types of mining. There are many abundant derelict mine lands in both China and Africa. Even for a low to moderate waste generation range of 1.5–2.0 kg per person per day, huge quantities of solid wastes are generated in both regions. A rapidly increasing challenge is the management of electronic wastes.

11.3.2   Liquid Effluents from Industrial Operations and Acid Mine Drainage

Manufacturing is the engine of the economy of China while agricultural processing is also a major component of the economies of both developing regions. These industrial sectors are typically associated with pollution of water resources by polluted effluents and drainage water, although increasingly effective regulations and their enforcement have brought many associated hazards under control. Expectedly, acid rain which is caused by intensification of combustion of fossil fuel for industrial production and gas flaring has contaminated water resources in industrial areas of China and Africa. Acid mine drainage is also a challenge because China and many African countries operate several mines in areas of moderate to high annual precipitation. Water demand has continued to grow with population growth, economic development, and changes in consumption patterns in both regions. It is estimated that water scarcity in Africa will reach dangerously high levels by 2025 (UN-Water, 2000). Demands for agriculture, industrial and domestic operations combined with seasonality of rainfall in both China and Africa have implications on both water quantity and quality. Although circumstances have improved in China due to regulatory and enforcement activities, since then, assessments of 200 major rivers in China (MEP, 2009) indicated that 20.8% of 409 monitored sections had water quality in the worst possible grade.

A more recent assessment (Tang et al., 2022) indicates that river basins of eastern China remain the most polluted areas due to intensive industrial activities. Wang et al. (2008) found that about 33% of the total waste discharges into Chinese rivers, lakes and the sea were from industries, most of which are rural in location. Furthermore, 80% of the discharges were untreated then. Ma et al. (2020) note that water scarcity in China is closely related to water quality in the sense that only a fraction of available water can meet intended uses. State-owned enterprises in China have been increasingly regulated since the past 25 years but privatization and growth of rural industries have posed challenges in enforcement and account for increasing emission of water pollutants (Wu et al., 1999).

Erosion from devegetated construction sites as well as poor agricultural practices have degraded surface water quality in China and Africa. A typical scenario is depicted in Fig. 11.7. Therein, rainwater leaches contaminants from waste piles that could be mining waste, domestic waste or byproducts of industrial processes, and transports them to water bodies. Exposed ground is also eroded frequently. Earthen material is conveyed to many lakes, streams, and rivers such that quite a large number of these water bodies are contaminated and are often coloured brown.

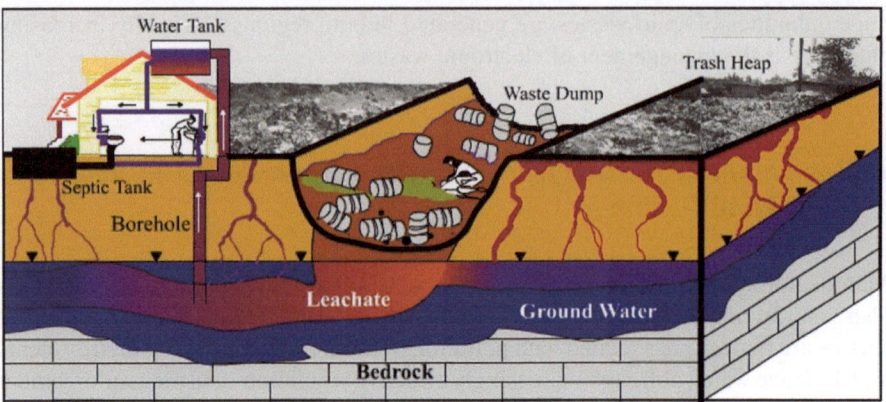

Fig. 11.7 Illustration of a scenario of contaminant leaching from a waste dump and its potential uptake by a domestic water borehole

11.3.3 Marine Pollution

Worldwide, dead zones in coastal oceans have developed around the mouths of more than 400 rivers systems with ecological impacts on more than 245,000 square kilometres (Diaz & Rosenberg, 2008). The dead zones result from the deposition of excessive amounts of chemicals and nutrients such that coastal marine life is endangered. Africa and China have extensive coastal seas that receive rivers that drain hinterland and coastal areas. The continental shelves of both regions are explored for oil and gas, and minerals using drilling rigs and geophysical methods that disturb marine life. There are many incidences of oil spills in the coastal ocean off the coasts of Nigeria, Angola and other oil-producing countries in Africa. Another category of marine pollution that is common to both China and Africa is marine plastics pollution (MPP). Garcia et al. (2019) observe that since the 1950s, the population of plastics has outpaced that of any other material. Plastic waste is ubiquitous, and large quantities are transported by wind and rivers into coastal oceans while another fraction is illegally dumped.

11.3.4 Groundwater Pollution by Chemicals

Although groundwater can be polluted on a regional basis through the natural dissolution of country rocks through physio-chemical processes, there is also superimposition of anthropogenic contaminants in some areas, especially urban and industrial areas where groundwater protection schemes are not implemented. Such areas are many in China and African countries. Sources of groundwater pollution are numerous and include waste dump sites as illustrated in Fig. 11.7, underground fuel storage tanks, septic systems, industrial impoundments, spill sites of contaminants including oil, leaking landfills, pipelines, and oil wells, applied pesticides and fertilizers. All these sources and contamination scenarios apply to China and most African countries. Jia et al. (2019) developed groundwater sustainability indexing system for 31 provinces and municipalities in China. Average scores of 59.5 out of 100 with a range from 47.3 for Tianjin to 72.9 for Tibet were recorded.

11.3.5 Air Pollution by Fine Particulates

Air pollution by dust emissions and its impacts on ecological systems of China and Africa were discussed in Sect. 11.2.3. In addition to those scenarios which were those of dust particles of the predominantly PM_{10} range (coarser particles), fine particulates in the $PM_{2.5}$ category are emitted from fuel combustion in coal-fired power plants, automobiles, and factories in both regions. This is a contributor to air pollution and is far more intense in China than in Africa. Huang (2019) estimated that air pollution caused more than 49,000 deaths in Beijing and Shanghai, China during the first half of 2021. In 2018, Greenpeace estimated that air pollution cost China as much as 6.6% of her economy. Earlier on, an official report of China's Ministry of Environmental Protection as indicated by Hernandez (2015), found that only 8 out of 74 Chinese cities that were monitored in 2014 met air quality standards. Since then, China has stiffened regulations on air quality with better enforcement systems. In Africa, growing industrialization, continuation of gas flaring, use of wood-fired cooking stoves, bush-burning and increase in the number of operational automobiles have also increased $PM_{2.5}$ concentrations in the atmosphere. You (2019) quoted UNICEF (2019) reports that indicated that deaths from outdoor air pollution in Africa increased by 57% from 1990 to 258,000 in 2017 with a resulting annual GDP loss of more than US$ 215 billion.

Air pollution in China and Africa has cascading damages on other ecological compartments. An example is its effects on rainwater acidity in the environs of Nanjing, a city that is close to the large industrial City of Shanghai, China. Studies of rainwater acidity in Nanjing by Tu et al (2005) indicated an average annual pH level of 5.15 within the range of 4.93–5.36. Clearly, air pollution was responsible for the observed acid rain. Acid rain is also prevalent in industrial areas of Africa. Particularly, the Niger Delta of Nigeria, Equatorial Guinea and some parts of Southern Africa have experienced acid rain. In the Niger Delta, gas flaring which is illustrated in Fig. 11.7 has seeded rainfall with sulphuric acid and other acid precursors, resulting in acid rain with negative impacts on coastal ecological systems.

11.4   Opportunities for China–Africa Collaboration

Regarding collaboration, many institutional frameworks exist for collaboration between China and Africa on the mitigation and adaptation to climate change and other environmental stressors. The major categories of such opportunities are described briefly below with a few examples.

11.4.1   Continental Regional Bloc Opportunities

The ongoing collaboration between China and the African Union Commission in various spheres of international development can be extended to cover climate change and environmental issues that are common to both regions. Among such issues are energy policies and technical systems with minding of global climate change, water security, large scale waste management, biodiversity management, marine pollution control and air quality control. At the African continental level, institutions such as the African Ministerial Conference on Water (AMCOW), Water and Sanitation for Africa (WAS) and Green Africa Foundation are collaborators with China. The direct unit within the African Union with the appropriate mandate is the Union's Department of Agriculture, Rural Development, Blue Economy and Sustainable Environment (ARBE). The collaboration can also be designed for partnership with each or any of the Regional Blocs of Africa. Appropriately, many declarations and initiatives that pertain to or address environmental sustainability targets have been scaled at the continental level. Among them are Africa's Agenda 2063 and initiatives such as the Africa Water Vision for 2025 (UN-Water, 2000), African Continental Green Recovery Action Plan 2021–2027, and initiatives of the African Development Bank. One of the African countries that is in the BRICS bloc with China, is South Africa which shares climate change and other environmental challenges with China.

11.4.2   Bilateral and Multi-Lateral Opportunities

China invests significantly in infrastructure development in many African countries. The China-led Belt and Road Initiative (BRI) is a US$4–8 trillion programme that is designed to bridge the existing infrastructure gap for acceleration of the economic growth of the Asia–Pacific area, Central Europe and Eastern Europe. Currently, it covers just the fringes of East Africa but will undoubtedly involve the sourcing of construction and other materials from Africa. It is an opportunity to test-run environmental stewardship initiatives at the bilateral and multi-lateral levels. Also, there is the instrumentality of bilateral and multi-lateral trade agreements.

11.4.3 Science Diplomacy Opportunities

Transactions on scientific collaboration to address challenges that are common to the collaborating regions or countries are generally less contentious than trade and security agreements and treaties. Science diplomacy which has had the function of reducing tension even among large countries with different political philosophies, is a useful instrument for promoting environmental sustainability in partner countries. Among the activities that can be designed and implemented within this context are the following:

(i)   Joint research and development on common environmental challenges, including climate change.
(ii) Joint workshops to address opportunities and challenges of current and prospective infrastructure development to support growing population.
(iii) Bi-directional visits and internships to promote understanding between partner countries.
(iv) Joint exhibitions and special programmes to promote the diffusion of knowledge and best available technologies and systems to address persistent and new socio-economic and environmental challenges.

China and many African countries are rapidly embracing science diplomacy as an instrument of international relations.

11.5 Broadening of China–Africa Collaboration into Worldwide Collaboration

Major environmental stressors may originate in one country, but their impacts tend to spread worldwide. This circumstance justifies the adoption of a universal approach to the tackling of major environmental challenges such as international circulation of air pollutants, desertification, biodiversity loss, ocean pollution, sea level rise and polar ice melting. Essentially, bilateral agreements between China and Africa at the African Union level or specific country levels on global-scale environmental challenges have greater potential for success if they fit into global arrangements. In the past few decades, United Nations agencies and their units such as the World Meteorological Organization (WMO), the United Nations Environment Program (UNEP) and the Intergovernmental Panel on Climate Change (IPCC) have developed and implemented programs that have helped in managing environmental risks worldwide. China and African countries must continue to be active participants in such programs.

11.6   Conclusions

China and Africa have many common environmental challenges, some of which derive from their cultural values and their need to rapidly develop and operate infrastructure to support large populations that presently exceed 1 billion people in either region. Both regions have the advantage of the possibility of developing and applying technologies to leapfrog the pollution that is associated with past approaches and technologies that have caused much global warming and other types of environmental challenges. Owing to the pressures associated with its higher levels of pollutant emissions and resulting higher pollution burden as well as increasing technological strength, China has gone further in the development and implementation of environmental management systems than most African countries. Regardless, there are opportunities for greater collaboration between China and African countries on environmental management with focus on international conventions/agreements; regulatory development and enforcement; exploration of policy options; environmental monitoring systems; research and technology deployment; technical guidance and science diplomacy; and formulation and provision of market incentives to support sustainable development.

References

AfDB. (2018). *Roadmap and Work Programme: Africa Nationally Determined Contributions (NDCs) Hub, Abidjan.* Cote D'Ivoire: Africa Development Bank.
AfDB. (2019). *Accelerating climate action in Africa.* Climate Change in Africa Report of the African Development Bank (AfDB) at the COP-25 Conference, Madrid, Spain, December 2–13.
Babugura, A. (2021). *Climate Change and the Global South: The case of Africa.* The Cairo Review of Global Affairs, Fall Edition.
Bae, S., Inyang, H. I., De Brito Galvao, T. C., & Mbamalu, G. E. (2006). Soil desiccation rate integration into empirical dust emission models for polymer suppressant evaluation. *Journal of Hazardous Materials, 132*, 111–117.
Diaz, R. J., & Rosenberg, R. (2008). Spreading dead zones and consequences for marine ecosystems. *Science, 321*, 926–927.
Epstein, E. R. (1999). Climate and health. *Science, 285*(5426), 347–348.
Friedlingstein, P., Jones, M. W., O'Sullivan, M., & Andrew, R. M. (2019). Global carbon Budget. *Earth Systems Science Data, 11*(4), 1783–1838.
Garcia, B., Fang, M. M., & Lin, J. (2019). Marine plastic pollution in Asia: All hands on deck. *Chinese Journal of Environmental Law, 3*, 11–46.
Han, B. Z., Ji, H., Guo, G., Wang, F., Shi, C., & Li, X. (2009). Chemical characterizations of PM10 fraction paved road dust in Anshan, China. *Transportation Research, Part D, 14*, 599–603.
He, B., Wu, J., Lu, A., Cui, X., Zhou, L., Liu, M., & Zhao, L. (2013). Quantitative assessment and spatial characteristic analysis of agricultural drought risk in China. *Natural Hazards, 16*, 155–166. https://doi.org/10.1007/s11069-012-0398-8
Hernandez, R. A. (2015). Prevention and control of sir pollution in China: A research agenda for science and technology studies. *SAPIENS, 8*(1), 1–8.
Huang, Y. (2019). The environmental challenges of China's recovery after COVID-19. *Time 2030*, February 2 (p. 4).

IMF. (2015). *Regional economic outlook: Sub-Saharan Africa: navigating headwinds*. International Monetary Fund (IMF), Washington DC., USA.

Inyang, H. I. (2017a). *Prospective and manifesting impacts of global climate change in Nigeria: Analysis, mitigation and adaptation*. The Text of the Keynote Lecture at the 10th Forum of Laureates of the Nigerian National Order of Merit, Abuja, FCT, Nigeria (p. 52).

IPCC. (2014). *The 2014 Climate Change Synthesis Report*. Intergovernmental Panel on Climate Change (IPCC), Geneva, Switzerland (p. 151).

Jia, X., O'Connor, D., Hou, D., Jin, Y., Li, G., Zheng, C., Ok, Y. S., Tsang, C. W., & Luo, J. (2019). Groundwater depletion and contamination: Spatial distribution of groundwater resources and sustainability in China. *Science of the Total Environment, 672*, 551–562.

Kan, H. (2009). Environment and health in China: Challenges and opportunities. *Environmental Health Perspectives, 117*(12), A530–A531. https://doi.org/10.1289/ehp.0901615

Lallanilla, M. (2013). China's top environmental concerns. *Live Science Newsletter*. https://www.livescience.com/27862-china-environmental-problems.html

Liang, F., Xiao, Q., Huang, K., & Gu, D. (2020). The 17-year spatio-temporal trend of $PM_{2.5}$ and its mortality burden in China. *Proceedings of the US National Academy of Sciences, 117*(41), 25601–25608.

Ma, T., Sun, S., Fu, G., Hall, J. W., Ni, Y., He, L., Yi, J., Zhao, N., Du, Y., Pei, T., Cheng, W., Song, C., Fang, C., & Zhou, C. (2020). Pollution exacerbates China's water scarcity and its regional inequality. *Nature Communications, 11*(650), 1–9.

MEPC. (2009). *The 2008 report on the state of environment in China*. Ministry of environmental Protection of China, Beijing, China. http://www.zhb.gov.cn/plan/zkgb/2008zkgb/

Njeru, G. (2022). Africa's inclusive green initiative. *Africa Report, 2022*, 37–39.

Oki, T., & Kanae, S. (2006). Global hydrological cycles and world water resources. *Science, 313*, 1068–1072.

Ritchie, H. (2018). Global inequalities in CO_2 emissions. In *Our world in data*, October, 16th Edition.

Tang, W., Pei, Y. H., Zhao, Y., Shu, L., & Zhang, H. (2022). Twenty years of China's water pollution control: experiences and challenges. *Chemosphere, 295*, 1.

Tu, J., Wang, H., Zhang, Z., Jin, X., & Li, W. (2005). Trends in chemical composition of precipitation in Nanjing: China during 1992–2003. *Atmospheric Research, 73*, 283–298.

UNICEF. (2019). *Silent suffocation of Africa*. A Report of the United Nations International Children's Emergency Fund (UNICEF), New York, USA (p. 13).

UN-Water. (2000). *The Africa Water Vision for 2025: equitable and sustainable use of water for socioeconomic development*. UN-Water Africa Report collaboratively developed by the Economic Commission for Africa, African Union and the African Development Bank, Addis Ababa, Ethiopia (p. 28).

Wang, M., Webber, M., Finlayson, B., & Barnett, J. (2008). Rural industries and water pollution in China. *Journal of Environmental Management, 86*, 648–659.

Westerdahl, D., Wang, X., Pan, X., & Zhang, K. M. (2009). Characterization of on-road vehicle emission factors and microenvironmental air quality in Beijing, China. *Atmospheric Environment, 43*, 697–705.

Wu, C., Mauer, C., Wang, Y., Xue, S., & Davis, D. L. (1999). Water pollution and human health in China. *Environmental Health Perspectives, 107*(4), 251–256.

You, Y. H. (2019). *Air pollution is starting to choke Africa*. Pollution-Earth Org-Past, Present and Future, August 27 (p. 6).

Part IV
China-Africa Collaboration in Telecommunications and Space

Chapter 12
Technology, Information and Power: the Role of CGTN Africa in China–africa Relations

Naledi Ramontja, Vhonani Petla, and Siphamandla Zondi

12.1 Introduction

Like many other countries, China's use of the mass media to drive strategic narratives has grown as it grew into the second-largest economy in the world, and its activities in Africa and elsewhere also grew astronomically in the past three decades. It has exponentially increased its quest for an advanced global image through using new technologies in its public and cultural diplomacy. This includes the expanded use of satellite broadcasting to advance China's national values and interests globally, suggests an assertive agenda toward nation branding in the face of the challenges of market disadvantages and the uncertainty of audience reception (Zhu, 2022) about China's rise and activities throughout the world.

State-owned media actors such as CGTN-Africa, Xinhua and People's daily play a central role in expanding China's media presence and its deliberate quest to craft a media narrative that supports its interests, including its rivalry with Western states. Since the early 2000s, China has consistently focused on maintaining a strong media and digital presence in Africa. Despite the volumes of aid and assistance invested in Africa, its image continues to be troubled by controversies, such as those about neocolonial designs. Despite this, new media and ICT initiatives followed the expansion in traditional mass media as part of China's strategic plan to seek greater engagements with Africa and African people. The Chinese media footprint in Africa and its engagement with Africans has increased in recent decades, despite the early existence of China's presence in African media. In the early 2000s, China's strategic media relations with Africa began to pick up in earnest with massive investments and the expansion of the African media sphere. As China intensified its cooperation with Africa, the need to manage the perception of these relations, of China in Africa

N. Ramontja (✉) · V. Petla · S. Zondi
Institute for Pan-African Thought and Conversation, University of Johannesburg, Johannesburg, South Africa
e-mail: nalediramontja@gmail.com; naledir@uj.ac.za

© The Author(s) 2024
M. Muchie et al. (eds.), *China-Africa Science, Technology and Innovation Collaboration*, https://doi.org/10.1007/978-981-97-4576-0_12

and of Africa in China, became necessary intensified. To a lesser extent, China's interest in Africa and its perception are affected by rivalry with Western interests in Africa, including expectations that China should be an alternative to the West and its narratives. This paper reflects on the use of technology, information, and power in Africa–China Relations through a case study of the Nairobi-based CGNTV-Africa and how China seeks to influence the media landscape in Africa to use media to shape global image.

12.2   Rise of China, Technology, and Power: Framing the Issues

Although the Chinese government promotes the use of the internet and independent news outlets to contribute to the global mediascape, it also uses the media to drive its narratives on international developments. China's efforts at actively strengthening its ties and advancing its image in developing and developed regions dates to the 1960s (Wu, 2012). Relations with Africa and media relations on the continent also intensified with a focus on independence movements and newly independent African countries committed to the Bandung agenda for intensified cooperation (Madrid-Morales, 2017). Chinese aid to Africa increased substantially to the point that by 1980 China was aiding more countries in Africa than the United States, a major donor power at the time. China's cultural revolution in the 1960s entailed the rejuvenation of Chinese culture, nationalism and foreign policy posture, informed and driven by the government's action to manage its mass (Lu, 2017).

In this period, media engagement between these two regions of the world remained relatively modest due to China's inward-looking focus during the cultural revolution. The media focused primarily on Asia, North America and Europe (Wekesa, 2013). In the post-1990 period, as relations between China and Africa grew, Beijing strengthened ties with Africa to enhance its influence and international image. Broadcasting became one of its voices to the world for showcasing its great power status (Wang, 2011). Unlike the Western media practice of promoting Western culture, ideology and policy ideas, China uses media engagement to signal mutual partnerships, non-interventionism and win–win relationships (Mosher and Farah, 2010). China advances its image as that of a leading world soft power (Monyae, 2021; Wu, 2012). Its growing media engagement with African audiences has drastically boosted its ability to drive narratives that favours its interests and image, projecting it as a benign force (Gagliardone, 2013). China is also projecting itself as championing Africa's artificial intelligence anchored on the Huawei infrastructure being expanded throughout Africa (Nkwanyana, 2021).

East Africa in particular, is not new to Chinese media interactions. For more than two decades, it has seen the rise of conventional and unconventional media strategies such as the *Perking Review* (now called the *Beijing Review*) (Wang, 2011). In fact, the region has long been a media hub for Chinese media, which was strengthened in the aftermath of the launch of the China–Africa Cooperation in 2000 (FOCAC) (Wang, 2011; Xinhua News Agency, 2021). The FOCAC serves as a formal framework for official relations and overall cooperation between China and Africa providing a structured platform for dialogue, policy coordination, and partnership initiatives. Within this framework, media engagement has significantly expanded, leveraging broadcasting platforms to enhance communication and cultural exchange (Wu, 2012).

The relationship also focuses on the infrastructure development of African state broadcasters such as the Zambia National Broadcasting Corporation (ZNBC) (Wu, 2012). This focus on media infrastructure is evident in the substantial financial support provided by China to various African countries. Between 2003 and 2012 the Chinese government offered US$4.5 million for TV and radio equipment for the government of Lesotho and donated media equipment worth US$8 million to Zambia; in 2008, China further offered US$4 million to FM Radio Expansion funds in cooperation with the Liberia Broadcasting System (LBS) to support its Radio broadcasting services, in the same year China funded the Malawian state broadcaster US$250,000 to develop a radio channel (Wu, 2012). Furthermore, China invested in Ethiopia's telecommunications sector by offering a multi-billion-dollar loan to Ethio-Telecom (Ethiopian telecom operator) to strengthen its access to mobile phones and the internet (Gagliardone, 2013). Media cooperation has developed significantly from this point, covering other aspects of the China–Africa bilateral media relationship (Wu, 2016).

This cannot be understood without grasping the following. Since the early 1960s, China's worldview has been shaped by Mao Zedong's 'Three World' Theory, where Mao categorized the world into three worlds; the First World has the United States, Russia, and other superpowers (Nye, 2019). The Second World consists of the socialist regimes in Europe and other developed countries. The Third World is the developing countries—including African countries (Calzti, 2022). The Maoist paradigm advocated for an alternative international order that would challenge the United States and its policies, promote world peace instead of war and the liberation of Third World countries from colonialism (Wang, 2011). Now, with that at the back of their mind, China has developed a foreign policy focusing on its global image, seen as critical in advancing this alternative way of managing world affairs (CCTV Africa, 2015). It sees its public diplomacy and nation branding as a tool for advancing a change in the international order. This makes the work of its media more profound and offers Africa an opportunity to participate in shaping new world orders.

12.3 The CGTN: A Platform for China's Contestation of the Africa Media Sphere

China's national broadcaster, China Central Television (CCTV), was founded in 1958 with just one channel. Today it produces over 40 channels dedicated to news, sports and films. In 2004 CCTV launched its first English language channel, CCTV

Table 12.1 Chinese media
platforms spreading in Africa.
Source Compiled by Authors,
using various sources

Xinhua	China daily
CGTN	CNC
CRI	People's daily
Weibo	WeChat
TikTok	

international, to broaden its scope (Wu, 2012; Fearon & Rodrigues, 2019). CCTV was later rebranded from CCTV to China Global Television Network (CGTN) to broaden its scope and global reach (Fearon & Rodrigues, 2019). Since its rebranding and the establishment of its foreign affiliates, CGNT has grown significantly, covering over 400 million users at the end of 2016 (Fearon & Rodrigues, 2019). CGTN's mission is to promote an alternative understanding of events across nations and continents to that presented by Western media through Chinese perspectives. The idea is to diversify the voices on world affairs (CGTN, 2017). In 2012 China Global Television Network-Africa (CGTN-Africa) was officially launched in Nairobi, Kenya, to disseminate news from both Chinese and African perspectives with an emphasis on enhancing Sino-Africa relations (Fearon & Rodrigues, 2019). CGTN presented itself as a '*new voice of Africa*' and China's biggest milestone in building connections in Africa with Africans (Yangui, 2014). Today Chinese state-owned corporations, including CCTV (now CGTN) and Xinhua news agency, are operating in Rwanda, Kenya, Nigeria, and Côte d'Ivoire. Its private media houses, such as StarTimes, and leading telecommunications companies, such as ZTE and Huawei, also operate in various parts of Africa (Jedlowski & Röschenthalerb, 2017).

Social media platforms play a significant role in China–Africa media relations. Table 12.1 shows several Chinese social media platforms across Africa to disseminate news and information and share content. The usage of these platforms remains diverse in Africa; they have gained more popularity and reportedly have more social impact.

China's internet-based mobile messaging App WeChat recently launched the mobile money service WeChat Wallet, which allows users to store bank cards and withdraw cash at the automated teller machines (ATM) of a partner, such as Standard Bank in South Africa (China Daily, 2016).This development marks WeChat's growing influence in the financial technology sector, particularly in Africa, where instant mobile money services are increasingly vital for financial inclusion and digital payment solutions.

The nature of content disseminated by CGTN-Africa reshapes China's international image while raising African voices, which have been marginalized and misappropriated by dominant Western media such as CNN, BBC, and France24 as well as the Middle Eastern platform, AlJazeera. To a larger extent, these have monopolized the African media market (Madrid-Morales & Gorfinkel, 2018). To this day, Africa's media freedom and presence is limited in the scope of the content. It is marred by Western perspectives driving an afro-pessimistic outlook on Africa (Madrid-Morales & Gorfinkel, 2018).

CGTN-Africa has five widely used programs: *Africa Live* is one of them—an hour broadcasting session that covers current affairs, business and sports news. The second is the *Faces of Africa*—a weekly documentary series on remarkable African literary and political icons and ordinary Africans. The third is *Talk Africa*—a talk show hosted once a week to discuss important events or activities for the week. The fourth is *Match Point*—discussing sports. The final one is *Global Business*, a show that covers African and global business news (Madrid-Morales, 2017; Gagliardone, 2013). Following the launch of CGTN-Africa, President Xi Jinping urged the television network to use the platform to tell Chinese stories and perspectives impartially (Fearon & Rodrigues, 2019). This assertion could only mean that CGTN is set to broadcast Beijing's worldview and perspectives to enhance its soft power.

CGTN-Africa strategically focuses on African voices created with the participation of local audiences in an almost ethnographic fashion (Umejei, 2018). Iginio Gagliardone (2013) argues that when CCTV opened its Nairobi broadcasting and production centre—what is now CGTN Africa—the key message it conveyed was that the station would promote positive journalism (Li, 2016; Gagliardone, 2013). This was about telling good stories about Africa, the positive narrative about Africa and about Africa–China relations as an alternative to the tendency to pick up negative stories about both in the Western media (Xinhua News Agency, 2021). To this end, the network focuses on inspiring and positive optics. It shares international satellite news production norms through its popular promotional slogan, '*see the difference*' (Fearnon & Rodrigues, 2019). The idea that Chinese media focuses on positive journalism is also the story promoted by the network itself. Its bureau chief, Song Jianing, once said in a media statement:

"We hope to strengthen a positive image of Africa in Africa and worldwide. If you take the case of Somalia, our journalists not only cover war and violence but also want to stress other aspects of life in Somalia. How life in the capital is improving, how the African Union is making a difference and even how a country like that has been able to send athletes to the Olympics" (quoted in Gagliardone, 2013: 32).

CGTN-Africa invested resources to ensure a continental reach in its content, coverage and audience. This meant ensuring that stories covered all five regions of Africa and that African journalists filed reports in various parts of Africa. It also ensured that continental events and meetings were covered prominently, even where there was no controversy. CGTN-Africa is then a tool China uses to strengthen its relations with Africa through positive reporting (Gravett, 2020). African stakeholders seem to appreciate the diversity of platforms and have welcomed the Chinese network alongside Western, Indian, Russian, Middle Eastern and other platforms.

12.4 CGTN and the Media Sphere in Africa

The launch of CGTN-Africa in Kenya is one of China's strong attempts to contest in the African media sphere, which its Western rivalries have long dominated. Today, there are three perceptions of Chinese's presence in the African media sphere, namely: China as Africa's neutral development partner, China as a hegemonic power and China as a competitor (Banda, 2009). As a competitor, China is seen as displacing local businesses with its own. As a development partner, it is seen as a neutral partner engaging in win–win cooperation with African states to achieve development goals. As a hegemonic power, China is framed with suspicions of plundering mineral-rich African countries (Gorfinkel et al., 2013; Banda, 2009). The African perspective in all three accepts China as a neutral partner in its development.

CGTN-Africa's content is broadcasted across Africa and in the world through a range of new and old technologies to accommodate its audience. New and old technologies, including radio, satellite (DSTV in South Africa), TV packages and social media platforms such as YouTube, Twitter and Facebook and its website, are used to disseminate information (Gorfinkel et al., 2013). To this day, CGTN-Africa has successfully promoted positive opinions of China and China–Africa relations through its two major channels, Talk Africa and Africa Live (Gagliardone, 2013). CGTN and CGTN-Africa both lack an element of disaster reporting; instead, both channels cover up on stories that could either reduce their viewership or harm their image (Fearon & Rodrigues, 2019).

The strategic increase of China's media in the African media sphere is regarded by many as an instrument to shape its soft power and win African audiences' hearts and minds. This is through its motto of *'positive journalism'*(Jedlowski, 2021; Wekesa, 2013). Many African scholars, amongst other Gagliardone (2013), argues that it is not practical to consider Africa a source of positive coverage, taking into account the challenges Africa is faced with, which includes the surge of ongoing violent conflicts, poverty, underdevelopment, the presence and resurgence of coups d'etats and military insurgents (Gagliardone, 2013). Analysis conducted by Madrid-Morales and Gorfinkel also challenges the idea that CGTN-Africa and other Chinese media houses focus on positive journalism in Africa (Madrid-Morales & Gorfinkel, 2018). This is done by revealing that the number of negative stories covered in *Africa Talk* aired by CGTN-Africa is almost the same as the number of the positive ones, by illustrating that only 38% of videos out of 500 Youtube covers positive stories while 27.6% covered the negative ones (Madrid-Morales, 2017). Other studies focused on the geopolitical and macroeconomic aspects of the media interactions by adopting Joseph Nye's soft power theory, emphasizing the power to shape others' preferences through appeal and attraction. These perspectives have led to wider debates around the transformation of China's soft power strategies and whether its strategies and programs are beneficial to both parties (Wu, 2012). Recent work explores the people-to-people interactions (Jedlowski, 2021), newsroom ethnographies (Gagliardone, 2013), Africa's coverage of Chinese news (Jedlowski & Röschenthalerb, 2017), the nature of the content produced, the response of audiences as well as the overall impact of the media engagements (Jedlowski & Röschenthalerb, 2017; Madrid-Morales & Gorfinkel, 2018). Most of these analyses position China at the centre of the picture while neglecting the role of African and African actors. This approach tends to produce a polarized understanding of China–Africa relations and media interactions (Jedlowski, 2021).

12.5 CGTN and Africans

The opening of CCTV-Africa (now CGTN) offices in Kenya opened to media engagements with Africans. It started with hiring local talent to cover local stories of ordinary
Kenyans broadcasted by GCTN. The broadcaster is contributing to the economy and
development of Africa by offering employment opportunities to local talent. Kenyan
news editors, technicians and reporters often undergo media training and workshops in China under FACOC to enhance their media expertise (Wekesa, 2013). This
contributes to the development of Africa and African talent. Its commitment to using
African voices and true African stories is described as a top feature that continues
to make Chinese media favourable in the eyes of Africans. The majority of African
audiences who watch CGTN indicate that they watch the channel because it offers
a unique perspective on local and international news, 46.9% indicated that CGTN's
reporting of African stories is much better than that of other international media
channels (Zhang & Mwangi, 2016). CGTN maintains its engagement with Africans
through its flagship program Talk Africa. Talk Africa is a weekly talk show that
keeps African audiences informed on political, social and current affairs. Amongst
the stories covered on the talk show if Chinese meet Africans, a show that encourages socio-cultural interactions amongst Chinese and Africans (Rhodes, 2012; Lee,
2020).

The most impactful engagement CGTN has made with Africans is through
the Faces of Africa film, which was launched in 2012 to capture African stories.
Faces of Africa has opened a space for a more positive portal of Africa and a
humane view of black Africans, it is all about the true representation of Africa and
what Africans have done for themselves, including the fight against colonialism,
apartheid, racial discrimination, poverty, hunger, inequality, and unemployment
(Madrid-Morales & Gorfinkel, 2018). Faces of Africa substitutes afro-pessimistic
narratives of hopelessness and destitution with positivist ideas of hopeful and rising
Africa.

African filmmakers directed 66% of the films, in this way, GCTN is promoting the
production of African stories by Africans, allowing Africans to tell their stories their
way. These films are created with Africans in mind and feature prominent Africans
who have contributed to the continent (Rhodes, 2012). The documentary series covers
stories from 28 of 54 African countries. The film includes, amongst others, the story
of Burkina Faso's revolutionary leader Thomas Sankara, the first president of Independent Ghana—Kwame Nkrumah, the pillar of modern Ethiopia—Haile Selassie,
South Africa's anti-apartheid activist and politician—Winnie Madikizela Mandela,
Nigerian's top literacy icon and Nobel Prize Winner Wole Soyinka and other less
known Africans who have and still continue to contribute to the development of
Africa (Madrid-Morales & Gorfinkel, 2018). Not only does the channel cover stories
of famous Africans but it also covers stories of ordinary Africans and the change
they make in their countries, for instance, stories of local peasants and how they
contribute to the economy of agrarian African countries such as Zimbabwe and
South Africa. The type of engagement Faces of Africa offers to ordinary Africans

proves that African stories are newsworthy, and that Africans are making efforts to develop their continent. Other films GCTN has produced target local audiences, show real-life engagements with Africans, and allow Africans to tell their stories. These include the *Chinese in Africa* film, *Passage to Africa* and *African Chronicles* (Madrid-Morales & Gorfinkel, 2018). All these films promote cultural awareness and general public engagement and understanding among Africans and Chinese.

Despite CGTN's Faces of Africa, several studies have been conducted to analyse the consumption of Chinese media in Africa and its impact on African communities. This included perceptions and thoughts, which revealed a reluctance to consume Chinese media, which creates scepticism that drives unfavourable perceptions. Firstly, students in Nairobi, where the CGTN Africa headquarters are situated, have acknowledged that they know of the platform's existence. However, very little engagement with the content exists in their area (Madrid-Morale & Wasserman, 2018). This was also similar to the use of Chinese applications such as WeChat and TikTok. The problem here is not necessarily the access to the Chinese media but rather the reluctance to engage it, but it is due to negative stereotypes associated with China and its motive in Africa (Madrid-Morale & Wasserman, 2017). These include that the role of the Chinese media seems to be directed at shaping positive perceptions to drive its own narrative (Wakesa, 2013).

Some African journalists and media organizations question the possible negative long-term consequences of editorial independence African media entities might face and overrepresentations of certain countries such as Kenya and South Africa with little to no coverage in other parts of the continent (Madrid-Morales & Wasserman, 2018). These factors influence perceptions about the Chinese Media in Africa and, to some extent, hinder the engagement of Africans with Chinese media.

There is a dichotomy that exists between decision-making, reporting and gatekeeping at CGTN-Africa. Some African journalists argue that this serves as a control mechanism to filter stories that are not favourable (Umejei, 2018). The delays in decision-making and receiving final authorisation to report stories means that reporters get to cover overdone stories. Other media organisations have covered these stories (Zhang & Mwangi, 2016). Other reporters claim that there are approximately four stages one has to go through to get approval before journalists are assigned to cover events (Umejei, 2018). On the people-to-people level, there still needs to be more in-depth cultural understanding and connection between the two (Gorfinke et al., 2013). This can be attributed to a lack of awareness and lack of sustainable interest in GCTN-Africa coverage. Despite this, CGTN occupies a unique space in the spectrum of the African media sphere as a neutral developmental partner.

12.6 Conclusion

From modest levels of the 1960s, the cooperation between Africa and China has grown astronomically, straddling areas of trade, investment, technology sharing, science and education. This has created a demand for China to tell its story and manage its conversation with Africa so that it does not happen only through its rivals in the West, who see the rise of China as a threat to their dominance of the African space. Technology has become a major catalyst of this growth in cooperation, both as goods traded and as means to facilitate cooperation. This study takes a critical look at the broadcast technology deployed by China through the CGTN in order to find new ways of looking at African stories and shape new narratives about Africa. It argues that this contributes to defining the evolving Africa–China relations and the continued involvement of China in the African media sphere. It shows that this is China's contribution to the rivalry with Western media houses for the telling of the story of China's relations with Africa, and in the process, the focus on positive news is contrasted with one on negative news to wonder what news might be that Africans want in the end. This illustrates how CGTN-Africa has influenced and shaped China's media perceptions in Africa. This is done not only by driving the Chinese narrative but also by promoting the ways of looking at African stories and African people and driving the concept of 'positive reporting' or 'positive journalism' in contrast to Afro-pessimistic and sinophobic views peddled in the Western media. Yet, positive journalism has markings of controlled news informed by strategic agendas that are not entirely clear but generally have to do with decisions in Beijing, which is concerning. China's presence in the African media sphere has opened up new opportunities for redefining Sino-Africa relations, including by finding something beyond the positive–negative binary in the current broadcasting landscape.

References

Banda, F. (2009). China in the African mediascape: A critical injection. *Journal of African Media Studies, 1*(3), 343–361.

Calzati, S. (2022). 'Data sovereignty' or 'Data colonialism'? exploring the Chinese involvement in Africa's ICTs: A document review on Kenya. *Journal of Contemporary African Studies, 40*(2), 270–285.

CCTV Africa. (2015). *About CCTV Africa.* http://cctv-africa.com/about-cctv-africa/

CGNT. (2017). CGTN Special. Available at: https://www.cgtn.com/special-list/2017.html [Accessed 31 August 2024].

China Daily. (2016). *WeChat banks on African money transfers.* https://www.chinadaily.com.cn/kindle/2016-12/11/content_27634199.htm

Fearon, T., & Rodrigues, U. M. (2019). The dichotomy of China global television network's news coverage. *Pacific Journalism Review, 25*, 1–20.

Gagliardone, I. (2013). China as a persuader: CCTV Africa's first steps in the African media sphere. *Ecquid Novi: African Journalism Studies, 34*(3), 25–40. https://doi.org/10.1080/02560054.2013.834835

Gorfinkel, L., Sandy, J., Van Staden, C., & Wu, Y. (2013). CCTV'S global outreach: Examining the audiences of China's 'New Voice' on Africa. *Media International Australia, 151*(1), 81–88.

Gravett, W. (2020). Digital neo-colonialism: The Chinese model of internet sovereignty in Africa. *African Human Rights Law Journal, 20*(1), 125–146.

Jedlowski, A. (2021). Chinese television in Africa. *Theory, Culture and Society, 38*(7–8), 233–250.

Jedlowski, A., & Röschenthalerb, U. (2017). China–Africa media interactions: Media and popular culture between business and state intervention. *Journal of African Cultural Studies, 29*(1), 1–10.

Lee, C. K. (2020). *The specter of global China.* University of Chicago Press.

Li, L. (2016). The image of Africa in China: The emerging role of Chinese social media. *African Studies Quarterly, 16*(3–4), 1.

Lu, X. (2017). *The rhetoric of mao zedong: Transforming China and its people.* University of South Carolina Press.

Madrid-Morales, D. (2017). China's digital public diplomacy towards Africa: Actors, messages and audiences. In K. Batchelor & X. Zhang (Eds.), *China–Africa relations: Building images through cultural co-operation, media representation and communication* (pp. 129–146). Routledge.

Madrid-Morales, D., & Gorfinkel, L. (2018). Narratives of contemporary Africa on China global television network's documentary series faces of Africa. *Journal of Asian and African Studies, 53*, 917–931.

Madrid-Morales, D., & Wasserman, H. (2017). Chinese media engagement in South Africa. *Journalism Studies, 19*(8), 1218–1235. https://doi.org/10.1080/1461670X.2016.1266280

Madrid-Morales, D., & Wasserman, H. (2018). How influential are Chinese media in Africa? An audience analysis in Kenya and South Africa. *International Journal of Communication, 12*, 2212–2231.

Monyae, D. (2021). *Reflections on the global village. Opinions and analysis.* Real African Publishers.

Mosher, A., & Farah, D. (2010). *Winds from the east: How the people's Republic of China seeks to influence the media in Africa, Latin America, and Southeast Asia.* Center for International Media Assistance.

Nkwanyana, K. (2021). *China's AI deployment in Africa poses risks to security and sovereignty.* Australian Strategic Policy Institute. Available at: https://www.aspistrategist.org.au/chinas-ai-deployment-in-africa-poses-risks-to-security-and-sovereignty/

Nye, J. (2019). Soft power and public diplomacy revisited. *The Hague Journal of Diplomacy, 14*(1–2), 7–20.

Rhodes, T. (2012). *China's media footprint in Kenya—Committee to Protect Journalists.* [online] Committee to Protect Journalists. Available at: https://cpj.org/2012/05/chinas-media-footprint-in-kenya/. Accessed 25 April 2022.

Umejei, E. (2018). Chinese media in Africa: Between promise and reality. *African Journalism Studies, 39*(2), 104–120.

Wang, J. (2011). *Soft power in China. Public diplomacy through communications. Palgrave Macmillan series in global public diplomacy.* Palgrave Macmillan.

Wekesa, B. (2013). Emerging trends and patterns in China–Africa media dynamics: A discussion from an East African perspective. *Ecquid Novi: African Journalism Studies, 34*(3), 62–78. https://doi.org/10.1080/02560054.2013.845592

Wu, Y. S. (2012). *The Rise of China's state-led media dynasty in Africa.* South African Institute of International Affairs.

Wu, Y. (2016). China's media and public diplomacy approach in Africa: Illustrations from South Africa. *Chinese Journal of Communication, 9*(1), 81–97.

Xinhua News Agency. (2021). *Confucius Institutes, 48 Confucius Classrooms established in Africa:* White paper.

Yanqiu, Z. (2014). *Understand China's media in Africa from the perspective of constructive journalism.* Communications and Public Diplomacy.

Zhangb, J., & Mwangi, J. M. (2016). A perception study on China's media engagement in Kenya: From media presence to power influence? *Chinese Journal of Communication, 9*(1), 71–80. https://doi.org/10.1080/17544750.2015.1111246

Zhu, Y. (2022). *Media power and its control in contemporary China: The digital regulatory regime, national Identity, and global communication.* Springer Nature.

Chapter 13
China–South Africa Collaboration in Astronomy

Y.-Z. Ma, D. A. H. Buckley, X. Chen, C. Cui, S. Dong, J.-X. Hao, Y.-C. Liang, J. Liu, B. Peng, Z. Shen, A. R. Taylor, and P. Vaisanen

13.1 Background

13.1.1 General Motivation

Astronomy captures the imagination of people everywhere and touches a fundamental human desire to understand the Universe that surrounds us and human's position in it. It provides an ideal way to attract young students to scientific and technic studies and developing human capacity for the knowledge-based economy for the future. The fundamental knowledge of the Universe is a common culture of all nations and all ethnic groups. In recent years, there have been increasing astronomy telescopes and projects being constructed and taking place in South Africa and China to exploit the cosmos in the two countries. These projects, standing at the cutting-edge of modern astrophysics, require a lot of technological and scientific expertise to involve. However, although the research teams of the two countries have been fast developing individually, there is still a lack of human capacity in terms of scientific analysis and technological development and the linkage between science and technology. More importantly, many of the scientific research and technological

Y.-Z. Ma (✉)
Department of Physics, Stellenbosch University, Matieland, Stellenbosch 7602, South Africa
e-mail: mayinzhe@sun.ac.za

D. A. H. Buckley
South African Astronomical Observatory, P.O. Box 9, Observatory, Cape Town 7935, South Africa

Department of Astronomy, University of Cape Town, Private Bag X3, Rondebosch 7701, South Africa

Department of Physics, University of the Free State, P.O. Box 339, Bloemfontein 9300, South Africa

© The Author(s) 2024   227
M. Muchie et al. (eds.), *China-Africa Science, Technology and Innovation Collaboration*, https://doi.org/10.1007/978-981-97-4576-0_13

development between the two countries have strong, internal synergy, and a collaboration will certainly strengthen the research team, make up the shortage of expertise and educate young scientists. From 2016, initiated by National Research Foundation and National Astronomical Observatory of China, astronomers of the two countries started to build collaborations relations and have established several projects that have world-wide impact, as illustrated below.

13.1.2 Major Facilities in China

China has been heavily investing its astronomy research, with strategic aims to strengthen the optical, radio astronomy and multi-messenger astronomy. Its major

X. Chen · C. Cui · J.-X. Hao · Y.-C. Liang · J. Liu
National Astronomical Observatory, Chinese Academy of Sciences, 20A Datun Road, Beijing 100101, People's Republic of China

X. Chen · J.-X. Hao · Y.-C. Liang · J. Liu
University of Chinese Academy of Sciences, Beijing 100049, People's Republic of China

X. Chen
Center of High Energy Physics, Peking University, Beijing 100871, People's Republic of China

Department of Physics, College of Sciences, Northeastern University, Shenyang 110819, Liaoning, People's Republic of China

C. Cui
National Astronomical Data Center, Beijing 100101, People's Republic of China

S. Dong
Kavli Institute for Astronomy and Astrophysics, Peking University, Beijing 100871, People's Republic of China

B. Peng
CAS Key Laboratory of FAST, National Astronomical Observatories, Chinese Academy of Sciences, Beijing 100101, People's Republic of China

Hebei Key Laboratory of Radio Astronomy Technology, Shijiazhuang 050081, Hebei, People's Republic of China

Z. Shen
Shanghai Astronomical Observatory, Chinese Academy of Sciences, Shanghai 200030, People's Republic of China

Key Laboratory of Radio Astronomy, Chinese Academy of Sciences, Nanjing 210008, People's Republic of China

A. R. Taylor
Inter-University Institute for Data Intensive Astronomy, Cape Town, South Africa

P. Vaisanen
South African Astronomical Observatory, P.O. Box 9, Observatory, Cape Town, South Africa

Southern African Large Telescope, P.O. Box 9, Observatory, Cape Town, South Africa

organizational centre, the National Astronomical Observatory China (hereafter NAOC), is the largest astronomy research institute in the system of the Chinese Academy of Sciences, which has 600 affiliated staff. Aiming at the forefront of astronomical science, NAOC conducts cutting-edge astronomical studies, operates major national facilities, and develops state-of-the-art technological innovations. Applying astronomical methods and knowledge to fulfil national interests is also an integral part of the mission of NAOC. The Purple Mountain Observatory (PMO) and the Shanghai Astronomical Observatory (SHAO) are separate CAS institutes, running in parallel of the research projects. Besides NAOC, there are four subordinate units across the country: the Yunnan Observatories (YNAO), the Nanjing Institute of Astronomical Optics and Technology (NIAOT), the Xinjiang Astronomical Observatory (XAO) and the Changchun Observatory.[1]

There are a few facilities being run country-wide in China, operated by the three major observatories. The Large Sky Area Multi-Object Fiber Spectroscopy Telescope (LAMOST), also known as the Guoshoujing Telescope, which is equipped with large FOV optics and up to 4000 fibres on the focal plane, was put into scientific use in September 2013.

The world's largest single-dish, Five-hundred-meter Aperture Spherical Telescope (FAST), has started to operate since 2018. The FAST has three outstanding features: the unique karst depression in Guizhou as the telescope site, the active main reflector of 500 m diameter which corrects the spherical aberration on the Earth, the light focus cabin driven by suspension cables. FAST was constructed by 2018 covering frequency range of 70 MHz–3 GHz with capacity for upgrading to 8 GHz (Fig. 13.1), equipped with a 19-beam receiver in 2018 and being fully operational since January 2020 with first high priorities on pulsar searching, pulsar timing, HI mapping in M31 galaxy, FRB (Fast Radio Burst) etc.

The 21-Centimeter Array (21CMA) has been running since 2006, which is designed to study the cosmic Epoch of Reionization, Construction was completed on the Chinese Solar Radio Heliograph (CSRH). In addition, TianLai project, which is a 21-cm intensity mapping (radio) survey was built and has been running since 2010. The aim of the Tianlai project is to measure the 21-cm intensity mapping from neutral hydrogen and also measure the fast radio bursts.

The Chinese VLBI Network (CVN) includes the Shanghai TianMa-65 m, the Shanghai Sheshan-25 m, the Urumqi-26 m, the Kunming 40 m and the Miyun 50 m radio telescopes. FAST is also ready to join this network at L-band.

In addition, China has been actively cooperating with scientists around the world, by participating in building the Thirty-meter-telescope (TMT) International Observatory. China also initiated the Square Kilometre Array (SKA) international project with 10 other countries. The CAS South American Center for Astronomy, also known as the China–Chile Joint Research Center for Astronomy, was established in Santiago, Chile in 2016. China has a South American station located in Argentina that runs its Satellite Laser Ranging (SLR) project. In addition, the China–Argentina 40 m

---

[1] http://english.nao.cas.cn.

Fig. 13.1 FAST established in 2016, Pingtang county of Guizhou province China

Radio Telescope is under construction. In addition, NAOC also engages in collaboration through the Chinese-French Origins International Associated Laboratory and the East Asian Core Observatories Association.

13.1.3 Major Astronomy Facilities in South Africa

The South African Astronomical Observatory (SAAO), originally established as the Royal Observatory in 1820 in Cape Town, is a National Facility of National Research Foundation (NRF) of South Africa. The SAAO is the national centre for optical and infrared astronomy in South Africa, and the premier optical astronomy facility on the continent of Africa. It operates multiple telescopes near the small town of Sutherland, about 400 km north-east of Cape Town. Some of the telescopes are owned by NRF/SAAO, such as the 1.9 m, an older 1 m telescope, and Lesedi, a new 1 m. There are over 20 other facilities on the Sutherland observing plateau owned by other international institutions, many of which have time accessible to the SA community.

In addition to the SAAO, the Boyden Observatory, operated by the University of the Free State, has several telescopes up to 1.5-m in diameter, including fast GRB follow-up telescopes, Watcher and Bootes, both part of global networks.

The largest facility at the SAAO is the Southern African Large Telescope (SALT), a 10 m-class optical telescope, the largest single optical telescope in the southern hemisphere. SAAO is tasked with operating it with a team of astronomers and engineers funded by the SALT Foundation. South Africa is the largest shareholder in the SALT Foundation, using approximately half of the time on the telescope currently, with other partners from Europe, North America, and India sharing the other half. SALT boasts a competitive suite of multi-mode astronomical instrumentation in optical and near-infrared spectroscopy and imaging. Observation programs cover a wide range of topics, utilizing two spectrographs and an imaging camera. Observations are fully queue-scheduled and undertaken by dedicated operations staff.

South Africa has been heavily invested in research of astronomy, particularly in radio astronomy. Its strategic plan of radio astronomy research is strongly aligned with the human capacity development in physics, computational programming, mathematics, and the knowledge economy which can be derived from. With this in aim, the following projects have been established in South Africa over the past 5 years[2]:

13.1.3.1   KAT-7 Dish

The dish array of seven 12-m dishes was primarily built as a precursor to the 64-dish MeerKAT radio telescope array as a demonstration telescope. It is considered a compact radio telescope, with all antennae lying within an area only 200 m across. The configuration is mainly for observing nearby galaxies, which emit radio waves on a large scale.

13.1.3.2   MeerKAT

The South African pathfinder for SKA, is located on the SKA site in the Karoo, and is a pathfinder for SKA-mid technologies and science. It was designed by engineers within the South Africa Radio Astronomy Observatory and South African industries, and most of the hardware and software was sourced in South Africa. It comprises 64 antennas, each 13.5 m in diameter, equipped with cryogenic receivers. The array configuration has 61% of the antennas located within a 1 km diameter circle, and the remaining 39% distributed out to a radius of 4 km. Construction of the SKA precursor MeerKAT (Fig. 13.2), which is a 64-element array of offset parabolic antennas, has been completed in July 2018 near the SKA central site on the South African Karoo plateau. Each antenna has a primary reflector with an effective diameter of 13.5 m and a secondary reflector with a diameter of 3.8 m, covering 350 MHz–13.8 GHz (up to 20 GHz in the future). Its optical path is designed in an offset Gregorian layout. It ensures excellent reception sensitivity and imaging quality. The MeerKAT is hosted and operated by the South African Radio Astronomy Observatory (SARAO), started

---

[2] https://www.sarao.ac.za/science/.

Fig. 13.2 Chinese SKA delegation at MeerKAT site in Jan. 2017 South Africa

its large science surveys in 2019, which will be integrated into SKA1-MID around 2026. Together with 133 SKA1 reflector antennae, the SKA1_Mid will be formed.

13.1.3.3 Hirax

The Hydrogen Intensity and Real-time Analysis eXperiment (HIRAX) is an interferometric array of 1024 6-m (20ft) diameter radio telescopes, operating at 400–800 MHz, that will be deployed at the Square Kilometer Array site in the Karoo region of South Africa. The array is designed to measure redshifted 21-cm hydrogen line emission on large angular scales, in order to map out the baryon acoustic oscillations, and constrain models of dark energy and dark matter.

13.1.3.4 Hera

Hydrogen Epoch of Reionization Array, consists of 350 dishes of 14-m size, operating at frequency 50–400 MHz, aiming to detect the power spectrum of the 21-cm line at the epoch of reionization. It is an American, South African and British collaboration to build a telescope capable of making a solid detection of the Epoch of Reionisation (EoR) redshifted hydrogen power spectrum signature, as well as conducting initial EoR science and launching this new scientific field of the observational cosmic dawn.

13.1.3.5   Przim

The "Hibiscus" antenna measures the first absorption trough of mean 21-cm brightness at redshift 20. Probing Radio Intensity at high-Z from Marion, is a low-frequency radio telescope which collects information about the universe during the "Cosmic Dawn", the period a few hundred million years after the big bang when the first stars in the universe formed. The light from these first stars is too dim for optical telescopes to view, therefore they have never been measured directly. PRIZM was designed to make this measurement and to help determine when the first stars and galaxies formed.

13.1.4   Joint Project: Square Kilometre Array

The Square Kilometre Array is the world's largest radio telescope being built on earth, and it will provide a lot of ground-breaking astronomy research. For this reason, South Africa and China have jointly contribute to the Square Kilometre Array project. The SKA-mid array located in South Africa is an array of several thousand dish antennas (around 200 to be built in Phase 1) to cover the frequency range 350 MHz–14 GHz. It is expected that the antenna design will follow that of the Allen Telescope Array using an offset Gregorian design having a reflector of 18 m × 15 m. It covers a wide science area including Cosmology, HI galaxy, EoR, Cosmic magnetism and time-domain astronomy. It offers a tremendous opportunity for South Africa and China astronomers to collaborate from now to the next few decades.

13.2   Scientific Research

13.2.1   Cosmology and Radio Astronomy

Cosmology is a vibrant field of research, with new insights about the nature of the universe developing quickly. The past decade has seen a significant growth in cosmological observations that have placed increasingly tighter constraints on the cosmological model and the basic parameters that describe it. Chinese and South African research teams have contributed substantially to the extension of the current knowledge base.

It is now firmly established that about 85% of the gravitating matter in the Universe is invisible, i.e., made of the elusive dark matter, which so far, thwarted all detection attempts despite intense searches. Observations of Type-Ia supernovae and the analysis of anisotropies in the cosmic microwave background (CMB) have revealed the need for a significant dark energy component which contributes a staggering 70% to the total cosmic energy budget at the present time. Measurements of the CMB anisotropies have also constrained the primordial spectrum of fluctuations produced by inflation.

While we have an excellent phenomenological model a more fundamental picture is largely missing, considering both the early-time Universe where very high-energy processes are relevant and the late-time universe, where we are in the curious position of living in a universe that is to 95% dark. The big questions around the birth of the universe and the dark sector remain, despite our advances on the observational front. Understanding the nature of dark matter and dark energy will require both theoretical and experimental leaps. Deciphering the physics of the first instant of life of our Universe is evoking large experimental efforts and new ideas on the emergence of the space–time as we experience it here and now.

Observational progress in multiple directions has continued at a rapid pace in the last few years. Cosmic Microwave Background observations continue to develop quickly, with results from Planck, ACT, SPT, and others continuing to improve. Studies of neutral hydrogen at intermediate and high redshifts will revolutionize our understanding of the first stars in the universe and the fabric of space. The intensity mapping of neutral hydrogen at low redshifts will probe a large cosmic volume of the Universe which essentially measure more modes of fluctuations than previous observations. Currently, there are several South African and Chinese radio and optical surveys, complemented by large computational facilities, which are intended to provide accurate measurement on galaxies, neutral hydrogen and CMB intensity and polarization features. All of them aim at deepening our understanding of the Universe. The three major objectives of the cosmology collaboration are:

1. To bring together established experts, postdoctoral fellows and young students working at the cutting-edge of cosmology and astrophysics to discuss their research. The list of South African and Chinese surveys we show above share a lot of common scientific objectives and even share similar technologies and data analysis techniques. Therefore, we organized workshops for people who work on the different survey projects a unique opportunity to interact with each other and seeking collaborative room between the two to improve their research.

2. To teach and train SA and Chinese graduate students in conducting research in astronomy, which is a fast-developing area. To be an excellent researcher one must have a broad horizon of knowledge at the international level and be able to work with different international researchers from different background in a multi-disciplinary and multi-wavelength framework. This is, what South African and Chinese students require and need further training. The series of workshop we organised taught and trained students and postdocs in the frontier knowledge and expertise of modern astrophysics. In addition, the collaboration made young people know more about each other, and encourage their mutual interactions and collaboration.

3. To attract young graduate and undergraduate students to work in this field by offering them the opportunity to travel to and collaborate with researchers from the other country. For SA students, it is a great opportunity and great interest for them to study astronomy and be able to travel and work in China. Vice versa, for

Chinese students, traveling, collaborating, and working in SA provided a unique opportunity for them to do better research, but also enhance their English level, and make new networks. Therefore, opening this bilateral collaboration window of cosmology for SA and China attracted young people to study astronomy.

13.2.2 Optical Astronomy and Open Skies

Through several bilateral workshops, we brought together the Chinese and South African astronomers working specifically on optical and infrared observational topics. This should also be seen as paving way for a fruitful bi-lateral collaboration in astronomy in the BRICS context. A specific goal was to introduce the respective astronomical observational communities to each other. An important objective was to share experience also in instrument/telescope development for mutual benefit. In addition to holding the main event, a scientific workshop, and facilitating other trips, the objectives can be divided into the following aspects: (a) Starting new sustainable research collaborations. (b) Finding ways to share time on research facilities. (c) Strengthening each other's telescope and instrument development.

Overall, collaboration between Chinese and SA astronomers during this program grew significantly. There have been several collaborative observing proposals and data sets taken at Sutherland (SALT, 1.9-m, IRSF). The most active collaborations have included active Chinese participation on an SA-led Large Science Program on SALT regarding transients, on supernovae in general, on polarimetry, and on AGN. Details below.

The beginning of the collaboration in November 2016 was already a successful one. A Chinese-South African observing program was submitted to SALT during the visit, and at least one active collaboration started, with the Chinese astronomer (Jirong Mao) visiting in 2017 to observe on the SAAO telescope facilities in collaboration with an SAAO scientist (Steve Potter). Later on this collaboration was crucial in also taking scientific advantage of the major astrophysical event of the decade, the electro-magnetic follow-up of the GW170817 gravitational wave/neutron star merger event using SALT and other SAAO facilities. Two of the several SAAO-related papers stemming from the event were co-authored with SAAO astronomers and Jirong Mao. The Chinese-South African program is continuing to search for optical counterparts of gravitational wave events in 2019 using the facilities in South Africa and China.

13.2.2.1 SALT Transients Collaboration

Collaboration started between Dr. David Buckley, SAAO, and Prof. Subo Dong at Kavli Institute for Astronomy and Astrophysics at Peking University (KIAA). This broadened the scope of the transients program led by Buckley to include unusual supernovae, with a focus on those discovered from the ASAS-SN global network. The team has classified more than twenty supernova candidates and posted the findings at the Astronomer's Telegram and IAU Transient Naming Services. The collaboration has resulted in seven papers thus far (others in preparation).

13.2.2.2   Development of BRICS Transient Program

A BRICS concept proposal PI'ed by SAAO astronomers arising from discussions with Chinese colleagues during the workshop in Lijiang was submitted in 2018. This led to further developments. Buckley continued collaboration with several astronomers involved in transients in China, leading to development of the BRICS Flagship Program on Optical Transient Network (BRICS-OTN): Jifeng Liu, Dong Xu, Roberto Soria (all NAOC) and Jirong Mao (Yunnan Obs of CAS). These led to an amalgamation of various BRICS transients initiatives, including the Chinese "Sitian" project, into a very ambitious BRICS-OTN proposal submitted in August 2019.

13.2.2.3   Collaboration on HXMT ("Insight")

With J. Thomas, SA astronomers have been in contact with IHEP astronomers involved in the HXMT X-ray mission concerning coordinated X-ray/optical studies of black hole transients, in particular MAXI_J1820 + 070. People involved are: Shu Zhang and Shuang-Nan Zhang, Institute of High Energy Physics, CAS (IAHP).

13.2.2.4   AGN and Reverberation Mapping

The Lijiang workshop led to the initiation of a partnership of Hartmut Winkler from University of Johannesburg, with the group of Jian-Min Wang from the Institute of High Energy Physics at the CAS in Beijing. Winkler visited Wang in Beijing immediately after that workshop. Wang's group runs an international AGN reverberation mapping programme, mainly with the Chinese 2.5 m telescope at the Lijiang Observatory. That site regularly experiences very adverse weather conditions during the period June–August, and that made it particularly desirable to obtain spectra at Sutherland over this period. Hartmut Winkler, with the help of Francois van Wyk and Bynish Paul, observed AGN reverberation targets for 4 weeks in 2018 and again for 5 weeks in 2019. The Chinese team sent their PhD student Bo-Wei Jiang to South Africa and SAAO for 4 weeks in August this year. In 2019 the programme was also widened to include photometry with the SAAO 1.0 m telescope where the Balmer emission line light curve is determined through narrow filters—this latter work also saw the participation of Michael Hlabathe, Encarni Romero Colmenero and Steve Potter from SAAO.

The collaboration has so far led to one published paper and one submitted paper in Astrophysical Journal, with the analysis and write-up of further targets in progress.

Hartmut Winkler and Jian-Min Wang have also (with Thaisa Storchi-Bergmann from Brazil) submitted a proposal for 3-year funding under the BRICS framework.

13.2.2.5   Sharing of Observational Resources

A detailed plan to facilitate usage of Chinese and South African observing facilities has been discussed between SAAO and National Observatories of China (Petri Vaisanen, Subo Dong, Jirong Mao, Dong Xu, Ramotholo Sefako, David Buckley). The plan is to allocate approximately 5% of Sutherland observing time on SAAO telescopes to Chinese astronomers needing access to the Southern sky, and likewise a same amount of time for SA astronomers on the 2-m and 1-m class Xinglong and Lijiang telescopes. An advanced draft of an MoU exists between the SAAO and NAOC Directors. There has also been discussion on usage of reciprocal SALT and LAMOST access.

13.2.3   Computational Astrophysics, Big Data and Virtual Observatory

The emergence of global data networks, coupled with tremendous advances in imaging technologies and automation, has led to an explosion in the detection of transient phenomena in astronomy. In the next decade, large-scale all-sky surveys across the electromagnetic spectrum will produce enormous amounts of data and provide unprecedented discovery opportunities. This field is now a driving force expanding the frontier of modern astrophysics. Multi-wavelength observations of all types of transient signals, from nearby Solar System objects to the most distant and energetic sources in the Universe, are revealing previously unknown phenomena and unlocking the nature of their cosmic sources. Astrophysical phenomena radiate over a broad range of the electromagnetic spectrum, each window providing complementary physical insights, and multi-wavelength astronomy (the merging and joint analysis of such several data sets) is a critical requirement to achieve the science goals set out by modern astronomy programmes.

Beyond the multi-wavelength explorations of the electromagnetic spectrum, the field of multi-messenger astronomy today has made a quantum leap, with the inclusion of other domains such as cosmic rays, neutrinos, and gravitational wave astronomy. There is tremendous scope of synergy of these with traditional astrophysics, including optical and radio astronomy, thereby adding another dimension to combining large data sets to extract the maximum information about our Universe. Both of the SKA and the Rubin Observatory LSST will have a huge impact on the study of the structure and the variability of the Universe on unprecedented scale, across the radio and optical domains, respectively. Such projects represent a new era of mega-data production that will render conventional research and collaboration methods, as well as current data and visual analytics tools, ineffective. The SKA and LSST are driving the most significant big data challenges of the coming decades, and the required data solutions pave the way for the telescope network as unique and strategic programme under the leadership of astronomers within both countries.

13.2.4 Radio Astronomy, Geodesy and VLBI

13.2.4.1 Astrophysical Masers and Massive Star-Formation

Massive star-forming regions are energetic and highly dynamical regions. Many molecular species are found associated with these star-forming regions, in particular: hydroxyl, methanol, and water. Many of these species have transitions that exhibit maser emission. Some of these maser transitions are found exclusively towards massive, e.g., class II 6.7 and 12.1 GHz methanol masers. All of these astrophysical masers are variable, some even exhibit periodic variability. Since Galactic masers have high brightness temperature and their emission spots are very compact, they are good astronomical targets for VLBI observations. The workshop investigated possible projects to be carried out by taking advantage of the fact that China and South Africa are situated in the northern and southern hemispheres, respectively.

13.2.4.2 Astrometry

Both Shanghai Astronomical Observatory (SHAO/CAS) and HartRAO carry out experiments in absolute astrometry. SHAO/CAS has an ongoing observing program called VLBI Ecliptic Plane Survey (VEPS) whose purpose is to increase the number of phase reference calibrators along the ecliptic plane, while HartRAO focuses on accurate astrometry of very southern sources. In the area of high frequency astrometry HartRAO is organizing K-band Celestial Reference Frame (CRF) observations while SHAO/CAS is developing X/Ka-band astrometry experiments. Both facilities are well-placed, and are participating in, international VLBI imaging experiments studying calibrator sources from catalogues produced by the VLBA BESSEL project. SHAO/CAS and HartRAO offer all researchers access to the entire sky. We discussed joint efforts to strengthen the collaboration on the densification of the standard S/X-band CRF and its extension at high frequency such as K-band and X/Ka-band during the workshop.

13.2.4.3 Geodesy

HartRAO and SHAO/CAS are heavily involved in international geodetic research. Both facilities are erecting new VLBI2010 VGOS antennas for expanded geodesy programmes. Joint geodetic activities have already been conducted within the framework of the IVS. China hosted the 8th IVS conference and HartRAO hosted the 9th IVS conference in March 2016. There is considerable room for growth in collaboration between SHAO/CAS and HartRAO in the field of geodesy discussed at this workshop.

Collaborative projects within the fields of Global Navigation Satellite Systems (GNSS), Lunar Laser Ranging (LLR) and Satellite Laser Ranging (SLR) and intercontinental time transfer were considered. The Chang'E-4 lunar lander was equipped with a corner cube retro-reflector array; this enabled both optical and radiometric ranging to the Chang'E-4 lander.

Combination of these techniques will allow inter-technique reference system and technique range accuracy comparisons.

13.2.4.4   Lunar Geology and Lunar Radio Ranging

The National Astronomical Observatories (NAOC/CAS) constructed a radio solar imager array, with baselines of 5–50 km operating at frequencies between 600 MHz and 8 GHz. Already NAOC/CAS and NRF have signed an agreement for collaboration between this new Chinese facility and South Africa's MeerKAT. MeerKAT is a 64-element interferometer with a maximum baseline of 8 km and can operate at frequencies up to 15 GHz. The geological structure of the moon, its evolutionary history, sub-surface structure and volcanic dynamics, is not fully understood. It is proposed that the MeerKAT and solar radio imager be used to map the lunar radio temperature distribution at different depths in the frequency range from 300 MHz to 15 GHz.

Lunar radio ranging experiments, where radio signals are transmitted and received from the Chinese lunar rover Chang'E-3, improve the Earth-Moon distance measure to a resolution of millimeters. This new kind of space geodetic technique, built upon the success of geodetic programmes in both China and South Africa, greatly aids in lunar dynamical studies. Joint lunar radio ranging experiments (using radio telescopes from both countries), for geodetic, geodynamic, lunar dynamic, selenodetic studies are proposed.

13.2.4.5   An African Correlator

Chinese astronomers participate in AVN planning. SHAO/CAS operates a correlator for the CVN, which can also involve the correlation for the East Asia VLBI Network (EAVN) and IVS, while HartRAO is installing a high-speed computer with the view of developing an African correlator for the AVN. SHAO/CAS expertise in this area is instrumental in the success of HartRAO's correlator. Once completed each facility can act as a backup to the other or augment data correlation when the other is over-utilized. We discussed synergies in this area between SHAO/CAS and HartRAO during the workshop.

13.2.5 FAST and MeerKAT Synergy

Based on FAST and MeerKAT, we should be able to detect several thousand of new pulsars, study details of the HI gas in nearby and distant galaxies. With significant improvements in sensitivity and resolution by combining the two, the survey area of their common sky is still large in probing the low HI column density gas (cosmic web), low mass galaxies and HI intensity mapping for cosmological BAO measurements. With the joint efforts and funding of the National Science Foundation of South Africa and the National Astronomical Observatories of the Chinese Academy of Sciences, scientists from China and South Africa have conducted extensive and in-depth technical and scientific research on SKA project, signed relevant technical cooperation agreements, held a series of academic seminars, and realized bilateral exchange of visits and stable cooperation.

13.3 The Actual Collaboration

13.3.1 Cosmology and Radio Astronomy

13.3.1.1 Establishing a Joint Centre for Computational Astrophysics

The scientists from NAOC and UKZN had the strong synergetic interests in cosmology and radio astronomy, in particular using the South Africa's Radio Astronomy facilities such as SKA and MeerKAT to conduct cosmology research with 21-cm surveys. To facilitate this collaboration, scientists from the two sides established the NAOC-UKZN Computational Astrophysics Centre in November 2016. During the five-year cycle of 2017–2021, our major focus has been to employ a few joint postdoctoral fellows working on joint projects between the two sides. Before the end of each year, we posted job advertisement of Joint Postdoctoral fellows on the American Astronomical Society website and interviewed the subsequent candidates correspondingly. Then we identified one or two potential candidates and offered them the position. They have to stay in South Africa to work with the supervisor for 2 years and then relocate to China. In this way, the postdoc fellows carried out the actual research program from the two sides and effectively bridge the connection between the two countries.

This model has been working out successfully. So far, the following postdoctoral fellows were employed in this metric.

Dr. Yi-Chao Li, 2016–2018
Dr. Di-Fu Shi, 2018
Dr. Denis Tramonte, 2018–2019
Dr. Anthony Walters, 2018–2020
Dr. Cheng Cheng, 2017–2022

Dr. Prabhkar Tiwari, 2018–2020
Dr. Ramij Raja, 2021–2023

Most of the postdoctoral fellows are very productive during their tenure periods, with one or two publications completed. In the following, we highlight some of the scientific achievements from this collaboration.

13.3.1.2 Scientific Collaborations

21-cm Intensity mapping: Measuring the 21-cm emission from neutral hydrogen is a novel way to map out the large-scale structure of the Universe. The ultimate goal is to use 21-cm intensity maps to measure the baryon acoustic oscillation feature and deduce the dark energy equation of the state, which is indeed the motivation for the MeerKAT large-sky survey project and SKA-I. In the last several years, we have been developing the computational pipeline to analyse the mock SKA-I data (Harper et al., 2018, MNRAS) and FAST data (Hu et al., 2021, MNRAS; Yohana, Ma, Li, & Chen, 2021, MNRAS). With a joint team between NAOC and UKZN, we showed that robust neutral hydrogen power spectra can be obtained with different/complementary foreground removal methods (Yohana, Ma, Li, & Chen, 2021, MNRAS; Yohana, Li, & Ma, 2019, RAA). We further showed the prospect of using intensity maps to constrain primordial non-Gaussianity (Li & Ma, 2017, PRD). We recently expanded our collaboration to the Brazilian BINGO telescope (The BINGO Collaboration, 2021, A&A, arXiv:2107.01633; arXiv:2107.01634).

Apart from mocking SKA, we have also collaborated on real data. Of course, MeerKAT and FAST intensity maps are still pre-mature to use, and SKA construction just started. Instead, we collaborated with Australian astronomers to use Parkes telescope data. In Tramonte and Ma et al. (2019 MNRAS, 2020 MNRAS), we achieved 12.5-sigma C.L. detection of the aggregated signal of 48 k galaxies, which is the first detection of direct stacking of galaxies on 21-cm intensity maps. We further deduced the neutral gas fraction parameter (Omega_HI) of dark matter halo and gave an upper limit of HI in filaments, which are quantities of importance.

13.3.1.3 Theoretical Cosmology

The measurements of H_0 from high-redshift CMB and low-redshift distance ladder (e.g. Cepheids) have gradually diverged in the past 8 years. In Gong, Ma, Zhang, Chen (2015 PRD), we did not find much tension between the two, but it has been rising since. By 2019, the 1% precision of H_0 from local Type-Ia supernovae and Planck-derived H_0 have reached more than a 4-sigma discrepancy. In Dai, Ma, He (2020, PRD-RC), we proposed a novel "holographic dark energy" to reconcile this tension. The model gracefully reduced acceleration early and increased at a late time to capture the local high value of H_0 while keeping the CMB observable unchanged. This model is generically different from the standard LCDM paradigm and is verifiable/falsifiable with the soon DESI survey data. Along the time, we have probed the other interacting dark matter and dark energy models (Xu, Ma, & Weltman, 2018, PRD; Cheng et al., 2020, PRD) as alternative explanations.

13.3.2 Optical Astronomy and Open Skies

Collaboration began between Prof. David Buckley, SAAO, and Prof. Subo Dong at Kavli Institute for Astronomy and Astrophysics at Peking University (KIAA) following a visit by South African astronomers to China in 2017. This widened the scope of the existing SALT large science programme on transients program to include unusual supernovae, particularly those discovered from the ASAS-SN global network and other types of transients. The team has classified more than twenty supernova candidates and posted the findings at the Astronomer's Telegram and IAU Transient Naming Services. The collaboration has resulted in seven papers thus far, covering a broad range of topics including supernovae, tidal disruption events and exoplanet systems discovered via gravitational microlensing. One highlight was the discovery of SN 2019bkc, which was then the most rapidly declining hydrogen-poor supernova (Chen et al., 2020). The spectroscopic observation by SALT played an important role in its classification. SN 2019bkc could not be satisfactorily interpreted by existing supernova models, and its discovery has inspired several new theoretical and observational investigations on fast-evolving stellar explosions.

Transient science collaboration was expanded with the GW170817 gravitational wave event, where Dr. J. Mao (Yunnan Observatory) and Dr. Dong Xu (NCAO) participated in the follow-up observations of the kilonova optical counterpart, the first electromagnetic counterpart of a gravitational wave event, conducted by South African astronomers (including Profs. D. Buckley, P. Vaisanen & S. Potter, amongst others) on SALT and other SAAO telescopes, which led to two major papers being published.

A more recent collaboration between Mao & Buckley, together with an international team, involved the SALT and VLT Spectro polarimetric observations of the— ray burst, GRB191221A, leading to two papers covering the observational results and their interpretation, plus a theoretical prediction concerning detection of Zeeman features in GRB afterglows, respectively.

In November 2016, a delegation of 10 Chinese astronomers visited SAAO. A two-day science workshop was held at SAAO/Cape Town, and a site visit to Sutherland was organised for the delegation together with several senior SAAO/SALT astronomers. In addition to introductions and science talks and collaboration during this visit, the main event of the whole program was planned; it was decided to be held in 2017 in China.

The main event of the program was held in 2017 in China, the "Open Skies from China to South Africa – Sharing Resources and Building Collaborations in Optical and Infrared Astronomy" conference in Lijiang, Yunnan Province, on Oct 30–Nov 1, 2017. Apart from about 30 Chinese astronomers, the scientific workshop was attended by 19 South Africans, including senior astronomers, postdocs, students, and a handful of engineers. In addition to science talks and discussion, there were topical discussion sessions on observatory cooperation, and a visit to the Lijiang Station of the Yunnan Observatory and its 2.4 m telescope.

In addition, a smaller (7) group of the SA delegation visited the Nanjing Institute of Astronomical Optics & Technology (NIAOT) to specifically plan instrumentation collaboration. They also visited the Purple Mountain Observatory. And still another group (3) visited the Xinglong Station of the National Astronomical Observatories (NAOC), as well as the NAOC headquarters in Beijing. All these visits, conducted during the week before the Lijiang meeting in October 2017, and directly after the Lijiang meeting in November 2017, included formal talks and plenty of active and positive collaboration.

The workshop also supported the standalone visits of Hartmut Winkler to the Institute of High Energy Physics at the Chinese Academy of Sciences in Beijing, after the Lijiang workshop, as well as an extended science collaboration trip of Moses Mogotsi also after the Lijiang workshop. The program also partially supported David Buckley's visits to Peking University and NAOC after the workshop. We intend to finalise and sign the SAAO/NAOC MoU on time sharing in the near future.

13.3.3   Computational Astrophysics, Big Data and Virtual Observatory

The science goals of the BITDN build on past collaborations and bilateral STI programmes between South Africa and China in both transient and time domain astronomy and data intensive survey science. An example of this is the collaboration between two of us (Buckley, Dong) which widened the scope of the SALT transients programme to include unusual supernovae, particularly those discovered from the ASAS-SN global network, and other types of transients. The team has classified more than twenty supernova candidates and posted the findings at the Astronomer's Telegram and IAU Transient Naming Service. The collaboration has resulted in seven papers thus far, covering a broad range of topics including supernovae, tidal disruption events and exoplanet systems discovered via gravitational microlensing. One highlight was the discovery of SN 2019bkc, which was then the most rapidly declining hydrogen-poor supernova. The spectroscopic observation by SALT played an important role in its classification. SN 2019bkc could not be satisfactorily interpreted by existing supernova models, and its discovery has inspired a number of new theoretical and observational investigations on fast-evolving stellar explosions.

Fig. 13.3 The South Africa–China bilateral workshop on Radio Astronomy, Geodesy and Space Science, Pingtang County, Guizhou province

13.3.4   Radio Astronomy, Geodesy and VLBI

The China–South Africa Bilateral Workshop on Radio Astronomy, Geodesy and Space Science was held June 1–6, 2017 in Kedu Town of Pingtang County with more than 50 participants from both countries (Fig. 13.3). The main topics of the workshop include molecular masers and massive star-formation, cosmology, astrometry and space geodesy, lunar radio science, as well as VLBI network and correlator. In addition, there are talks on the topics such as the SKA, the FAST and the MeerKAT. Based on the common interests, more than 20 collaboration projects have been proposed during the meeting, which strengthen the bilateral collaboration in the field of radio astronomy, geodesy and space science as well as related technology between the South African and Chinese astronomy communities.

13.3.5   FAST and MeerKAT Synergy

In January 2017, the Chinese SKA delegation, consisting of experts and scholars from the Ministry of Science and Technology (MOST), the Chinese Academy of Sciences (CAS) and China Electronics Technology Group Corporation (CETC), visited the headquarters of the South African National Research Foundation (NRF) in Bituo, the SKA Observatory in Cape Town. They inspected the construction of SKA precursor MeerKAT in Karoo (Fig. 13.2). Deputy project manager of FAST Project Bo Peng and chief scientist of SKA South Africa Fernando Cimilo are committed to establish MeerKAT-FAST coordination mechanism, especially the complementary cooperation between China's relatively strong pulsar team and South Africa's relatively strong neutral hydrogen power. Deputy Director of SKA Office of CETC Feng Wang and SKA Antenna Project manager in South Africa William Easterhuyse delivered and received the first SKA prototype antenna at SKA Site in South Africa, together with chief technologist of SKA South Africa Justin Jonas, to study the incorporation of CETC54 antennas into AVN (African VLBI Network) construction plan. Science and Technology Counsellor of the Chinese Embassy in South Africa Mr. Wei Huang, Deputy Director-General of the South African Department of Science and Technology Mr. Daan Du Toit, and Deputy Director of the National Research Foundation Mr. Nithaya Chetty, attended the discussion at the NRF headquarters. The two sides planned to develop a bilateral astronomy institute or laboratory based on the existing personnel exchange program of the South African Ministry of Science and Technology, the science popularization program, and the bilateral international seminar program of the South African NRF, to comprehensively coordinate astronomy research and relevant high technology under SKA framework.

On June 15–16, 2017, the China–South Africa FAST and MeerKAT collaboration meeting was held in Cape Town, which was mainly to prepare for the bilateral scientific cooperation seminar on large scientific installations to be held in Guizhou in a year (Fig. 13.4). The two sides exchanged in-depth information on the development and early science of South Africa's SKA pathfinder MeerKAT, the construction and commissioning progress of China's FAST, and topics of bilateral interest including neutral hydrogen, cosmic magnetic field, transient objects, pulsars and VLBI. More than 20 participants (Fig. 14.4) including NAOC, Purple Mountain Observatory, Peking University, SKA South Africa Observatory, UCT, WCT, UKZN and Rhodes University attended. Claude Carignan of the University of Cape Town and Bo Peng of NAOC co-chaired the conference.

In June 13–15, 2018, we organized a 2.5 days workshop in Pingtang county of Guizhou province. The main purpose is to explore the synergies between the different SKA pathfinders (including MeerKAT) and FAST. There are more than 100 attendees in the workshop, entitled "FAST-MeerKAT and SKA Pathfinders Synergies" (Fig. 13.5), including 17 postdocs and 37 post-graduate students. There are 26 oral talks as well as 4 posters presented, the topics involved progress reports on pulsars, understand evolution of the Universe and Galaxies by HI observations, cosmic magnetism and galactic ISM, instrumentation, and data handling. A conference proceeding (Fig. 13.6) was published including 23 papers.

In June 2020, in order to implement the spirit of the remarks made by General Secretary Xi Jinping and President Cyril Ramaphosa at July 2018 High-level Dialogue between South African and Chinese Scientists, encouraging scientists from both sides to continue the efforts to build the SKA project, the bilateral seminar between China and South Africa was held by video to discuss scientific cooperation. Seminar was convened by the Chinese embassy in South Africa Long Shen, more than 30 people attended, including SKA China office Director of Qi-an Wang, Chinese SKA chief scientist Xiangping Wu, scientific innovation of South Africa's deputy director general of Dan du Kuyt, Radio Astronomy Observatory in South Africa (SARAO) Director Robert Adam, chief scientist of the SARAO Fernando Camilo, and representatives of SKA scientists, participated online. Prof. Xiangping

Fig. 13.4 Preparing for FAST-MeerKAT synergy in Cape Town, June 2017

Fig. 13.5 Conference photo of FAST-MeerKAT and SKA pathfinders synergies

Fig. 13.6 Conference photo in front of the FAST telescope (left panel) and the conference proceedings (right panel) in Guizhou province, June 2018

Wu pointed out that FAST and MeerKAT are both medium frequency radio observation facilities, SKA in China and South Africa can cooperate in data sharing, joint VLBI observation and SKA Regional Center construction. Fernando Camillo introduced the technical features and latest research progress of MeerKAT, as well as the recent scientific research plan and suggested key directions for cooperation. The two sides indicated that future cooperation could be conducted in establishing joint research centers.

13.4   Future Outlook

As shown above, China–South Africa collaboration in astronomy has started and fast developed in the past 5 years, in cosmology and radio astronomy, big data science and open skies, VLBI and also the optical astronomy. Several major observational initiatives are being explored in both countries, namely:

13.4.1   The SAAO Intelligent Observatory (South Africa)

This project aims to network the SAAO's suite of telescopes into an intelligent transient follow-up machine where targets are selected by event brokers and follow-up observations automatically scheduled. This will support follow-up of LSST transient and variable sources, an area of strategic importance with South Africa's involvement in the project. This is both for research but also to support the wider US and Chilean LSST communities as one of South Africa's in-kind contributions to the 10 years survey project is access to its facilities for follow-up.

13.4.2   The Sitian Project (China)

This ambitious proposed future strategic project, led by NAOC, is to develop a network of globally distributed wide-field 1-m telescope (~ 70 of them) enabling the entire celestial sphere to be monitored for transient and variable object with a cadence of ~ 1 h. This will be a potential game changer for the discovery of fast transients.

Both projects have overlapping science goals and now form part of a BRICS Flagship programme, the BRICS Intelligent Telescope and Data Project. In support of these science goals, a new 2nd generation SALT instrument is being developed, namely RSS-Dual, which will incorporate the existing RSS spectrograph with a newly developed red optimized spectrograph, enabling efficient coverage of the entire spectral region from 390 to 900 nm by the two "arms" of RSS, in the blue and red pectral regions. A driver for this is again transient and survey science.

13.4.3   The VLBI Network

China and South Africa hold common goals in developing VLBI networks, satellite laser ranging facilities, lunar radio & laser ranging facilities. Closer relations in the field of VLBI where networks of telescopes are required. Both South African and China will benefit from linking to the respective Northern and Southern hemisphere networks and sharing expertise in VLBI technology and development (e.g. new high frequency receivers). This strengthened relationship will also add value to the development of the African VLBI Network.

13.4.4   Radio Astronomy Research

Each country is investing heavily in radio astronomy facilities and research. China is fast developing a leading space programme requiring Southern Hemisphere facilities to track and monitor spacecraft and probes. Below we list the value-add to Chinese-South African cooperation in Astronomy:

(a) Both China and South Africa are founder countries and members of the SKAO. South Africa has successfully built and operated the unique SKA-mid precursor, MeerKAT. On the other hand, China constructed the 21 Centimeter Array (21CMA) in west of China in 2016, the first low-frequency array dedicated to the detection of the epoch of reionization. Both MeerKAT and 21CMA work in a compensating manner such that the whole SKA frequency band can be well covered. SKA is recognized to be a high-priority project during the South Africa–China Scientists' High Level Dialogue attended by the two Presidents. More can be expected from bilateral SKA collaboration.

(b) Both countries will benefit from the exchange of students and researchers leading to greater research output and new research fields. In particular in the field of star-formation and pulsars, where both countries have expertise, by working together we can become leading global players.

(c) Each country shall benefit from access to the respective Northern and Southern skies from an observing perspective, e.g. MeerKAT and FAST collaboration.

(d) By collaborating we will be able to develop the exciting field of lunar geology that will lead to manned missions and potential lunar mining etc. Likewise, our shared knowledge will ensure success in the fields of lunar radio and laser ranging and the research associated with this.

13.4.4.1   Computational Capacities Will Be Co-developed in Two Countries

Projects undertaken by large global collaborations, in which large amounts of observing time are devoted to major key science programs that create vast data sets, is becoming the new paradigm. This mode of observing combined with the new instrumental capacities is driving an exponential growth in the rate of data confronting researchers. As illustrated in Fig. 13.7, China and South Africa are at the forefront of this challenge. New and upcoming major facilities, in particular the SKA and its pathfinders, present the biggest data challenges of the coming decade. Developing the systems to support effective analysis of vast amounts of data has become an

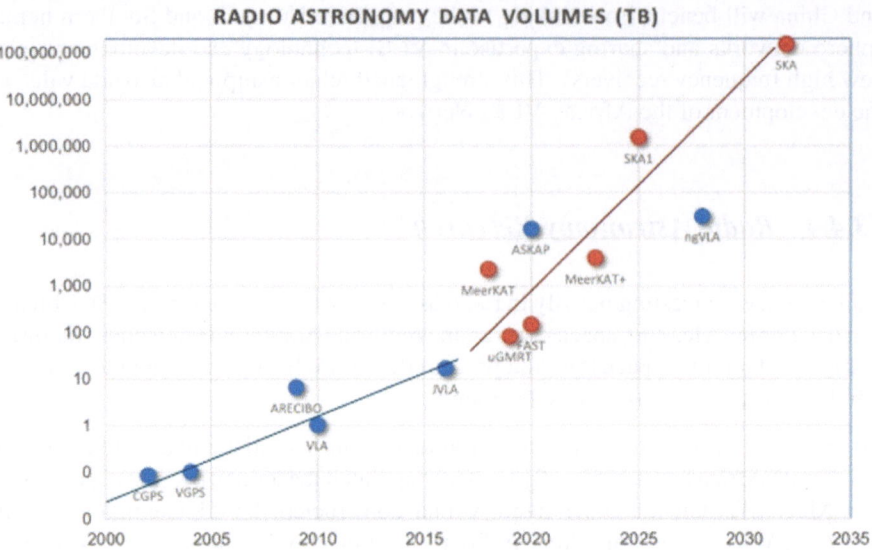

Fig. 13.7 Data volumes per day for large projects on radio astronomy facilities as a function of time. Current and planned facilities are indicated in red. (Taylor, Private communication)

integral part of how we approach modern science and technology. We must develop the necessary skills and tools that astronomers will need, many of which can be borrowed by, and applied in, other fields of science and engineering.

13.4.4.2   FAST-MeerKAT Synergy Will Be Strengthened

China's FAST passed national acceptance and inspection in January 2020, has been then officially opened to the international astronomical society in March 2021. Construction of South Africa's MeerKAT radio telescope array was completed in 2018, there are 10 key science projects setting for independent observations until SKA-1, including pulsars and testing the theory of gravity. Currently, MeerKAT is the most sensitive interferometry, FAST is the world's largest single-aperture tele-scope. The two facilities complement each other, and can contribute to breakthroughs in cutting-edge science on pulsars, neutral hydrogen, gravitational wave detection, galaxy formation and evolution, and the epoch of re-ionization etc.

The cooperation based on FAST and MeerKAT has made substantial progress in personnel exchanges and scientific research. Dr. Yinzhe Ma is a professor of UKZN, who is very active in academic activities and exchanges, plays a very important role in the cooperation. Dr. Ming Zhu has joined the MeerKAT's galaxy Survey working group, planning to perform MeerKAT + FAST observation experiments to explore two data joint imaging processing methods of single antenna and interference array, and obtain high dynamic range images. In September 2020, MeerKAT launched its first global call for applications, receiving 113 PIs applications from 18 countries and officially announcing the judging results on February 16, 2021. A total of 57 projects were approved, including 42 in Category A and 15 in Category B. China obtained 3 category A projects, accounting for more than 5% (3/57) of MeerKAT approved projects, and the priority ratio was more than 7% (3/42).

The SKA is a super-scientific installation that brings scientists from both countries together, where China and South Africa are both formal members. South Africa is one of the two site countries, the SKA cooperation between China and South Africa will be close. The establishment of MeerKAT-FAST coordination mechanism is conducive to the radio astronomy between China and South Africa, which is beneficial to scientific research capacity of both sides. With the emergence of the new international cooperation pattern of FAST and MeerKAT, radio astronomy in China and South Africa will surely usher in a new and great development and welcome the arrival of the SKA era.

The Joint Platform for Astronomy between China and South Africa (JPACS) is proposed to be created based on MeerKAT and FAST's early scientific research, to carry out cooperation in frontier astronomy, high and new technologies and personnel training. The high-performance antenna and receiver technology cooperation will raise the level of the related electronics, information technology, ultra weak signal detection and processing, data transmission, storage, sharing, to provide an opportunity to lay a foundation for international talents (radio astronomy research and technical teams).

13.5   Concluding Remarks

The China–South Africa collaboration in astronomy has been started and driven forward by a group of diligent scientists from the two countries in the past 6 years. With appropriate funding support, this initiative will propel the two countries to the scientific forefront of the exploration of the mystery of the Universe.

Acknowledgements The China–South Africa Astronomy collaboration has been strongly supported by several institutions since 2016, so here we acknowledge the support from Department of International Cooperation Ministry of Science and Technology of People's Republic of China (MOST), Bureau of International Cooperation Chinese Academy of Sciences (CAS), National Natural Science Foundation of China (NSFC), National Remote Sensing Center of China (NRSCC), SKA China Office. We would also like to thank National Research Foundation and University of KwaZulu-Natal for distinctive kick-off support for the joint collaboration. In particular, we thank Sui-Jian Xue (NAOC薛随建), Gong-Bo Zhao (NAOC 赵公博), Prof. Nithaya Chetty (NRF/Wits), Mr. Takalani Nemaungani (DSI) and Mr. Guo-Xuan Dong (NSFC 董国轩) for their significant propelled work.

Funding Y.-Z. Ma acknowledges the support from National Research Foundation with Grant No. 150580, No. 159044, No. CHN22111069370 and No. ERC23040389081 and Dr. C. Cheng (程程) and Dr. W.-M. Dai (戴卫明) for helping to shape this document. S.D. was supported by the National

Key R&D Program of China (Grant No. 2019YFA0405100) and the National Science Foundation of China (Grant No. 12133005).

References

Chen, P., Dong, S., Stritzinger, M. D. et al. (2020). The most rapidly declining type I supernova. *The Astrophysical Journal Letters, 889*(1), 2019bkc/ATLAS19dqr.

Cheng, G., Ma, Y.-Z., Wu, F., Zhang, J., & Chen, X. (2020). Testing interacting dark matter and dark energy model with cosmological data. *Physical Review D, 102*, 043517.

Dai, W.-M., Ma, Y.-Z., & He, H.-J. (2020). Reconciling hubble constant discrepancy from holographic dark energy. *Physical Review D, 102*, 121302 (**Rapid Communication**).

Gong, Y., Ma, Y.-Z., Zhang, S.-N., & Chen, X. (2015). Consistency test on the cosmic evolution. *Physical Review D, 92*, 063523.

Harper, S., et al. (2018). Impact of simulated 1/f noise for HI intensity mapping experiments. *Monthly Notices of the Royal Astronomical Society, 478*, 2416.

Hu, W., et al. (2021). 1/f noise analysis for FAST HI intensity mapping drift-scan experiment. *Monthly Notices of the Royal Astronomical Society, 508*, 2897–3290.

Li, Y.-C., & Ma, Y.-Z. (2017). Constraints on primordial non-gaussianity from future HI intensity mapping experiments. *Physical Review D, 96*, 063525.

Tramonte, D., & Ma, Y.-Z. (2020). The neutral hydrogen distribution in large-scale haloes from 21-cm intensity maps. *Monthly Notices of Royal Astronomical Society, 498*, 5916–5935.

Tramonte, D., Ma, Y.-Z., Li, Y.-C., & Staveley-Smith, L. (2019). Searching for HI imprints in cosmic web filaments with 21-cm intensity mapping. *Monthly Notices of the Royal Astronomical Society, 489*, 385–400.

Xu, X., Ma, Y.-Z., & Weltman, A. (2018). Constraining the interaction between dark sectors with future HI intensity mapping observations. *Physical Review D, 97*, 083504.

Yohana, E., Li, Y.-C., & Ma, Y.-Z. (2019). Forecasts of cosmological constraints from HI intensity mapping with FAST, BINGO & SKA-I. *Research in Astronomy and Astrophysics, 19*, 186.

Yohana, E., Ma, Y.-Z., Li, D., Chen, X., & Dai, W.-M. (2021). Recovering 21-cm signal from simulated FAST intensity maps. *Monthly Notices of the Royal Astronomical Society, 504*, 5231–5243.

Chapter 14
Technology Transfer and Technological Spillovers from Chinese Tech Giant in North African Countries: The Case of Huawei in Algeria

Tin Hinane El Kadi and Abdelkader Djeflat

14.1 Introduction

South-South cooperation (SSC) has attracted the attention of development economists for decades with the hope they bear in building radically different relations between nation States. While the South-South cooperation charter dates back to the Groupe of 77's meeting in Algiers in 1967 (Muhr & De Azevedo, 2018), there has been a widespread sense recently that the time is ripe for putting SSC once again at the centre stage of world politics and economics to redraw the architecture of the global order (Grey & Gills, 2016).

China, as an emerging economy with an exceptional development path (Naughton, 2006), has shown strong support for South-South cooperation and commitment in recent years to increase investments in other developing countries through its Belt and Road initiative (BRI). It has also signalled through white papers and discourses its willingness to share with African countries its rich experience in technological catch-up and transfer of 'best practice' (Morais, 2008). For instance (Mu & Lee, 2005), the China–Africa Science and Technology Partnership Program (CASTEP) was launched in 2009 to establish a robust S&T partnership with African countries. In a recent move, China's engagement in Africa's digital sector materialized under the China–Africa Partnership Plan on Digital Innovation passed in August 2021. This partnership pledges to share digital technologies with African countries to promote digital infrastructure, connectivity and to help expand Internet access in Africa's

T. H. E. Kadi
International Development Department, London School of Economics and Political Science (LSE), London, UK

A. Djeflat (✉)
University of Lille, Lille, France
e-mail: adjeflat@gmail.com

remote areas. It constitutes perhaps the single most important foreign initiative to help the continent's digital transformation.

With the highest number of internet users worldwide, estimated at around one billion (Tomala, 2022) and a thriving digital economy, China has expanded its global digital footprint since the early 2000s. Chinese information and communication technology (ICT) multinational corporations (MNCs) have built the backbone infrastructure used by millions of internet users across developing countries (Gagliardone, 2019; Oreglia, 2012). China stands to become an even more critical actor in the ICT industry in the foreseeable future through the Digital Silk Road (DSR), the digital component of Beijing's multibillion-dollar BRI (Hillman, 2021).

While China's BRI is best known through its two central paths: the "Belt", which refers to the overland routes, known as "the Silk Road Economic Belt", and the road, which refers to the sea routes, or the "Maritime Silk Road", the BRI's digital dimension is significant. In 2016, China's State Council published the "13th Five-Year Plan for National Informatization," dedicating an entire section to the creation of an "online Silk Road" and encouraging the full participation of Chinese tech firms (CCCPC, 2016). In May 2017, speaking at the first BRI forum in Beijing, Chinese President Xi Jinping announced that big data would be integrated into the BRI to create the "Digital Silk Road of the twenty-first century".[1]

The digital Silk Road brings to BRI countries advanced telecommunication infrastructures such as fibreoptic cables, 5G networks, data centres, satellites, as well as e-commerce hubs and smart cities. It aims to deepen digital cooperation, develop common technology standards, and improve the efficiency of security systems across BRI countries (Shen, 2012). The Digital Silk Road aims to build what Beijing calls "a community with a shared future in cyberspace" (Oreglia et al., 2021; Wei, 2019). This ambitious plan is primarily driven by China's tech giants, most notably Huawei and ZTE (Zhongxing Telecommunication Equipment), who can deliver high-quality fibre optic cables at much lower costs than their European and US competitors, namely Ericsson, Nokia and Cisco. Huawei's ICT equipment generally sell 5–15% cheaper than its main international competitors, Ericsson and Nokia, while ZTE has in the past offered equipment up to 30–40% cheaper (Cisse, 2012; Hart & Lind, 2020; Joo et al., 2016). In 2012, Huawei overtook the Swedish-headquartered company Ericsson in revenue to become the world's largest telecommunications equipment vendor (Lee, 2012).

With the announcement of the Belt and Road Initiative (BRI) in 2013 and the Chinese government's 2016 Arab Policy Paper, China has shown renewed interest in the region. Beijing has committed to increasing investments in high-value-added sectors and to boost cooperation in science and technology with countries across North Africa and transfer technology through academic exchange and R&D activities (Cheney, 2019).

---

[1] Keynote speech delivered by the Chinese President Xi Jinping at the opening ceremony of the Belt and Road Forum (BRF) for International Cooperation in Beijing, capital of China, May 14, 2017 (Xinhua/Wang Ye).

With its strategic location, connecting Asia, Africa, and Europe through the Suez Canal, North Africa holds a central position in China's BRI (Abdel Ghafar & Jacobs, 2019). All of the five North African countries have signed memorandums of understanding to join the BRI, and the region is home to several hallmark BRI projects. From Mao Zedong's "Three World" theory to Xi Jinping's Belt and Road Initiative, China and North Africa have developed and sustained strong relations rooted in a shared experience of colonial domination (Pairault, 2017). In recent years, the region became host to several hallmark DSR infrastructure projects, including data centres and smart cities built by Chinese firms (El Kadi, 2019). Between 2004 and 2020, trade between China and North African countries grew from $4.9 billion to $33 billion, nearly seven times more in just seven years[2], Between 2005 and 2016, Chinese companies carried out contracts around $ 22 billion (Ghanem & Benabdallah, 2016).

While North African countries have different political economies, four of them (Algeria, Tunisia, Egypt and Morocco) middle-income status and have in common growing numbers of tech-savvy young people, a relatively high rate of internet penetration, and proximity to the EU market (World Bank, 2019b), making the region a strategic hub for the DSR. Like elsewhere on the continent, after rapid socio-economic progress in the aftermath of independence, the 1980s and 1990s were characterized by rapid deindustrialization and a rise in inequalities across North Africa. The region's economies are stuck in low added value sectors, and for some of them, primary-commodity exports with chronic unemployment representing a distinctive feature of North Africa (Azmeh & Elshennawy, 2020; Kabbani, 2019). In this context, the DSR is perceived by North African governments as an opportunity to accelerate digital transformations and escape the middle-income trap while creating quality jobs for millions of unemployed people in the region. Although commentaries on the BRI's implications abound (Ahrens, 2013; Hillman, 2021; Kennedy, 2006), China's ambition to extend its digital footprint has so far been marginalized in both academic and policy circles, despite its importance.

Existing writings on China's global digital expansion have predominantly focused on the potential threat this could represent to the West's hegemony over the Internet (Hillman, 2021) and, more specifically, to Washington's supremacy over the cyberspace with the fear that Beijing would reap the economic, political and intelligence advantages that once flowed to Washington (Segal, 2018). Mainstream accounts have expressed concerns regarding the reproduction of China's Internet model abroad, a model that has been dubbed as digital authoritarianism, or even more vaguely as "digital Leninism".

A major problem with some of this literature is that it tends to marginalize domestic agency and depict China as an "all-powerful actor able to transpose its model into poorer nations, convincing them through attractive loans, while Beijing advances its masterplan. An emerging body of empirically grounded studies has challenged this assumption, showing wide variations between companies depending on local

---

[2] "China–Africa Trade," Johns Hopkins School of Advanced International Studies China Africa Research Initiative, 2020, http://www.sais-cari.org/data-china-africa-trade.

context, sectors, and ownership type, i.e., whether a company is private or state-owned (Brautigam, 2009; Oya & Schaefer, 2019; Calabrese & Tang, 2020). In the digital sphere, emerging evidence suggests that China's engagement supports nationally rooted visions of the Internet, with Chinese companies expanding by forming various partnerships with local actors and adjusting domestic needs and preferences (Gagliardone, 2019). To date, empirical research on the developmental contribution of Chinese digital multinationals, especially with regard to technology transfer remains scant (Agbebi, 2019; Tugendhat, 2021; Demena et al., 2019). Using the case study of Huawei in Algeria, one of the largest recipients of Chinese capital in North Africa, this chapter investigates the technological spillovers emanating from the Chinese tech giant. Going beyond the West's grim picturing of China's presence in Africa as a new form of colonialism, this chapter seeks to tell an empirically rich and nuanced account of the Chinese contribution to digital transformation of North African countries through the Algerian experience and more specifically answer a key question: to what extent Huawei has been able to transfer technology through knowledge spillover to local.

14.2 Technology Transfer and Spillover Through MNCs Operations in the South: Literature Review

Several contributions have highlighted the issue of technology transfer in relation to building capacity in the host country (Kim & Dahlman, 1992; Mowery & Oxley, 1995). Kim and Dahlman (1992) argue that many of the 'late-industrialisers' of the postwar period, such as South Korea, Taiwan (China), and Hong Kong (China), initially exploited foreign sources of relatively mature technologies. Such technology could be transferred through channels that relied on *arm's length* transactions, such as licensing, turnkey plants, and capital goods imports. The economies that have benefited most from inward technology transfer had national systems of innovation that include public policies strengthening their 'national absorptive capacity" (Cohen & Levinthal, 1990). Similarly, at firm level, the exploitation of external technology requires the creation within the firm of some 'absorptive capacity', an ability to understand an externally sourced technology and apply it internally. Inward technology transfer during the postwar period supports the conclusions that the mix of channels through which an economy obtains technology from foreign sources (FDI, arm's length, licensing etc.) but most of all the overall effort to exploit foreign sources of technology.

But broadly speaking, technology transfer, seen as the dissemination of technical knowledge and know-how embodied in products, processes, and management (Wahab et al., 2011), seem to be primarily through foreign direct investment (FDI) long been regarded as a significant engine of technological upgrading and capacity building (Globerman, 1979; Markusen & Venables, 1999; Amsden, 2001; Saggi, 2002; Blalock & Gertler, 2008; Fu et al., 2011). While technology transfer takes on several modes and shapes, the attention of scholars seem to privilege spillovers as an important channel, when it comes to FDI (Hoi Quoc and Pomfret, 2011) Technology spillovers reflect the unintended transfer of technology, while technology transfer has a more intentional/deliberate connotation (Smeets, 2008). Similarly, knowledge transfer implies a broader, more general type of knowledge, while technology transfer is narrower and more targeted (Holm et al., 2020). For the sake of this chapter, these terms are used interchangeably.

There are several mechanisms through which a host country can reap the benefits of inward technology. At macro-economic level, technological benefits generally assume the form of spillovers, or external effects, that go beyond such direct sources of economic benefit as higher-productivity, higher-wage manufacturing jobs.' Important sources of spillovers are 'reverse engineering, which may result in the development of similar products for indigenous manufacture and export, and skills acquisition through 'learning by using'. The extent of these spillovers depends on several factors: the age of transferred technology, the channel of transfer and the level of indigenous technical capabilities. Potential spillovers are greatest for the most up-to-date technologies, but exploitation of these spillovers is limited by the controls imposed by the transferring firm, and by the level of indigenous technical capabilities (Mowery & Oxley, 1995).

The basic premise underlying the existence of technology spillovers is that foreign invested firms are technologically superior to local ones. Thus, their interaction is assumed to lead to technology transfers which, in turn, lead to productivity gains (Saggi, 2002; Javorcik, 2004). Given the lower technology base within developing economies, these spillovers may help local industries build up their domestic technological capabilities and catch up with the international technology frontier (Lall, 1992; Ning & Wang, 2018). When multinational firms venture abroad, they cannot always maintain full control over their managerial and technological knowledge, allowing local ecosystems to learn from them. These instances are known as knowledge spillover (Crespo & Fontoura, 2007). Demonstration effects are recognized in the literature as the most powerful channel for technology spillovers: local firms (customers, suppliers, subcontractors) observe and potentially adopt new practices and techniques by collaborating with multinational firms, something that leads to productivity gains (Blomstrom & Kokko, 2001; Crespo & Fontoura, 2007). A second important mechanism of spill over is labour mobility; this is when local employees trained by foreign firms move to domestic firms, taking with them new managerial and technical knowledge (Görg & Strobl, 2005).

The empirical evidence on the transfer of technology and spillovers through FDI by Western MNCs are, 'at best' mixed. In their seminal study on technology spillovers in Morocco, Haddad and Harrison (1993) found that if domestic and foreign firms compete to capture the same market, then the latter does not have incentives to promote technology linkages. In some instances, foreign firms were found to operate as enclaves with little connection to the local economy (Liu et al., 2009; Aitken & Harrison, 1999; Matija & Knell, 2018). Measures adopted by foreign companies to limit technology transfer, and spillover include protecting their intellectual property, trade secrecy, hiring mainly foreign workers, and preventing labour turnover by providing wages that are significantly higher than local industry averages (Liu, 2008; Liu et al., 2009). In other instances, research showed that foreign subsidiaries did more harm than good to the local economy by capturing the domestic market and crowding out local competitors without engaging in any meaningful form of technology transfer (Amendolagine et al., 2013). This is often explained by the wide technology gap between Northern companies and their counterparts from the South (Sazali et al., 2011).

Consequently, Chinese firms emanating from a developing nation raise new hopes. Unlike Ericsson and Cisco, who are dispatched from rich countries, Chinese tech giants like Huawei could thus represent a greater opportunity for host middle-income countries to extract new technology and, like China, upgrade to producing high-tech equipment and services for the digital era. This is substantiated in several contributions. According to the literature, the smaller the technological gap between the country from which the MNC is dispatched and the host economy, the more likely the foreign subsidiary will transfer technology and know-how that can more readily be absorbed and applied locally (Glass & Saggi, 2002: 497). South-South investments are thus assumed to generate more meaningful technology transfer than investments from industrialized countries, increasing the likelihood of technological learning and innovation (Takii, 2005; UNCTAD, 2012; Kubny & Voss, 2014).

14.3   Methodology

From a methodological point of view, this work uses a qualitative case method, felt more appropriate in this situation (Cunningham et al., 2017). It draws from fieldwork research carried out through in-country interviews in Algeria's ICT ecosystem conducted between July 2021 and February 2022. Interviewees included Huawei representative in Algiers and employees, Algerian subcontractors, customers, former assembly line managers and workers, Algerian ICT students and students receiving training from the tech giant. They also included, policymakers, ICT experts and university faculty/researchers. It also draws on the lead author's observations during employment at the Algiers office of Huawei between 2015 and 2016. This work experience has allowed for a better understanding of the dynamics shaping the localization of Chinese firms and the bargains around knowledge transfers. In addition, a large documentary search was conducted in both Latin and Chinese language, thanks to the fluency of the lead author in Chinese language. Among the obstacles met the difficult of access to several relevant documents, which are not publicly disclosed for reasons ranging from commercial secrecy to national security.

14.4 Huawei Localization in Algeria and Technology Transfer and Spillover: Main Findings

This section analyses technology transfer and the extent to which Huawei's presence in Algeria generates such spillovers. It looks first at Huawei's activities in North Africa in general and then analyses the two main components: technology transfer through training and technology spillover through labour mobility.

14.4.1 Huawei and Its Activities in North Africa

Founded in Shenzhen in 1987, Huawei has grown exponentially to become the world's leading OEM (Original Equipment Manufacturer) by moving into new markets as it began constructing telecommunications networks for phone carriers, making mobile devices for consumers, and providing a myriad of cloud, big data, and other services to other firms (Li & Kee-Cheok, 2017). In 2020, Huawei had more than 197,000 employees, operated in 170 countries and regions, and estimated that its equipment served over 3 billion people worldwide.[3] Huawei adopted a more strategic approach to R&D and strove to boost its patents by increasing considerably its investments in R&D, and particularly after adopting appropriate policies to boost "indigenous innovation" in strategic areas (Shen, 2012; Kennedy, 2006). In 2020 alone, its R&D spending reached a staggering amount of $22.3 billion (Kirton, 2021). Finally, a less-recognized factor behind Huawei's success lies in the firm's capacity to adjust to disparate cultural, political, economic, and institutional settings in different regions around the world (El Kadi, 2022). Thus, Huawei has played a significant role in North African digital transformation: in Tunisia, with the deployment of broadband networks and the training of thousands of Tunisian ICT engineers, in Egypt, with the project of the first systems for cloud computing and artificial intelligence in Africa and an Open Lab in Cairo and in Morocco with the construction of $10 billion smart city in Tangier.[4] In Algeria, Huawei played an important role in the country's telecom industry with its price-competitive equipment and managed to open its first mobile phone factory on the African continent in Algiers and in training thousands of Algerian students, staff members, and subcontractors.

Since the drop in oil prices in 2014, Algeria has adopted a series of measures aimed at diversifying its economy and moving towards higher value-added activities. Among other things, it has invested heavily in upgrading network infrastructure and has undertaken several digital initiatives. Internet penetration rate in the country reached 60.6% in 2020.[5] By the same date, mobile broadband access is correlated with a surge in mobile-cellular subscriptions and the expansion of 3G and 4G

---

[3] Huawei, "Our Company," Huawei, https://www.huawei.com/en/corporate-information.

[4] The North Africa Post: https://northafricapost.com/44913-mohammed-vi-tangier-tech-city-project-moving-forward-with-new-partnership-agreements.html.

[5] https://datareportal.com/reports/digital-2022-algeria.

networks coverage. Although the recent reduction of the gap in terms of connectivity is significant, the region's internet penetration remains just slightly above the world average, estimated at 57%, (World Bank, 2019a) something that creates important opportunities and demand for ICT infrastructure provided by OEMs like Huawei.

After extensive negotiations between Huawei and the Algerian government, the decision to build the factory was made. Algeria was severely impacted by the 2014 hydrocarbon price collapse because 60 percent of its budget is based on oil and gas. In order to conserve foreign currency, the government decided to cut back on imports and implement an import substitution strategy. The import of 900 products, including cell phones, was prohibited by the government in 2018. The government introduced a number of policies aimed at foreign companies for increasing local manufacturing in its quest to diversify the national economy and increase added value, notably, in order to localize production, the Algerian government entered negotiations with a number of cell phone producers including Samsung of South Korea which pledged to open a smartphone assembly plant in 2018 with a production capacity of 1.5 million units annually the creation of 400 direct jobs.[6]

Huawei complied with the government's request to localize production out of concern that Samsung would take market share away from it. At the factory's opening ceremony in 2019, Huawei representatives emphasised that the plant would be outfitted with the newest technologies and would be used to transfer cutting-edge technologies and production techniques. Huawei's smartphone factory in Algeria is one of the most important Digital Silk Road projects in Northern Africa and the most compelling example of Huawei's localization strategy, making it the first of its kind outside of China and in Africa. A joint venture between Huawei and the Algerian company AFGO-Tech, the factory has a monthly production capacity of 15,000 smart devices. Around 100 people work in the production unit, including 18 engineers who were trained in Shenzhen in Huawei's production methods. The Y7 prime smartphone was the first model that was put together as part of the plan. This product, in the opinion of the Chinese tech giant, was best suited to the needs of the Algerian market.

The Chinese company managed to establish a solid presence and is currently the leading network equipment provider, in terms of market share, in the Algerian telecom industries through its subsidiaries, Huawei Telecom Algeria. It employs more than 500 workers with about 83% of the staff made up of local employees and the remaining 30% of Chinese and other foreign engineers.[7] Huawei played a key role in the upgrade of 3G and 4G and is likely to continue playing a key role in the country's move to 5G. It signed a contract with Sonatrach, Algeria's state-owned

---

[6] APS (2017) https://ambalg-sofia.org/samsung-launches-first-smartphone-assembly-plant-in-alg eria/.

[7] Express DZ: https://www.express-dz.com/2019/03/20/huawei-telecommunications-algerie-dev oile-sa-strategie-2019/.

oil company, to upgrade its digital systems by providing cloud services and big data applications.[8]

The Oued Smar factory, like other Huawei facilities, plays a crucial strategic role in the company's internationalization. Through a supply chain that combines Chinese inputs into the assembly of the finished product in Algeria, it ensures ongoing access to the promising 43 million-person market in Algeria. Prior to the opening of the factory in December 2018, Huawei had a market share of about 6% of Algeria's phone market; by August 2020, it had increased to 12.34%, outpacing Condor, a domestic manufacturer in Algeria. Huawei Algeria disputes these figures, claiming that it holds nearly 18% of the mobile phone market in the country, just behind Samsung.

Concurrently to these projects, Huawei has partnered with several government institutions, universities, research institutes and local firms to expand its footprint in Algeria. The firm's public relations have stressed Huawei's role as not just the one of a turnkey project provider but as the one of a development partner willing to adjust and accommodate local development needs. Through training Algerian students and opening its manufacturing plant, the firm managed to embed itself in the local ecosystem and consolidate its image among policymakers. We will examine that in more depth in the following section.

14.4.2 Technology Transfer Through Training Initiatives

In Algeria, our fieldwork shows training initiatives are multi-faceted and include several components: training of new recruits, training within universities, training of subcontractors.

Firstly, for high-tech firms, providing regular training is crucial to ensure that employees are competent in new products and processes, which are vital to their functioning. Interviewees in Algeria indicated that new graduate intakes at Huawei go through rigorous training and induction within the first weeks of their employment. This training continues throughout their employment, with mandatory tests undertaken at different stages of their careers. Local engineers and managers are also reported going through training programs when they were first hired. The training covered technical and soft skills and continued throughout their employment period, with mandatory tests undertaken at different stages of their careers. In addition, Huawei also sends their local employees abroad for further training. A key motive driving many young engineers to work with the corporation, as reported in field interviews, is the learning opportunities they are given to upgrade their skills and know-how through training abroad and through observation.

---

[8] https://e.huawei.com/fr/publications/global/ict_insights/201902271023/Success-Story/201904170833.

Secondly, beside industry-specific training, Huawei's enterprise business has been particularly dynamic in establishing cooperation agreements with universities for training students in the region. Huawei counts two types of partnerships with local universities through their two academies: the *Huawei Authorized Network Academy* (HANA) and the *Huawei Authorized Information and Network Academy* (HAINA). According to the tech giant, the objective of these academies is to participate in capacity building to enable the digital transformation of local economies by connecting talents to local industry. More specifically, these academies aim to promote certifications in Huawei technologies among ICT university students. These certifications attest that their holders are competent in using and maintaining the technologies of a specific manufacturer. Students then go on to find employment with mobile operators, OEMs, or other firms that use these technologies. Alternatively, some graduates join channel partners who sell and install OEM's equipment for customers such as governments and large corporations. More specifically, in Algeria Huawei has ramped up efforts to create ICT academies across the country in recent years. Through a partnership with the ministry of higher education signed in 2021, Huawei launched five major ICT labs within leading universities: these are the National Institute of Post and Information and Communication Technologies of Algiers and Oran, the University of Saida, the University of Sciences and Technologies Houari Boumediene of Algiers, and the National School of Computer Science of Algiers. The labs are equipped with high-performing computers and cutting-edge equipment to be used for students' training. Participating universities also have access to ICT courses taught by accredited Huawei instructors who train both students and future instructors. The Chinese company claims to have trained over 2500 young Algerians in 2020 in various fields related to ICT. Besides ICT academies, the firm has launched a myriad of Social Corporate Responsibility (CSR) programs, such as the "Seeds for Future" scholarship, which takes some of the brightest students to Huawei's headquarters in Shenzhen and offers them exposure to cutting-edge technologies and immersion in Chinese culture. Huawei also organizes large scale ICT competitions within and across countries. In 2019–2020, Algeria's student teams won first place in the ICT Global Contest.[9]

Thirdly, Huawei offers training to local subcontractors for installing, troubleshooting, and maintaining the equipment they sell to customers. Usually, OEMs would have different subcontractors covering the various regions of the country where ICT infrastructure is being rolled out. Beyond training in universities, Huawei has provided regular training to subcontractors and workers. Local engineers and managers at the two Chinese firms, both on in-house and leased contracts, are reported going through training programs when they were first hired. The training covered technical and soft skills and continued throughout their employment period, with mandatory tests undertaken at different stages of their careers. They also send their local employees abroad for further training. A key motive driving many young

---

[9] https://aptantech.com/2020/11/18/algeria-egypt-nigeria-teams-among-winners-at-huaweis-glo bal-ictcontest/.

engineers to work with Chinese MNCs, and Huawei in particular, is the learning opportunities provided by the companies.

Finally, they provide training to customers, like mobile operators and big firms, on the use of purchased equipment. In this respect, Huawei guarantees that major key players remain attached to the corporations' standards and branding and better equipped to resist attractions from other competitors.

14.4.3   Technology Spillovers Through Managerial Practices and Labour Mobility

Fieldwork findings suggest that the majority of high-level positions were filled with Chinese managers reaching 920 in 2015, third largest behind the French (1963) and the Lebanese (1146) (Hanniche & Bellache, 2020). Algerian workers were placed under the supervision of 20 Chinese experts sent from the company's various factories during the first few months of operation to ensure strict adherence to Huawei's standards. But due to growing labour costs in China, Huawei has, in recent years, localized a bigger share of their labour in North Africa, including in managerial positions.

This being said, interviewed workers at the plant highlighted the importance of the training provided and its role in boosting management skills in the manufacturing sector. In fact, the factory started operating with about 40 workers, among which 18 local engineers were sent to China to observe Huawei's factories and learn about production processes. Later the factory expanded to 140 workers as extra production lines were added. The plant is equipped with the latest generation equipment and uses the most innovative technologies and all of Huawei's know-how according to a Huawei representative. This means that local employees and engineers in particular could have access to latest technologies within a relatively short time and could upgrade their skills and know-how allowing thus for leap-frogging.

As trained workers at multinationals move to domestic firms or start their businesses, technology may be disseminated from MNCs to other firms within the same industry. Within the instances of labour mobility, former employees interviewed highlighted they had grasped a great deal of managerial knowledge that they could apply to their new work.

Similarly, interviewed start-up owners in Algiers who had previously worked for Huawei highlighted that they had gained significant managerial knowledge. Managerial knowledge tends to be overlooked in the literature on knowledge spillovers from FDI. Yet, just as technology, management practices can have a significant impact on a firm's productivity (Fu et al., 2011). The rate of integration becomes thus an opportunity for the MNC to enhance spillover both within its eco-system and outside.

This implies that more and more employees should be given the opportunity to hold managerial position in Chinese MNCs. Labour mobility should not be viewed as a threat by Chinese firms but rather as a vehicle to expand Chinese management style to other companies and to other sectors in the country.

14.5 Discussion

Huawei's presence in Algeria is a good example of how flexible Chinese IT companies are when they venture outside, both before and after the Digital Silk Road. Aspects of what is universal are always present in particular circumstances, notwithstanding the evident constraint of generalizability on other Chinese actors from a single case study. Unpacking Huawei's tremendous dynamism in North Africa requires an understanding of large-scale training, CSR initiatives, and, more recently, manufacturing and R&D. With the help of this localization strategy, the Shenzhen based company was able to establish close relationships with a wide range of ecosystem participants and position itself as an active participant in the region's transition to the digital economy. Huawei has responded to Algeria's demands for greater value addition from foreign firms by embarking on a range of initiatives that promote technology transfer, such as large-scale training and manufacturing. We find that these initiatives have resulted in managerial knowledge spillovers even though the transfer of more technical knowledge has remained limited. These managerial skills spillovers, which specifically refer to the adoption of foreign management practices by domestic firms, have been recognized as powerful tools for promoting innovation and increasing productivity (Javorcik, 2004; Ning & Wang, 2018).

The spread of managerial information from Chinese multinational firms operating in North Africa could be significant if localization practices spread more widely among Chinese and local companies. First of all, localization enhances the integration of the foreign firm into the economic structures of the host country by promoting the employment of local human resources. Increased interactions with local players, particularly suppliers and subcontractors, means more chances for them to pick up knowledge from the international company. Second, localization fosters further spillovers at the managerial level by giving local staff members specialized training and first-hand experience with managing huge projects. The most seasoned local workers may then transfer these new talents deeper into the home economy through labour mobility (Auffray & Xiaolan, 2015).

Opportunities for managerial knowledge spillovers remain, with important implications for productivity gains in the digital sector. Moreover, with its energetic efforts in organizing ICT competitions, providing scholarships to students and grants to promising startups, Huawei may have a greater footprint on skill building than its competitors.

By bridging the cultural and linguistic divide between Chinese management teams and their local partners and employees, a good localization plan could also enhance the environment for managerial knowledge spillovers. Thus Huawei, had all the chances to bring the cultural and linguistic gap in Algeria, between its managers and its partners and employees. This was done through its training activities directed towards local employees. Thus the 140 workers trained locally and then sent to Huawei were put in the cultural and linguistic bath in China. This contributed to making communication a lot easier and interpersonal relationship more frequent and intense going sometimes beyond work matters. In this respect Huawei managed to build bonds of trust in the relation and made it easier for local employees to acquire the technology.

The hiring of local managers who will deal directly with the firm's local stakeholders and workers is likely to boost communication and interpersonal interactions. This should make it easier for coworkers to become trusted, which will improve mentorship possibilities and learning environments, both of which are crucial for management knowledge spillovers. It's also feasible that regional managers are more suited to teach their associates and partners new knowledge.

However, following field investigation, the extent to which these knowledge transfer initiatives contribute to technological upgrading remains questionable. In an empirical study on Huawei's role in human capital development in Nigeria, Agbebi (2019) finds that the Chinese firm has contributed to skill building through its numerous training programs, including training activities targeted toward its local employees, suppliers, and customers, through organizing ICT competitions and providing scholarships to local students. In contrast, Tugendhat (2021) finds in a study looking at Huawei's training centres in Nigeria and Kenya that Huawei's presence fell short of offering meaningful opportunities for knowledge transfers that could promote technological upgrading in the two countries. He argues that international equipment vendors limit the scope of the knowledge they are willing to share with local employees and actors.

This is also the case in North African economies and Algeria in particular, in our view. Chinese tech players have helped to build sustained production capacity in telecommunications hardware and software, including a reservoir of well-trained local engineers. Yet, they perceive the risks of imitation are high and make deliberate efforts to hinder any meaningful understanding of the deeper functioning of their cutting-edge technologies. Some interviewees commented, however that by bringing more Chinese staff locally than other foreign competitors, Chinese firms like Huawei and ZTE provided more opportunities for exposure to engineers working on cutting-edge technologies and learning the latest practices and standards (Mattli & Tim, 2003).

While the Chinese corporation has pledged to contribute to upgrade local capacities through outsourcing, the rate of technological integration in the factory continues to raise questions, though as it did in other parts of Africa (Rwehumbiza, 2021). The assembly line's reliance on imported SKD (Semi Knocked Down) and CKD (Completely Knocked Down) kits, which are produced in China before being exported for the final stages of assembly, raised concerns among the experts interviewed. Thus, there is little room for technology transfer and little room for value addition in the manufacturing process. Authorities in Algeria called the practice "fictitious production" and "disguised import," which has become standard among various manufacturers.

These initiatives have not resulted in the meaningful transfer of technology and technical knowledge but have created instances of managerial knowledge spillovers that could support innovation in Algeria in the future.

14.6 Conclusion

Although the expansion of digital technologies in North Africa could drive economic and social development, this trajectory is not automatic. We will go along Tsui's questioning 'Do Huawei's Training Programs and Centres Transfer Skills to Africa?' (Tsui, 2016) in view of some limitations of knowledge transfer in Huawei's training centres (Tugendhat, 2021). Without pro-active policies aimed to maximize the gains from the digital economy, digital infrastructure built as part of the digital silk road risk increasing the technological dependence of North African countries, weakening their capacity to learn, innovate and move into competitive positions within the global knowledge economy. On the other hand, this is also the result of the absorptive capacity (Cohen & Levinthal, 1990) of local engineers and local firms. Furthermore, while joint venture requirements have been effective tools for technology transfer in practical terms (Blomström & Sjöholm, 2001), the experience of Huawei's factory in Algiers shows that these are unlikely to result in meaningful learning opportunities without more extensive local content requirements.

This study bears a variety of policy ramifications that might be relevant to nations beyond Algeria. If the region's leaders are serious about the digital economy stirring up growth and creating jobs, it is imperative to stop adopting the posture of mere consumers of tech products and services and start acting more like potential producers. This strategy may well also be the safest for protecting national data and ensuring cybersecurity.

In the coming years, North African governments, and Algeria in particular ought to ensure that cooperation agreements with China entail comprehensive knowledge and technology transfer mechanisms, including cooperation in research and development, and that Chinese investments yield quality jobs for the region's young population.

BRI countries should implement a set of digital industrial policies that encourage technology localization and productive linkages to reverse present trends. Governments would benefit from putting in place regulations that protect and assist the expansion of local businesses, simplifying their integration into intricate production networks (Ernst & Linsu, 2002), in addition to increasing investment in human resources and investing in domestic R&D skills. By drawing on China's own development history, regulations might help to guarantee that up-and coming tech leaders have the financial means and the necessary protection from intense international competition they need to successfully seize domestic markets and join and advance within global value chains. By increasing the quantity and calibre of exchanges, digital industrial strategies should also aim to promote learning from international digital enterprises. To do this, one strategy would be to mandate consortium bidding between domestic and foreign businesses. The tasks would need to be divided between the successful bidders with explicit rules for technology transfers and clear compensations for each party.

Local public institutions also have a role to play in making local businesses visible to international businesses and compatible with them. Partnerships with both Chinese and non-Chinese businesses operating in the country could be facilitated by reducing information asymmetries and by building databases of available local companies and their skills. In the spirit of South-South exchange, greater regional cooperation could enable smaller economies to reap the full benefits of international digital initiatives like the Digital Silk Road. Smaller developing countries may find the concept of a regional digital policy, similar to the one governing the European Digital Single Market, advantageous (Azmeh et al., 2019). To boost their bargaining power with major tech corporations, African countries should further their regional integration. It should be possible to level the playing field for all African countries by moving past fragmented bilateral commercial deals with China and its tech giants, and doing so would ultimately increase opportunities for local agencies to design institutions that support inclusive digital development. The African Continental Free Trade Area (AfCFTA) adopted in March 2018 may constitute a favourable setting for pushing ahead the idea of a single African digital market.

References

Abdel Ghafar, A., & Jacobs A. (2019). *Beijing calling: Assessing China's growing footprint in North Africa*, 18 September. https://www.brookings.edu/research/beijing-calling-assessing-chinasgrowing-footprint-in-north-Africa/

Agbebi, M. (2019). Exploring the human capital development dimensions of Chinese investments in Africa: Opportunities, implications and directions for further research. *Journal of Asian and African Studies, 54*(2), 189–210. https://doi.org/10.1177/0021909618801381

Ahrens, N. (2013). *Huawei in China's competitiveness: Myths, reality and lessons for the United States and Japanese. Center for Strategic and International Studies.* https://www.csis.org/programs/japan-chair/japan-chair-archives/chinas-competitiveness-myths-realities-and-lessons-united

Aitken, B. J., & Harrison, A. E. (1999). Do domestic firms benefit from direct foreign investment? Evidence from Venezuela. *American Economic Review, 89*(3), 605–618. https://doi.org/10.1257/aer.89.3.605

Amendolagine, V., Amadou, B., Nicola, D. C., Francesco, P., & Adnan, S. (2013). FDI and local linkages in developing countries: Evidence from sub-Saharan Africa. *World Development, 50*, 41–56. https://doi.org/10.1016/j.worlddev.2013.05.001

Amsden, A. H. (2001). *Challenges to the west from late-industrializing economies*. Oxford University Press.

Auffray, C., & Xiaolan, F. (2015). Chinese MNEs and managerial knowledge transfer in Africa: The case of the construction sector in Ghana. *Journal of Chinese Economic and Business Studies, 13*(4), 285–310. https://doi.org/10.1080/14765284.2015.1092415

Azmeh, S., & Abeer, E. (2020). North Africa's export economies and structural fragility: The limits of development through European value chains. LSE Middle East Centre, No. 42. http://www.lse.ac.uk/middle-east-centre/publications/paper-series

Azmeh, S., Foster, C., & Echavarri, J. (2019). The international trade regime and the quest for free digital trade. *International Studies Review, 22*(3), 671–692. https://doi.org/10.1093/isr/viz033

Blalock, G., & Gertler, P. (2008). Welfare gains from foreign direct investment through technology transfer to local suppliers. *Journal of International Economics, 74*(2), 402–421. https://doi.org/10.1016/j.jinteco.2007.05.011

Blomstrom, M., & Kokko, A. (2001). Foreign direct investment and spillovers of technology. *International Journal of Technology Management, 22*(5–6), 435–454. https://doi.org/10.1504/IJTM.2001.002972

Blomström, M., & Sjöholm, F. (2001). Technology transfer and spillovers: Does local participation with multinationals matter? *European Economic Review, 43*(4–6), 915–923. https://doi.org/10.1016/S0014-2921(98)00104-4

Brautigam, D. (2009). *The Dragon's gift: The real story of China.* Oxford University Press.

Calabrese, L., & Xiaoyang, T. (2020). *Africa's economic transformation: The role of Chinese investment in Africa.* http://cdn-odi-production.s3-website-eu-west-1.amazonaws.com/media/documents/DEGRP-Africas-economic-transformation-the-role-of-Chinese-investment-Synthesis-report.pdf

CCCPC. (2016). *The 13th five-year plan for economic and social development of the People's Republic of China—2016–2020.* https://en.ndrc.gov.cn/newsrelease_8232/201612/P020191101481868235378.pdf

Cheney, C. (2019). *China's digital silk road: Strategic technological competition and exporting political illiberalism council on foreign relations, net politics (blog),* September 26. https://www.cfr.org/blog/chinas-digital-silk-road-strategic-technological-competition-and-exporting-political

Cisse, D. (2012). Chinese telecom companies foray into Africa. *African East-Asian Affairs.* https://doi.org/10.7552/69-0-94

Cohen, W., & Levinthal, D. (1990). Absorptive capacity: A new perspective on learning and innovation. *Administrative Science Quarterly., 35*, 128–152.

Crespo, N., & Fontoura, M. P. (2007). Determinant factors of FDI spillovers—What do we really know? *World Development, 35*, 410–425.

Cunningham, J. A., Menter, M., & Young, C. (2017). A review of qualitative case methods trends and themes used in technology transfer research. *The Journal of Technology Transfer, 42*, 923–956. https://doi.org/10.1007/s10961-016-9491-6

Demena, B. A., & Van Bergeijk, P. (2019). Observing FDI spillover transmission channels: Evidence from firms in Uganda. *Th World Quarterly, 40*, 1708–1729. https://doi.org/10.1080/01436597.2019.1596022

El-Kadi, T. H. (2019). Travail, Algériens, Chinois et mythes culturalistes [Work, Algerians, Chinese and Cultural Myths]. In Lamine Khan (ed.) Chihab Edition, Algiers.

Ernst, D., & Linsu, K. (2002). Global production networks, knowledge diffusion, and local capability formation. *Research Policy, 31*, 1417–1429. https://doi.org/10.1016/S0048-7333(02)00072-0

Fu, X., Pietrobelli, C., & Soete, L. (2011). The role of foreign technology and indigenous innovation in the emerging economies: Technological change and catching-up. *World Development, 39*, 1204–1270. https://doi.org/10.1016/j.worlddev.2010.05.009

Gagliardone, I. (2019). *China, Africa, and the future of the internet: New media, new politics.* Zed Books.

Ghanem, D., & Benabdallah, L. (2016). *The China syndrome.* Carnegie Middle East Center, November 18th. https://carnegie-mec.org/diwan/66145

Glass, A. J., & Saggi, K. (2002). Multinational firms and technology transfer. *Scandinavian Journal of Economics, 104*, 495–513.

Globerman, S. (1979). Foreign direct investment and spillover efficiency benefits in Canadian manufacturing industries. *The Canadian Journal of Economics, 12*(1), 42–56. https://doi.org/10.2307/134570

Görg, H., & Strobl, E. (2005). Spillovers from foreign firms through worker mobility: An empirical investigation. *Scandinavian Journal of Economics, 107*, 693–709.

Gray, K., & Gills, B. K. (2016). South-South cooperation and the rise of the Global South. *The World Quarterly, 37*, 557–574. https://doi.org/10.1080/01436597.2015.1128817

Haddad, M., & Harrison, A. (1993). Are there positive spillovers from direct foreign investment? Evidence from panel data for Morocco. *Journal of Development Economics, 42*, 1–74. https://doi.org/10.1016/0304-3878(93)90072-U

Hanniche, B., & Bellache, J. (2020). China–Algeria: Opportunities and challenges of a growing relationship. *Financial and Markets Review, 7*, 88–107.

Hart, M., & Lind, J. (2020). *There is a solution to the Huawei Challenge, Center for American Progress, October 14.* https://www.americanprogress.org/article/solution-huawei-challenge

Hillman, J. E. (2021). *The digital silk road: China's quest to wire the world and win the future.* New York: Harper Business. https://www.amazon.com/Digital-Silk-Road-Chinas-Future/dp/0063046288

Hoi Quoc, L., & Pomfret, R. (2011). Technology spillovers from foreign direct investment in Vietnam: Horizontal or vertical spillovers? *Journal of the Asisa Pacific, 16*, 183–201.

Holm, J. R., Bram, T., Østergaard, C. R., Coad, A., Grassano, N., & Vezzani, A. (2020). Labor mobility from R&D-intensive multinational companies: Implications for knowledge and technology transfer. *The Journal of Technology Transfer, 45*, 1562–1584. https://doi.org/10.1007/s10961-020-09776-8

Javorcik, B. S. (2004). Does foreign direct investment increase the productivity of domestic firms? In search of spillovers through backward linkages. *American Economic Review, 94*(3), 605–627. https://doi.org/10.1257/0002828041464605

Joo, S. H., Oh, C., & Lee, K. (2016). Catch-up strategy of an emerging firm in an emerging country: Analyzing the case of Huawei vs. Ericsson with patent data. *International Journal of Technology Management, 72*, 19–42.

Kabbani, N. (2019). *Youth employment in the middle east and North Africa: Revisiting and reframing the challenge.* Brookings Institution, 26 Feb. https://www.brookings.edu/research/youth-employment-in-the-middle-east-andnorth-africa-revisiting-and-reframing-the-challenge/

El Kadi, T. H. (2022). *How Huawei's localization in North Africa delivered mixed returns.* Carnegie Endowment for International Peace.

Kennedy, S. (2006). The political economy of standards coalitions: Explaining China's Involvement in High-Tech Standards Wars. *Asia Policy, 2*, 41–62. https://www.nbr.org/publication/the-political-economy-of-standards-coalitions-explaining-chinas-involvement-in-high-tech-standards-wars

Kim, L., & Dahlman, C. (1992). Technology policy for industrialization: An integrative framework and Korea's experience. *Research Policy, 21*, 437–452.

Kirton, D. (2021). Huawei posts 3.2% rise in profit in 2020, as revenues decline from outside of China. *Reuters*, March 31. https://www.reuters.com/article/us-huawei-tech-results-idUSKBN2BN0XA

Kubny, J., & Hinrich, V. (2014). Benefitting from Chinese FDI? An assessment of vertical linkages with vietnamese manufacturing firms. *International Business Review, 23*, 731–740. https://doi.org/10.1016/j.ibusrev.2013.11.002

Lall, S. (1992). Technological capabilities and industrialization. *World Development, 20*, 165–186. https://doi.org/10.1016/0305-750X(92)90097-F

Lee, C. (2012). Huawei Surpasses Ericsson as World's Largest Telecom Equipment Vendor. *ZDNet.com*, 24. https://www.zdnet.com/article/huawei-surpasses-ericsson-as-worlds-largest-telecom-equipment-vendor

Li, R., & Kee-Cheok, C. (2017). Huawei and ZTE in Malaysia: The localization of Chinese transnational enterprises. *Journal of Contemporary Asia, 47*(5), 752–773. https://doi.org/10.1080/00472336.2017.1346697

Liu, X., Chengang, W., & Yingqi, W. (2009). Do local manufacturing firms benefit from transactional linkages with multinational enterprises in China? *Journal of International Business Studies, 40*, 1113–1130. https://doi.org/10.1057/jibs.2008.97

Liu, Z. (2008). Foreign direct investment and technology spillovers: Theory and evidence. *Journal of Development Economics, 85*, 176–193. https://doi.org/10.1016/j.jdeveco.2006.07.001

Markusen, J. R., & Venables, A. (1999). Foreign direct investment as a catalyst for industrial development. *European Economic Review, 43*, 335–356.

Matija, R., & Knell, M. (2018). Why is there a lack of evidence on knowledge spillovers from foreign direct investment? *Journal of Economic Surveys, 32*(3), 579–612. https://doi.org/10.1111/joes.12207

Mattli, W., & Tim, B. (2003). Setting international standards: Technological rationality or primacy of power? *World Policy, 56*, 1–42.

Morais, S. (2008). *Opportunity New Yor City: A performance-based conditional cash transfer.* International Poverty Centre Working Paper, no. 49. https://www.researchgate.net/public ation/46463120_Opportunity_NYC_a_PerformanceBased_conditional_Cash_Transfer_Prog ramme_A_Qualitative_Analysis

Mowery, D. C., & Oxley, J. E. (1995). Inward technology transfer and competitiveness: The role of national innovation systems. *Cambridge Journal of Economics, 19*, 67–93.

Mu, Q. M., & Lee, K. (2005). Knowledge diffusion, market segmentation and technological catch-up: The case of the telecommunication industry in China. https://doi.org/10.1016/J.RESPOL.2005.02.007

Muhr, T., & De Azevedo, M. L. (2018). The Brics development and education cooperation agenda. *International Relations, 3*, 517–534.

Naughton, B. J. (2006). *The Chinese economy: Transitions and growth.* MIT Press.

Ning, L., & Wang, F. (2018). Does FDI bring environmental knowledge spillovers to developing countries? The role of the local industrial structure. *Environmental and Resource Economics, 71*, 381–405. https://doi.org/10.1007/s10640-0170159-y

Oreglia, E. (2012). *Africa's many Chinas. University of California-Berkeley.* http://www.ercolino.eu/docs/Oreglia_Proj_AfricasManyChinas.pdf

Oreglia, E., Ren, H., & Liao, C. C. (2021). The puzzle of the digital silk road. In N. Kassenova & B. Duprey (Eds.), *Digital silk road in central Asia: Present and future.* Harvard University.

Oya, C., & Schaefer, F. (2019). *Chinese firms and employment dynamics in Africa: A comparative analysis.* SOAS, University of London.

Pairault, T. (2017). La Chine au Maghreb: De l'esprit de Bandung à l'esprit du capitalisme (China in the Maghreb: From the spirit of Bandung to the spirit of Capitalism). *Revue De La Régulation.* https://doi.org/10.4000/regulation.12230

Rwehumbiza, D. A. (2021). Huawei's linkages with local firms in Tanzania: Idiosyncratic benefits and risks. *Global Business and Organizational Excellence, 40*, 20–35. https://doi.org/10.1002/joe.22076

Sazali, A. W., Che Rose, R., & Idayu Wati Osman, S. (2011). Defining the concepts of technology and technology transfer: A literature analysis. *International Business Research, 5*(1), 61. https://doi.org/10.5539/ibr.v5n1p61

Segal, A. (2018). *When China rules the web technology in service of the state.* September/October, 10–18.

Shen, H. (2012). *Across the great (fire) wall: China and the global internet.* PhD Dissertation. The University of Illinois https://www.ideals.illinois.edu/handle/2142/97589

Smeets, R. (2008). Collecting the pieces of the FDI knowledge spillovers puzzle. *The World Bank Research Observer, 23*(2), 107–138. https://doi.org/10.1093/wbro/lkn003

Takii, S. (2005). Productivity spillovers and characteristics of foreign multinational plants in indonesian manufacturing 1990–1995. *Journal of Development Economics, 76*(2), 521–542.

Thomala, L. L. (2022). Number of internet users in China 2008–2021. *Statista.* https://www.sta tista.com/statistics/265140/number-of-internet-users-in-china

Tsui, B. (2016). *Do Huawei's training programs and centers transfer skills to Africa?'* Policy Brief No. 14, China Africa Research Initiative: John Hopkins SAIS.

Tugendhat, H. (2021). Connection issues: A study on the limitations of knowledge transfer in Huawei's African training centers'. *Journal of Chinese Economic and Business Studies.* https://doi.org/10.1080/14765284.2021.1943194

UNCTAD. (2012). *Technology and innovation report 2012: Innovation, technology and south south collaboration.* New York and Geneva: United Nations.

Wahab, S. A., Raduan, C., & Suzana, I. W. (2011). *Defining the Concepts of Technology and Technology Transfer.* https://doi.org/10.5539/ibr.v5n1p61

Wei, H. (2019). *Xi: Internet a joint global responsibility.* World Internet Conference, October 21. https://www.wuzhenwic.org/2019-10/21/c_562803.htm

World Bank. (2019b). *Individuals using the internet (% of population).* World Bank, Washington, DC. https://data.worldbank.org/indicator/IT.NET.USER.ZS

World Bank. (2019a). *Internet penetration rates—World average.* Washington, DC. https://data.worldbank.org/indicator/IT.NET.USER.ZS

Part V
Evolution of China-Africa Collaborations in Science, Technology and Innovation

Chapter 15
Evolution of China–Africa Cooperation in Science, Technology and Innovation: A Way to Capacity Building

Xianlan Lin, Xiao Zhang, and Siqian Wang

15.1 Introduction

It is believed by historians that China–Africa relations started in the Han Dynasty or even earlier (Anshan, 2021, pp. 97–109). Historical records of China–Africa exchanges before the Ming Dynasty are rare, and it is not until Zheng He who commanded expeditionary voyages to the west that China–Africa relations got into the public eye and were pushed to a new high in ancient times. Scientific and technological exchange has been a constant feature of China–Africa relations since ancient times. At present, a more institutionalized and refined cooperation of both sides in STI is an important driving force that leads the development of China–Africa relations.

The academic community has formed some foundations on the study of scientific and technological cooperation between China and Africa. Some scholars focus on China–Africa scientific and technological cooperation in specific fields, such as agricultural science and technology (Rui et al., 2021), science technology engineering and mathematics (STEM) (Taling, 2021), and digital fields (Heng & Zhemeng, 2021); while others focus on scientific and technological cooperation between China and specific countries in Africa, such as Kenya (Hongzhong & Ruizhou, 2019) and Ethiopia (Yonghong & Yuanfei, 2016). Some scholars examine China–Africa scientific and technological cooperation from the perspective of cooperative research papers and patents (Mammo & Swapan, 2019), while others study the cooperation from a strategic perspective (Yonghong et al., 2012, 2019). Some summarize the development process of China–Africa science and technology cooperation (Tao, 2011), some sort out the models of cooperation (Zhaokun, 2019), and some analyze its

X. Lin
School of Public Administration, Hubei University, Wuhan, China

X. Zhang · S. Wang (✉)
China-Africa Innovation Cooperation Center, Wuhan, China
e-mail: caicc@caicc.net.cn

© The Author(s) 2024
M. Muchie et al. (eds.), *China-Africa Science, Technology and Innovation Collaboration*, https://doi.org/10.1007/978-981-97-4576-0_15

forms and policies (Xiao, 2013). To analyze the existing research results in general, most of them focus on the cooperation between China and Africa at the level of "science" and/or "technology", but there is not much of full discussion on China–Africa cooperation under the new situation with the theme of how scientific and technological innovation can promote economic and social development.

Based on existing studies, this paper extends the scope of China–Africa scientific and technological cooperation to China–Africa cooperation in STI, and gives the term "cooperation" a relatively broad concept, which includes not only cooperation in scientific and technological knowledge exchanges, but also direct aid in funds, materials, personnel and technologies, as well as the form that targeting at capacity building through innovative cooperation institutions in STI. What is the historical foundation of China–Africa cooperation in STI? After years of evolution, has a clear development direction of China–Africa cooperation in STI been formed? In the face of new opportunities and challenges, where will China–Africa cooperation in STI step towards in the future? Focusing on these core issues, this paper aims to summarize and refine the evolution trend of China–Africa cooperation in STI by sorting out its historical evolution and the latest endeavours in the new era. Analysis of new opportunities and challenges that China–Africa cooperation in STI faced is also going to be carried out, for the purpose of putting forward countermeasures and suggestions for further deepening such cooperation.

15.2 Historical Evolution of China–Africa Cooperation in STI: From Mutual Exchanges to Direct Aid

China–Africa cooperation in STI from the Han Dynasty to the founding of the PRC is dominated by mutual exchanges, and that from the founding of the PRC to the late twentieth century features direct aid.

15.2.1 Ancient China–Africa Cooperation in STI Characterized by "Mutual Exchanges"

China–Africa cooperation in STI in ancient times took the form of mutual exchange of needed technologies. The period from the Western Han and Eastern Han Dynasties to the Sui and Tang dynasties saw incipient but overall limited scientific and technological exchanges between China and African counties (Tao, 2011, pp. 113–126). After the Silk Road was opened up in the Western Han Dynasty, scientific and technological communication between the Han Dynasty and North African countries under the rule of the Roman Empire gained further development marked by mutual exchanges and the use of China's medicinal herbs and Africa's aromatic and medicinal plants, which had boosted the advancements of medical technology of both sides (Deming, 1997, pp. 105–108). From the Northern and Southern Song dynasties to the early Ming Dynasty, the scientific and technological exchanges between China and Africa flourished as economic and trade exchanges grew. China's printing, compass and gunpowder spread to the North and East African regions (Leften, 2006, pp. 204–205). Ambassadorial visits made by Zheng He during his voyages to the west promoted commerce, trade as well as scientific and technological exchanges between China and African countries. After the fleets commanded by Zheng He arrived in East Africa, a large number of African medicinal materials were loaded on board while China's rich geographical knowledge and excellent shipbuilding techniques were imparted to the locals (Niane, 1992, pp. 538–539). In the sixteenth century, China–Africa cooperation in STI was weakened but not completely cut off by Western colonialism. Since the seventeenth century, many Chinese people were trafficked to the northern, southern and eastern African regions, and their offspring engaged in jobs in relation to Chinese traditional medicine, engineering and agriculture, among others in Africa, promoting technological exchanges between the two sides (Tao, 2011, pp. 113–126).

15.2.2 Modern China–Africa Cooperation in STI Characterized by "Direct Aid"

After the PRC established diplomatic relations with African countries successively, the focus of China–Africa cooperation in STI gradually shifted to China's development aid to Africa, especially direct aid in terms of funds, materials, personnel and technologies. Since the founding of the PRC, the Chinese government proposed the "Five Principles of Peaceful Coexistence", the fundamental norms guiding international relations including China–Africa relations. At the Afro-Asian Conference (also known as the Bandung Conference) held in 1955 in Bandung, Indonesia, they were extended and upgraded to the ten principles guiding international relations which were widely recognized by third-world countries. Many African countries including Egypt and Sudan established diplomatic relations with the PRC after the Bandung Conference. China helped African countries by supporting their national liberation movements, promoting peaceful settlement of conflicts and disputes and providing necessary aid, which drives China–Africa relations to a deeper level. Correspondingly, with the support of African countries, the PRC regained its legitimate seat in the United Nations (UN) in 1971. The establishment of mutual trust politically had laid a more solid foundation for China–Africa cooperation in STI. Depending on political mutual trust, China pushed its cooperation in STI with African countries one step further by increasing foreign direct aid to Africa.

Official cooperation in STI between China and African countries goes back to the 1970s when cooperation programs were administrated by China's Ministry of Science and Technology (MOST) and government authorities of science and technology of

African countries. By entering into bilateral intergovernmental agreements on scientific and technological cooperation, establishing joint committees for intergovernmental scientific and technological cooperation with African countries, and sending science and technology diplomats to South Africa and Egypt, China and African countries carried out extensive cooperation in STI in areas of mutual concern (Xiao, 2013, pp. 142–146). For example, during the three decades from 1949 to 1979, there were more than 180 China–Africa cooperation programs in agricultural science and technology. China helped African countries construct agricultural technology experiments and promotion stations and develop agricultural processing projects, spreading China's technologies in growing rice, cotton, tea, tobacco and mulberry, and breeding silkworms to Africa. Most of the aid programs were non-reimbursable, with technologies unilaterally transferred by China to Africa unconditionally (Jiali, 2005, pp. 11–14). As to infrastructure construction, the Tanzania-Zambia Railway built with China's technical, financial and manpower assistance is a milestone of China–Africa cooperation in STI in the second half of the twentieth century. As the largest single item foreign-aid project ever undertaken by China (Yanting, 2019, p. 20), the railway helped people in southern African regions fight colonialism and racialism and was dubbed the "Freedom Railway" by the locals. Since the railway was open to traffic, regions along the railway gained rapid development. Inland regions in southwestern Tanzania far from the Indian Ocean, eastern regions of Zambia and other poverty-stricken regions saw the formation of new towns and industrial and agricultural production areas along the railway, leading to an urban and industrial development corridor completely free from legacies of colonial domination (Yanting, 2019, pp. 21–22). The railway is of great political and strategic significance. In tandem with large-scale construction projects in Africa, China made endeavours to improve living standards of African people through technical assistance and cooperation. Examples include the China–Egypt crop straw gasification technology demonstration project that conveys gas to farmers' homes through pipelines, the China–Zimbabwe solar energy utilization technology demonstration project that enables the locals to effectively utilize solar energy, and the CBERS (China–Brazil Earth Resources Satellite) data receiving station in South Africa that provides technical support for environmental monitoring, disaster prevention, and agricultural yield estimation in southern African regions (Xiao, 2013, pp. 142–146). Besides, China has made ongoing efforts to promote cooperation in STI by sending doctors, medical teams and engineering and technical personnel and providing technical training for the locals. The cooperation programs and results yielded build a strong foundation for all-around and in-depth development of China–Africa cooperation in STI in the new era.

15.3   New Endeavours to Promote China–Africa Cooperation in STI in the New Era: Focusing on Capacity Building

Cooperation in STI has been one of the themes of China–Africa relations since the beginning of the twenty-first century. Different from previous "mutual exchanges" based and "direct aid" based cooperation, China–Africa cooperation in STI in the new era prioritizes "capacity building", which is consistent with the needs of Millennium Development Goals and Sustainable Development Goals of the United Nations. China–Africa cooperation in STI in the new era is represented by the China–Africa Science and Technology Partnership Program (CASTEP) 1.0 and the China–Africa Science and Technology Partnership Program (CASTEP) 2.0, which are aimed at empowering China and African countries to achieve transformation and development by themselves through joint development and in-depth cooperation in STI.

15.3.1   The First Stage of Exploration: CASTEP 1.0

The CASTEP originates from the Forum on China–Africa Cooperation (FOCAC) established in 2000 and joined by China, 53 African countries with diplomatic relations with China and the African Union Commission. FOCAC, a mechanism for collective dialogue on South-South cooperation between China and African countries, aims to promote equity, mutual benefit and equality-based consultation, enhance mutual understanding, broaden consensus, strengthen friendship and promote cooperation. At the opening ceremony of the Fourth Ministerial Conference of the FOCAC held on November 8, 2009 in Sharm el Sheikh, China's Premier Wen Jiabao proposed the CASTEP, in a bid to join hands with African countries to achieve joint and sustainable development by carrying out efficient and pragmatic cooperation and giving full play to the key role of STI in economic and social development.

The first phase of the CASTEP focuses on cooperation programs in STI needed by African countries and of significant value for their development so as to help them enhance capacity. The CASTEP is guided by several principles. The first principle is to achieve mutual benefits and win–win outcomes, that is, experience, information, knowledge and other tangible and intangible resources in favour of scientific and technological development are shared on the basis of mutual respect for sovereignty to enhance STI capacity and ultimately achieve sustainable economic and social development for both sides. The second principle is to engage in demand-driven cooperation in STI, with full consideration of both the requirements of African partners and the features and advantages of China's scientific and technological development. The third principle is to define priorities, that is, projects in STI are properly prioritized according to characteristics and requirements of African countries so that those significantly contributing to economic and social development are implemented with priority in a step-wise way. The fourth principle is to play the guiding role of governments, that is, government authorities use policy and financial incentives to encourage and enable enterprises, higher learning institutions and scientific research institutes to get involved by contributing capital or otherwise to create synergy. The program prioritizes scientific and technological cooperation in livelihoods-related areas of mutual concern, including water resources, food, health, energy and environment, and other areas where science and technology are expected to significantly drive national economic and social development. Cooperation is also carried out in fundamental research, applied research, technology demonstration, transfer of scientific and technological achievements and other aspects. Special attention is paid to the improvement of African countries' capabilities of making policies, managing technology, promoting the development of technology industry, and improving livelihoods with technology. The majority of funds required for the execution of the program comes from the Chinese government, with capital from relevant international organizations, enterprises and other private investors being brought in. The MOST and other government authorities of China are leading agencies responsible for coordinating relevant departments in China and maintaining communication with African countries and regions for joint advancement of the program. Besides, the advisory committee, secretariat and other mechanisms are established to guarantee that the program runs well.

On November 24th, 2009, the "CASTEP 1.0" was officially launched in Beijing (MOST of PRC, 2009). Since its inception, China and African countries joining the program carried out substantial cooperation in talent training, policy coordination, technical service, joint research and technology demonstration as they knew that "teaching one how to fish is better than giving him fish". Remarkable results have been achieved in promoting a new pragmatic, efficient and dynamic partnership in science and technology between China and African countries. Examples include the donation of equipment to African researchers, which was initiated on the day when the program was launched. According to the action plan, equipment and facilities for scientific research were donated by China's MOST to African researchers returning to their home countries after completion of long-term scientific research in China so that relevant research and development endeavours were continued without interruption, helping African countries improve scientific and technological conditions and furthering cooperation between China and African counties. For another example, in 2011, China's MOST compiled the *China–Africa Science and Technology Partnership Program (CASTEP) Applicable Technology Manual* (MOST of PRC, 2011) which elaborates the development and current applications of 46 specific technologies of different categories in relation to agriculture, forestry, energy saving and emission reduction in industrial production, energy saving and emission reduction building construction, energy saving and emission reduction in commercial & civil activities, the supply of clean energy, waste disposal and utilization, water resource utilization, resource and environment, disaster prevention and mitigation, and health, etc. It also specifies contact information of technology providers so that technologies needed are accessed by African countries.

15.3.2 The Second Stage of Exploration: CASTEP 2.0

At the 18th National Congress of the Communist Party of China (CPC), China advocated that a community with a shared future should be built. That means a nation should take into account reasonable concerns of other countries while pursuing its national interests for the purpose of achieving shared development and new-type global development partnership which features balance, equality, collaboration, joint responsibilities and enhanced common interests of humankind. As a part of the community with a shared future, China and African countries share strong complementarity, extensive mutual interests and reciprocal demands. In response to the new round of technological revolution and industrial transformation across the globe, the two sides need to further enhance cooperation in STI by diversifying contents, areas and modes of innovation cooperation to find a way to innovation-driven development. To this end, the "CASTEP 2.0" was launched by China in light of fruitful results of the "CASTEP 1.0.", in a bid to better respond to new circumstances and mutual demands of both sides and play a more prominent role in sustainable development of China and African countries and in the building of a community with a shared future.

In 2016, Chinese Foreign Minister Wang Yi attended the Coordinators' Meeting on the Implementation of the Follow-up Actions of the Johannesburg Summit of the FOCAC and made a work report where he announced the official launch of the "CASTEP 2.0" (Yi, 2016). Compared with the "CASTEP 1.0" launched 7 years ago, the upgraded version emphasized its objectives to comprehensively improve the level of China–Africa cooperation in STI, provide strong support for the establishment and development of China–Africa comprehensive strategic and cooperative partnership, enhance and consolidate the China–Africa community with a shared future, and promote innovation-driven development of China and African countries. Its guiding principles give more prominence to "people orientation" and "capacity building", and training of scientific and technological talents is placed at the core of China–Africa cooperation in STI, in order to give full play to their key role in shaping the future of Africa by making them more motivated and more creative. The program also emphasizes efforts to be made to help African countries enhance STI capacity and gradually realize innovation-driven sustainable development. The "CASTEP 2.0" shows superiority and improvement in both design and implementation of policies.

In terms of policy design, the "CASTEP 2.0" has been incorporated into the top-level design of the "Belt and Road" Science, Technology and Innovation Cooperation Action Plan, and brought into line with the overall diplomacy blueprint for building China–Africa comprehensive strategic and cooperation partnership.

Chinese President Xi Jinping proposed to create the "Silk Road Economic Belt" in a speech delivered at Nazarbayev University during his visit to Kazakhstan in September 2013, and build a "21st Century Maritime Silk Road" at the Indonesian parliament in October 2013, so as to give play to the role of the Belt and Road Initiative (BRI) as a platform to build a community with a shared future for mankind. In September 2016, China's MOST, along with other ministries, issued the *Special*

Plan on Advancing Cooperation of Science and Technology Innovation in the Belt and Road Construction (MOST of PRC, 2016). At the Belt and Road Forum for International Cooperation held in May 2017 in Beijing, Xi Jinping proposed the "Belt and Road" Science, Technology and Innovation Cooperation Action Plan, which consists of the Science and Technology People-to-People Exchange Initiative, the Joint Laboratory Initiative, the Science Park Cooperation Initiative and the Technology Transfer Initiative. Echoing the "Belt and Road" Science, Technology and Innovation Cooperation Action Plan, the "CASTEP 2.0" promotes sustainable development of China and African countries and the building of a community with a shared future by advancing pragmatic cooperation with African countries in talent training, science and technology policy, scientific research platform, technology transfer, science park, research and demonstration, resource sharing, and innovation and entrepreneurship, etc. In the speech delivered at the opening ceremony of the Johannesburg Summit of the FOCAC in December 2015, Xi Jinping proposed to upgrade the new-type China–Africa strategic partnership to a forcomprehensive strategic and cooperation partnership and announced the "10 Major Cooperation Plans" to be executed together with African countries. At the Beijing Summit of the FOCAC held in September 2018, China announced that the "Eight Major Initiatives", succeeding the previous "10 Major Cooperation Plans", would be implemented in close cooperation with African countries in the next 3 years and beyond to help them achieve independent and sustainable development at a faster pace. To promote social development cooperation, China called attention to cooperation in STI and knowledge sharing and proposed to continue the implementation of the "Belt and Road" Science, Technology and Innovation Cooperation Action Plan and the "CASTEP 2.0". Priority has been given to STI helpful for livelihood improvement and national economic and social development. China also undertook to work with African countries to promote the implementation of the "Science, Technology and Innovation Strategy for Africa", aiming to help African countries enhance STI capacity.

Moreover, China stressed the enhancement of science and technology related people-to-people exchange and cooperation and expressed its willingness to encourage and support technology transfer to African countries. In response to the trend of the times and needs of development of new technologies such as artificial intelligence and quantum computing, China is willing to make the best use of its own advantages to provide assistance to African countries in operating system, network security, big data, blockchain and other applications to the best of its ability. To meet development requirements of African countries, higher learning institutions, scientific research institutes, enterprises and other players in China and African countries are encouraged to join hands in key areas of mutual concern, including joint laboratories, high-level joint research and training of science and technology talents. In addition to discussions on pragmatic cooperation in the construction of science parks, China is also willing to continue its support for the construction and development of the "Sino-Africa Joint Research Center" and the Square Kilometer Array (SKA), which is an international large scientific project for the construction of radio telescope and a flagship project of science and technology on the African continent.

In terms of policy implementation, the advancement of the "CASTEP 2.0" features greater flexibility, diversity, pragmatism and efficiency. At the Beijing Summit of the FOCAC held in 2018, Xi Jinping proposed the construction of a China–Africa innovation cooperation center as an important move to advance the "CASTEP 2.0", improve the "Belt and Road" technology transfer network, promote China–Africa cooperation in STI, and build a closer China–Africa community with a shared future in the new era. In response to the call of Xi Jinping, China's MOST and Hubei Provincial People's Government launched the construction of China–Africa Innovation Cooperation Center (CAICC) in May 2020. In an early phase, CAICC focuses on cooperation with Kenya, Ethiopia, South Africa, Egypt, Ghana, Nigeria and other African countries and establishment of sub-centers in 1–2 countries in eastern, western, southern and northern African regions, followed by continuous expansion of the territory of cooperation. Its priorities include cooperation in technology transfer, innovation and entrepreneurship in fields like agriculture, medicine and health, resources and environment, energy, space, and information and communication, etc. Against the backdrop of global spread of COVID-19 pandemic and formidable challenges to African public health system, CAICC pays special attention to cooperation in STI with African countries in public health, particularly in the prevention and control of infectious diseases. It aims to develop a technology transfer, innovation and entrepreneurship network that covers China and African countries, promote innovation and business startups by young people in China and African countries, accelerate training, transfer and demonstration of applicable technologies, and boost social and economic development of African countries. In December 2021, China–Africa Innovation Cooperation Conference co-organized by CAICC was successfully held in Wuhan, Hubei, and contracts were signed for 15 cooperation programs in STI involving many African countries, covering joint construction of platforms, technological research and development, commercialization of science and technology achievements, and talent training. This further promoted the building of an extensive network of China–Africa cooperation in STI. Besides, substantial results have been yielded in bilateral and multilateral cooperation between China and African countries in STI under the "CASTEP 2.0" supported and joined by Egypt, South Africa, Ethiopia, Kenya and many other African countries. For example, a number of joint research platforms construction has been launched and promoted, including the Kenya–China Joint Laboratory for Crop Molecular Biology, China–Ethiopia Joint Laboratory for Leather Industry, China–Egypt Joint Laboratory for Renewable Energy, and China–South Africa Joint Research Center for the Development and Utilization of Mineral Resources. Discussions on science park cooperation with Egypt and South Africa have also commenced, thanks to China's abundant experiences in constructing high-tech zones and science park. Proactive efforts, including organization of China–South Africa Hi-Tech Exhibition and construction of platforms like China–Arab States Technology Transfer Center and China–Africa Innovation Cooperation Center, have been made to build a network for China–Africa cooperation in technology transfer (Xia, 2018).

15.3.3 Cases of China–Africa Cooperation in STI Promoting Capacity Building of African Countries

Promoting the realization of the United Nations' 2030 sustainable development goals is the direction of capacity building for African countries. Many cases can prove that current China–Africa cooperation in STI is moving towards the above goals, with its focuses on the application and promotion of advanced technologies in Africa through practical cooperation methods such as technology transfer and personnel training, so as to help African countries strengthen their capacity building.

Eliminating all forms of poverty in the world is the primary goal of the United Nations 2030 Agenda for Sustainable Development. China–Africa cooperation in STI in the field of agriculture helps African countries accelerate the elimination of regional poverty by mastering advanced agricultural technologies. In Madagascar, the African Center of the China National Hytrid Rice Research and Development Center helped local people overcome the drought that occurred at the end of 2019. On one hand, the center selected and bred hybrid rice varieties and cultivation techniques suitable for local needs; on the other hand, it also trained local growers to master advanced technologies. In Burkina Faso, the Chinese expert team built a 50-hectare lowland rice demonstration area in its Central and Western Regions. In addition to providing agricultural production materials and participating in the whole process of sowing and harvesting, more importantly, the Chinese expert team also organized field technical training for the popularization of new technologies. As a result, the rice in the demonstration area in Burkina Faso is of high quality, and the yield increases by 2–3 times per year, which sets an example of capacity building in rice planting for other areas within the country. In Guinea-Bissau, the China-aided agricultural group compiles a series of Chinese-Portuguese bilingual teaching materials such as "Key Points of Rice Planting Technology" for the training of local technical talents (Hongjiang et al., 2022). In Burundi, the China-Aided Agricultural Technology Demonstration Center guides 22 villages in 14 rice-growing provinces in Burundi, providing rice farmers with technology and guidelines. The 49 trainees trained by the Center have then provided rice planting training for more than 1500 farmers across the country. After such a cooperation sustaining for seven consecutive years, rice production in Burundi has increased from an average of 3 tons per hectare to 10 tons, achieving of rapid growth and contributing to the country's food security (Li, 2022).

In addition to the cooperation in STI in the field of agriculture, China and African counties have also deepened the cooperation in STI in terms of digital technology and aerospace technology, which help African countries transform their economic development patterns and build their capability of achieving modernization in their own ways. Taking the promotion of economic transformation by digital technology as an example, based on technology and experience exchanges with China, many localized mobile payment networks have grown rapidly in African countries, such as M-Pesa in Kenya, Opay in Nigeria, and Zapper in South Africa, etc. Successful cases such as those mentioned above have jointly promoted the development of digital transformation in Africa. Taking the promotion of modernization by aerospace technology as another example, the cooperation between China and Africa in various forms such as satellite import and export, joint construction of aerospace infrastructure, sharing satellite resources, and joint research and development of satellites has made aerospace technology play a more and more significant role in the economic and social development and in the improvement of people's production and life of African countries. Specific areas of action include but are not limited to climate monitoring, breeding, disaster prevention and mitigation, and communications (Jiabao, 2022).

15.4   New Opportunities and Challenges for China–Africa Cooperation in STI

Today's world is experiencing great changes that have not been seen in a century. "Great changes" means the reshaping of the international order. Developing countries are more deeply involved in the wave of major changes than ever before. The international trend of accelerated development of multi-polarization in the world and increasingly balanced international patter has an increasingly far-reaching impact on China and the vast number of African countries, which also provides new development opportunities for China–Africa cooperation in STI.

First, the accelerated reconstruction of the international political and economic pattern has provided a new window of opportunity for the in-depth promotion of China–Africa cooperation in STI. A large number of emerging markets and developing countries are rising and embracing greater space and potential for development. According to a survey by the Development Research Center of the State Council of PRC, emerging markets and developing countries are likely to surpass developed countries in total economic output and contribute to about 60% of global economy and investment by 2035 (Yiming, 2020). As the largest developing country and the fastest-growing nation in the group of emerging markets and developing economies, China will have access to a more promising market for technology transfer and technological application by advancing cooperation with Africa, the continent with the largest number of developing countries, in STI.

Second, the continuous and in-depth development of a new round of scientific and technological revolution and industrial reform has also brought new development opportunities to China–Africa cooperation in STI. The integration of multiple technological breakthroughs based on digital and intelligent technologies, a predominant feature of the ongoing Fourth Industrial Revolution, is driving the revolution at an unprecedented speed to a depth and breadth never seen before. The new revolution improves global industrial chain network, enhances the division of labor in the international industrial chain and reduces inter-regional transaction costs, facilitating further advancement of China–Africa cooperation in STI. The formation of a global industrial chain network makes productions more flexible and helps China and geographically remote African countries make the best of technology to create new forms of business and new sources of economic growth.

Third, with the increasingly close relationship and interdependence between different countries, the common problems faced by the survival and development of human society are also increasing. The key of solving such interrelated issues lies in STI, particularly in intergovernmental cooperation in STI. This trend makes China–Africa cooperation in STI even more relevant. For instance, in the context of global economic recession caused by COVID-19 pandemic, developing countries, especially those in Africa, are suffering wide-reaching negative consequences due to weak economic base and limited medical support capability. It, therefore, calls for STI to achieve effective epidemic control, rapid post-pandemic recovery and economic rebound, which is exactly what China–Africa cooperation in STI can contribute against the new backdrop.

China–Africa cooperation in STI sees new development opportunities in tandem with new challenges in the new era. From the perspective of international politics, currently in the turbulent period of international political development, the intensification of the game between countries and the complexity and variability of the international political pattern will inevitably interfere with the cooperation in STI between China and African countries. From the perspective of technological change, while digital and intelligent technologies bring new development opportunities, they also produce a more significant "capacity gap", that is, the differences between different groups in accessing digital resources, processing digital resources, and creating digital resources are increasing. From the perspective of the global environment, the increasingly prominent global issues such as climate change and environment deterioration, food and energy crisis, and the widening gap between the rich and the poor have posed challenges to different countries, especially to developing countries.

Developing countries, who are technologically inferior, have to learn advanced technologies while dealing with risks brought by new technologies and other issues usually expected to be encountered by developed countries. For China and African countries, how to break through the limitations of their own economic and technological development levels and truly achieve the strategic goal of innovation-driven transformation and development in the changing international political and economic environment is an important challenge for China–Africa cooperation in STI.

15.5   Response Strategies for Future China–Africa Cooperation in STI

In response to new challenges and opportunities outlined above, capacity building-oriented is still the future direction of China–Africa cooperation in STI. Specifically, the following endeavors are recommended to further China–Africa cooperation in STI in the future.

Firstly, it's imperative to enhance the strategic significance of China–Africa cooperation in STI and increase the consensus of both sides on development. China and African countries share the missions of accomplishing innovation-driven transformation and development, lifting low-income groups out of poverty, and improving the standard of living and quality of life of people on both sides. China's development experience shows that science and technology are the primary productive force, a strong support for economic growth and social prosperity, and a leading force driving sustainable and coordinated development. The key for African countries to speed up industrialization and modernization lies in drawing momentum from improved capacity and level of science and technology to turn technological revolution into an engine for economic growth. China–Africa cooperation in STI is an endeavour made by developing countries to achieve indigenous development and a source of strength for building the China–Africa community with a shared future.

Secondly, China and African countries have to further expand their cooperation in the fields of new technologies. China and African countries need to carry out cooperation in STI in line with the trend of the times, and make the best of important opportunities offered by the new round of technological revolution and industrial transformation to catch up with developed countries. In particular, enhancing the development of digitalization is integral for China–Africa cooperation in STI. Although the cooperation between China and African countries in the promotion of digitalization has achieved certain results, there is still a lot of room for expansion in the breadth and depth of such a cooperation. The endeavour to cooperate in the fields of new technology will inevitably face more difficulties, which need to be dealt with through strategic cooperation between China and African countries by deepening their mutual trust and by innovative cooperation methods.

Lastly, mechanisms of China–Africa cooperation in STI need to be diversified and improved. In addition to government-led innovation cooperation between China and African countries, more flexible and diverse cooperation mechanisms like nongovernmental and inter-regional cooperation may be developed. Although China–Africa cooperation in STI has established a relatively complete official cooperation mechanism at national level and is operating effectively, the continuous enrichment of cooperation forms will help to further tap the potential of cooperation. In the construction of non-governmental cooperation mechanism on one hand, some existing international experiences can be used for reference, and the role of market mechanism in promoting China–Africa cooperation in STI can be strengthened. While mobilizing non-governmental forces to participate more, it is also significant to establish a long-term and stable mechanism to better combine different advantages of both official cooperation and that of the civil one. In the construction of inter-regional cooperation mechanism on the other hand, new explorations have already been made. The China–Africa Innovation Cooperation Center recently founded in Wuhan of Hubei is a valuable example of inter-regional cooperation, which can promote more flexible cooperation at the local level. In the future, it is necessary to continue strengthening inter-regional cooperation in STI between China and African countries, which will help to further mobilize the initiative of cooperation entities of both sides, fully tap the potential and space for cooperation in STI at the grassroots level, and carry out more distinctive, targeted, pragmatic and efficient cooperation programs.

References

Anshan, L. (2021). Forty years of study on the history of China–Africa relations in ancient times. *Social Science Front, 2*, 97–109.

Deming, Z. (1997). Medical exchanges between China and African countries in ancient times. *Chinese Medical Journal, 2*, 105–108.

Heng, W., & Zhemeng, L. (2021). Africa's digital economy development and China–Africa digital cooperation: Opportunities, challenges and responses. *Journal of Zhejiang Normal University, 6*, 20–26.

Hongjiang, W., et al. (2022). China's poverty reduction experience contributes to Africa's sustainable development. *Xinhua News Agency*, March 29th. https://www.yidaiyilu.gov.cn/xwzx/hwxw/231080.htm

Hongzhong, H., & Ruizhou, Z. (2019). Study on agricultural science and technology exchanges and cooperation between China and Kenya. *Chinese Agricultural Science Bulletin, 29*, 151–158.

Jiabao, L. (2022). China–Africa space cooperation benefits the African continent. *People's Daily* (Overseas Edition), September 19th. https://www.yidaiyilu.gov.cn/xwzx/hwxw/277547.htm

Jiali, L. (2005). Considerations on strengthening China–Africa agricultural cooperation. *World Agriculture, 5*, 11–14.

Leften S. S. (2006), A global history: from prehistory to the 21st century (volume 1). Translated by Xiangying, W. Beijing: Peking University Press.

Li, J. (2022). China-aided Burundi agricultural technology demonstration center helps Burundi increase crop yield. *Xinhua News Agency*, May 22nd. https://www.yidaiyilu.gov.cn/xwzx/hwxw/245412.htm

Mammo, M., & Patra, S. K. (2019). China–Africa science and technology collaboration: Evidence from collaborative research papers and patents. *Journal of Chinese Economic and Business Studies*. https://doi.org/10.1080/14765284.2019.1647004

MOST (PRC). (2009). Launch ceremony of China–Africa science and technology partnership program (CASTEP) & donation of equipment to African researchers held in Beijing. November 30th. http://www.most.gov.cn/ztzl/lhzt/lhzt2010/gjkjhzlhzt2010/201003/t20100301_76042.html

MOST (PRC). (2011). China–Africa science and technology partnership program (CASTEP) applicable technology manual. The Ministry of Science and Technology (MOST), People's Republic of China, Beijing.

MOST (PRC). (2016). The notice on issuing the Special Plan on Advancing Cooperation of Science and Technology Innovation in the Belt and Road Construction. September 14th. http://www.scio.gov.cn/xwfbh/xwbfbh/wqfbh/35861/36653/xgzc36659/Document/1551346/1551346.htm

Niane, D. T. (ed.) (1992). *General history of Africa: Africa from the twelfth to the sixteenth century.* Beijing: China Translation Corporation.

Rui, W., Huijie, Z., Tianjin, C., & Shuai, Z. (2021). Status quo and countermeasures for China–Africa agricultural science and technology cooperation. *Journal of Agriculture, 8*, 98–103.

Taling, T. R. (2021). China–Africa cooperation on science technology engineering and mathematics: Challenges and prospects. *International Journal of Research and Scientific Innovation, 8*(9), 120–126.

Tao, W. (2011). Analysis on the development history and features of China–Africa cooperation in science and technology. *World Outlook, 2*, 113–126.

Xia, L. (2018). Building a "Belt and Road" community of STI: fruitful results of cooperation in STI between China and Asian and African countries. *Science and Technology Daily*, December 18th. http://ip.people.com.cn/n1/2018/1218/c179663-30472981.html

Xiao, W. (2013). Analysis on situation and suggestions on policy regarding China–Africa science and technology cooperation. *Forum on Science and Technology in China, 8*, 142–146.

Yanting, D. (2019). *Friendly cooperation between China and member states of the East African community*. China Social Sciences Press.

Yi, W. (2016). Work report of Chinese Foreign Minister Wang Yi at the coordinators' meeting on the implementation of the Follow-up Actions of the Johannesburg Summit of the Forum on China–Africa Cooperation. July 30th. https://www.fmprc.gov.cn/wjbz_673089/zyjh_673099/201607/t20160730_7478472.shtml

Yiming, W. (2020). Great changes in the past century, high-quality development and building a new development pattern. *Journal of Management World, 12*, 1–12.

Yonghong, Z., Tao, W., & Hongxiang, L. (2012). China–Africa science and technology cooperation: Strategic significance, policy direction, and mechanism system. *World Outlook, 5*, 52–71.

Yonghong, Z., Tao, W., Tao, W., et al. (2019). *A study on the strategic background of China–Africa scientific and technological cooperation*. China Social Sciences Press.

Yonghong, Z., & Yuanfei, G. (2016). Mechnism and content of China–Ethiopia cooperation in science and technology. *Journal of Southwest Petroleum University (social Sciences Edition), 3*, 40–49.

Zhaokun, H. (2019). Research on China–Africa science and technology cooperation mode and promotion strategy. *Scientific Management Research, 1*, 102–105.

Chapter 16
Science, Technology and Innovation (STI): Synergic Cooperative and Collaborative Relations Between China and Africa

Manir Abdullahi Kamba

16.1 Introduction

The international community and multilateralism have become highly active, coordinated, and robust in a rapidly global world with increasing science, technology, and innovation development. According to Umesh, Sai Vinod and Sivakumar (2021), the effectiveness of these initiatives depends on the point of regional and international cooperation in developing countries. There are, however, little substantive information and knowledge gaps on how international cooperation and activities can serve as tools for managing science, technology, and innovation of Sustainable Development Goals (SDGs).

China–Africa science, technology, and innovation development initiatives are timely in addressing innovation and development research gaps, South-South development platforms, and cooperation in shaping innovative national development evidence policies, priorities, programs, and benefits in developing countries. International bilateral or multilateral cooperation on innovation and development has been evolving rapidly since the late twentieth century to meet the increasing needs of vulnerable populations, moving toward achieving sustainable development goals (SDGs).

In addition, to address the challenges of STI in Africa, collaborative diplomacy and policies on research and innovation have become more significant than ever in saving lives, improving public well-being, and enhancing and providing long-lasting benefits to the world's poorest and most developed countries.

While the news has often been reported on the increasing Chinese involvement in Africa, there is been very little literature on its effects on STIs. Furthermore, it is

M. A. Kamba (✉)
Bayero University, Kano, Nigeria
e-mail: manirung@yahoo.com

© The Author(s) 2024
M. Muchie et al. (eds.), *China-Africa Science, Technology and Innovation Collaboration*, https://doi.org/10.1007/978-981-97-4576-0_16

291

still unclear exactly how the Chinese approach differs from the Western approach, as the difference between the "horizontal" and "vertical" approaches is not practical or operational. There is a need for a deeper analysis of China's cooperation and efforts at improving STI at large-scale development in Africa.

16.2 Science Technology and Innovation Challenges in Africa

According to Schumpeter's economic theory, as explained by Mêgnigbêto (2019), development is a historical process of structural changes substantially driven by innovation (Schumpeter, 2004). Therefore, in any given situation, a low level of scientific production means a low level of novelty and knowledge production, resulting in a low level of innovation. In this regard, evidence has shown that Africa is among the poorest regions of the world in terms of wealth production; it also faces the most significant challenges: poverty, hunger, armed conflicts, poor governance, population growth, access to education, and health care. The African region needs to produce more knowledge to take up these challenges. So, there is a need to understand how much knowledge produced by Africa-based researchers helps the region in wealth creation.

One of the key challenges of STI systems research focusing on developing countries is therefore to understand real differences, rather than benchmarking against an ideal type of innovation system (Cozzens & Sutz, 2014; Egbetokun et al., 2009). Some of the ideas and concepts which have emerged in the innovation systems community have been derived mainly from specific experiences in rich countries and cannot be used as universal templates. While the systemic perspective and interactive learning are widely relevant, the agencies involved, and the patterns of interaction will differ and there will be variations in the nature and degree of system external actors (Cassiolato et al., 2002).

In other words, researchers focusing on countries in Africa cannot uncritically adopt analytical frameworks of innovation developed in the North. This is an analytical point in itself, but there are also normative reasons for avoiding the uncritical transfer of concepts, although this is being recently addressed by scholars from the region and those working on the region (Oyelaran-Oyeyinka & Gehl Sampath, 2010). The scale of poverty, as well as the relative importance of the informal sector with an infrequency of 'decent jobs' require a narrower focus on the relationship between innovation and economic growth on one hand, and between productive development and competitiveness on the other, be combined with an understanding of the prerequisites for inclusive development (Johnson & Andersen, 2012). Needless to say, in today's globalized economy, inclusive development challenges are numerous, as are the advantages associated with what seems to be a borderless transfer of knowledge. In addition, the innovation policy lens needs to focus on Africa's recent economic growth success, the impact of the rising consumer class, and the potential impact of the resulting structural change driven by key sectors such as construction (notably with the inflow of direct foreign investments from China) as well as the ICT and services sector broadly.

The complex challenges cannot be grasped without strengthening the political economy dimension of innovation systems research. There has been fairly limited research that asks how politics, power and interests influence technology and innovation policy and practice in Africa (Bell, 2009). While the political economy dimension is widely relevant and underexplored in innovation studies, it is particularly relevant where 'new leaderships have largely tended to carve out renter opportunities and created increasingly monopolistic and autistic governance practices' (Karuri-Sebina, Sall, Maharajh, & Segobye, 2012, 492).

Siyanbola, Adeyeye, Olaopa, and Hassan (2016), explained that the concept of STI in development is geared towards creating and exploiting new knowledge. While knowledge accumulation is pivotal to the development process, there is no agreement on the trajectory of the process. Notably, Metcalfe (2000), indicated that it is difficult to predict the outcome and consequences of the exploitation of knowledge in development. Despite this, Siyanbola et al. (2016), reiterated that scientific developments may open up new opportunities for generating new ideas and knowledge for practical use and application of technology. On the other hand, it is driven by the need and search for the underpinning natural principles for solving development challenges. In addition, there are differences in knowledge and how knowledge is accumulated and applied to solve societal challenges among countries. These differences are in the strength and depth of their institutional structures for generating and applying knowledge.

As the knowledge economy proliferates, African countries have made a bold attempt to turn around their development fortunes by adopting the Monrovia Strategy in July 1979, the Lagos Plan of Action (LPA) for the Economic Development of Africa (1980–2000), and Final Act of Lagos in April 1980. The LPA was a visionary, far-reaching, and unprecedented blueprint to foster the continent's collective self-reliance and sustainable development. Subsequent attempts at charting Africa's development have drawn inspiration from that visionary framework.

According to Khan (2022), advocacy regarding the importance of science, technology, and innovation (STI) came early in the life of the African Union, founded in 2002 in Durban, South Africa, as a member of the Organization of African Unity (OAU). Therefore, this assessment's most appropriate point of departure is referencing the African Union seminal document "On the Wings of Innovation". The Science, Technology, and Innovation Strategy for Africa 2024 (STISA–2024) places science, technology, and innovation at the epicentre of Africa's socio-economic development and growth. This strategy follows the prior "Africa's Science and Technology Consolidated Plan of Action (CPA)" (African Union, 2005) which sets out five programs to improve policy conditions and build innovation mechanisms, including science and technology policies and their measurement through the African Science, Technology and Innovation Indicators (ASTII) initiative. The CPA reaffirmed the 1980 Lagos Plan of Action target of 1% for Gross Expenditure on Research and Development to GDP (GERD: GDP). The work on indicators gained substance through founding the African Observatory on Science, Technology, and Innovation (AOSTI) and the African Innovation Outlook series. African and international donors supported these moves.

The implementation of the CPA influenced the role that science, technology, and innovation play in Africa's socio-economic development. These influences were translated into policy instruments at various levels to achieve transformative and emancipatory goals using building institutions and implementing programs. The situational analysis of STI in Africa summarized below builds on evidence generated by the survey, "Science, Technology and Innovation Policy-making in Africa: An Assessment of Capacity Needs and Priorities and the Environment Scan," which supported the review of the CPA.

(a) ***Increased recognition by African leadership and the public of the critical role STI plays in economic growth and human development***

Recent political, and policy statements and instruments underscore increased investment in STI to achieve sustainable socio-economic growth, reduce poverty and food security, fight acute communicable and non-communicable diseases, and stem environmental degradation. The evidence-backed launch by the regional networks as implementation mechanisms for the CPA R&D flagship programs in biosciences, biotechnology, biosafety, laser technology, mathematical sciences, water, and energy, as well as programs related to measuring STI support evidence-based policymaking.

(b) ***Insufficient Funding for STI***

Recent statistics from UNESCO and ASTII show that the current level of investment in R&D by Africa as a continent (of which international donors fund more than half) puts Africa at a strategic disadvantage. Most STIs achieve the 1% of GDP target agreed upon by the AU member states as desired minimum expenditure on R&D.

(c) ***Organizational Capacity by Entities Responsible for STI Policymaking***

Most entities responsible for STI policymaking in Africa operate in isolation from other policy agencies. They have weak links to the private and education and research sectors and African and international Policy Research Think Tanks. Again, they do not have easy access to empirical material and current knowledge in STI policymaking. Furthermore, they ignore the inter-sectoral linkages and policy mixes, making their institutional outputs much less reliable.

(d) ***Infrastructure to Support Innovation***

To support innovation and facilitate competitive business activities requires infrastructure such as broadband internet access, essential telecommunication services, reliable electricity supply, water, good transportation networks, laboratory facilities, and tax systems that support private sector innovation. The AU Program on Infrastructure Development for Africa (PIDA) revealed different levels of infrastructure ready to support innovation in African economies. However, it reflects Africa's low scores in many major classifications or indices, such as the world's leading universities, and competitiveness index.

(e) ***Inadequate Expertise in STI Policy Development***

Minimal evidence-based policy development takes place in Africa. Many officials involved in or responsible for drafting policy documents do not have the necessary skills or training and no evidence-based policymaking experience. Moreover, in most countries, institutions responsible for STI policy do not have appropriate libraries or easy access to relevant information sources for policymaking purposes.

(f) *Emergence of African Civil Society Organizations and Think Tanks Dedicated to Raising Awareness of STI*

Civil society organizations and think tanks are championing the use of African indigenous knowledge to support sustained economic growth and inform public attitudes and understanding of the relevance and importance of STIs. At the same time, they contribute to the STI policy debate in biosafety, climate change, biodiversity, environment regulation, and ICT, and most contributions are not by evidence.

(g) *Bilateral and Multilateral Cooperation*

Bilateral and multilateral partnerships have shaped STI development in Africa (e.g., the European Union–Africa Joint Strategy, the India–Africa Science and Technology Initiatives, and the China–Africa Science and Technology Partnership). However, these interventions and cooperation mechanisms are not promoting African ownership, accountability, and sustainability.

(h) *Scientific Output*

Africa is registering an increasing number of scientific publications as well as the acquisition of capital goods. For example, Tunisia reported a tenfold increase in scientific publications between 1990 and 2010, while Uganda achieved over 1200% growth during the same period. About 18 African countries achieved a fourfold increase in imports of capital goods between 2000 and 2011. Steady investment in STI, expansion of R&D institutions, and political support may account for this surge in both technology acquisition and the number of scientific papers published.

From the preceding analysis, although Africa is reorienting its development policies to include STIs at various levels, its STI capacity is still deficient. Only 12 African countries out of 141 countries were surveyed and ranked among the top 100 innovation achievers in the 2015 Global Innovation Index. Only one of 31 African countries surveyed in the 2016 Network Readiness Index was among the world's top 50 network-ready countries. Africa has poor STI infrastructure, a small pool of researchers, low patronage of science and engineering programs, weak intellectual property frameworks, and low scientific output relative to the rest of the world. Africa remains disadvantaged in overall STI efforts due to the low investments in STI capacity development. It accounts for about 5% of the global gross domestic product but only 1.3% of global spending on research and development (UNESCO, 2015). Indeed, about 84% of the African countries surveyed in 2022 were ranked Low or Very Low in capacity development outcomes (Khan, 2022).

Africa Capacity Report 2017 shows that STI capacity is one of Africa's biggest challenges. A survey of 44 African countries undertaken by ACBF in 2016 to assess capacity needs in STI priority areas showed that African countries consider training as a High or Very High priority area in STI Investment in STI development is meagre in Africa. African countries are taking a short-term approach to developing STI skills, as evident in low public spending on research, development, and scientific infrastructure. This situation has not been improving to a large extent, as indicated in the report of Khan, 2022. Most African countries have a weak institutional capacity to develop and sustain STIs since few public institutions have adequately qualified human resources in science and engineering. African countries have a weak capacity to retain the few qualified scientists and engineers it has, and the migration of skilled African scientists and other experts, the "brain drain" has further depleted Africa's STI capacity. For instance, from 2007 to 2011, the number of tertiary-educated Africans who had migrated was estimated at 450,000 (UN-DESA and OECD, 2013).

So, Africa incurs a net loss in skilled human capital with the critical technical skills to foster Africa's sustainable development. Zimbabwe (43%), Mauritius (41%), and the Republic of the Congo (36%) recorded the highest proportions of educated persons living in OECD countries. Burundi, Algeria, Mauritania, Chad, and Guinea are the top five African countries least able to retain their top talent (WEF, 2014). So, Africa's training institutions subsidize other developed regions since training is costly. Another critical challenge is the lack of accurate data to target STI policies and strategies. The lack of a robust standard set of African STI indicators has constrained the continent's ability to make evidence-based STI decisions. The weak capacity to manage the data affects the ability to update the STI policies and strategies and to determine how much to allocate to build STI capacities and frameworks.

16.3   Africa Science and Technology Innovation Initiatives

Science, Technology, and Innovation (STI) underpin almost every aspect of human existence. Therefore, optimizing the benefits of STIs is an increasing priority for Africa's governments and people. Developing STI initiatives and strategies, which give due consideration to Africa's environment and concerns, is one of the most effective weapons for winning the struggle to reduce and eliminate absolute poverty in Africa. In recent years, African governments, scientists, policymakers, private sector actors, and many crucial civil society organizations are beginning to re-discover the importance of STIs in driving technical and economic progress. STI fosters productivity growth, social welfare, and sustainable development. The vision for a significant improvement in the physical quality of life, competitiveness, and overall prosperity in Africa can only be achieved and sustained through a sincere commitment to STI development.

Let us compare traditional societies with contemporary ones. It becomes clear that norms and processes of creating a meaningful environment for living have to be in line with science-based knowledge. In the last few decades, significant advancements have been made in the global economy resulting in a departure from traditional production processes. The increasing needs of society in the face of limited resources have propelled renewed thinking toward efficiency.

Therefore, it requires improved knowledge of science and technology and a revolution in the entire knowledge system toward a culture of innovation. Evidence in the economic literature indicates that science, technology, and innovation can play a significant role in the economic growth of a given nation. In recent years, progress in STI has increased productivity in several developed countries, including Africa, indicating greater efficiency in using labour and capital. The rise in productivity resulted from improved managerial practices, organizational change, science, technology, and innovation in producing goods and services. More so, increased investments in information and communications technology (ICT) have led to improved provision of quality of capital and labour as well as witnessing the rising skills of the average worker in African economies. Furthermore, technological change obtained through the returns to research and development (R&D) and other knowledge-based investments and spillovers from innovation also contributes significantly to growth.

Noting the critical role of Science, Technology, and Innovation in Africa's development agenda, the 23rd Ordinary Session of African Union Heads of State and Government Summit, in June 2014, adopted a 10-year strategy called Science, Technology and Innovation Strategy for Africa (STISA-2024). The strategy aims to support the AUA Agenda 2063, with science, technology, and innovation as critical enablers for achieving sustainable development goals (SDGs). The agenda stresses the need for diversification of sources of growth and a solid need to sustain Africa's current robust economic performance so that a large section of people are out of poverty. The STISA also calls for social transformation and economic competitiveness through human capital development, innovation, value addition, industrialization, and entrepreneurship.

Due to the cross-cutting nature of science, technology, and innovations, the STISA-2024 strategy is a blueprint for meeting the knowledge, technology, and innovation demands in various A.U.A.U. economic and social sector development frameworks. The STISA-2024 is anchored on six areas aiming at contributing to the achievement of the vision of the African Union:

(a) Eradication of Hunger and Achieving Food Security;
(b) Prevention and Control of Diseases;
(c) Communication (Physical and Intellectual Mobility);
(d) Protection of our Space;
(e) Live Together-Build the Society;
(f) Wealth Creation.

The strategy further defines four mutually reinforcing pillars that will serve as the prerequisite conditions for its success:

(a) Building and upgrading research infrastructures;
(b) Enhancing professional and technical competencies;
(c) Promoting entrepreneurship and innovation;

(d) Providing an enabling environment for STI development in the African continent.

At the first African Ministerial Conference on Science and Technology (AMCOST), the participating countries committed themselves to developing and adopting a standard set of indicators. One of its key outcomes was the adoption of NEPAD's African Science, Technology, and Innovation Indicators (ASTII) in 2005 to contribute toward a better quality of STI policies at the national, regional, and continental levels. ASTII aims to support and strengthen the capacity of Africans to develop and use STIs.

More specifically, ASTII aims to:

- Develop and promote the adoption of internationally compatible STI indicators;
- Build human and institutional capacities for STI indicators and related surveys;
- Enable African countries to participate in international programs for STI indicators; and
- Inform African countries about the state of STI in Africa.

16.4   China Science, Technology, and Innovation Initiatives

The definition of Science and Technology (S&T) policy in China differs from that in the West. In China, the meaning is more inclusive than its Western equivalent. In the West, it usually means 'high-technology' and its derivatives. However, it has a more inspiring quality in China, which means 'an invocation of national economic good and the rapid achievement of that good. In the case of China, the S&T policy aims to accelerate the commercialization of technology, integrate S&T with the economy, and promote its people's living and health standards. It also supports sustainable development, safeguards national security, enhances innovative capability, encourages the creative passion of scientists, and expands international S&T cooperation and legislation and regulation, including Intellectual Property Rights (IPR).

In January 2006, China initiated a 15-year "Medium- to Long-Term Plan for the Development of Science and Technology." The MLP calls for China to become an "innovation-oriented society" by 2020 and a world leader in science and technology (S&T) by 2050. It commits China to develop capabilities for "indigenous innovation" (*zizhu chuangxin*) and to leapfrog into leading positions in new science-based industries by the end of the plan period. According to the MLP, China will invest 2.5% of its increasing gross domestic product in R&D by 2020, up from 1.34% in 2005; raise the contributions to economic growth from technological advances to more than 60%; and limit its dependence on imported technology to no more than 30%. The plan also calls for China to become one of the top five countries in the world in the number of invention patents granted to Chinese citizens and for Chinese-authored scientific papers to become among the world's most cited. In all likelihood, the MLP will significantly impact the trajectory of Chinese development. Thus, it warrants careful attention from the international community.

"Indigenous innovation" has become a top priority in the past 5 years. As "guiding principles for science and technology undertakings," China's National Medium- and Long-Term Program for Science and Technology Development for 2006–2020 lists "indigenous innovation, leapfrogging in priority fields, enabling development, and leading the future." The document states that "core technologies cannot be purchased "in areas critical to the national economy and security. "China must "master core technologies in some critical areas, own proprietary intellectual property rights, and build several internationally competitive enterprises." The plan calls for boosting China's gross R&D spending to 2.5 percent of GDP by 2020, for science and technology to account for 60 percent of the economy, and cutting dependence on imported technology to 30 percent.

With these developments, China aims to become an innovation powerhouse by 2020, according to a newly adopted national scientific and technological innovation plan during the 13th Five Year Plan (2016–20). The plan says China will advance its global ranking of innovation competence with its combined efforts to enhance innovation, build essential science innovation parks, and attract top-tier science and technology researchers.

According to the plan, China will begin implementing critical scientific and technological innovation projects. It emphasizes areas contributing to China's industrial upgrading and the new economy, including modern agriculture, clean and efficient energy, and fifth-generation mobile telecommunication. With Beijing and Shanghai pioneering the effort, several innovation zones are coming up in China. This initiative in science and technology innovation will always keep a close eye on its application. The government of China has repeatedly highlighted the importance of innovation, supporting and encouraging mass innovation and business startups. Innovation is vital to the economy, shifting from being driven by investment and manufacturing to being more consumption- and service-based. In this regard, business startups in China with innovative ideas and investments have flourished. Some recent homegrown innovations have significantly changed people's lives in mobile messaging apps, i.e., WeChat, which offers a user experience that rivals international competitors.

The plan also has the potential to single out measures to tackle obstacles that have long been hindering science and technological innovation in China. These brought about the synergy between industry-academia and research institutions, including a more systematic technology transfer from research institutions to companies. Accelerating reform of the management system for science and technology, better coordination to improve resource distribution and intellectual property rights protection. "The case for overhauling science, research systems and generating greater enthusiasm among science and technology researchers is robust if we mean to enable true breakthroughs in innovation,"

Liu Bing stated that technological innovations are needed to boost China's economic vitality and international competitiveness, which demand adequate financial inputs, integration with market demand, and education that encourages new ideas.

Evidence has shown that in the past 15 years, China has launched many initiatives to boost science, develop high-tech industries, and reduce dependence on foreign technologies. The success stories can lie in the 973 Programs initiated. For example, it supports 175 chief scientists focusing on "strategic needs" in agriculture, energy, information, and health. Furthermore, the 863 Programs, known as the State-High Tech Development Plan, aim to reduce China's dependence on imported advanced technologies, resulting in the development of China's Shenzhou spacecraft and Loongson computer processors. The Torch Program, meanwhile, promotes the development of high-technology industrial zones.

Sharma (2021) noted that the *UNESCO Science Report*, released on 11 June 2021, highlighted that despite the tremendous progress made by China in science, technology, and innovation since 2015. It shows that China's manufacturing sector has become technologically sophisticated but is still dependent on some core foreign technologies like semiconductors and the acquisition of foreign technology companies (Sharma, 2021). Its ambition of becoming an innovation-driven economy may have reached a crossroads, with a more arduous path ahead.

16.5   The Need for Synergic Cooperation and Collaboration in Africa

Africa is the poorest continent, according to World Bank, 2006; Mabogunje, 2004; Bigsten & Durevall, 2008. The African continent faces overwhelming challenges that hamper its development. These include inadequate infrastructure, poor access to essential services, widening inequality, rising unemployment, and conflicts. The continent also lacks the basic infrastructure to develop new technologies and innovation. In addition, the continent lacks a clear policy and capacity to adapt, adopt, absorb and diffuse imported technologies for her advancement. To address these developmental challenges in Africa. Governments, multilateral institutions, and developmental partners globally converged together and set up the Millennium Development Goals (MDGs) and the Sustainable Development Goals (SDGs) as the roadmap to African development.

Given the above situation, one could ask whether there is hope and a chance for Africa to leapfrog for development. If there is, what would be the essential issues that can guarantee sustainable development in Africa? Despite recognizing the role of science and technology (S&T) as the global engine of development. To this extent, applying STI can increase production systems' efficiency and enhance industrial competitiveness. Therefore, a nation's competitive edge is the speed at which it can identify, utilize and diffuse new knowledge.

The expansion of the global economy, which leads to the growth and more complexity of market activities in Africa, has increased competition and a scarcity of resources. This situation has made cooperation, collaboration, and partnerships inevitable. It can serve as an official platform where China and African economies can cooperate and collaborate mutually to benefit and boost STI, facilitating the flow of economic resources under less stringent rules. Therefore, cooperation and collaboration are developing to favour partners within a group so that the strengths of the strong countries can overcome the weaknesses of weaker countries. Similarly, the weaker countries can learn from the stronger countries regarding STI development initiatives. In this context of discussions, China is seen as a more significant player in the partnerships because it has more STI capacity and economic engagement than Africa.

16.6 Lessons to be Learned by Africa

16.6.1 Lesson 1: Planning of Technological Progresses and Innovations

The Soviet Union model, the Manhattan Project, the Apollo Project, and the Star Wars Project are clear examples and proof that, to some extent, innovations or technological progress planning is possible. Provided that there is firm determination from stakeholders, sufficient resources, and support by the government. Presently, China is running many Apollo-like projects, such as the 863, 973, and 13th 5-year plans. Most of these projects aim to catch up in strategic and selected industries, such as those identified in the "Made-in-China 2025" strategy.

The Chinese version of Quantitative Easing (QEQE) generated massive funds channelled towards expanding R&D, hiring overseas Chinese experts and foreign experts (especially in strategic sectors such as advanced material, electronic chips, computing, aviation, biotech, and AIAI and robotics). This driving force complements the two pieces of China's technological puzzle. In addition to importing high-tech capital goods, acquiring foreign patents and licenses, and merging with or buying out foreign high-tech companies, mainly state-owned Chinese corporations. These government initiatives matched the unprecedented amount of fiscal and financial resources made available in the post-2008 period.

16.6.2 Lesson 2: Size Matters

Population and land mass matter because technological innovation is intrinsically about capturing a new phenomenon through identifying or creating new combinations of component technologies. Together with improvements in the structure, design, or method of an existing technology that delivers better performance in incremental innovations.

China's status as a world factory is a lesson to be reckoned with by African countries because the supply chains of various industries, consisting of thousands of component technology suppliers, are now clustered in the country. Moreover,

Research and Development (R&D) activities in China are becoming more effective in identifying and creating new combinations of technologies, as evidenced by the many global industrial giants setting up R&D centres in China. Similarly, in searching for the best supply chain capability to commercialize innovations, high-tech industries increasingly look to China for manufacturing solutions; complementary innovations by Apple's products are typical examples. This development has made China's internal market stand worth 1.4 billion consumers with a per capita income of $8000 + in nominal US dollars and $15,000 + in P.P.P. U.S. dollars. It has also been a blessing to innovation in many ways.

The vast China market size has promise that is sufficient enough for a return to cover the costs of acquiring foreign technology or undertaking a cumulative learning process to catch up with foreign frontier technologies. In a similar vein, the potential size of return also enables Chinese industries to afford the pursuit of a costly combination of technologies, such as CPU chips, nuclear energy, aviation, and space technologies. In some cases, the market size is also the key to unlocking the "learning curve" effects, as in the cases of the Chinese solar and electric vehicle industries.

More importantly, technologies by themselves do not recognize physical borders. The African market is as large as China's and can also attract not only the best technologies but also the best scientific and technological talents globally, as evident from the experience of China.

16.6.3 Lesson 3: Capacity Building

The Chinese tradition of emphasizing education is crucial for its technological development, besides the China government's spending 20 percent of its budget on education. Even Chinese households are investing heavily in education, reaching levels equivalent to 50 percent of the government's education budget. Globally, data has shown that China has the highest number of students studying overseas, and the number of these students returning to China has increased. All these measures endow China with an educated and disciplined labor force to work with newly introduced technologies and the R&D capability to learn and eventually innovate on top of existing tech.

Some argue that China's strict control of information flows—most notable in the Great Firewall—will block the exchange of ideas and thus stifle innovation. However, China's recent phenomenal scientific and technological achievements show the contrary. For innovation, the most significant need is the freedom to access and communicate well-defined scientific and technological information, knowledge, and ideas to inspire innovation and technological progress. The Chinese government provides generous funding to researchers in China to interact and collaborate with their global counterparts. Today, the exponential growth of Chinese researchers' publications, scientific discoveries in international academic journals, conference proceedings, and patent rights registration prove that China's strategy has worked well for them.

16.6.4 Lesson 4: Chinese Model of Technological Capability

Development in China's rise of technological capability is taking a path that is drastically different from the Soviet Union model and the newly industrialized Asian economies. China's technological capabilities are reshaping the global economic structure from two perspectives. First, China has accumulates a high level of capabilities in medium-level technology, such as machinery, infrastructure construction, modern logistics, electronics, and renewable energy equipment. These technologies are precisely what the developing world needs to improve economic and social well-being quickly.

Second, as China's technological capability is high-tech, many of today's so-called high-tech industries will become commoditized at an accelerated speed. In the past 10 years, China has witnessed commoditized computers, smartphones, modern metro subways, and even high-speed rail, from which low-income countries and social groups have benefited the most. Shortly, some of the advanced technologies are enjoyed today by people in developed countries, such as innovative and clean energy systems, new energy vehicles, autonomous driving cars, automation and robotics, advanced medical equipment, and medicines. It will become increasingly more affordable to developing countries, a process that China will primarily drive. This development is one of the key drivers behind the Belt and Road Initiative.

16.7 China–Africa Cooperation and Collaboration as the Game Changer

Cooperation and Collaboration on STI between Africa and China have evolved into a multi-layered set of relations. These had grown from colonial times, when the research was mainly extractive, through the emphasis on Africa in the 1990s—developing from individual ties among scientists to formal governmental involvement.

Mapping out the policies, literature, and reports is safer to say that China–Africa Cooperation and collaboration is timely and is a good initiative in the right direction. Africa is one of the largest developing countries and the continent with the most developing countries in the world. Similarly, and considering both sides,' i.e., their development, history, background, and characteristics, Vision 2035 of China, the UN 2030 Agenda for Sustainable Development, the 2063 Agenda of the African Union, and the national development strategies of African countries. China and Africa can jointly formulate this China–Africa Cooperation to determine the directions and objectives of mid- and long-term cooperation and promote a closer community with a shared future for China and Africa.

16.8   Benefits Derived by Africa from STI Cooperation and Collaboration with China

The cooperation and collaboration between China and Africa can be an essential instrument to ensure that STI plays its full role in the sustainable development of African countries. The cooperation will help strengthen their STI capacities, foster the development of national and regional innovation ecosystems, generate homegrown research and innovation, and take them to market. It also aims to assist African countries in building their national and regional capacities in intellectual property rights and technology-related policies, as well as facilitate the transfer of appropriate technologies and, in the process, accelerate African integration into the knowledge-based economy.

In light of the benefits of cooperation and collaboration that African countries will enjoy, the following objectives guide the cooperation between the two continents. The new entity has five specific objectives:

- To strengthen STI capacity in African countries, including the capacity to identify, integrate, develop, and absorb the deployment of technologies and innovations, as well as the development of the capacity to address and manage intellectual property rights issues;
- To promote the development and implementation of African STI initiatives;
- To strengthen partnerships among STI-related public entities and with the private sector;
- To promote cooperation and collaboration among stakeholders involved in STI, among African countries, as well as other countries;
- To promote and facilitate the identification, utilization, and access to appropriate technologies by the African countries and their transfer.

Given the above objectives, the China Africa Cooperation and Collaboration may be well-placed to play an essential role in providing scientific and technical knowledge, whose current under-provision has significant implications for sustainable development in African countries and at the global level. Indeed, African countries' sustainable development depends on the ability to acquire scientific and technical knowledge and apply it to solve the many challenges they face.

To this extent, resolving growth issues in the African economy becomes necessary to gain the most from FDI's ultimate objective. Lingering growth problems will continue to hinder adequate investment allocations in the region. Therefore, without specifically outlining the core issues facing development in Africa, resolving them will be a recipe for underdevelopment. For example, there is a need for oil-producing nations to go beyond crude oil activities with a lower market price to processing activities with a higher market price. One such simple way to do that is cooperation and collaboration with China because China knows what it wants from Africa and, as a result, it can comfortably deal with Africa.

On the other hand, African economies in context need to know what they want as investment programs from China. Therefore, it understands what will attract investments that will resolve its growth issues other than going for anything. The situation has a long-run effect of collapsing the domestic activities and the exporting sectors. Diversification programs will have a more significant impact, and the non-oil sectors will be well developed.

Today, most countries are experiencing heavy financial debts from donor organizations (the IMF and World Bank). The most viable approach for expanding infrastructural development in the African region is planning to adopt and implement the Mutual or Pooled Growth Model (MGM/PGM) in cooperation with China. As the name implies, this strategy requires that China and African economies should initiate a deposit plan depending on the agreement of both countries. China and Africa can pull a fixed amount of funds to be invested in a viable structural program or project in a partner country (say Africa) for at least 2 years. Additionally, the African community can leverage China's economic interest to attract investment resources to bridge the enormous infrastructural gaps facing development in the African region.

Moreover, later pull another for the remaining country (say China) for a similar or different project/s. After the 2 years, both countries will account for the funds to avoid miss-appropriateness. This type of financing or investment will serve as a unique investment because, unlike the IMF funding system, it has no interest commitment, reducing the debt-to-GDP ratio for partner economies.

Africa is an endowed region with diverse resource potential, hence a suitable place to site processing and manufacturing industries. Africa's manufacturing sector remains underdeveloped, yet China is an industrial hub. Africa's lack of adequate technical competencies is delaying the region's industrial programs. *China–Africa cooperation* can serve as a forum that can help Africa to close the technical gaps and boost its industrial development. The cooperation can be used as a framework to transfer technical resources to support the different sectors of the African economy. Through government policies, young and women entrepreneurs can be supported by giving special incentives for at least 1 year, technical skills, and training to sustain activities. The government, through its machinery, can also protect young vision from the competition and allow them to grow into substantial exporting industries and create more job opportunities for reducing unemployment in the region.

Lastly, African countries need to reform their investment policies by observing China's past partnership engagements and programs with the ASEAN community and other rising economies; this will help Africa to gain wisdom and help shape future engagement with China. Most such engagements have had a couple of successes and failures to learn. In the future, the African community will be able to develop good economic deals with China via trade and investment. Again, the U.S. has had long-standing trade and investment history with China.

16.9 Conclusion

The analysis in this chapter shows that despite the importance of STI, Africa is lagging far behind other countries in various STI indicators, including R&D, human resource capacity, patents, and innovation. For Africa to catch up, it must embrace the ongoing Technological Revolution in China while ensuring that the net effect on the labour market and productivity is positive. It will also significantly contribute to eradicating poverty and fostering economic growth in Africa. The Africa–China cooperation and collaboration have the potential to catalyse the ongoing initiatives on STI in Africa. In general, the significant alignment of China's initiatives will significantly contribute to increasing the African country's access to knowledge, especially if both cooperate with the various African organizations and promote synergies across them and between multilateral and regional initiatives. Playing an orchestrating role by harnessing the various African initiatives may produce a helpful rationalizing and ultimately a boosting effect on the ground. A feedback process between China and African STI initiatives will ensure such cooperation. This will entail dialogue across institutions to harness the potential that exists. STISA-2024 (p. 46) highlights this principle, which states that "establishing a healthy, vibrant and sustainable Innovation Ecosystem requires clear communication and knowledge sharing between all innovation stakeholders. It will reduce the duplication of effort, increases research and innovation excellence, and properly utilizes scientific and technological knowledge to address societal challenges through innovative products, services, processes, business models, and policies". In light of the complex development cooperation landscape, multilateral institution building (and its financing) remains challenging. Promoting complementarity of efforts between China and African initiatives will undoubtedly help achieve more effective results on the ground, thus showing the added value of institutional development at the multilateral level in the STI area, which the African countries have long pursued.

References

African Union. (2005) *The conference of the African Union meeting at its 5th Ordinary Session in Sirte*. Libyan Arab Jamahiriya from 4 to 5 July.

Bell, M. (2009) *Innovation capabilities and directions of development*. STEPS Working Paper 33, Brighton: STEPS Centre.

Bigsten, A., & Durevall, D. (2008). *The African economy and its role in the world economy*. Nordiska Afrikainstitutet. Current African issues

Cassiolato, J. E., Szapiro, M. H. S., & Lastres, H. M. M. (2002). Local system of innovation under strain: The impacts of structural change in the telecommunications cluster of Campinas, Brazil. *International Journal of Technology Management, 24*, 680–704.

Cozzens, S., & Sutz, J. (2014). "Innovation in informal settings: reflections and proposals for a research agenda" Innovation and Development. *Taylor & Francis Journals, 4*(1), 5–31.

Egbetokun, A. A., Siyanbola, W. O., Olamade, O. O., Adeniyi, A. A., & Irefin, I. A. (2009). Innovation in Nigerian small and medium enterprises: Types and impact. *J. E-Commerce in Organisations, 7*(4), 1.

Johnson, B., & Andersen, A. D. (red.) (2012). *Learning, Innovation and Inclusive Development: New perspectives on economic development strategy and development aid*. Aalborg: Aalborg Universitetsforlag. (Globelics Thematic Report, Vol. 2011/2012).

Karuri-Sebina, G., Sall, A., Maharajh, R., & Segobye, A. (2012). Fictions, factors and futures: Reflections on Africa's impressive growth. *Development, 55*(4), 491–496.

Khan, M. J. (2022). The status of science, technology and innovation in Africa. *Science, Technology & Society, 27*(3), 327–350.

Mabogunje, A. (2004). *Framing the fundamental issues of sustainable development*. http://www.start.org/links/cap_build/advanced_institutes/institute3/p3_documents_folder/Mabogunje.doc

Mêgnigbêto, E. (2019). Synergy within the West African Triple Helix innovation systems as measured with the game theory. *Journal of Industry-University Collaboration, 1*(2), 96–114. https://doi.org/10.1108/JIUC-03-2019-0008

Metcalfe, J. S. (2000). Co-evolution of systems of innovation. In *Paper presented at the Volkswagen Foundation Conference "Perspectives and Challenges for Research and Innovation,"* Berlin, June 8–9.

Oyelaran-Oyeyinka, B., & Gehl Sampath, P. (2010). *Latecomer development: Innovation and knowledge for economic growth*. Routledge.

Schumpeter, J. A. (2004). *The theory of economic development*. Transaction Publishers.

Science. (2024). *Technology and innovation strategy for Africa 2024 (STISA-2024)*.

Sharma, Y. (2021). China obstacles emerging on road to an innovation-driven economy. *The University World News Magazine*. https://www.universityworldnews.com/fullsearch.php?mode=search&writer=Yojana+Sharma

Siyanbola, W., Adeyeye, A., Olaopa, O., Hassan, O. (2016). Science, technology and innovation indicators in policy-making: The Nigerian experience. *Humanities & Social Sciences Communications*. https://www.nature.com/articles/palcomms201615

Umesh, P., Sai Vinod, M. S., & Sivakumar, N. (2021). Integration of human values in stakeholder engagement for CSR—Illustrations from Indian public enterprises. *Public Enterprise, 25*(1–2), 39–56. https://doi.org/10.21571/pehyj.2021.2512.04

UN-DESA, & OECD. (2013). *World migration in figures*. A joint contribution by UN-DESA and the OECD to the United Nations High-Level Dialogue on Migration and Development, New York, October 3–4. https://www.oecd.org/els/mig/World-Migration-in-Figures.pdf

UNESCO. (2015). *Towards 2030: UNESCO science report*. Paris.

WEF (World Economic Forum). (2014). *Matching skills and labour market needs: Building social partnerships for better skills and better jobs*. Geneva.

World Development Report. (2006). *Equity and development (English)*. World development report. Washington, DC: World Bank Group. http://documents.worldbank.org/curated/en/435331468127174418/World-development-report-2006-equity-and-development

Chapter 17
Overview of China–Africa Collaborations: Opportunities and Criticisms

Peter Chihwai

17.1 Introduction

There is much debate in the academic world about whether China–Africa relations are mutually beneficial, or they are one-sided in favour of China. There is also growing concern and interest from the West on China–African collaborations, regarding the relations largely as unequal and unorthodox (Eisenman, 2012). This chapter will unearth the benefits gained from such relations so far, the criticism levelled against China–Africa relations and seek to establish the win–win relationships prevailing thus far, with a greater emphasis on the more benefits Africa gained from the relationship.

China–Africa relations started flourishing in the late 1990s based on mutually beneficial standpoints which were to be established, maintained, and enhanced and remain sustainable. Ordinarily, it was known that such symbiotic relations existed normally between the South, developing nations, and the North, developed countries. In the North–South relations, it was generally empirically proven that the North benefited immensely from the South. The West-developed nations like the USA, UK, and France were the generally imposing known colonizers since the eighteenth and nineteenth centuries in the scramble for Africa era. This was an era of partitioning Africa and exploiting resources including the slave trade. With the advent of China–Africa relations emerging, it became apparent that a new superpower was emerging. That sparked political and intellectual scrutiny from the global village. This could have been partly because resources from Africa were now going to be shared and Africa became the new political and economic battlefield (Alden, 2005).

What resources does Africa have that call for such contentions? Africa has 54 countries, and each of them accounts for the United Nations votes. China is the

P. Chihwai (✉)
Department of Tourism and Integrated Communication, Faculty of Human Sciences, Vaal University of Technology, Vanderbijlpark, South Africa
e-mail: peterc1@vut.ac.za

© The Author(s) 2024   309
M. Muchie et al. (eds.), *China-Africa Science, Technology and Innovation Collaboration*, https://doi.org/10.1007/978-981-97-4576-0_17

emerging new giant beside the usual dominant Britain, France, and America in this continent endowed with resources. Africa accounts for 17% of the world's population, 9.6% of global oil output, 90% of platinum supply, 90% of cobalt supply, two-thirds of the world's manganese, 35% of uranium and 75% cotton supply. Above all, Africa has 54 votes in the United Nations General Assembly. This shows the importance of Africa to the world, both the West and East. Similarly, this also shows the potential conflict that may arise between the West neo-imperists' former African colonizers with the potential East emerging China. A school of thought emerging is whether Africa–China collaboration in trade or other scenarios is neo-imperialism or neoliberalism (Van der Wath, 2004; Mooney, 2005; Gumede, 2012).

China has however made tremendous strides in developing Africa. One in every five big infrastructures being built in Africa is being done by China. Since 2006, China has invested 2 trillion dollars in infrastructure development in Africa. Part of the infrastructure built includes the 6200 km railway from Ethiopia to Djibouti, build African Union Building Headquarters in Addis Ababa, Parliament building in Zimbabwe. In mutual benefits fashion, China has gained geographical influence over Africa, has sold its currency on the continent, and has spread its culture. Although the USA is a huge investor in Africa, China is Africa's biggest trading partner (Botha, 2006; Ndubisi, 2007; Manji, 2007; Brautigam, 2007; Prah, 2007; Weizhong, 2008). This chapter aims to present the mutual benefits established so far between China and Africa. It will seek specifically to establish to what extent Africa has benefited more from the relations. Briefly, the chapter will also highlight the criticism levelled against the collaboration between the two countries.

The chapter outline will be as follows, Section 17.1 will deal with the introduction. Section 17.2 will deal with the literature review. Section 17.3 will deal with methodology. Section 17.4 will deal with findings, and discussions whilst the last section deals with recommendations and conclusions.

17.2 Literature Review

China–Africa economic and trade cooperation, began in 1955 (Wenping, 2008). Although China and Africa have different degrees of development, the economic and trade cooperation between them is highly complementary. Over the past half-century, especially since the establishment of the Forum on China–Africa Cooperation in 2000, China–Africa economic and trade cooperation has gradually expanded in scale and broadened in areas of cooperation (Taylor, 2008).

China's funding assistance to Africa has significantly been a handful to achieve the Sustainable Development Goals (SDGs), especially in the economy (Fang, 2020). Foreign direct investment from China has greatly assisted in the realization of affordable and clean water (SDG 7). Decent work and economic growth (SDG 8) as well as Industry, Innovation, and infrastructure (SDG 9) were also achieved. Responsible Consumption and Production SDG 12 through the same mechanism was achieved, too. Climate Action (SDG 13) has received tremendous progress, and Life on Land (SDG 15) has shown significant improvements (Fang, 2020).

According to Achu (2019), FOCAC has been instrumental in achieving SDGs in Africa and the Africa Union's Agenda 2063. Between 2000 and 2015, the average annual GDP of 38 African countries surged by 4.9 per cent. The first FOCAC meeting in Beijing set the following objectives: promotion of industries, infrastructure linkages, smooth flow of trade, environmental development, capacity building, healthcare, personnel exchange, peace, and security (China Africa Research Initiative (CARI), 2018).

According to Botha (2006), the Forum on China–Africa Cooperation (FOCAC) was formed in 2006 to fulfil certain obligations with values embedded in such an alliance. The agreed core values to be observed were common development, solidarity, sincerity, and equality. Borovska (2011) comments that such relationships would bring trade parity between Africa and China and other critics see potential risks associated with such partnerships. It is necessary to interrogate whether such values are being observed.

According to Hoffman (2007), both political and economic factors determine trade patterns. The most applied theories of trade are relative factor abundance theory, gravity trade theory, and political trade. Political trade is a combination of generally accepted views that some American, European, and African observers regularly use to explain close economic ties between two democratic countries and two authoritarian countries. In 1994, during the State of the Union address, President Bill Clinton explicitly pronounced that 'Democracies don't attack each other, they make better trading partners and partners in diplomacy'.

Researchers do not agree on which model Africa–China relations hinge on. Some support the economic model, some support the political model, and others support a mixture of both. But, even those that agree on one side do not agree on how exactly that chosen model explicitly can be used to explain the relations between China and Africa. The researchers who advocate for the political model are dissatisfied with the usual traditional economic explanations. The concept of relative abundance theory also known as the Heckscher–Ohlin (H–O) model, is supported by researchers such as Wang (2007), Herbst and Mills (2009). H–O (also known as neoclassical trade theory) predicts that the relative abundance or scarcity of a country's fixed factor endowments (resources, labour, and capital) compared with those of its trade partner determines what it will sell and what it will buy. A country will export goods produced with its abundant endowments and import goods produced with locally scarce inputs. According to H–O, once trade barriers are removed fixed relative factor endowments will determine trade patterns. Herbst and Mills agree that in Africa 'the market, not a grand strategy, is the Chinese motivation'. They suggest that relative factor abundance theories have proven particularly valuable in explaining China–Africa trade because China's relatively scarce natural resource endowments mean it must 'lock up as many raw materials as possible. Indeed, many African countries' relatively abundant factor endowments are natural resources and they do tend to dominate those countries' exports to China. In the real world, the original abundance relative theory is seen as inapplicable by researchers.

According to Eisenman (2012), the Gravity trade theory has been used as a scape-goat to cover up for the weaknesses found in China and African trade relations of what is happening in real life. The gravity theory of trade predicts that the distance between any two countries (measured in shipping distance) and the size of their respective economies (measured in GDP) will be the principal determinants of the quantity of trade between them. Other researchers and advocates note that the gravity model can be made increasingly robust with the inclusion of social and political explanatory variables such as population size, common borders, and common language making it more robust to many researchers for its political and social variables linkages. In their examination of democracies, autocracies, and international trade, Mansfield and et al. (2000) for instance, found that 'the gravity framework is quite successful in explaining the flow of interstate commerce'.

According to, the new trade theory, sometimes known as increasing returns to scale, has proven most helpful in elucidating bilateral trade between two developed countries with similar technological and factor endowments. These conditions are not found between China and Africa. This model was a modification of the relative abundance theory. Caught in between, the concept of increasing returns to scale is crucial for China since it would have spent massive sums on government-initiated and funded industrial policies.

Political trade suggests that two democratic countries trade better and higher than two different countries such as an autocratic one and a democratic one. Bliss and Russett (1998), suggested that for democratic states the causal mechanism is rooted in the perceived state's security concerns. Countries in this political trade thrive on national interests and on behalf of private interests. In countries deemed unfriendly, unstable, unreliable, and hostile, trade is discouraged, or potential enemies are avoided. In political trade, countries can enter relationships of economic interde-pendence for absolute gains, without worrying as much about the hazard of relative gains as they might with non-democratic partners (Eisenman, 2012).

According to Sandrey (2009), China and South Africa's diplomatic and economic relations are increasingly deepening, calling for research to establish whether such relations have exploited benefits. As benefits accrue, so do the challenges also. There is a need to interrogate such opportunities and challenges with empirical evidence.

Another school of thought attributes China–Africa relations to diplomacy and economic interests as driving forces for such relationships. Taylor (2008) and Shelton (2008) argue that the current China–Africa relations are mainly commercial and strengthen their position by declaring China's interest as based on three economic fronts which are China's growing demand for oil: Africa's raw materials and natural resources, and Africa as a ready-market for China's finished products. In part, Mooney (2005), concurs with the aforementioned citing raw materials as the main driving force for China's forging of relations with Africa. Van der Wath (2004) concurs with China's interest in abundant resources specifically for the African south of the Sahara region. On the diplomatic front, Giry (2004), views China's relations with Africa as asserting herself as a soft superpower nation seeking international recognition as opposed to the West's rise to superpower status through aggression.

Many scholars viewed this partnership as a win–win scenario and an opportunity for Africa's growth and development (Ndubisi, 2007; Manji 2007; Brautigam, 2007; Prah, 2007; Weizhong, 2008). Botha (2006) cites the South African mining sector as a perfect opportunity for its growth with China's involvement whereas Manji (2007) views China as very strong in areas of rural development and technology, areas which Africa greatly needs assistance. Sustainable rapid economic growth and industrialization seem to be the core business of China than ideological philosophies (Brautigam, 2007; Prah, 2007). Shinn, (2005) findings on the Horn of Africa development with China's assistance were that China should be viewed as a real development partner and not a threat. Skills transfer, aid, debt relief, scholarships, training, and provision of technical specialists to African countries are all benefits that are being enjoyed by Africa from China (Alden, 2005; Baah, 2007; Jauch, 2007). China adopted a no-strings-attached approach to Africa unlike Western countries like the USA and Britain which attach strict conditions like governance and economic liberalization through institutions like the IMF and World.

17.3 Methods

This section deals with the methods applied to evaluate data. Critical document review and multiple case studies were employed. They are effectively used in qualitative types of research mostly.

Critical document review and multiple case studies were utilized in the analysis of China–Africa relations. According to Bowen (2009), document analysis is a systematic procedure for reviewing or evaluating documents. These documents include both printed and electronic versions. The electronic version is material that is computer-based and internet-transmitted. Examples given by Bowen (2009) of critical documents include journals, annual reports, and meetings. Like other analytical methods in qualitative research, document analysis requires that data be examined and interpreted to elicit meaning, gain understanding, and develop empirical knowledge (Corbin & Strauss, 2008; Rapley, 2007). Supporting the above sentiments (Stake, 1995; Yin, 1994), asserts that document analysis is particularly applicable to qualitative case studies and intensive studies producing rich descriptions of a single phenomenon, event, organization, or program. Nhamo (2020), purports that public and private institutions have valuable data sources like pictures, articles, and media statements that can respond qualitatively.

17.4 Results, Discussion, and Findings

This section deals with the results of trade between China and Africa, between South Africa and China, a multiple case study explorations of relations between China and African countries such as Ethiopia and South Africa as well as the application of some economic political theories following the results, findings leading to further discussions.

Figure 17.1 shows the total trade between Africa and China. African imports from China were slightly below Africa's exports to China. African exports to China decreased in 2009 and overtook imports between 2011 and 2013. From 2014 to 2017 imports from China seem to be constantly rising from 2004 except in 2016. However, in 2018, African exports to China started overtaking Africa's imports from China. Trade between China and Africa increased fourfold between 1995 and 2015, which contributed immensely to African countries' development. Xinhua (2018) supports China's contribution to Africa and mentions that from 1978 to 2017, the China–Africa trade has increased more than 200 times, significantly contributing from US\$ 765 million to US\$ 170 billion. The study found that on a country-by-country basis, the balance of trade between China and country resource exporters tends to favour the African country, whilst China's exports dominate its commerce with non-resource exporting African trade partners from Figs. 17.1, 17.2, 17.3, 17.4, 17.5, 17.6. This resonates with Eisenman's (2012) similar findings. Comparing China's trade with Egypt to its trade with Libya till 2011 helps to illustrate the leading role factor endowments (capital, labour, and resources) play in determining China–Africa trade patterns while holding regime type constant. In both cases, China's comparative advantage in consumer goods and capital equipment drives steep growth in its exports. However, the presence of an export commodity (oil) in Libya results in a moderate Libyan trade surplus, while the lack of resources in Egypt has led to an imbalanced trade relationship in China's favour as shown in Fig. 17.3, 17.4, 17.5 and 17.6. The study found that in a bid to further boost trade, China introduced eligibility for zero-tariff treatment, from 190 to over 440 items for the least developed countries that enjoy diplomatic relations with her. The finding resonates with a similar study by Burke et al. (2008). This shows the willingness of China to assist the least developed countries and should be seen as a progressive move designed to boost bilateral trade and enhance diplomatic relations.

The study finds real trade between China and Africa South of the Sahara. Capital goods contribute to most imports from China in the Southern Africa region. The Belt Road initiative and flourishing infrastructure projects in Africa contribute also to such capital goods imports. Consumer goods were ranked second between 2007 and 2017. The figure also shows that consumer goods occupy an important share of the Sub-Saharan African country's imports from China. Intermediate goods from China ranked third in the same period which was an increase from \$US5 Billion and maximum in 2013 at USD\$11.8 billion proving manufacturing sector growth. This concurs with Regissahui's (2019), assertion of China's massive contribution to Africa in general whilst the low import from China of raw materials largely is due to its abundance in the region. In line with Eisenman's (2012) findings, China–Africa trade patterns are dichotomous whereby surplus trade is in favour of African countries that export resources to China are experienced such as the Democratic Republic of Congo, Gabon, and Libya. On the other side countries with little or no resources to export to China suffer trade deficits. Such countries include Ethiopia, Benin, and Morocco. China's exports to Africa are diversified whilst Africa's exports to China

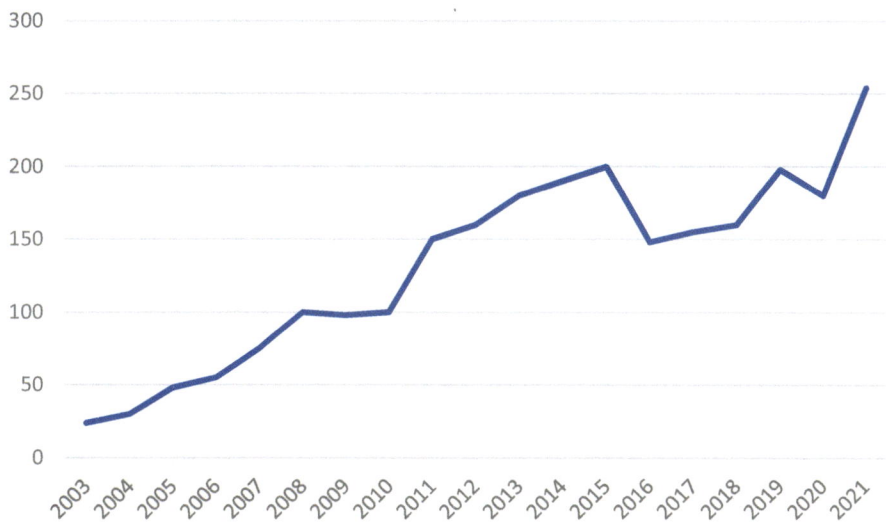

Fig. 17.1 Total trade between China and Africa (US$ Billion). *Source* Author compiled data from China customs

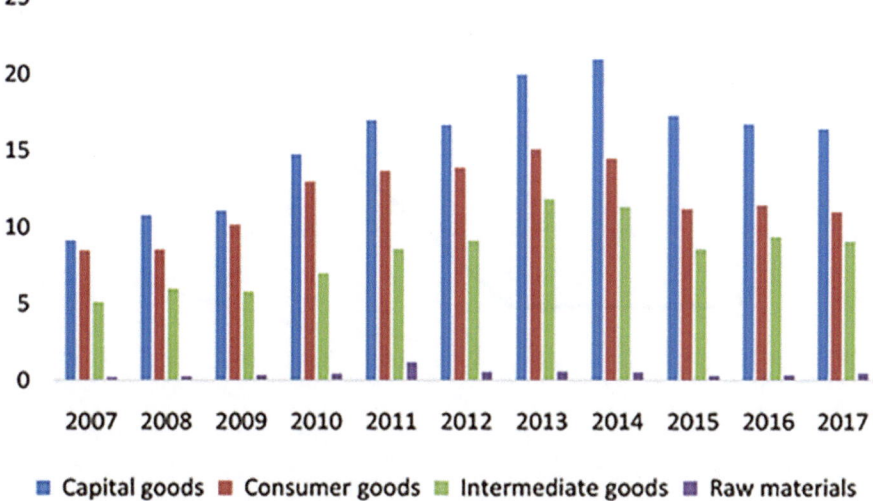

Fig. 17.2 Sub-Saharan African countries' imports from China from 2007 to 2017 in US$ Billion. *Source* Author compiled data based on Regissahui (2019), World Integrated Trade Solution (WITS) database (2019)

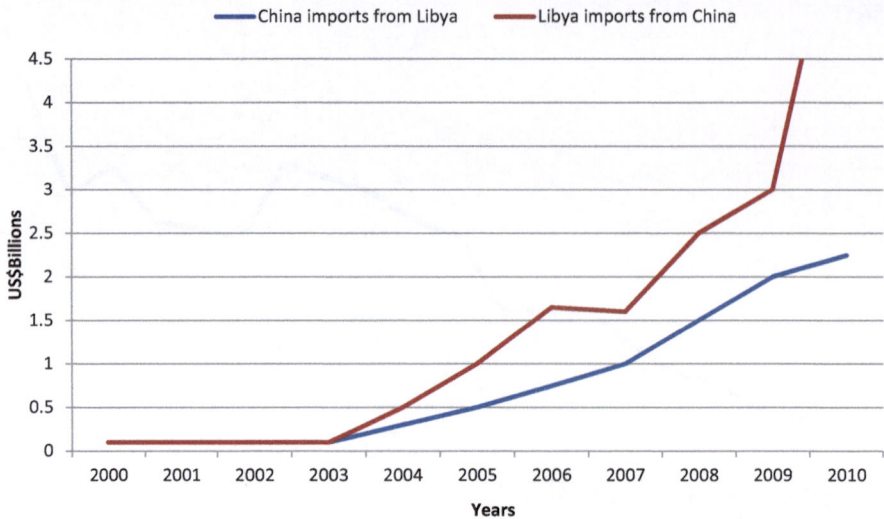

Fig. 17.3 China-Libya trade patterns. *Source* Author compiled data based on Eisenman (2012)

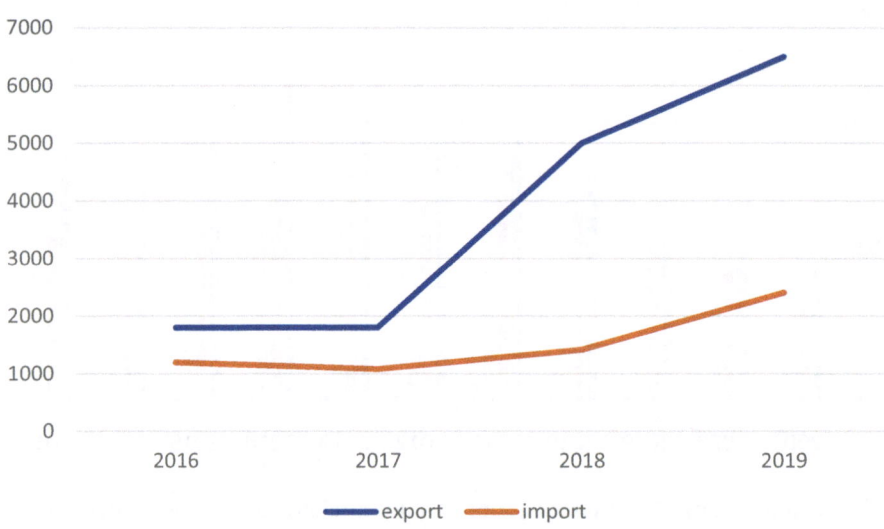

Fig. 17.4 Libya exports to and imports from China. *Source* Author compiled data based on Trading economics

are concentrated in narrow resource products such as crude oil (contributing two-thirds of total export value to China) being the most exported since 2000. Africa South of Sahara, the top 5 exported products are mineral products, base metals (including oil), precious stones and metals, wood products, and textiles and clothing—alone makeup 90% of its total purchases. By 2009, nearly 80% of China's exports from

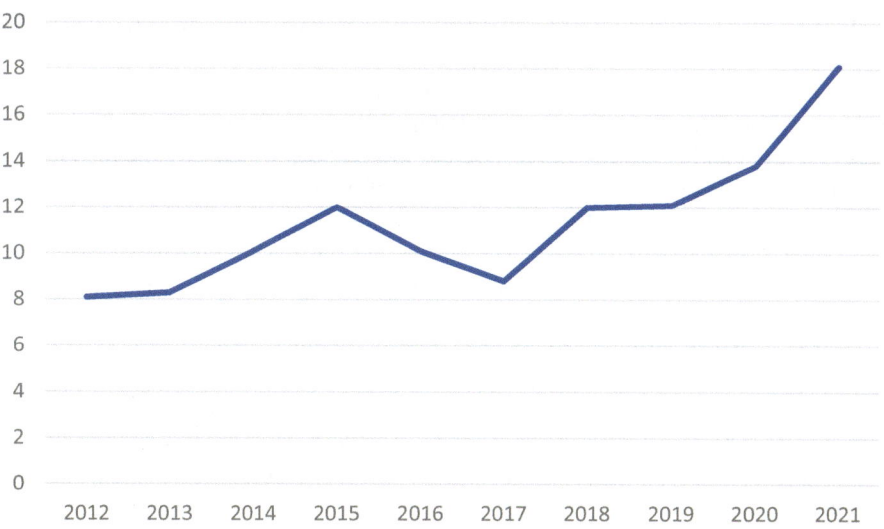

Fig. 17.5   China's export to Egypt (US$ billions). *Source* Author compiled data based on Trading Economics

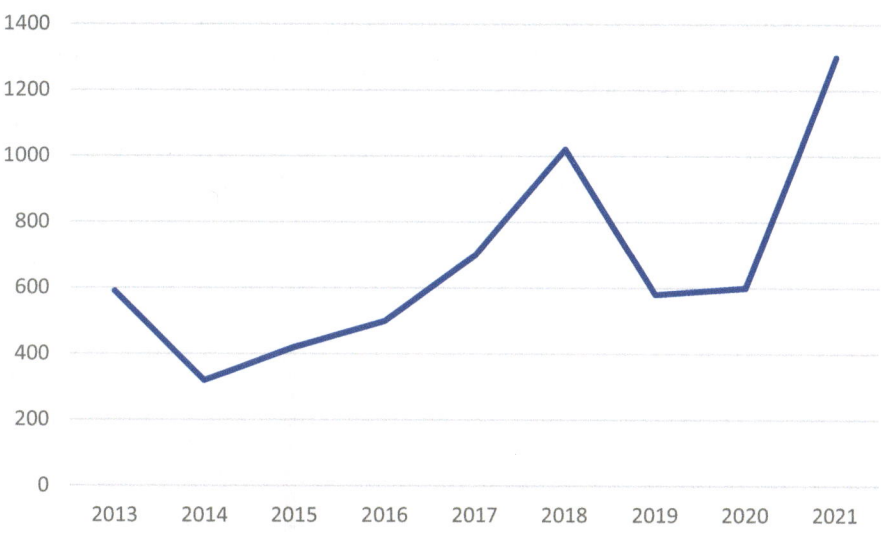

Fig. 17.6   Egypt's export to China (US$ million). *Source* Author compiled data based on Trading Economics

Africa were metals and petroleum products. Crude oil is Africa's top seller and since 2000 has made up over two-thirds of the total export value to China. Iron ore and platinum are also important African exports to China.

In Figs. 17.3 and 17.4, Libya has an advantage in trade with China because it has oil, so the balance of trade is even.

As shown by Fig. 17.4, between 2016 and 2017 there was a balance of exports and imports to and from China respectively but imports from China increased especially between 2018 and 2019. Egypt has nothing more to offer than Libya which has oil and that explains the difference in the balance of trade.

Figure 17.5 shows that China's exports to Egypt increased between 2012 and 2015 then dropped in 2016 up to 2017 the lowest ever then ascended gradually until 2021 up to 18 billion. This shows the symbiotic relations between the two nations between 2012 and 2021.

Figure 17.6 shows Egypt's exports to China as sluggish with a sharp drop in 2014 and a steady rise from 2015 up to 2018 then dropped in 2019 and 2020 possibly due to COVID-19 restrictions. There was a sharp upward trend in 2021. Egypt could improve its exports to China through government and private sector initiatives.

Figure 17.7 shows partly, Africa South of Sahara with a case of South Africa–China trade patterns.

Figure 17.7 shows that there is a progressive improvement from 1998 in imports from China to South Africa until 2022 although COVID-19 affected the trade moderately. South Africa has been lagging slightly in terms of exports to China. Between 2010 and 2012 there seems to be slight progress going forward with South African exports lagging. According to Botha (2006) at the early stage the China–South African trade seemed to favour China slightly more than South Africa. However, subsequently the China–South Africa trade became more balanced.

Figure 17.8 shows positive trade between South Africa and China between 2013 and 2021 where there is a steady rise from just above 8 billion in 2013 to nearly 14 billion in 2021. That is a positive result for China's South African trade.

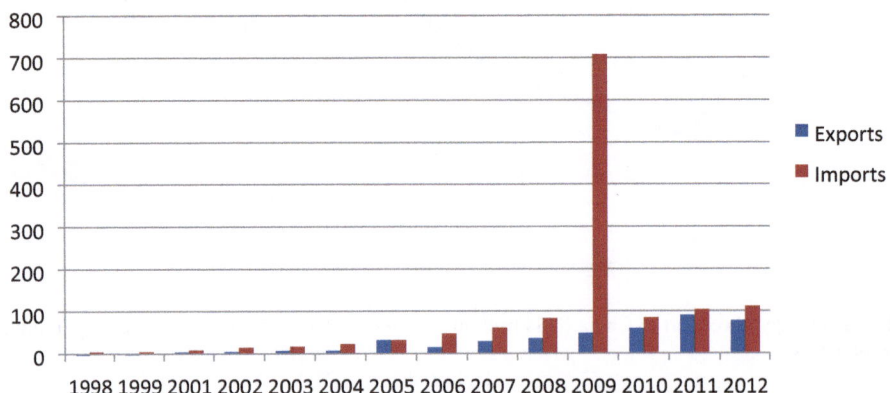

Fig. 17.7 South Africa–China exports and imports (US$ millions). *Source* Author compiled data based on South Africa Revenue Services

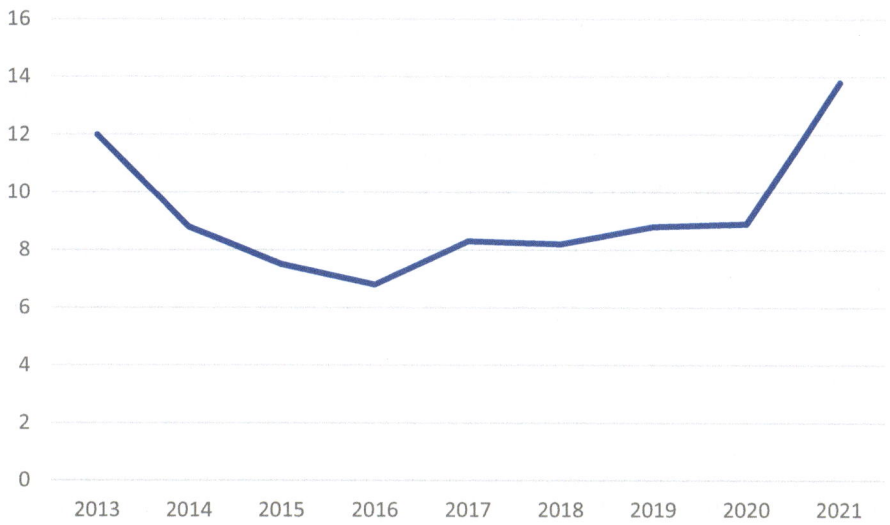

Fig. 17.8 Trade between China and South Africa (US$billions). *Source* Author compiled data based on Comtrade Trading Economics

As can be deduced from Fig. 17.9 South Africa has benefited from Chinese exports because it proves that such commodities have shortages in South Africa. February exports by China to South Africa seemed highest in the year 2021.

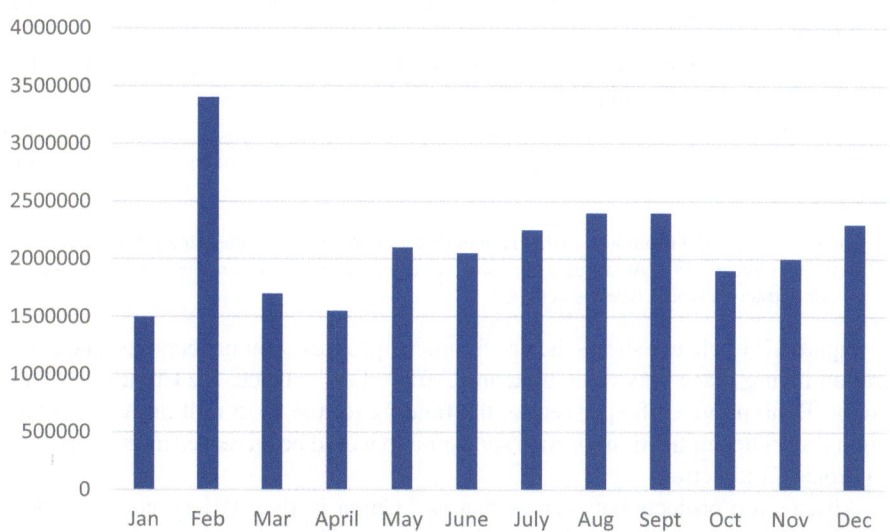

Fig. 17.9 China's exports to South Africa in 2022. *Source* Author compiled data based on China Customs (US$Thousands)

To address some trade concerns between China and South Africa and to promote sustainable symbiotic relationships, the formation of the South Africa–China Partnership for Growth and Development (PGD) in 2007 seemed to be the one to solve that puzzle. Under this agreement, South Africa had to increase its value exports to China, increase the export of manufactured and processed agrarian products, and establish a balanced investment flow.

Collapsing trade distance following China's African trade was necessitated by the Go global strategy adopted by China which saw China's heavy investment in shipping and port construction in Africa during the 1900s and 2000s. The building of super-sized cargo shipping fleets was a positive step in enabling such trade as well. Africa's infrastructure was enhanced.

Infrastructure development of roads, ports, ships, and railways has blocked African countries from competing in the Chinese market as well as in their home market in this labour-intensive sector. That leaves the Sino-Africa collaborations skewed in favour of China. Eisenman (2012) supports that notion by concluding that infrastructure development naturally makes China impose its influence on the destination of its products and equally where its suppliers will be located.

China is investing in almost every part of Africa as seen in Fig. 17.10. Countries with resources needed by China trade more with China as compared to those without resources needed. For example, in Nigeria, there is oil which China needs, therefore, trade with Nigeria is higher at about 18% whilst Angola equally has oil and mineral resources pegged at about 8%. Mozambique, without many resources, has China's investment at about 3%. China's investment in Africa after the 2018 FOCAC summit was $60 billion to the continent. Diverse countries' development proves that China is not restricted to a single country and is interested in the development of the whole continent.

Figure 17.11 shows how diverse China's investments in Africa are. China has invested in almost every major sector of the economy, not just in mineral extraction.

African leaders are refuting the claim of Africa as China's second continent (French, 2015), as President Cyril Ramaphosa (2018) co-chair of the FOCAC summit stated:

> "In the values that it promotes, in the manner that it operates, and in the impact that it has on African countries, FOCAC refutes the view that a new colonialism is taking hold in Africa, as our detractors would have us believe."

Figure 17.12 clearly shows the development, progress, mutual benefits, and cooperation taking place between China and Africa. For instance, the Chinese Export–Import Bank provided 85 per cent of the funding for the $475 million Addis Ababa Light Rail. The implication for such sponsorship would be increased trade promotion on project completion.

Africa's top projects between Africa and China in five African countries have had their railway systems funded by China: Kenya, Ethiopia, Angola, Djibouti, and Nigeria. Kenya's largest infrastructure project since independence, the Mombasa-Nairobi Standard Gauge Railway, was funded by China at an estimated cost of R57.2 billion. In Ethiopia, China has funded two railway projects: Addis Ababa Light Rail

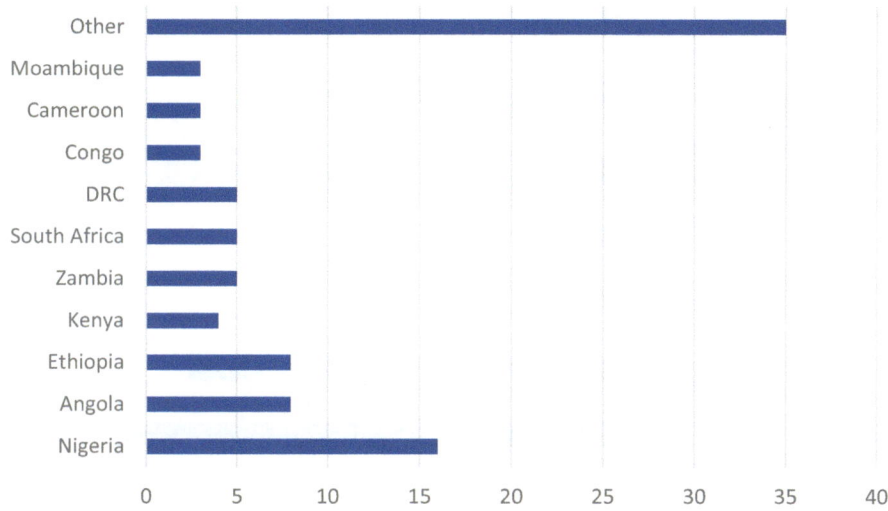

Fig. 17.10 Chinese Investment Distribution in Africa in US$billions. *Source* Author data based on Chinese Investment Tracker 2018

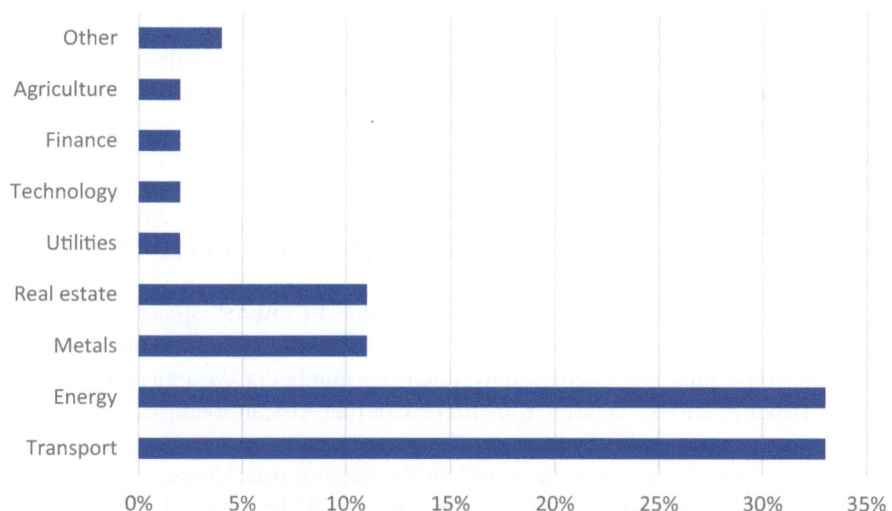

Fig. 17.11 Sectoral investment distribution in Africa by China. *Source* Author compiled data based on the Chinese Investment Tracker

Transit and Ethiopia-Djibouti Railway. Lobito-Luau Railway in Angola and Abuja-Kaduna Railway in Nigeria were also funded by China. The African governments would not have afforded to build these big railway projects which are necessities in enabling trade between and among states. It is recommended that such sterling efforts should continue building Africa.

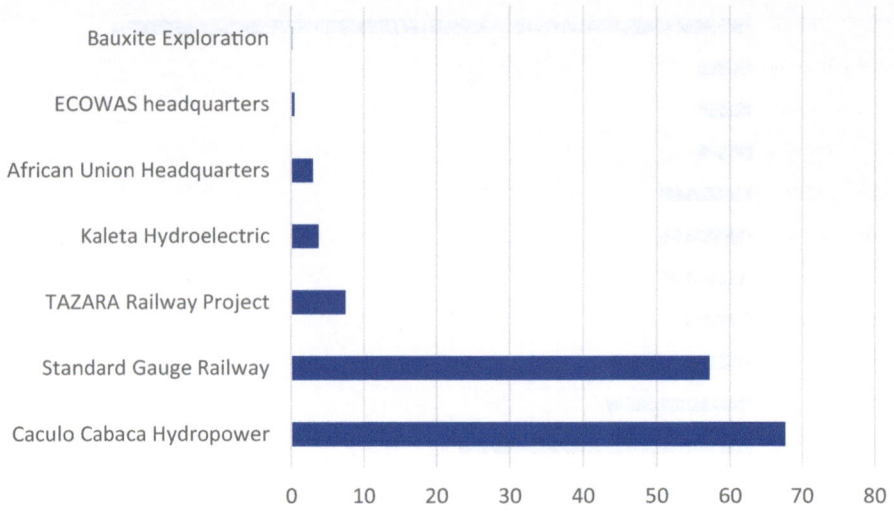

Fig. 17.12 Major projects are undertaken in Africa (US$ billions). *Source* Author compiled data based on CARI

The building of an R3 billion African Union headquarters is located in Addis Ababa. Ethiopia was fully funded and built by China. This is a milestone achievement that African Union is for all 54 African states and the ability of China to persuade all African states to build and fund a project for them is not easy. It is the diplomacy, commitment, and ability to establish good relationships across different African cultures, and it is something African countries can adopt in their economic expansion across the globe. This implication of such is the unification of Africa enabling Sustainable Development goals of people to people meeting with the Chinese, and it must prevail.

In March 2018, the West African regional bloc ECOWAS signed a deal with China to build their headquarters in Abuja for R475.7 million ($31.6 million). Africa benefited from China and it managed to convince a number of Western countries and establish trade relations. A lesson from this reveals that African countries individually or collectively in two or three groups may copy such strategies and approach different organisations and groups across the world such as the European Union, to get financial systems and aid without any strings attached. That could be achieved in a gentle and fairer brotherhood partnership.

In 2017, Ghana and China agreed to an R150 million bauxite exploration deal aimed at further exploiting the West African country's vast solid mineral deposits. If Ghana could exploit its mineral deposits, it would not have invited China. Due to Ghana's weaknesses in technology, capital, and skilled manpower to explore, China was the answer to their plight. The deal was struck which would benefit China and Ghana in a friendly and fairer way, unlike in the colonial era where resources were forcibly taken out of Africa without their knowledge and consent.

The study found that Angola and China signed a construction deal for the Caculo-Cabaca Hydropower Project in Dondo, Angola, worth R67.7 billion and is set to produce 2172 megawatts of electricity. A similar project is ongoing at the Kaleta hydroelectric facility in Guinea, worth R3.8 billion, with China funding 75 per cent of the project. The finding resonates with similar findings by Business Insider (2018). The implication of this undertaking proves China's interest in the energy sector and testimony to the diversification of China's interests in energy which do not only lie in mineral extraction, contrary to popular belief both politically and in academic circles.

The study found that Nigeria and China signed a deal to build an oil refinery in Edo State for R30.1 billion. China and Nigeria need each other because Nigeria is Africa's largest oil producer which aims to move from pumping 2 million barrels a day to 3 million barrels a day by 2023. China's oil production is reducing at the same time due to natural depletion and other geological challenges but targets that by 2030 it will be importing crude oil of up to 80%. This foresight by the two nations proves that these bilateral relations are mutually beneficial.

The study found that China is responsible for several projects in Zambia including the China National Building Material which was launched by President Lungu. The project is worth R7.5 billion (Business Insider, 2018). The TAZARA railway project is another project China has been involved with in Zambia, and this cement factory is yet another project bearing testimony to China's interest in the construction and transport sectors and as a developmental partner meeting its obligations at FOCAC summits.

The study found that before President Robert Mugabe was ousted, China presented the former head of state with an R2.1 billion donation gift for building a new Parliament. China–Zimbabwe relations started during the liberation struggle when the Zimbabwe African National Union received support during the liberation war. The relations gave Zimbabwe a political impetus when it got its independence in 1980, as well. The Look East Policy adopted by former President Robert Mugabe deepened bilateral relations. The donation building came after several successful projects had been running between Zimbabwe and China. It is no surprise that it was a big thank you. The study finds that China did several projects in Zimbabwe such as a $200 million purchase of a 92 per cent stake in Zimasco Consolidated Enterprises Limited by Sinosteel Corp, China's biggest chrome importer. The Chinese have also invested in diamond mining through Anhui, a joint venture between the government's Zimbabwe Mining Development Company, the military, and the Chinese company. China Machine-Building International Corporation signed a $1.3 billion agreement to mine coal and build thermal power generators in Zimbabwe. This finding resonates with Mano's, 2016 findings on the same. According to Africa Portal (2014), some critics say Zimbabwe–China relations were adopted by Zimbabwe to cover up for its criticized domestic policies. However, it is worth recognizing that China–Zimbabwe symbiotic relations date far back and the gift of building a Parliament in Zimbabwe does not come as a major surprise. The findings imply that good relations with China ideologically also achieve the best results.

The study found that energy seems to be China's second priority project as the figure above shows. The reason for China's interest in energy projects in Africa could be attributed to the fact that it is the world's second-largest consumer of energy and the concern of the world with environmental damage due to excessive carbon dioxide emission (Wang, 2008). However, the energy initiatives greatly benefit the host nations, improving their respective Gross Domestic Product. China's decreasing role in Africa's oil and gas sector has been decreasing since 2018 (French, 2018). The electric power sector has been fast-growing through the construction of large hydro- and coal-fired power stations across the continent. According to Boston University's China Energy Finance (2021), by 2020 China invested USD29.7 billion USD worth of power generation capacities in Africa. They also added 9.3 billion USD for the transmission systems that are often linked to these power plants, which far exceeds its 11.2 billion USD in the oil exploration and extraction area.

According to the China Africa Research Initiative (CARI) (2018), China is attracted to Africa by its natural resources and export markets, while African leaders hope Chinese engagement brings economic development. From the above figures, the study found that China indeed is interested in some of Africa's natural resources whilst Africa is not so much interested in importing raw materials from China which implies that Africa could improve value addition to its goods for export.

Africans' reactions to Chinese involvement have been met with mixed feelings. Government officials have supported Chinese involvement, while other elements of African societies criticize China including academics (Gumede, 2014), for what they see as an exploitative, neo-colonial approach. From the figures above it is difficult to conclude that the relations are exploitative because the figure above shows billions coming to Africa, whilst other scholars (Watson, 2020) allude to the fact that China offers no-strings-attached loans to Africa as opposed to the exploitative Western countries who set conditions like governance matters. Supporting further findings in this study, Hoffman (2007), asserts that China has its trade relations with Africa based on its 1950s international and nations relations norms which include mutual respect for sovereignty and territorial integrity, mutual non-aggression, mutual non-interference in internal affairs, equality and mutual benefit, and peaceful coexistence. These principles, if they are abiding by them, it implies that relations are not exploitative in China's favour.

China faces scepticism as it tries to establish sustainable development and win–win commercial deals with a range of soft-power tools to improve its image. Big and small businesses between Chinese companies and African companies have signed these commercial win–win deals in all the meetings and conferences they have met, with heads of state included in some instances. The implication of China supporting its small businesses to export to Africa should be adopted by Africa.

CARI (2018), believes that the United States of America and Chinese goals in Africa do not conflict but the presence of these giants in Africa should be taken advantage of by African states. Whilst the two big giants fight for economic and political domination over Africa, Africa is striving in the process because better conditions will be set by any of the two giants, especially China, and Africa being poor will go for the cheaper and viable options.

The study found that all bilateral alliances between South Africa and China were beneficial. The two governments have formed three alliances for trade proliferation and trade balance through the formation of FOCAC, PGD, South Africa–China Economic & Trade Cooperation Forum, and Joint Economic and Trade Committee (JETC) which were all established to tackle the challenges in the relationship between China and South Africa. The three seem to have made progress although duplication of effort seems to be inherent in the trio's discharge of their duties. If differences were highlighted and resolved through these joint groups and trade continue, that shows agreement on how to trade and achieve mutual benefits. In the figure above we see sometimes, imports from China exceed Africa's exports to China. Billions of United States dollars up to 18 billion, according to the figure above, are seen flooding to Africa for proving the symbiotic relations existing therein.

The lenience and tolerance of China are also shown when they give loans to African countries although the latter struggle to repay the loans citing economic difficulties. Zambia has struggled over decades to pay back the loan extended to them on the TAZARA railway line, citing the non-viability of the project up to now. It is anticipated that Kenya's SGR railway line will not be able to return the loan extended to them by China. Ethiopia has begun debt restructuring initiatives with China as well (Mkonza, 2022). China has extended its loan terms to South Africa's largest parastatal of R37 billion loan to Eskom but some critics still feel that the loan conditions are tilted in favour of China.

Mansfield et al. (2000) suggest that there is less trade between a democratic and an autocratic government by between 15 and 20%. Criticizing this finding, Eisenman (2012), argues that their model does not yield determinate predictions about whether trade between autocratic pairs is more likely than between mixed pairs. The academic literature then has a problem in answering the notion of whether two autocratic states trade well in balanced trade. Besides Mansfield et al. findings being regarded as biased and inconsistent, Penubarti and Ward (2000), argue that their results are both inefficient and have biased estimates, claiming that Mansfield et al. only uses a loose-fitting gravity model to determine bilateral trade. Barma and Ratner (2013) used China as a case study, in qualitative research and concluded that China is supportive of illiberal capitalism, where markets are free, but politics are not. They argue that by forming and nurturing bilateral and multilateral arrangements, the Chinese government is creating an alternative international structure anchored by illiberal norms. They contend that Chinese illiberalism presents a real long-term geopolitical challenge, it is easily exportable, and it is dangerously appealing to a disaffected world. Furthermore, Barma and Ratner link China's political illiberalism with its economic relations by arguing that Beijing leverages its mercantilist strength in the international system to attain its national interests and argue that this trend is nowhere more evident than in Africa.

The study finds that China's oversupply of capital and material determined China's desire to fulfil Africa's demand mutually. The sentiment is equally shared by Chinese President Xi Jinping and the Vice-Minister of the Overseas Chinese Affairs Office of the State Council Minister. He stated that the most important thing is to turn the challenge into an opportunity by moving out of this overcapacity based on its development strategy abroad and foreign policy (He Yafei, 2014).

In a brotherhood fashion and mutually beneficial and understanding manner when Ethiopia failed to repay its SGR loan and management fees to China as scheduled, China agreed that the loan would be paid in 15 to 30 years after negotiations. Ethiopia has faced similar problems to Kenya and fell behind in scheduled repayments for the loan and even Kenya was given a grace repayment period by China (Chen, 2019).

The study also found that politics and economic trade are closely linked as evidenced between Zimbabwe and China. The finding resonates with the findings of Eisenman (2012), who alludes that political and economic considerations determine trade patterns, particularly between developing countries like China and African states. The big question requiring further analysis is why and how political variables influence China's trade with Africa. Economies of scale and technological development are a product of a centrally administered national industrial policy born of a political process rather than an efficient market mechanism. The role of politics in trade is becoming increasingly more prominent between developing economies with autocratic political regimes. Eisenman (2012) laments the absence of the role of political variables in determining trade, despite many scholars having recently included the political variable in their new bilateral and multilateral trade models. The implications of this model in this study are the role of politics in the number of goods that can be traded, and which types of goods can be traded under certain political variables. For example, raw material versus manufactured consumer goods and capital equipment, many researchers believe they determine the China African economic and political ramifications prevailing today.

17.5   Conclusions

This chapter dealt with different perspectives on China–Africa collaborations. It also dealt with China's contributions to China and Africa's contribution to China. Bilateral and multilateral collaborations formed between China and Africa such as the FOCAC, JET, PDP AFRODAD were mutually beneficial because there were developments brought in by China such as infrastructure development, roads, and railways built in Africa, and the GDP of African nations were improved due to trade, skills transfer, access to affordable capital goods, consumer goods, and intermediate goods. China also benefited by establishing a market in Africa, redistribution of its massive population elsewhere, geopolitical dominance in Africa, and access to abundant resources in Africa. In practical terms, there seems to be a win–win situation. The study implies that theories of trade should factor in the political variable in the conceptualization of new models which specify how politics determine trade, the quantities of goods to be traded under certain political conditions, and the type of goods to be traded under specific political conditions so that China–Africa relations could be more easily evaluated. The study implies that China–Africa collaborations should continue and be sustained because they are based on mutual benefits. China's interest in energy implies that there will be development in all sectors of African economies increasing the GDP. Moreover, the projects that Africa shelved for years can now be expedited and bring positive results. The study implies that there is no neo-colonialism in the trade. There is an evidential acceptable balance of trade that replaces Western countries' models of lending favourably. It is recommended to strengthen and sustain the Africa–China collaboration.

References

Achu, F. N. (2019). Pro-environmental behaviour of attendees at a major sport event in Cameroon. *Geo Journal of Tourism and Geosites, 27*(4), 1307–1320.

Alden, C. (2005). *Leveraging the Dragon: Toward 'An Africa that can say No in Africa'.* South Africa Institute of International Affairs. February.

Barma, N., Ratner, E., & Weber, S. (2013). The mythical liberal order. *The National Interest,* (124), 56–67.

Bliss, H., & Russett, B. (1998). Democratic trading partners: The liberal connection, 1962–1989. *the journal of politics, 60*(4), 1126–1147.

Borovska, H. (2011). The China-Africa marriage. *Copanhagen Business School.*

Botha, I. (2006). *China in Africa: friend or foe?: China's contemporary political and economic relations with Africa* (Doctoral dissertation, Stellenbosch: University of Stellenbosch).

Brautigam, D., & Gaye, A. (2007). Is Chinese investment good for Africa?. *Council on Foreign Relations Online Debate, 20.*

Burke, C., Naidu, S., & Nepgen, A. (2008). *Scoping Study on China's Relations with South Africa.* Nairobi: AERC (mimeo)

Chen, X., & Mao, X. (2019). A Landmark in China–Africa Friendship: The China- Aided African Union Conference Center. *South-south Cooperation and Chinese Foreign Aid, 35–48.*

China–Africa Research Initiative. (2018). *Data: China–Africa trade.*

Corbin, J. & A. Strauss (2008): *Basics of qualitative research.* London, UK: Sage.

Eisenman, J. (2012). China–Africa trade patterns: Causes and consequences. *Journal of Contemporary China, 21*(77), 793–810.

Fang, Y. (2020). Influence of foreign direct investment from China on achieving the 2030 Sustainable Development Goals in African countries. *Chinese Journal of Population, Resources, and Environment, 19,* 1–8.

French, H. W. (2015). The Plunder of Africa: How Everybody Holds the Continent Back.

Giry, S. (2004). China's Africa strategy out of Beijing. *New Republic, 15,* 19–23.

Gumede, W. (2012). FPC Briefing: Challenges facing South Africa-China relations. *The Foreign Policy Centre* [Online]. http://fpc. org. uk/fsblob/1465. pdf (Accessed: 15 October 2012).

Gumede, V. (2014). Socio-economic transformation in post-apartheid South Africa: Progress and challenges. *The future we chose, 278–296.*

Herbst, J., & Mills, G. (2009). Commodities, Africa and China.

Hoffman, F. G. (2007). Conflict in the 21st Century: The Rise of Hybrid Wars. *Potomac Institute for Policy Studies.*

Jauch, H. (2007). Between politics and the shop floor. *Transitions in Namibia, 50.*

Manji, F., & Marks, S. (Eds.). (2007). *African perspectives on China in Africa.* Fahamu/Pambazuka.

Mano, W. (2016). *Engaging with China's soft power in Zimbabwe: Harare citizens' perceptions of China–Zimbabwe relations.*

Mansfield, E. D., Milner, H. V., & Rosendorf, B. P. (2000). Free to trade: Democracies, autocracies, and international trade. *The American Political Science Review, 94*(2), 305–321.

Mkonza, K. C. (2022). *Human Rights Diplomacy or Transactional Diplomacy? an Analysis of South Africa's Foreign Policy, 1994 to 2020*. University of Johannesburg (South Africa).

Mooney, P. (2005). China's African safari. *Yale Global, 3*(1), 2005.

Ndubisi, O.N. (2007), "Relationship quality antecedents: the Malaysian retail banking perspective", *International Journal of Quality & Reliability Management, 26* 8, 829–845

Nhamo, G. (2020). Higher education and the energy sustainable development goal: policies and projects from University of South Africa. *Sustainable development goals and institutions of higher education*, 31–48.

Penubarti, M., & Ward, M. D. (2000). Commerce and democracy. *Center for Statistics and the Social Sciences working paper, 2*(6), 1–36.

Prah, K. K. (2007). China and Africa: Defining a relationship. *Development, 50*(3), 69–75.

Rapley, J. (2007). Understanding development: theory and practice in the third world (3rd edition). *Boulder*, Colo: Lynne Rienner.

Regissahui, M. H. J. (2019). Overview of the China–Africa trade relationship. *Open Journal of Social Sciences, 7*, 381–403. https://doi.org/10.4236/jss.2019.77032

Sandrey, R. (2009). The impact of China-Africa trade relations: The case of Angola.

Shelton, G. (2008). South Africa & China: A strategic partnership? In C. Alden, D. Large, & R. Soares de Oliveira (Eds.), *China returns to Africa: A rising power and a continent embrace*. Columbia University Press.

Shinn, T. (2005). New sources of radical innovation: research-technologies, transversality and distributed learning in a post-industrial order. *Social Science Information, 44*(4), 731–764.

Stake, R. (1995). *Case study research*. thousand oaks, CA: Sage.

Taylor, S. J. (2008). *Modelling financial time series*. world scientific.

Van der Wath, K. (2004). "Enter the dragon: China's strategic importance and potential for African business" in *convergence, 5*(4): 72–75

Wang, J. Y. (2007). What drives China's growing role in Africa?

Wang, Y. (2008). Public diplomacy and the rise of Chinese soft power. *The annals of the American academy of political and social science, 616*(1), 257–273.

Watson, B. W., Badouel, E., & Niang, O. (2020, October). Foreword of Proceedings of CARI 2020. In *CARI 2020-Colloque Africain sur la Recherche en Informatique et Mathématiques Appliquées* (1–12).

Weizhong, X. (2008). Sino-African relations: New transformations and challenges. In Gerrero & Manji (Eds.), *China's new role in Africa and the South*. Fahamu and Focus on the Global South.

Wenping, H. (2008). China's perspective on contemporary China–Africa relations. In C. Alden, D. Large, & R. Soares de Oliveira (Eds.), *China returns to Africa*. Columbia University Press.

World Integrated Trade Solution (WITS). (2019). *Database*.

Xinhua, M. A., & Jun, X. I. E. (2018). The progress and prospects of shale gas exploration and development in southern Sichuan Basin, SW China. *Petroleum Exploration and Development, 45*(1), 172–182.

Yin, R. K. (1994). Discovering the future of the case study. Method in evaluation research. *Evaluation practice, 15*(3), 283–290.

Chapter 18
Evolution and Prospects of China–Africa Cooperation in Science, Technology and Innovation

Yebo Wang, Xu Pan, Siqian Wang, and Yu Pang

18.1 Introduction

Science, Technology and Innovation (STI) is an important method for solving global problems. STI cooperation offers an important way in which countries can establish mutually beneficial, win–win partnerships. China's development experience shows that STI development provides important underpinning for poverty elimination and the improvement of independent development capability, and powerful support for economic growth and social prosperity, acting as a dominant force leading sustained, coordinated development. China is the largest developing country in the world, while Africa is the continent with the largest number of developing countries. Similar historical encounters and a shared historical mission have tightly linked China and Africa together. China–Africa relations have consistently maintained vigorous development momentum, forming a distinctive path of win–win cooperation (The State Council Information Office of the PRC, 2021). STI cooperation has been a consistent and important component of China–Africa cooperation, and is playing an increasingly prominent role in South-South cooperation and the development of China–Africa relations. As China–Africa relations continue to develop, China–Africa STI cooperation has flourishing and borne fruit, providing strong impetus to both sides' economic and social development.

Y. Wang · Y. Pang (✉)
National Center for Science and Technology Evaluation, Ministry of Science and Technology, Beijing 100098, China
e-mail: pangyu@ncste.org

X. Pan
Institute of Ecological Protection and Restoration, Chinese Academy of Forestry, Beijing 100091, China

S. Wang
Hubei International Science and Technology Exchange Center (China-Africa Innovation Cooperation Center), Wuhan 430071, China

© The Author(s) 2024
M. Muchie et al. (eds.), *China-Africa Science, Technology and Innovation Collaboration*, https://doi.org/10.1007/978-981-97-4576-0_18

18.2   Evolution of China–Africa Cooperation in Science, Technology and Innovation

China–Africa STI cooperation has continued from ancient times to the present day. A comprehensive analysis of the history of the development of China–Africa STI cooperation suggests its evolution can be divided into four stages.

18.2.1   Stage I (Spontaneous Interaction of Traditional Science and Technology, Before the Founding of the PRC)

One of the world's four ancient civilizations, ancient China was a world leader in science, technology, culture and other fields. As the land and maritime Silk Roads "opened" China's technological achievements successively spread to Europe and Africa, the Africa's technological achievements also spread to China. With the formation of the Silk Road, commercial interactions developed and flourished (He et al., 1995), for instance, Chinese medicinal herbs and African aromatic medicines promoted bilateral development of medical technology. Chinese silk and porcelain were also introduced in Africa, and some regions of Africa began to learn Chinese silk-weaving and porcelain-making techniques. After the formation of the Maritime Silk Road in the eighth century, Chinese paper-making technology was introduced in North Africa, and Egyptian doctors welcomed traditional Chinese medicine. During the Middle Ages, technology exchange became an important feature of both the Eastern and Western worlds, with over 60 kinds of medicinal herbs imported from China to Africa, and three major Chinese inventions (printing, compass and gunpowder) reaching the North and East African regions. After the seventeenth century, many Chinese were brought to Africa to engage in traditional Chinese medicine, engineering and agriculture, promoting the exchange of medical technology and tea cultivation technology between China and Africa (Wang, 2011).

China–Africa technology cooperation at this stage possessed two main characteristics. Firstly, the major one focused on mutual exchange of medicine and medicinal materials. Chinese anatomy and pulse science were introduced in Africa, while Egypt's embalming and brain surgery technology were introduced in China. Significant trade in medicinal materials promoted improvements in medicine and medical technology on both sides. Secondly, traditional technology was spontaneously exchanged through trade and commerce. Chinese silk-weaving and porcelain-making technology, along with the four great Chinese inventions, were introduced in Africa. Meanwhile glass production and sugar-making technology invented by the Egyptians was introduced in China, which promoted economic development and cultural prosperity on both sides. Until the founding of the People's Republic of China (PRC) in 1949, there was no mechanism for China–Africa science and technology cooperation, and the interaction was, overall, spontaneous.

18.2.2 Stage II (Government-Led Technical Assistance, from the Founding of the PRC to the 1980s)

After the founding of the PRC, the international environment deteriorated severely. Within the context of the bipolar world system created by the United States and the Soviet Union, China implemented an independent, peaceful foreign policy, actively developing relations with African countries. Modern science and technology cooperation relations between China and Africa began in 1955 at the first Asian-African Conference, Bandung Conference, which adopted a resolution promoting cultural cooperation and exchange between Asian and African countries, and proposed "mutual technical assistance to the maximum extent practicable" (Ministry of Foreign Affairs, Republic of Indonesia, 1955). In 1956, China and Egypt signed a "cultural cooperation agreement" and began exchanges of professors, young students and scientists. Between 1963 and 1965, Chinese Premier Zhou Enlai visited Africa three times, proposing first the "Five Principles for China's Relations with African and Arab Countries" and then the "Eight Principles for Foreign Aid", which covered the content of China's technical assistance to Africa, clearly defining four basic principles for this assistance. The first principle was equality and mutual benefit; the second, the meeting of actual needs; the third, encouragement of technology transfer, and the fourth, non-specialization in the treatment of science and technology personnel. These principles provided a clear direction for China–Africa cooperation in science and technology (Xie, 2002).

Thus, at this stage, China–Africa cooperation in science and technology moved from being spontaneous to being government-led, as is apparent from three perspectives. First, the cooperation was grounded in the common aspirations of China and Africa in terms of political interests. Initially, China both actively expressed support for African national liberation movements, provided practical aid to independent African countries, as its capacity allows. Since the Bandung Conference, the technical assistance and cooperation agreements signed between China and some African countries have been distinctly political in tone, rather than clearly emphasizing interests. Second, cooperation was to be equal and mutually beneficial. The "Eight Principles for Foreign Aid", proposed by Zhou Enlai during his visit to Africa, clearly stated that Chinese technicians would not seek special treatment, guaranteed that African technicians would master technologies, and included a principle of equal cooperation putting China and Africa on an equal footing. Third, China has actively provided assistance to Africa, strengthening South-South cooperation. During the three decades from 1949 to 1979, over 180 China–Africa cooperation projects in agricultural science and technology have been undertaken. These have included China's provision of assistance to African countries for construction of experimental agricultural technology stations and agricultural processing projects, disseminating Chinese rice, cotton, tea, sericulture and tobacco cultivation technologies to Africa, all in the form of free aid. Through selfless assistance to Africa in science and technology, a close partnership between China and Africa has been established. The Tanzania-Zambia railway in particular, which China helped Tanzania and Zambia build, has become a landmark in China's science and technology (ST) assistance to Africa, serving to strengthen the friendship between China and Africa.

18.2.3 Stage III (Exploration of Mutually Beneficial Win–Win Mechanisms, from the 1990s to 2009)

After 1980, China adopted a policy of reform and opening up, and gradually adjusted the system for China–Africa relations in ST. In 1983, China established four principles for China–Africa economic, and ST cooperation: "equality and mutual benefit, pragmatism, diversity in form, and common development" (Li, 2005). Meanwhile, China gradually brought its system of ST assistance for Africa into line with international standards and strengthened the concept of two-way cooperation. Five notable changes in China–Africa ST relations occurred during this period: first, a greater emphasis upon economic benefit and demonstration effects in terms of project selection; second, the adoption of a contracting system for project implementation, with a focus on unity of responsibilities, rights and interests; third, the long-term follow-up of project management, covering the entire process from completion and handover to ongoing cooperation, fourth, strengthened technical and management cooperation to consolidate the achievements of ST assistance; and fifth, the exploration of new areas of ST cooperation. China–Africa ST relations had thus developed further. According to incomplete statistics, over 2000 China–Africa ST cooperation projects were implemented between 1978 and 1990, involving petroleum, chemicals, textiles, agriculture, animal husbandry, medicine and other fields. By this point, China–Africa ST relations had gradually changed from an "aid-based" format, to "mutually beneficial and win–win" cooperation, marking completion of the transformation of China–Africa ST relations towards sustainable development.

After 1995, China and Africa continued to expand and deepen their ST cooperation in fields including agriculture, forestry, energy, machinery, environmental protection, communications, and satellite technology. For instance, during the third meeting of China–Egypt Joint Committee on Intergovernmental Cooperation in Science and Technology in 1998, China and Egypt extended their cooperation to high-tech fields such as metrology, environmental protection, biology, information technology, and resource exploration and development. In 1999, an agreement on ST cooperation formally signed by China and South Africa established the Joint Commission on Science and Technology Cooperation, which covers areas of basic science and high technology including agriculture, biology, chemistry, medicine, minerals, information technology, traditional medicine and space technology.

In 2000, within the context of South-South cooperation, China and African countries jointly initiated the establishment of a mechanism for collective dialogue, the Forum on China–Africa Cooperation (FOCAC), in order to further strengthen friendly cooperation between China and African countries under the new circumstances, to jointly address the challenges of economic globalization and seek common development, and to gradually develop and improve China–Africa ST relations in terms of overall planning and formulation of specific policies.

18.2.4 Stage IV (All-Round Sustainable Development in STI Cooperation, 2009 to Present)

In April 2009, at the China–Africa Science and Technology Policy Exchange Conference, both sides discussed the Overall Plan of Action for African Science and Technology reaching out to China's ST policies, with the aim of enhancing mutual understanding of these policies and promoting cooperation. This meeting made preliminary preparations for the establishment of a mechanism for the development of China–Africa ST relations. At FOCAC's Fourth Ministerial Conference, in November 2009, Chinese Premier Wen Jiabao proposed the "China–Africa Science and Technology Partnership Plan" (FOCAC Sharm, 2009), marking the final formation of the development mechanism for China–Africa ST relations. Since 2009, first the China–Africa Science and Technology Partnership Plan, then the China–Africa Science and Technology Partnership Plan 2.0, have been launched. Under the framework of these plans, China and African countries have established a new type of pragmatic, efficient and dynamic ST partnerships, mainly implementing joint research, technology demonstration, ST policy planning, technology training, high-level ST talents training, donation of scientific research equipment and so on, providing strong support for the development of a China–Africa comprehensive strategic partnership, and the consolidation and reinforcement of China and Africa as a community with a shared future.

Entering the new era, Chinese President Xi Jinping proposed principles for China's Africa policy—sincerity, real results, amity and good faith, and the principles of pursuing the greater good and shared interests—providing the fundamental guidelines for China's cooperation with Africa, and charting its course. In 2015 and 2018, FOCAC led China–Africa cooperation to unprecedented heights, as both sides' leaders unanimously decided to build a closer community with a shared future, and further promote their cooperation on the "Belt and Road" Initiative, creating a new milestone in the history of China–Africa relations (The State Council Information Office of the PRC, 2021). In particular, the proposal of Eight Initiatives, and of the FOCAC-Beijing Action Plan (2019–2021), at FOCAC 2018 further strengthened the Forum as a project brand with a focus upon health, investment, trade, industrialization, infrastructure, agriculture and food security, climate change, peace and security, human resources, the digital and maritime economy and other areas. It is a key to improve China–Africa cooperation in science and technology (FOCAC 8th Ministerial Conference, 2021). At this Forum, President Xi Jinping also proposed the establishment of the China–Africa Innovation Cooperation Center, an initiative important to the implementation of the China–Africa Science and Technology Partnership Plan 2.0, the improvement of technology transfer network rollout along the

Belt and Road, the promotion of STI cooperation between China and African countries. This also represents an important element in the building of a closer China–Africa community with a shared future in the new era. Jointly sponsored by the Ministry of Science and Technology of China, Hubei Provincial Government, and supported by the science and technology authorities of African countries, the China–Africa Innovation Cooperation Center was officially inaugurated and began operation in February 2021.

China has signed cooperation agreements with 53 African countries and the African Union Commission, and intergovernmental STI cooperation agreements with 16 African countries, injecting fresh impetus into the development of China–Africa STI cooperation. In 2021, FOCAC's 8th Ministerial Conference adopted the Dakar Declaration, Dakar Action Plan (2022–2024), the Declaration on China–Africa Cooperation on Combating Climate Change and the China–Africa Cooperation Vision 2035. These included agreements on numerous matters. First, the sides agreed to further strengthen local government exchanges, support the establishment of more friendly relations at provincial and municipal levels, improve the mechanism of the Forum on China–Africa Local Government Cooperation, and promote deeper and more practical local cooperation between them. Second, they agreed to continue to strengthen bilateral and multilateral mechanisms for STI cooperation in traditional areas such as agriculture, public health, medical and health care, and education, to improve the depth and intensity of cooperation, formulate and implement the China–Africa Partnership Plan on Digital Innovation in Africa, and actively explore and promote cooperation in applications of new technology including cloud computing, big data, artificial intelligence, the Internet of Things and the mobile Internet. Third, they agreed cooperate in order to improve the China–Africa technology transfer and innovation cooperation network, to convene a China–Africa innovation cooperation and development forum, and to support the establishment of joint China–Africa laboratories, partner institutes and centers for STI cooperation. Fourth, the sides arrived at a consensus on 11 points concerning cooperation in science, technology and knowledge sharing, including the strengthening of STI strategic interaction and policy communication, the joint construction of a multi-level science, technology and humanities exchange system, and the joint implementation of technology transfer and cooperation in innovation and entrepreneurship, thus fully leveraging the platform effects of the "China–Africa Innovation Cooperation Center" and "China–Africa Joint Research Center", the strengthening of cooperation in the field of space technology and nuclear science research between China and Africa on an ongoing basis (FOCAC 8th Ministerial Conference, 2021).

Thus, by this stage, the top-level design, mechanisms and policies of China–Africa STI cooperation had been continuously enriched and improved, and the areas of cooperation further consolidated and expanded, with China actively strengthening its communication and coordination with Africa on STI strategies, sharing experiences and achievements in science and technology development, and promoting the exchange and training of STI talents, and the transfer and innovation in technology and entrepreneurship between the two sides. Together, China and African countries have also constructed a number of high-level joint laboratories, and established the China–Africa Joint Research Center and China–Africa Innovation Cooperation Center. In recent years, China has also assisted Africa in the cultivation of a large number of STI talents through projects such as the Alliance of International Science Organizations Scholarship for Young Talents in the Belt and Road Region, Chinese Government Scholarships, the Talented Young Scientist Program, and the International Youth Innovation and Entrepreneurship Program. Breakthroughs in the area of space cooperation have included the cooperative use of China's remote sensing data in the fields of disaster prevention and mitigation, radio astronomy, satellite navigation and positioning, and precision agriculture, and the sides' joint participation in the Square Kilometer Array Project, an international project in the field of astronomy. China has also supported the construction of a satellite assembly and test center in Egypt, and assisted Algeria and Sudan in the launch of their first satellites (State Council Information Office of the PRC, 2021).

18.3   Mechanisms of China–Africa Science and Technology Cooperation

After over half a century of development, contemporary China–Africa STI cooperation relations represent the concentrated efforts and wisdom of generations of Chinese and African people. The significance of STI cooperation in overall China–Africa cooperative relations has been increasing. The sides have established a series of cooperation mechanisms, including bilateral cooperation, multilateral forums, a leaders' summit, ministerial meetings, and senior official committees, along with other mechanisms of cooperation in various fields and day-to-day liaison mechanisms. Amongst these, three categories of China–Africa STI cooperation mechanisms can be identified: multilateral, bilateral and specialized. Cross-cutting at the multi-sectoral level, these form a pattern of multiple structures (Zhang et al., 2012).

18.3.1   Multilateral Mechanisms

The main multilateral mechanisms supporting China–Africa STI cooperation relies are FOCAC, the China–Arab States Cooperation Forum (CASCF), and South-South Cooperation (SSC) framework.

18.3.1.1   The Forum on China–Africa Cooperation (FOCAC)

FOCAC exists as a platform for collective dialogue between China and African countries in the context of South-South cooperation. Since its establishment in 2000, it has held eight meetings, emerging an important platform for collective dialogue, an effective mechanism for practical cooperation between China and Africa, and an important driving force both for the expansion of China–Africa ST cooperation in terms of areas, and for its upgrading in terms of level (Wang et al., 2012). The first FOCAC, in 2000, determined that the scope of China–Africa ST cooperation should include not only traditional areas, such as agriculture, transportation and health care, but also high-end fields such as information technology. The possibility of additional ST cooperation in the form of engineering project cooperation, and management cooperation was also identified. It was reaffirmed that the goal of strengthening basic and applied ST cooperation between China and Africa was the promotion of Africa's upgrading of local technology and economic development (FOCAC Beijing Summit, 2000a, 2000b). In 2003, China further proposed the strengthening of cooperation between China and Africa in the complementary fields of science and technology, the enhancement of technology transfer to Africa, and the promotion of new forms of ST cooperation such as science and technology training courses.

In 2006, the Chinese government released China's African Policy, which consolidated the basic principles of China–Africa ST relations policies (Ministry of Foreign Affairs of the PRC, 2006), as follows: (1) The principles of China–Africa ST cooperation are mutual respect, complementarity of strengths and sharing of benefits. (2) China–Africa ST cooperation should cover fields including agricultural biotechnology, solar energy utilization technology, geological exploration and mining technology, and new pharmaceutical research and development. (3) Forms of China–Africa ST cooperation should not only include applied research, technology development and achievements transfer, but should also create new forms of cooperation, such as practical technology training courses and technology aid demonstration projects for African countries. (4) The goal of China–Africa ST cooperation is to "actively help disseminate and utilize Chinese scientific and technological achievements and advanced technologies applicable in Africa", while practicing the principles of "sincerity, friendship and equality; mutual benefit, reciprocity and common prosperity; mutual support and close coordination; learning from each other and seeking common development" in China's relations with Africa.

18.3.1.2   The China–Arab States Cooperation Forum (CASCF)

China and the League of Arab States established the China–Arab States Cooperation Forum (CASCF) at the end of January 2004. The CASCF possesses mechanisms which provide new platforms for dialogue and cooperation between China and the Arab states on the basis of equality and mutual benefit, including ministerial meetings, a senior officials' committee, entrepreneurs' conferences, civilizational dialogue seminars, friendship conferences and energy cooperation conferences. Through these, it is consolidating and expanding the mutually beneficial cooperation between the two sides in politics, economics and trade, ST, culture, education, health, energy, environmental protection, forestry, agriculture, tourism, human resource development and publishing. Through CASCF, China is training thousands of senior officials and technicians for Arab countries, including member states of the Arab league in Africa, every year (Chang, 2010).

18.3.1.3 The South-South Cooperation Framework

As developing countries, China and African countries have been dominant forces in South-South cooperation. In recent years, China has been exploring effective ways to deepen this cooperation, attaching great importance to the joint promotion of South-South cooperation in its collaboration with international organizations including the UN's Food and Agriculture Organization, World Food Programme and International Fund for Agricultural Development. For example, by 2012, China had aided the construction of over 60 informatization projects, including fiber optic telecommunications transmission networks, e-government networks and radio and TV FM transmitters, etc. Fiber optic backbone transmission network projects in Cameroon, Tanzania and other countries have also broadened the application of fiber optic cables in African countries (State Council Information Office of the PRC, 2014).

Relying on the multilateral cooperation mechanisms described above, China and Africa have signed a series of cooperation agreements and issued various cooperation policies, including the Beijing Declaration of the Forum on China–Africa Cooperation (October 2000), the Program for China–Africa Cooperation in Economic and Social Development (October 2000), FOCAC's Addis Ababa Declaration (December 2003), FOCAC's Addis Ababa Action Plan (2004–2006) (December 2003), China's African Policy (January 2006), FOCAC's Beijing Summit Declaration (November 2006), FOCAC's Beijing Action Plan (2007–2009) (November 2006), CASCF's 2008–2010 Action Implementation Plan (May 2008), CASCF's 2010–2012 Action Implementation Plan (May 2010), FOCAC's Sharm El Sheikh Declaration (November 2009), and FOCAC's Sharm El Sheikh Action Plan (2010–2012) (November 2009), embodying the rich contents of China–Africa ST cooperation.

18.3.2 Bilateral Mechanisms

The signing of bilateral cooperation treaties and agreements has been the main mechanism for ST cooperation between China and African countries. Since the foundation of the PRC, China has signed intergovernmental agreements on ST cooperation with 16 African countries, establishing mechanisms for ST cooperation and implementing exchanges and cooperation of various types.

In terms of signing cooperation agreements, in the 1950s, China signed an official agreement on cultural cooperation with Egypt, opening the way for bilateral technology cooperation between the two nations. Since then, China has signed bilateral technology cooperation agreements with 16 African countries, including Guinea, Nigeria, Somalia, Kenya, the Congo (DRC), Mali and Sudan, at different times. Since 2000, the bilateral cooperation agreements in place between China and Lesotho,

Nigeria and Algeria have reflected new concepts of ST cooperation, providing China–Africa bilateral ST cooperation with additional contents. Most of the above bilateral technology agreements have been focused on ST cooperation, and provide platforms and mechanism for this (Zhang et al., 2012).

Meanwhile, Joint Committee Meetings have been established with some countries (Wang, 2013), strengthening communication and coordination on major ST issues at bilateral and regional levels, providing an effective dialogue mechanism for bilateral cooperation, providing a mechanism for regular bilateral joint research, and playing an important role in promoting the development of bilateral relations in the research fields. In addition, ST diplomats have also been stationed in South Africa and Egypt, reflecting the importance of ST exchanges and cooperation in China–Africa relations. This has promoted the in-depth, pragmatic and efficient development of bilateral ST cooperation.

18.3.3   Specialized Mechanisms

At the end of 2009, the China–Africa Science and Technology Partnership Program (CASTEP) became the first dedicated mechanism for China–Africa ST cooperation. This was followed by the China–Africa Agricultural Cooperation Forum, established in 2010, and the China–Africa Partnership Plan on Digital Innovation, proposed in 2021.

The CASTEP is led and funded by the Ministry of Science and Technology of China together with other Chinese ministries. In order to meet both sides' common requirements and new needs in terms of development, China decided to further enhance the functions of this program and has implement an upgraded version, the CASTEP2.0. This is expected to play a more important role in the China–Africa joint promotion of sustainable development and the building of a community with a shared future. CASTEP2.0 focuses on attracting enterprises and private investment, together with participation from international organizations and other social funds, and proposes eight ways in which to develop China–Africa ST relations, including the exchange and training of ST talents, cooperation in STI policies, the establishment of research cooperation platforms, the construction of collaborative networks for technology transfer, cooperation in science parks, joint research and technology demonstration, the sharing of ST resources, and cooperation in innovation and entrepreneurship.

Until November 2021, CASTEP's achieved significant outcome. Firstly, there has been a flourishing in terms of China–Africa ST and people-to-people exchanges. The Ministry of Science and Technology of China has supported over 300 young African scientists to implement scientific research in China, and technology training courses in China for nearly 2000 African technicians and government officials from 47 African countries. In addition, a series of branded activities such as the "China InnoTour for African Young Scientists" and "Young Scientists Exchange Program" have been arranged to further broaden channels for exchanges between Chinese and African scientists and technicians. Secondly, China–Africa joint research in key areas has achieved practical results. To date, China and many African countries have established joint research platforms, implementing joint research programs in areas including agriculture, light industry, new energy, health care, resources and the environment. It has provided to over 130 bilateral joint research projects during the past 10 years. The two sides have cooperated in the "Square Kilometer Array" international large-scale science project, jointly participating in the international organization and governance of the Observatory, telescope construction and operation, scientific research and technology development. A large number of joint research achievements have received international recognition, and are now widely used. Thirdly, China–Africa cooperation in technology transfer, innovation and entrepreneurship has made positive progress. The Ministry of Science and Technology of China has implemented over 30 ST aid projects in Africa, providing strong support for African countries' technology application. China has established science park cooperation with South Africa and Egypt, and launched the China–South Africa cross-border incubator. Cooperation in innovation and entrepreneurship has become an important way of the development of China–Africa cooperation.

Within this, agricultural cooperation has been the focus of China–Africa ST cooperation. China is a largely agricultural country, and Africa is one of the world's major areas of agricultural development. With huge populations in both of the regions, agriculture provides the basis for development and stability. Thus China–Africa agricultural cooperation has formed the main content of cooperation between the sides for many years, and also the main area of their ST cooperation. Since the launch of FOCAC, China has sent large numbers of agricultural experts to dozens of African countries, established several agricultural demonstration centers, and trained thousands of agricultural management personnel and agricultural technical personnel to assist African countries. The China–Africa Agricultural Forum, launched in August 2010, provides a collective mechanism for dialogue on agricultural cooperation between China and Africa within the FOCAC framework. It reflects the great importance that China and Africa attach to the issue of food security, and offers another special platform for China–Africa ST cooperation in the area of large-scale agriculture.

In the present world, the accelerating development of a new generation of digital technologies is driving changes in production, creating new spaces for human life, and expanding the arena of national governance. In addition, during the COVID-19 pandemic, the digital economy bucked the trend to become an important driver of global economic recovery. At the Extraordinary China–Africa Summit on Solidarity Against COVID-19 in 2020, President Xi Jinping emphasized China's willingness to work with African countries to expand cooperation in the digital economy, smart cities, 5G and other new business models. Against this background, China announced the China–Africa Partnership Plan on Digital Innovation at FOCAC 2021. This proposed six cooperation initiatives. Firstly, the strengthening of digital infrastructure and opening up of the information arteries of economic and social development. Secondly, the development of a digital economy, and promotion of development integrating digital technologies into the real economy. Thirdly, the rollout of digital education and elimination of "talent bottlenecks" in digital innovation. Fourthly, an increase in digital inclusivity, serving the ordinary people of Africa. Fifthly, the creation of digital security, and improvements in Africa's digital governance capacity. Sixthly, the establishment of a cooperation platform promoting digital progress through communication (Deng, 2021).

18.4 China–Africa STI Cooperation Promotes Economic and Social Development

In recent years, as China–Africa STI cooperation has flourished, China has been actively implementing the "Belt and Road" Science, Technology and Innovation Cooperation Action Plan, promoting the China–Africa Science and Technology Partnership Program, sharing China's STI achievements and development practices with African countries, and providing assistance to African nations with economic development, livelihood improvement, social governance and environmental improvement. At the same time as increasing its provision of aid to Africa, China has affirmed its adherence to the concept of giving fish, but also teaching people how to fish, thereby turning the provision of aid to African countries with their independent development into a new form of aid. China has been assisting Africa with STI capacity building and enhancing Africa's independent innovation capability, leading many African countries to speak highly of Chinese aid to Africa. At the 2021 China–Africa Innovation Cooperation Conference, during the opening ceremony and in live interviews, various African government officials, including the Egyptian Minister of Higher Education and Scientific Research, the South African Minister of Higher Education, Science and Innovation, and the Kenyan Deputy Minister of Education, together with the ambassadors of Gabon, Tanzania, Benin and Mali to China, expressed desire for further strengthening of China–Africa cooperation in the fields of digital technology, ecology, artificial intelligence, cultural exchange, and innovation and entrepreneurship, enable China's innovation achievements to benefit greater numbers of African people.

18.4.1 China–Africa STI Cooperation is Boosting African Regional Economic Development

Chinese enterprises' extensive participation in China–Africa STI cooperation has imbued it with vitality.

18.4.1.1 Promoting Africa's Industrialization

Industrialization is a prerequisite for Africa's inclusive and sustainable development, and key to creating jobs, eliminating poverty and improving living standards. In recent years, technology transfer from Chinese private enterprises, especially from manufacturing enterprises, to Africa's local market has gradually led to the development of related industrial chains, nurturing numerous local upstream and downstream enterprises. These investments have had notable technology spillover effects in the host countries, supporting their industrialization. Of the over 150 Chinese enterprises in South Africa, high-tech enterprises, including Huawei, ZTE, Hikvision, Dahua, BGI, StarTimes, Goldwind, Longyuan, account for nearly one-third. A report from McKinsey has concluded that, in Africa as a whole, over 10,000 Chinese enterprises, 90% of which are private enterprises, are in operation. Nearly half of these Chinese companies in Africa have brought new products or services to local markets, and over a third have brought new technologies, promoting the economic and social development of African countries in areas such as telecommunications, cell phones and new energy, in particular (Jayaram et al., 2017). For example, Huawei's 4G technology has improved the level of network services and improved the quality of mobile financial services in Kenya, and even the East Africa region as a whole.

18.4.1.2 Promoting Employment in Africa

According to Ernst and Young's Africa Attractiveness Report, since 2005, Chinese companies have created over three times more jobs in Africa than the United States, making China an important contributor to job creation in African countries (EY, 2017). The McKinsey report revealed the high localization rate for Chinese companies' employment in Africa, of around 89%, with about 2/3 of Chinese companies providing technical training and guidance to African employees. For example, Hisense South Africa employs over 800 local workers, accounting for 90 percent of their total workforce, and has indirectly created more than 5000 local jobs. In addition, Chinese companies have also significantly raised employee's income levels. In Ethiopia, for example, the average salary of African workers in the textile enterprises of Eastern Industrial Park is about 50% higher than that paid by local enterprises (Liang, 2020).

18.4.1.3 Deepening Infrastructure Cooperation

China is supporting Africa's prioritization of infrastructure development in economic revitalization, encouraging support Chinese enterprises' adoption of multiple modes of participation in infrastructure construction, investment, operation and management in Africa. Since the establishment of FOCAC, Chinese enterprises have assisted African countries with the construction and upgrading of over 10,000 km of railroads, nearly 100,000 km of roads, 1000 bridges and 100 ports, 66,000 km of power transmission lines, 120 million kilowatts of installed power capacity, 150,000 km of communication backbone networks, and network services covering nearly 700 million user terminals. Built and operated by Chinese enterprises, Kenya's Mombasa–Nairobi Standard Gauge Railway, the first modern railway to be built in that nation in a century, is known as the "Road of Friendship", "Road of Cooperation" or "Road of Win–Win", in this new era of cooperation between China and Africa. It has transported 5.415 million passengers and 1.308 million TEUs of containers, contributing 1.5% to Kenya's economic growth.

18.4.2 China–Africa STI Cooperation is Promoting Livelihood Improvement in Africa

18.4.2.1 Developing a Technology Cooperation Network

The China–Africa Joint Research Center has developed a cooperation network platform in East Africa. Based in Kenya, this extends across Tanzania, Madagascar and Ethiopia, and has maintained long-term, stable STI cooperation with nearly 20 universities and research institutions in more than 10 African countries, representing an upgrade from infrastructure aid to "soft" cooperation encompassing science and technology, management and education. Based upon the distribution and geographical characteristics of African resources, the Center has also established 5 sub-centers, dealing with, respectively, Biodiversity and Conservation, Water Resources Management, Geographic Science and Remote Sensing, Microbiology and Epidemic Disease Control, Modern Agriculture Demonstration, with bases of cooperation in Kenya, Tanzania, Ethiopia and other countries with whom China has long-standing friendly relations. Through these platforms, over 60 Chinese scientists have conducted scientific research in Africa, making important contributions to the solution of the major problems confronting African countries' economic and social development, such as food shortages, environmental pollution and infectious disease epidemics, while also upgrading these countries' levels of science and technology, and capacity for related talent training. Commenting on the China–Africa Joint Research Center, Kenyan Vice President William Ruto said that technological innovation is helping Kenya solve problems such as poverty, and is important for achieving economic prosperity and ensuring the people's happiness, and that Chinese aid brought Kenya to occupy a central position in African scientific research and innovation.

18.4.2.2 Boosting Africa's Agricultural Development

China is actively sharing agricultural development experience and technology with African nations, supporting improvements in agricultural production and processing, and promoting their construction of agricultural industrial chains and development of trade. Since 2012, 7456 African agricultural trainees have received training in China, while courses for over 50,000 people have been delivered locally, and 23 agricultural demonstration centers have been constructed, through projects such as the 100 Agricultural Experts' Assistance for Africa and the Agricultural Expert Group for Africa. To date, China has established agricultural cooperation mechanisms with 23 African countries and regional organizations, and signed 72 bilateral and multilateral agricultural cooperation documents. In 2019, China and Africa held the first China–Africa Agricultural Cooperation Forum, established the China-AU Agricultural Cooperation Committee, and began formulation of the China–Africa Agricultural Modernization Cooperation Plan and Action Plan.

18.4.2.3   Expanding Cooperation in the Digital Economy

China is actively helping African countries bridge the "digital divide", with the rapid development of China–Africa cooperation in the "digital economy" bearing fruit in areas spanning digital infrastructure construction and digital societal transformation, as well as the application of new technologies such as the Internet of Things and mobile finance. Chinese enterprises have participated in a number of submarine cable projects connecting Africa with Europe, Asia and America; cooperated with mainstream operators in Africa to achieve almost full telecommunication service coverage; constructed over half of the wireless site and high-speed mobile broadband networking in Africa, laid over 200,000 km of optical fiber, helped 6 million households gain access to broadband Internet, and served over 900 million African people. Chinese enterprises have also actively participated in the construction of public service platforms in Africa, including those for electronic payments and smart logistics, achieving win–win cooperation in interconnection.

18.4.2.4   In Strengthening the Foundations of Health

China has practiced the concept of "people first, lives first", helping African countries cope with diseases and epidemics, constructing public health systems, and promoting to build a China–Africa health community of a shared future. The dispatch of Chinese medical teams to African countries has been one of the China–Africa cooperation projects of longest history and greatest effectiveness, with China sending a total of 23,000 medical personnel to Africa from 1963 to 2021. China has focused on helping African countries to strengthen the development of specialist medicine, training 20,000 medical personnel of all kinds for African countries. To date, China has helped 18 African countries establish 20 medicine centers for specialties including cardiology, critical care medicine, trauma and lumpectomy, and has established cooperation mechanisms with 45 African hospitals in 40 African countries. China is supporting African countries' improvements in port hygiene and quarantine capacity, dispatching disease control experts to African countries' CDC to provide technical support. In particular, Chinese medical teams and scientists have played an active role in helping African countries deal with Ebola epidemic in West Africa, and with the COVID-19 pandemic.

18.5 Prospects for China–Africa STI Cooperation

In recent years, China–Africa STI cooperation has been flourishing, driven by the existing high-quality cooperative projects, cooperation centers and fields cooperation, with collaborating enterprises' investment in R&D incentivized by innovation funds and development funds. The establishment of a China–Africa partnership addressing climate change has provided a hub for STI cooperation concerning economic restructuring and industrial transformation, and as the implementation of joint research and exchange programs has advanced, the quality and relevance of cooperation has continuously improved. All of this attests to a tendency towards an orientation of multi-directional linkage.

However, the world is undergoing profound changes, unseen in a century, in the context of a complex and changeable international situation, and the ebbing and flowing of the COVID-19 pandemic, profound impacting development in all countries. In particular, since the dawn of the twenty-first century, the emergence of the Fourth Industrial Revolution, as typified by artificial intelligence, 5G telecommunications technology and the Internet of Things, has shifted the focus away from traditional geopolitical competition towards competition in the high-tech field and in global relations of production, creating rare opportunities for developing countries to participate in global industrial chains, supply chains and value chains. Despite facing stubborn challenges, further deepening in China–Africa STI cooperation is anticipated to emerge as an important means by which two-way interaction and win–win cooperation can be achieved, and the technology gap bridged, promoting the multipolarization of the world's STI and the common progress of humankind. To this end, continued expansion of the depth and breadth of STI cooperation between China and Africa, especially in the following five respects, remains crucial.

Firstly, the top-level design of innovation cooperation should be strengthened, and the related consensus further consolidated. Both China and Africa are facing new development situations, environments and stages. At present, China is expediting the establishment of a new pattern development. Taking the domestic cycle as its major element, while incorporating the mutual promotion of domestic and international cycles, this is creating a new, higher-level, open economic system. China should strengthen STI strategic cooperation and policy communication with the African side, while both sides adopt innovation-driven development, jointly leveraging STI's leading role in economic transformation and sustainable development as fully as possible.

Secondly, the foundations of cooperation should be strengthened, and innovation cooperation further expanded. In response to both sides' needs in terms of science, technology and industry development, and especially in response to the pandemic, China and Africa should actively support the creation of STI alliances, bases and networks, for China–Africa cooperation in key areas including health, sustainable development, agricultural science and technology, and digital technology and engineering technology, in order to provide further STI solutions for overcoming the pandemic and promoting economic recovery.

Thirdly, technology transfer and commercialization should remain the focus, alongside further promotion of the outcomes from innovation. The China–Africa Innovation Cooperation Center should be actively leveraged, close linkages between China–Africa innovation resources and demand should be promoted. The exchange activities between Chinese and African enterprises and institutions should be supported in technology transfer, demonstration of applications, innovation and entrepreneurship. The leading role of STI in economic development and livelihood improvement should be fully leveraged to help African countries to reduce poverty and increase incomes.

Fourthly, supporting policies should be improved, and the efficiency of innovation cooperation should be further enhanced. At present, the need remains for further strengthening of China–Africa STI cooperation policy's supporting systems and service system at the medium- and micro-levels. As China–Africa cooperation has developed in depth, the diversification of its subjects of cooperation and broadening of its fields of cooperation have become more prominent. Different subjects of STI cooperation have different needs in terms of interests, while different mechanisms for sharing benefits exist in different fields. Further promotion of China–Africa STI cooperation through subdivision of the management of subjects and fields of cooperation, and establishment of supporting policy systems and service systems, therefore remains a fundamental task.

Fifthly, non-governmental exchanges should be broadened and the network of partners for STI cooperation should be further expanded. From a global perspective, many countries have established strong non-governmental mechanisms for STI cooperation with Africa. For example, U.S. private foundations, such as the Bill and Melinda Gates Foundation, the Exxon Mobil Foundation, and the Coca-Cola Africa Foundation, play an important role in U.S.-Africa STI cooperation. Japan relies mainly on JICA in STI cooperation with Africa, while India and the UK focus on the promotion of STI cooperation with Africa via market mechanisms. By contrast, China–Africa STI cooperation relies mainly upon inter-governmental promotion. In the future, it is important to expand the people-to-people exchanges and exert advantages of no-government sectors in the China–Africa STI cooperation, which will enhance its competitiveness.

References

Chang, H. (2010). The growing road of "China–Arab states cooperation forum." *Arab World Studies, 6*(3), 1.

Deng, L. (2021). *Remarks by assistant foreign minister Deng Li at China–Africa Internet Development and Cooperation Forum.* Ministry of Foreign Affairs of the People's Republic of

China PhysicsWeb. https://www.fmprc.gov.cn/mfa_eng/wjbxw/202108/t20210825_9134689. html. Accessed: 26 January 2023.

EY. (2017). *EY's Attractiveness Program Africa: Connectivity redefined*. EY PhysicsWeb. https:// assets.ey.com/content/dam/ey-sites/ey-com/en_za/topics/attractiveness/reports/ey-aar-2017-connectivity-redefined.pdf. Accessed: 26 January 2023.

FOCAC Beijing Summit. (2000a). *Beijing declaration of the forum on China–Africa cooperation*. China.org.cn PhysicsWeb. http://www.china.org.cn/english/features/focac/185148.htm. Accessed: 26 January 2023.

FOCAC Beijing Summit. (2000b). *Program for China–Africa Cooperation in Economic and Social Development*. China.org.cn PhysicsWeb. http://www.china.org.cn/english/features/focac/185 182.htm. Accessed: 26 January 2023.

FOCAC Sharm. (2009). *Forum on China–Africa Cooperation Sharm El-Sheikh Action Plan (2010– 2012)*. Sardc Physics Web. https://www.sardc.net/en/southern-african-news-features/forum-of-China-Africa-cooperation-sharm-el-sheikh-action-plan-2010-2012/. Accessed: 26 January 2023.

FOCAC 8th Ministerial Conference. (2021). *Dakar declaration of the eighth ministerial conference of the forum on China–Africa cooperation*. FOCAC Dakar Physics Web. http://www.focac.org. cn/focacdakar/eng/hyqk_1/202112/t20211222_10474202.htm. Accessed: 26 January 2023.

FOCAC 8th Ministerial Conference. (2021). *Forum on China–Africa cooperation dakar action plan (2022–2024)*. FOCAC Dakar Physics Web. http://www.focac.org.cn/focacdakar/eng/hyqk_1/202112/t20211222_10474206.htm. Accessed: 26 January 2023.

He, F. C., & Ning, S. (1995). *General history of Africa: The ancient*. East China Normal University Press.

Jayaram, K., Kassiri, O., & Sun, I.Y. (2017). *The closest look yet at Chinese economic engagement in Africa*. McKinsey & Company PhysicsWeb. https://www.mckinsey.com/featured-insights/mid dle-east-and-africa/the-closest-look-yet-at-chinese-economic-engagement-in-africa. Accessed: 26 January 2023.

Li, J. L. (2005). Some thoughts on strengthening China–Africa agricultural cooperation. *World Agriculture, 05*, 11–14.

Liang, Y. M., & Li, G. (2020). The impact of Chinese enterprises' investment in Africa on local inclusive growth—A case study of Ethiopia. *Globalization, 6*, 14. https://doi.org/10.16845/j. cnki.ccieeqqh.2020.06.007

Ministry of Foreign Affairs of the PRC. (2006). China's Africa policy paper. *People's Daily,* 01-03.

Ministry of Foreign Affairs, Republic of Indonesia. (1955). *Final Communiqué of the Asian-African conference of Bandung (24 April 1955)*. Asia-Africa speak from Bandung, Djakarta: The Ministry of Foreign Affairs, Republic of Indonesia (pp. 161–169).

State Council Information Office of the PRC. (2014). *China's Foreign Aid (2014)*. The State Council, The People's Republic of China Physics Web. https://english.www.gov.cn/archive/white_paper/ 2014/08/23/content_281474982986592.htm. Accessed: 26 January 2023.

The State Council Information Office of the PRC. (2021). *China and Africa in the New Era: A Partnership of Equals*. The State Council, The People's Republic of China PhysicsWeb. http://english.www.gov.cn/archive/whitepaper/202111/26/content_WS61a07968c6d0df 57f98e5990.html. Accessed: 26 January 2023.

Wang, T. (2011). Analysis on the development history and features of China–African Science and Technology Relationship. *World Outlook, 2*, 117–130. https://doi.org/10.13851/j.cnki.gjzw. 2011.02.009

Wang, T., & Zhang, Y. C. (2012). New point for China–Africa relationship—Review of "China–Africa science and technology partnership program." *Journal of Southwest Petroleum University (social Sciences Edition), 14*(2), 78–83.

Wang, X. (2013). Analysis on situation and suggestions on policy regarding China–Africa science and technology cooperation. *Forum on Science and Technology in China, 8*, 142–146. https:// doi.org/10.13580/j.cnki.fstc.2013.08.023

Xie, Y. X. (2002). *History of contemporary diplomacy of China (1949–2001)*. China Youth Press.

Zhang, Y. H., Wang, T., & Li, H. X. (2012). China–Africa science and technology cooperation: Strategic significance, policy direction, and mechanism system. *Global Review, 5*, 20. https://doi.org/10.13851/j.cnki.gjzw.2012.05.002

Chapter 19
China–Africa Multifaceted Collaboration

Sureyya Yigit

19.1 Introduction

"The essence of neo-colonialism is that the State which is subject to it is, in theory, independent and has all the outward trappings of international sovereignty. In reality its economic system and thus its political policy is directed from outside."

Kwame Nkrumah

China's presence in Africa has become particularly important in recent years due to the economic agreements signed with several countries and the cooperation in aid and development in infrastructure, health, education, and humanitarian assistance. China has pledged not to interfere in political affairs but has tried to establish relations of mutual benefit. China desires a friendly and multipolar international climate to conduct what it defines as its peaceful rise. For this reason, it has fostered the rhetoric of foreign relations based on trust and mutual benefit, equality and respect between countries and cooperation. Located within this framework of action is its policy of foreign aid for development towards Africa. The lack of political conditionality for granting Chinese development aid turns relations with it into an alternative model of cooperation for development. Due to its staunch defence of national sovereignty, Chinese foreign aid is extremely attractive to African countries, which are still highly susceptible to foreign intervention due to their colonial heritage.

The relationship between power and knowledge can define concepts such as development (Yigit, 2022). In this way, the conception of development and its practical application has much to do with the control and discipline from which the concept is generated. Therefore, one can analyse China's contemporary concept of development, which does not envisage a domain separate from its foreign policy concerns and

S. Yigit (✉)
School of Politics and Diplomacy, New Vision University, Tbilisi, Georgia
e-mail: samarkand2020@yahoo.com

© The Author(s) 2024
M. Muchie et al. (eds.), *China-Africa Science, Technology and Innovation Collaboration*, https://doi.org/10.1007/978-981-97-4576-0_19

actively mobilises historical discourses of geopolitics—respect for sovereignty, non-interference in political affairs and anti-hegemonism—and the language of commonality and reciprocity such as solidarity, friendship, and anti-imperialism, to justify its current approach.

19.2   Areas of Inquiry

The structure of this research identifies three main objectives, which can be summarised as:

(i)   Identifying Chinese development aid and its political scope;
(ii) Evaluating the implications of China's presence in Africa for the rest of the world;
(iii) Analysing the underlying reasons for Western criticism of Chinese foreign aid;

The general objective of this work is to demonstrate the importance of China's cooperation in Africa, with the primary aim being to present the importance of the relationship between these two territories, China due to its economic growth and global influence, and Africa as the region with minor economic progress compared to the rest of the world. Therefore, the goal is to expose the relationship between the two and their particularities, especially in Chinese assistance to Africa. The specific objectives include the following:

- Identify the areas of cooperation between China and Africa, whereby African countries benefit;
- Demonstrate the progress or improvements these countries have experienced in agricultural development, education, health, infrastructure, and humanitarian assistance;
- Analyse the cooperation strategies that China has conducted in Africa, highlighting its typical characteristics.

The areas of cooperation and investment projects to be studied are the following: agricultural development, education, health, infrastructure, and humanitarian assistance. For their part, the African states in which China has had the most considerable trading influence are South Africa, Nigeria, Angola, Egypt, and the Democratic Republic of the Congo, as they are the top five trading partners for Beijing (Mureithi, 2022).

Chinese cooperation in Africa has been continuing for decades. It has covered different areas of development that have been key for Africa—one of the regions with the most negligible human development in the world—to advance. This has been significant as Chinese investment represents a strategic advance in crucial sectors for the development of any region, primarily through investments in infrastructure, which is essential for job creation and revitalisation of the African economy. Another area of cooperation is agriculture to achieve food security for the population living in African countries. The African case is overly complex due to the multiple internal and external factors that affect it; thus, investigations of this continent's current evolution are

Table 19.1   African poverty factors. *Source* Open Sources

Political instability	Many governments and violent opposition groups in African countries are in constant conflict, which dampens prospects for long term political stability. It fails to establish minimum democratic guarantees for societal development
Wars	Internal conflicts between violent and terrorist groups that are part of irregular governments, adversely affect the population
Epidemics	Epidemics in African countries are common and constant due to the intense exploitation of resources. Some of the most significant have been AIDS, Ebola, and Covid-19
Weather conditions	Africa presents different natural environments that have been seriously affected by the intensive exploitation of resources. This results in climatic consequences such as floods and droughts which negatively affect agricultural production
Low diversity of economic activity	As the main economic activity, many countries are dependent on the primary sector. Many find themselves in extreme poverty without the possibility of engaging in other productive activities
Increase in external debt	Due to internal crises, many countries have had to borrow significantly; with South Africa, Guinea Bissau, Eritrea, Ghana, Togo, Sierra Leone, Gabon, Congo, Angola, Mozambique, Kenya, and Zambia all having debts totalling more than 70% of their GDP's

increasingly necessary. In order to understand the importance of China's assistance to Africa, it is necessary to identify the main challenges outlined in Table 19.1.

19.3   Political Conditionality and Chinese Cooperation

China–Africa relations have a long history, and with the establishment of the China–Africa cooperation Forum, relations have become closer. Both sides have made progress in cooperation in various fields. A new type of strategic partnership has been established between China and Africa, which excludes any political conditions. Political conditionality emerged and gained prominence at the same time as the coming down of the Berlin Wall in 1989. In the post-cold war era, it was primarily applied by many international organisations, most famously by the European Union (EU), to encourage ex-socialist countries to establish and maintain democracy and protect human rights. When countries emerged from authoritarian pasts that had lasted many decades and were undergoing a transition to democracy during the cold war, international organisations did not implement a policy of political conditionality. As evidence of this, we can highlight the examples of Greece, Portugal, and Spain as good examples of countries that had endured half a century of dictatorship in some cases.

The cold war witnessed the maintenance of trade relationships and detente with the socialist bloc, which was intended to encourage liberalisation and improvements in human rights. Financial assistance to the underdeveloped south was non-political.

When one looks at such policies implemented by international organisations, there are no references to democratic norms such as human rights.

It was only after the end of the cold war that political conditionality was exerted externally to influence the domestic protection of human rights. The cold war attitude of international organisations in their multilateral and western states' bilateral relations consisted primarily of rational mutual gain. In the aftermath, however, rationality as objective analysis gave way to the normative dimension of political conditionality. A normative role allows, permitting, the whole array of tools and devices available to achieve the spread of liberal norms.

The most well-known of these norms become known as good governance. This notion was new as it first surfaced during the fall of the Berlin Wall in the World Bank's report on Sub-Saharan Africa (Kerandi, 2008). Here the crisis in the region was one of a crisis of governance. Given the feeble positive impact of previous policies implemented by international economic organisations such as the World Bank and the IMF, a change in emphasis was considered necessary. The central factors underlying the focus on political conditionality were threefold:

1. Acceptance of financial assistance remaining ineffective,
2. Meagre commitment towards reform of recipient governments,
3. Endemic corruption that remained in developing countries.

A novel approach towards political authorities was deemed necessary, whereby questions were directed at governments' capacity, ability, and willingness to govern effectively in not sectional but the common interest. There was a heightened awareness that the quality of a country's governance system was a vital determinant of the ability to pursue sustainable economic, social, and political development.

According to the World Bank's definitions of governance during the 1990s, one can highlight the three most prominent features that it encompassed:

(a) Form of political regime
(b) Processes by which authority is exercised managing a country's economic and social resources for development
(c) Capacity of governments to design, formulate, implement policies, and discharge functions.

The effectiveness of political conditionality and any other international policy ultimately and truly depends on whether they are applied consistently. Suppose one set of political conditions is applied to one state or a group of states and another to other parties, in that case, the ability of the policy to achieve the desired results becomes much more difficult. Third parties, if they sense, and here perception is vital even if no such duality exists, will be unwilling to comply with the demands made as there will be a lack of trust and expectation that the rewards offered will not be realised. Therefore, any contradictions or double standards associated with political conditionality inevitably decrease the chances of being a successful venture.

One of the factors that made political conditionality possible was, as implied, the ending of the cold war. Pursuing broader "milieu" goals that were impossible due to the international systemic constraints were lifted and norms such as the enlarge-ment of democracy as popularised by President Clinton's National Security Adviser, Anthony Lake, became achievable (Lake, 1994). International organisations, rather than having little choice but to support anti-communist, authoritarian regimes with horrendous human rights records due to their leader's political allegiance, could change their direction and make demands on them if they wished to continue their beneficial relationship.

The initial reaction to political conditionality came from the oldest school of thought in international relations theory, realism. Foreign policy of states had always been hierarchical, with the conception of national interest and the focus on main-taining or maximising power remaining paramount. Changing this alignment by promoting democracy and human rights to the top of the foreign policy determinants list was considered unrealistic. Realists reacted by stating that even if such a change was to be announced, it would only be rhetoric. Namely, human rights or democracy would only be masking other, more traditional interests.

It was reminded that political conditionality was a norm, a standard of behaviour, due to that fact, it was competing against other interests (Schimmelfennig, 2007). Further criticism was levelled at political conditionality as assertions were made that the international organisations and other donors only used this new policy in their interests; the genuine interests of third-party populations intended to be ensured through political conditionality would not be ultimate, that the developing states' citizens interests would not be increased through the use of political conditionality (Crawford, 1997).

As foreign aid budgets were reduced despite the benefits of the peace dividend after the cold war ended, this new direction was simply an excuse for the western states and influential international organisations to reduce their financial assistance to strategically or commercially unimportant states. The ideological straitjacket was ripped open and such actors could now direct their policies towards what was more to their interests rather than the military alliances to which they belonged. Counteracting this argument was the defence put forward by the Democratic Peace theory, primarily through Doyle, whereby the claim that democracies not going to war with each other made political conditionality the most viable policy (Richmond, 2008). Who could argue against the aim of the policy—which was the creation and maintenance of democracies, which would make the world a safer—less conflictual place? A distinction was made about political conditionality. Positive political conditionality consisted of a positive attitude; that of promising and delivering benefits or rewards. Negative political conditionality similarly offered a harmful type of behaviour; that of threatening and reducing or even terminating rewards.

Whilst such actions and intended policies reminded one of linkage politics, it was different since political conditionality was much broader in its objectives. General political reform, improvement in human rights, and the maintenance of the rule of law were involved and invoked. Furthermore, its second difference from linkage politics stemmed from the range of actors it was directed against. Here again, political conditionality cast its net much more expansive than linkage politics which is only concerned with communist states. Political conditionality knew no limits; it covered, in theory, at least the entire world; it was meant to be universal and not selective, as that would mean double standards.

European countries regarded that expressing concern about human rights abuses and widespread violations could no longer be seen as an intervention in the domestic affairs of a state since over one hundred states had signed the UN conventions on civil and political as well as economic and social rights (Macfarlane et al., 2004). Relying on this global situation, a powerful argument was made that such interest and action could not be deemed as external intervention. It violated the central pillar of the contemporary international system, namely sovereignty.

When political conditionality came to be applied by the European Community and later by the European Union, internal and external criticism arrived at the same principle of insufficient democracy. Third parties complained of the EU suffering from a lack of democracy and not possessing its bill of rights. It was forcefully asserted that such an actor could not ask other actors what it lacked. Equally, internal critics labelled Eurosceptics vehemently accused the EU of a democratic deficit. Until the EU solved its central problem, its credibility abroad could never be what it ought to be.

Different grey areas emerged regarding the judgment of the political conditionality criteria. One problematic issue concerned the definition and understanding of democracy. What exactly was democracy meant to imply? To some, it meant a multi-party system. For others with a communist past, the transition from a people's democracy rooted in orthodox Marxism pursuing social equality meant developmental democracy attempting to broaden popular participation in decision-making, advancing freedom, and flourishing individualism (Jankovic, 2016). Furthermore, whereas the demandeurs had enjoyed democratic governance for many decades, they had also witnessed precarious developments in the early days of their democratic establishment. The states attempting to meet the criteria imposed through political conditionality were eager for the democratic states to recall the difficulties associated with establishing such norms.

Within international organisations and the industrialised states, a schism in strategy emerged. There was consensus on trying to offer aid to the populations of the transition countries, especially the poorest sections. The divergence of opinion and strategy focused on what action to take when political conditionality was not met. The declared move was for aid to be suspended. Some argued that the termination of aid only punished the poorest for the sins committed by their political leaders (Carey, 2007). Theoretically, this reflected the preferences of asphyxiation, which advocated the blockage of economic flows, which was intended to halt unacceptable behaviour by some states (Lavin, 1996). The choice of oxygenising was the opposite, whereby only enhanced economic activity could be expected to induce positive political consequences.

Understanding that developing countries should have room to determine their policies for meeting such aims as the Millennium Development Goals (MDG) and utilising aid more effectively if they could genuinely rely on it as part of their long-term budget plans came to be accepted. The partnership was based on an open dialogue with mutual rights and responsibilities where each party would be committed to transparency. Regarding the use of conditions in reducing poverty it clearly stated that the terms for aid should support and not buy reform (Radelet, 2004). These incentives should encourage a genuine policy of reform conducted by the recipient states. Aid should not be seen or offered as a reward or payment in exchange for reform (Erbeznik, 2011). Terms and conditions of aid had to be strongly linked to benchmarks. Furthermore, conditions to strengthen broader public participation were included, especially among the poor.

Considering such actions, one can state that from the mid-1990s onwards, there was a significant evolution in aid relationships, which naturally has implications for the appropriate role of political conditionality. As mentioned, the MDG provided a new framework for development based on a different kind of partnership. Poverty reduction became paramount in terms of development assistance. This innovative approach emphasised inclusive development, prioritising poor people, with recipient governments taking policy responsibility. Secondly, despite the acceptance of macroeconomic policies remaining vital, widespread agreement focused on the effects of good governance in reducing poverty and conflict. The role of democratic, participatory processes in developing plans for reducing poverty was especially encouraged. An increasing acceptance of the view that better political and economic governance in terms of quality positively correlated with policy outcomes. Countries with excessive military spending and endemic corruption affected foreign investment and the delivery of public services. Due to these observations, states and international economic organisations attached conditions linked to political and institutional change in their roles as donors. Political conditionality attempting to buy reform from an unwilling recipient had rarely worked, and therefore such a philosophy had to change.

To summarize, regarding political conditionality, African states were included in this reinforcement strategy used by international organisations and other international actors in bringing about and stabilising change at the state level. After the dissolution of the Union of Soviet Socialist Republics (USSR), the promotion of democracy, human rights and the rule of law was within the remit of political conditionality. In applying political conditionality, the western international community primarily set the adoption of liberal-democratic norms for African states as conditions for rewards. These rewards were threefold:

- Political—in terms of military protection
- Social—international recognition or public praise by international originations
- Economic—financial assistance and trade liberalisation.

Therefore, given this situation, cooperation between China and Africa and cooperation between the two sides to become more diverse gathered more outstanding political support. Relations between China and African countries have endured for more than half a century, maintaining solid foundations and mutual support; and observing cooperative and close bilateral relations that positively impact the economic, social, and political fields yielding fruitful results. Relations between China and sub-Saharan

African countries present a strategic vision of importance to both parties. The cooperation strategy of China has profound economic, social, and commercial effects on the eagerness for its integration with the world economic system.

Looking into the future, China–Africa relations promise a prospect of pervasive cooperation (Muchie & Patra, 2020). From this, with the further development of China–Africa relations, the common interests of both sides may continue to increase in the coming years, and this international influence may also increase. The strategic importance of the new China–Africa strategic partnership has become increasingly prominent in recent years.

Due to its strong international presence, China has become a source of international cooperation. It plays a significant role in being active and constructive regarding different problems that affect the world today. One of the fundamental pillars of China that makes possible the ability to cooperate is Chinese banking entities, which support different aid projects. On the other hand, to improve its image in the international context, in recent years, China has dedicated itself to maintaining and deepening bilateral relations with other countries, conducting policies of cooperation and friendly exchanges (Yigit, 2022a). This is important since it advocates mutual relations based on peace, stability, and trust. The attributes of China's Cooperation in Africa present specific characteristics, of which one can identify four:

(i)   Principle of non-interference referring to the non-imposition of conditions in the formalisation of cooperation agreements and keeping China out of internal political affairs;

(ii) Promoting the self-determination of developing countries through their capacities aimed at self-efficiency. Countries should be free to choose their development model without the political intervention of China;

(iii) Peaceful coexistence between China and the host countries;

(iv) Adaptation of China to the real needs of the beneficiary country; without this requirement, cooperation would not have a definite goal and ultimately fail.

The types of aid China has been conducting in Africa are summarised in Table 19.2.

The fields of action are broad and complementary; at the same time, they form the basis by which China can conduct various activities since it allows for a wide range of actions. Thus, the project and technical cooperation areas are strategic for the beneficiary countries to achieve self-determination and self-development. For example, many African countries depend on agriculture for survival; therefore, technical assistance for improving such processes becomes vital.

19.4   South–South Cooperation

The 2008 Accra Agenda for Action declared that South–South cooperation for development aims to uphold the principle of non-interference in internal affairs, equality among developing partners and respect for their independence, national sovereignty,

Table 19.2 Types of Chinese Aid in Africa. *Source* Open Sources

Human resource development	This type of cooperation is conducted through multilateral and bilateral channels in research and training program exchanges between China and participating countries
Medical resources	Shipping of medical equipment, medical services, and free medicines. These services are provided in underdeveloped areas where local doctors are trained
Humanitarian aid	Offered to countries that have suffered severe natural or humanitarian disasters. Help is through materials or monetary aid. Volunteer staff is also sent if necessary
Projects	Civil projects financed by China and conducted in the recipient country. Includes materials and specialised personnel. Projects represent the proportion of China's broader spending on cooperative relations with Africa
Goods and materials	Goods, materials, equipment, personnel specialised in projects
Debt relief	Suspension of debt for certain countries
Technical cooperation	Dispatch of specialised technical personnel to recipient countries to undertake specific collaboration tasks especially in productive areas such as agriculture, energy, hydrocarbons, and industry

cultural diversity, identity, and local content (Sridhar, 2009). The traditional objective of South–South Cooperation has been to break with Western global domination, rising as a counter-model to imperialism during the Cold War. South–South Cooperation has provided Africa with the necessary political and psychological impetus for its emancipation and development by reducing its dependence on and intellectual domination by the West.

The widespread use of the concept of South–South Cooperation in defining China's current relations with Africa contrasts favourably with the traditional post-colonial image of European policies in Africa (Yigit, 2024). There has been a rapprochement between Chinese and African positions, which is in tune with the underlying politics and ideological imperatives of South–South Cooperation in challenging Western domination and defending the interests of developing countries based on solidarity. The mere fact of the global economic impact of the rise of China, and its commitment to Africa, has provided African countries with more economic opportunities for trade and investment, as well as greater room and political alternatives from which to challenge Western agendas and development aid prescriptions. In this sense, it may be asserted that one of the critical objectives of South–South Cooperation has already been fulfilled.

However, South–South Cooperation, as it arises in the relationship between China and Africa, has proven to be a challenge for African countries concerning relations with the North. This is because although South–South Cooperation seeks to break with the commercial chains that have dominated the global economy since colonial times—through which Africa's resources have been plundered and made dependent on the West—the continent must be wary that dependency does not simply shift to China. The innovative aspect of Sino-African relations is that China does not try to seize African natural resources with finances. However, it pays for it through infrastructure, and critical aid projects which become paramount. It becomes essential that African states ensure their national interests are defended and not compromised in their dealings with China. Nevertheless, China's involvement in Africa has allowed triangulation to flourish, which means that African states can seek relations with more than one external state, as reflected in Angola seeking China's help in negotiations with the International Monetary Fund in 2003 and thus expanding its capability to confront donors and investors (Power & Mohan, 2010). The definition of Chinese cooperation is complex, but a historical continuity can be observed from the founding of the People's Republic in 1949 to its current diplomacy, which seems to suggest that the guiding principles of Chinese foreign aid in the past have remained constant. China's State Council issued a white paper on China's Foreign Aid which became the framework from which aid for cooperation was articulated (State, 2011). It presents the essential elements for cooperation on which Chinese foreign management is based.

The first of the five basic principles are to promote a self-management mechanism, constantly improving its foreign aid work. For the implementation of development aid projects, China offers diverse ways, based on the development of host countries based on their presence in the white paper of 2011; China will assist host countries in laying the foundation for future development. It allows them to move onto the path of self-sufficiency and independent development. The lack of political conditionality detailed earlier occupies second place in the list of principles. China upholds the Five Principles of Peaceful Coexistence and the right of host countries to choose their development model independently. Thus, China dissociates itself from using foreign aid to interfere in the internal affairs of recipient countries or seek political privileges for itself. Third, adhering to the principle of equality, mutual benefit and joint development, China maintains that foreign aid will focus on practical effects, accommodate the interests of recipient countries, and strive to promote friendly and mutually beneficial bilateral relations through economic and technical cooperation with other developing countries. This typology is divided between complementation of projects, material goods, technical cooperation, human resource development, Chinese teams working abroad, emergency humanitarian aid, volunteer programs abroad and debt relief.

China is not familiar—or at least uncomfortable—with the notion of development cooperation policy as its independent policy, according to the historical evolution that it has had in Western countries since the 1950s. For this reason, its development aid is not managed by a single ministry. However, there is a complex network of delegations that deal with it. Moreover, aid for development cooperation does not come only from the central government. However, since 1971 all the provinces and the most critical municipalities have had economic and technical cooperation offices to help conduct cooperation projects. Also, maintaining a realistic approach is crucial, hence the fourth principle. China maintains that it does its best to tailor its aid to the actual needs of recipient countries and provides foreign aid within its capabilities and according to China's internal conditions. Finally, adapting to the times and historical context, and paying attention to reform and innovation, is the fifth principle for China. To this end, China adapts its foreign aid for development to domestic and international situations, pays attention to experiences and innovations in the field of foreign aid, and aims to adjust and accordingly proceed with required fine-tuning quickly (Trinidad, 2013).

It is difficult to ascertain the total amount of Chinese development aid due to a lack of official data as the Chinese government, unlike traditional donors, did not publish official statistical data on foreign aid until 2011 (Regilme & Hodzi, 2021). One thing is clear; however, China's reluctance to publish aid statistics does not mean that China does not know how much aid it is giving. Instead, aid is part of a tightly controlled government information system. However, at times declarations are made concerning this, as in 2008, when Chinese Premier Wen Jiabao announced that China had disbursed a total of more than $56 billion in aid to all developing countries between 1950 and 2012 (Zhang, 2014). The IMF estimates that China has signed more than a thousand loan commitments worth US$153 billion over 2000–19 (Mandon & Woldemichael, 2022).

The typology of aid to Chinese cooperation distinguishes three main types, also included in the 2011 white paper. First, interest-free loans are granted to construct large infrastructures such as dams or roads. They are delivered to developing countries which demonstrate better economic performance. Second, concessional loans— the Chinese commonly call preferential loans—are low-interest loans intended to help recipient countries develop productive projects that generate economic and social benefits. Finally, the white paper establishes subsidies to help recipient countries construct infrastructure such as hospitals, schools and low-cost housing, health systems, drinking water projects, or projects that benefit society. In addition, they are used in cooperation projects to develop human resources, technical cooperation, and emergency humanitarian aid.

It is evident from the economic point of view that China has a strong presence in African countries thanks to direct foreign investment and loans, which have been constant for over two decades. China has a presence in all African countries, both in the form of investment and with the functioning of Chinese Small and Medium-sized Enterprises (SMEs) in the territories of Africa. Additionally, the presence of Chinese SMEs is concentrated in the extractive industry, with the project's infrastructure outsourced to suppliers from China and Chinese labour (Ajakaiye & Kaplinsky, 2009). In turn, Chinese foreign trade to Africa benefits in three critical sectors: textiles, transportation, and infrastructure.

19.5   Chinese Cooperation in Africa

The last FOCCAC meeting, held in May 2021, resulted in an action plan—developed from the Beijing Declaration—which highlighted two main aspects. On the one hand, the fight to halt the pandemic created by Covid-19, for which China committed to supplying vaccines destined for African countries. In this respect, China constructed hospitals to care for Covid-19 patients (Rudolf, 2021). To the most affected countries, China suspended debt repayment. On the other hand, in order to maintain economic openness, officials and Chinese experts have remained in African countries in strategic positions to support the economy, especially in projects that are related to the BRI, in which 46 African countries and the AU Commission have signed cooperation agreements with China under the framework of this initiative (Addaney, 2021).

As can be seen, China's action in Africa is decisive concerning assistance to combat Covid-19 and to assist in the economic recovery of African countries most affected by the pandemic. At the same time, this allows China to recover part of the essential business activity for both parties. One can categorise four strategic implications of the new China–Africa partnership:

19.5.1 Peace Promotion

Throughout history, Western settlers have exploited African countries and, after independence, have imposed an unfair and unreasonable role in the international order. In addition, African countries also experienced political unrest, poverty, wars, and natural disasters. These factors slowed the economic development of African countries and reduced their international influence. The standard of living of the population remained relatively low. Since China and Africa maintained a high degree of coherence in handling affairs in the international community, the relationship between the two parties developed, demonstrating a positive example of South–South cooperation.

19.5.2 Sustained Economic Growth

As the scale of cooperation continues to expand trade and financial relations between China and African countries, the two sides are increasingly interdependent. Today, Africa has become an essential source of import resources and an export destination for Chinese commodities. Bilateral trade and investment between China and Africa continue to rise, and economic cooperation between China and African countries is increasingly plentiful. Since 2009, China has been Africa's largest trading partner (Shinn, 2019). The trade volume between China and Africa reached US$10.6 billion in 2000, increasing to US$126.9 billion in 2010 and reaching an all-time high in 2021 with $254 billion (Olander, 2022). The rapid growth in the trade volume between China and Africa has significantly impacted African countries and has gradually increased its impact on China's economic development. In this respect, the increase in trade is an essential part of the relations between China and Africa, which is established through economic assistance from China to African countries that require it, improving their conditions. At the same time, the increase in imports from Africa to China is also promoted, stimulating the development of African businesses from the sectors that have received technical assistance.

19.5.3   Maintenance of Sovereignty and Mutual Interests

In recent years, the international presence of African countries has further improved, and their role in the international community has become increasingly relevant in terms of peace and development. China–Africa relations have also become a salient point for the former. The Chinese government has attached immense importance to Africa's key role in China's diplomacy. China has proposed new measures and goals to develop the relationship between China and Africa further, maintaining the steady, rapid, and healthy development of friendship between the two sides, which has effectively enhanced the influence of China's international level of cooperation among developing countries.

19.5.4   Diversified Relationships to Achieve Win–Win Outcomes

With a multipolar world on the horizon and a globalising economy, the world is intricately connected, though, at the same time, problems such as asymmetric economic development continue unabated. For developing countries, the task of peace and development becomes even more critical. Therefore, strengthening South–South cooperation, promoting North–South relations, uniting for self-sufficiency becomes an attractive choice for developing countries. The African Unity for Renaissance and Knowledge Exchange series of conferences initiated by Mammo Muchie is a notable contributor to this endeavour (Nzewi & Maramura, 2021). In addition to South–South cooperation, there is a greater incentive to use their advantages to promote general social progress. The cooperation and friendship between China and African countries have become a potential model of South–South cooperation, which benefits both China and Africa. They may play a vital role in promoting peaceful global development.

19.6   Improving Education

Development cannot be complete if the beneficiary society does not advance its human capital. In this regard, China and the African countries that are part of cooperation have evolved in the field of education, beginning in the 1950s, with exchange programs for students and teachers. In the 1960s, scholarships were awarded to train

African professionals in different fields. These scholarships rose significantly after the millennium. However, the level of education in Africa continues to lag. Children who do not attend school, inequality in primary education, and the mismatch between supply and demand for vocational education are significant problems facing African education. The gap between the rich and poor in Africa is also unacceptably wide. In general, most education and training programs in Africa suffer from low-quality teaching, inequality, and exclusion at all levels. While the number of children with access to primary education has increased significantly, many do not attend school.

For countries aspiring for global leadership, a critical area of foreign policy is aid, and educational assistance is undoubtedly one of the crucial ways to cultivate a country's soft power. Africa is a key area for educational assistance from China. China's educational assistance through the granting of various scholarships and the construction of African schools has not only effectively promoted the development of education in Africa, but it has also become the promoter of education in China: an essential avenue for Africa's soft power.

For China, strengthening cooperation with Africa in education also possesses many benefits. On the one hand, improving the quality of the African population will help Chinese enterprises in Africa reduce the cost of human resources and promote deeper China–Africa cooperation in more significant fields such as investment, trade, production capacity and technology. On the other hand, compared to cooperation in infrastructure, education cooperation has a more decisive humanistic attribute, which will help the people of China and Africa to connect. It will facilitate the exchange and mutual learning of civilisations between China and Africa which will have a long-term strategic impact.

In this regard, China and Africa can have an excellent, mutually beneficial cooperation in education. On the one hand, Africa has a high demand for education and countering this, China has made many achievements in the field of education—especially in the field of primary education, such as literacy, the popularisation of compulsory education and the teaching of mathematics and science—and has accumulated many experiences and methods that African developing countries can use as a reference.

China grants significant importance to improving people's livelihoods and living standards, and education in Africa alongside assistance in infrastructure, agriculture, forestry, fishing, and energy production (Broich, Szirmai & Adedokun, 2020). Through technical assistance in Africa, China has raised African countries' scientific and technological levels in industrial production and management, energy, culture, education, medicine, and health (Yigit, 2013).

It is self-evident that education's importance is exceedingly high. It is also true that helping construct and maintain school buildings, providing teaching teams, training teachers, increasing the number of scholarships to study in China, supporting the development of education technical and vocational education, and training, African civil servants, are vital. All these activities promote the improvement of people's livelihoods in Africa and strengthen assistance to people's livelihood in areas such as health care and agriculture.

19.7 Support Medicine and Health

Medical and health care is one of the critical areas of aid for China. By assisting in constructing hospitals, providing medicines and medical equipment, dispatching medical equipment, training staff doctors, and developing disease prevention exchanges, China supports recipient countries in improving other medical and health conditions and disease prevention and control (Chen et al., 2019). Furthermore, it strengthens capacity building in public health, earning its medical assistance a good reputation in Africa.

When the Covid-19 pandemic broke out and spread rapidly with confirmed cases in many African countries, China provided aid not only in terms of epidemic prevention materials but also various forms of bilateral cooperation, such as professional training for medical personnel, to help fight the spread of the epidemic in Africa (Zhao et al., 2020). Over the past few years, China's bilateral assistance has ranged from emergency assistance to sustainable public health cooperation in the long term.

19.8 Infrastructure Construction

Strengthening infrastructure construction is not only the most effective way to boost the rapid growth of the national economy and improve the general level of national economic development but also a strategy to improve the population's living environment and achieve socially sustainable development. Strengthening infrastructure construction is key to ensuring that Africa's rich natural resources can be transformed effectively into becoming the driving force for economic development. It is also a prerequisite to help Africans get rid of poverty and improve their quality of life as soon as possible.

Lagging infrastructure is a significant problem that restricts the development of developing countries in Africa, with infrastructure such as electricity, water supply, roads and information technologies information and communications remaining precarious and insufficient (Asongu & Odhiambo, 2019). Building infrastructure in African countries requires funding, technology, and sustainable expert support. Africa's current information and communication technology situation are highly unbalanced, and the general level is low. The relative backwardness of technology has weakened Africa's economic development momentum and competitiveness and hampered the pace of infrastructure development. If African countries want to develop facilities for communication services and keep up with the information era, they urgently need to solve the existing technology gap and adjust to Industry 4.0 (Bongomin et al., 2020).

In terms of supporting the construction of infrastructures in Africa, such as supporting China's participation in international cooperation and the implementation of foreign aid, the China Export–Import Bank has played an active role (Zhang, 2020). Since its creation in 1994, the Bank has attached importance to supporting

China–Africa cooperation, providing loan-based development assistance, preferential loans, and investment loans abroad under effective risk control; it has played a vital role in relieving capital from insufficient development in African countries and gradually gaining the support of sustainable capital.

19.9   Increased Humanitarian Assistance

In recent years, earthquakes, typhoons, floods, droughts and other natural disasters and humanitarian disasters triggered by wars led the affected countries to endure severe material losses. China responded actively to the international community's calls, providing emergency relief supplies or timely cash assistance and dispatching rescue teams and medical teams as needed to help affected countries reduce the impact of the disaster and rebuild their homes as soon as possible.

The internationalisation of Chinese society begins with the provision of humanitarian aid supplies, as the China Foundation for Poverty Alleviation demonstrated in the tsunami in Indonesia (Li & Dong, 2018). Regarding aid to Africa, delivering humanitarian aid supplies is a temporary act of Chinese social organisations. It is also an area where Chinese social organisations conduct long-term projects. Temporality manifests itself in the rescue of public crises caused by various emergencies, as witnessed in 2001, with heavy rains and floods in Angola causing many deaths and displaced people. The Chinese Red Cross made donations to the Angolan Red Cross to help the victims of the floods in Angola.

Humanitarian assistance has been established as a crucial means of developing public diplomacy (Yigit, 2012). Given that Africa is a most significant concern to China, and China often provides humanitarian assistance to Africa, it has paid more attention to humanitarian aid work, the scale and speed of which are slowly increasing.

19.10   Conclusion

The benefits of aid to Africa cannot be measured in resources and financial flows alone. Africa has abundant natural and human resources, while China has applicable technology and experience. In addition, the economies of China and Africa are complementary. One can assert that cooperation between the two sides continues to possess enormous potential. China seeks mutual or bilateral interests, and in the case of Africa, it seeks mutual benefit through practical results, interaction, and progress for both parties. Beijing's rapprochement with Africa has emerged to jointly explore new ways of interacting to expand economic and commercial cooperation.

China and Africa share a relationship and constant long-term cooperation, a model of mutual assistance demonstrating a significant benefit among developing countries.

China's knowledge is helpful for African countries. It is a truism that China's development experience cannot be fully replicated in Africa, but both sides can learn from each other. China–Africa cooperation has always intended to be mutually beneficial bi-directional, either at the governmental or economic level since both are considered developing regions. Cooperation has come about thanks to highly organised joint efforts and the provision of specialised institutions. This allows the aid provided by China in Africa to be particular and represent support to areas where they are most needed.

In the final analysis, China's participation in Africa must be considered from all angles, both the tangible benefits it brings to Africa as well as China's concrete and strategic interests. Therefore, the balance of this relationship has been positive for Africa since it influences its development, and it has been positive for China since it has found a variety of allies in a comprehensive and diverse set in different African countries. The future of this relationship can continue to be one of deepening and development between the two if it succeeds in building a better functioning organisation and increasing the significant fields of cooperation.

It is impossible to conclude any study which focuses on African external relations, especially concerning China, without giving the final word to Mwalimu, where he declared more than half a century ago: "You don't have to be a Communist to see that China has a lot to teach us in development. The fact that they have a different political system than ours has nothing to do with it" (Robinson, 1970).

References

Addaney, M. (2021). *Cascading climate impacts and Africa's engagement with China's Belt and Road Initiative.*

Ajakaiye, O., & Kaplinsky, R. (2009). China in Africa: A relationship in transition. *The European Journal of Development Research, 21*(4), 479–484.

Asongu, S. A., & Odhiambo, N. M. (2019). Challenges of doing business in Africa: A systematic review. *Journal of African Business, 20*(2), 259–268.

Bongomin, O., Nganyi, E. O., Abswaidi, M. R., Hitiyise, E., & Tumusiime, G. (2020). Sustainable and dynamic competitiveness towards technological leadership of industry 4.0: Implications for East African community. *Journal of Engineering, 2020,* 1.

Broich, T., Szirmai, A., & Adedokun, A. (2020). Chinese and Western development approaches in Africa: Implications for the SDGs. In: *Africa and the sustainable development goals* (pp. 33–48). Springer.

Carey, S. C. (2007). European aid: Human rights versus bureaucratic inertia? *Journal of Peace Research, 44*(4), 447–464.

Chen, J., Bergquist, R., Zhou, X. N., Xue, J. B., & Qian, M. B. (2019). Combating infectious disease epidemics through China's Belt and Road Initiative. *PLOS Neglected Tropical Diseases, 13*(4), 1.

Crawford, G. (1997). Foreign aid and political conditionality: Issues of effectiveness and consistency. *Democratization, 4*(3), 69–108.

Erbeznik, K. (2011). Money can't buy you law: The effects of foreign aid on the rule of law in developing countries. *Indiana Journal of Global Legal Studies, 18*(2), 873–900.

Jankovic, I. (2016). Das tocqueville problem: Individualism and equality between democracy in America and Ancient Regime. *Perspectives on Political Science, 45*(2), 125–136.

Kerandi, A. M. (2008). Governance Agenda for Sub-Saharan Africa: Issues and challenges. *Federal Governance, 5*(1), 1–22.

Lake, A. (1994). A strategy of enlargement and the developing world. *Foreign Policy Bulletin, 4*(4–5), 91–94.

Lavin, F. L. (1996). Asphyxiation or oxygen? The Sanctions Dilemma. *Foreign Policy, 104*, 139–153.

Li, X., & Dong, Q. (2018). Chinese NGOs are "going out": History, scale, characteristics, outcomes, and barriers. In: *Nonprofit policy forum* (Vol. 9, No. 1). De Gruyter.

Macfarlane, S. N., Thielking, C. J., & Weiss, T. G. (2004). The responsibility to protect: Is anyone interested in humanitarian intervention? *Third World Quarterly, 25*(5), 977–992.

Mandon, P., & Woldemichael, M. M. T. (2022). *Has Chinese aid benefited recipient countries? Evidence from a meta-regression analysis.* International Monetary Fund.

Muchie, M., & Patra, S. K. (2020). China–Africa science and technology collaboration: Evidence from collaborative research papers and patents. *Journal of Chinese Economic and Business Studies, 18*(1), 1–27.

Mureithi, C. (2022). *Trade between Africa and China reached an all-time high in 2021* [online] Quartz. Available at: https://qz.com/africa/2123474/China-Africa-trade-reached-an-all-time-high-in-2021/. Accessed: 16 October 2022.

Nzewi, O. I., & Maramura, T. C. (2021). A big picture perspective of the decolonization of public administration debate in Africa: Looking back and looking forward. *South African Journal of Higher Education, 35*(5), 204–215.

Olander, E. (2022). China–Africa trade in 2021 amounted to $254 Billion, Breaking an All-Time Record. [online] *The China Global South Project.* Available at: https://chinaglobalsouth.com/2022/01/19/China-Africa-trade-in-2021-amounted-to-254-billion-breaking-an-all-time-record/. Accessed: 16 October 2022.

Power, M., & Mohan, G. (2010). Towards a critical geopolitics of China's engagement with African development. *Geopolitics, 15*(3), 462–495.

Radelet, S. (2004). *Aid effectiveness and the millennium development goals.* Center for Global Development Working Paper (39).

Regilme, S. S. F., Jr., & Hodzi, O. (2021). Comparing US and Chinese foreign aid in the era of rising powers. *The International Spectator, 56*(2), 114–131.

Richmond, O. P. (2008). Welfare and the civil peace: Poverty with rights? In: *Whose peace? Critical perspectives on the political economy of peacebuilding* (pp. 287–301). Palgrave Macmillan.

Robinson, D. (1970). *100 most important people in the world today.*

Rudolf, M. (2021). *China's health diplomacy during Covid-19: The Belt and Road Initiative (BRI) in action.*

Schimmelfennig, F. (2007). European regional organizations, political conditionality, and democratic transformation in Eastern Europe. *East European Politics and Societies, 21*(1), 126–141.

Shinn, D. H. (2019). China's economic impact on Africa. In: *Oxford research encyclopaedia of politics.*

Sridhar, D. (2009). Post-Accra: Is there space for country ownership in global health? *Third World Quarterly, 30*(7), 1363–1377.

State Council. (2011). *White paper on China's foreign aid.*

Trinidad, D. D. (2013). The foreign aid philosophy of a rising Asian power: A Southeast Asian view. In: *A study of China's foreign Aid* (pp. 19-45). Palgrave Macmillan.

Yigit, S. (2012). Olimpiyat, Yumuşak Güç ve Ölçekler. [online] ORSAM. Available at: https://www.orsam.org.tr/tr/olimpiyat-yumusak-guc-ve-olcekler/. Accessed: 16 Oct. 2022.

Yigit, S. (2013). *Chinese energy diplomacy in central Asia* [online] ORSAM. Available at: https://www.orsam.org.tr/en/chinese-energy-diplomacy-in-central-asia/. Accessed: 16 Oct. 2022.

Yigit, S. (2022). EU-Central Asian Civil Societal Relations: Unrealistic Expectations, Discouraging Results, Special Issue "EU-Asia Pacific social and cultural dialogue: involving civil society in interregional relations" of the Journal Cuadernos Europeos de Deusto/Deusto Journal of European Studies.

Yigit, S. (2022a). Mongolia-China-Russia Economic Corridor: Mongolian Benefits as China Looks North, 2. International Cappadocia Scientific Research Congress, Nevsehir, Turkey, June 2022.

Yigit, S. (2024). Pan-African Unity, in 3. International Cankaya Scientific Studies Congress February 28–29, 2024 / Ankara-TÜRKİYE, The Proceedings Book, Editor Prof. Dr. Gökhan ACAR, IKSAD Publications, Issued: 15.03.2024 ISBN: 978-625-8254-35-8, p.108–124

Zhang, D. H. (2014). *China's second White Paper on foreign aid: impressive growth in 2010–12*. Development Policy Centre.

Zhang, D. H. (2020). Chinese foreign aid and financing: An example of new development assistance? In: *New development assistance* (pp. 167–180). Palgrave Macmillan.

Zhao, Z., Li, X., Liu, F., Zhu, G., Ma, C., & Wang, L. (2020). Prediction of the COVID-19 spread in African countries and implications for prevention and control: A case study in South Africa, Egypt, Algeria, Nigeria, Senegal and Kenya. *Science of the Total Environment, 729*, 138959.

Sureyya Yigit Professor of Politics and International Relations at the School of Politics and Diplomacy, New Vision University, Tbilisi, Georgia. Senior Consultant to the Zhenskaya Demokratichyskaya Set Kyrgyzstana (ZDS) Women's Democracy Network in the Kyrgyz Republic since 2013 and a consultant to London-based Aeropodium since 2018.

Part VI
China-Africa Collaboration in Digital Technologies

Chapter 20
China–Africa Collaboration in Digital Technology: Present and Future Challenges

Mingfeng Tang, Xiaomeng Liu, Mammo Muchie, and Angathevar Baskaran

20.1 Background of China–Africa Digital Technology Collaboration

The rapid development of modern information technology has promoted the emergence of new technologies, new products and new commercial modes. In recent years, digital technology plays an important role in serving as the engine of economic development and has been becoming an indispensable technology to enhance economic development and international competitiveness for countries' strategic decision-making around the world. Being the world's second largest and populous continent, Africa has a lot of potential in developing digital economy. The Agenda 2063 of

M. Tang (✉) · X. Liu
Sino-French Innovation Research Center (SFIRC), Faculty of Business Administration,
Southwestern University of Finance and Economics, Chengdu, China
e-mail: tang@swufe.edu.cn

X. Liu
e-mail: 542389573@qq.com

M. Muchie
DSI/NRF SARChI Research Chair on Science, Technology and Innovation Studies, Tshwane
University of Technology, Pretoria, South Africa
e-mail: muchiem@tut.ac.za

A. Baskaran
Department of Political Science, Public Administration and Development Studies, Faculty of
Business and Economics & University of Malaya North-South Research Centre (UMNSRC),
University of Malaya, Kuala Lumpur, Malaysia

SARChI (Science, Technology and Innovation Studies), Tshwane University of Technology,
Pretoria, South Africa

A. Baskaran
e-mail: baskaran@um.edu.my

© The Author(s) 2024
M. Muchie et al. (eds.), *China-Africa Science, Technology and Innovation
Collaboration*, https://doi.org/10.1007/978-981-97-4576-0_20

Africa signifies African countries commit to developing their information communication technology (ICT) and digital economy (Adonu, 2021). China and Africa both are developing economies and have keeping cooperation relationships since 1950s (Liu & Luo, 2021). The inauguration of Forum on China–Africa Cooperation (FOCAC) in 2000 in Beijing further deepens mutual trust and equality-based cooperation between China and African countries to meet the challenges of economic globalization in search of common prosperity.[1] Since 2000, China–Africa cooperation has extended from financial aid and trade to all fronts (e.g. telecommunication, green energy, infrastructure investment, technical support and digital economy). For example, in 2013, China put forward 'One Belt and One Road' initiative. In December 2015, South Africa became the first African country to sign 'One Belt and One Road' cooperation document with China. Africa has become one of the most important and actively participated regions to respond to 'One Belt and One Road' initiative in the world. There are broad prospects for China–Africa joint construction of 'One Belt and One Road'. According to the latest data released by the General Administration of Customs of China, in 2021, the total bilateral trade volume between China and Africa reached a new high. China has maintained the status of Africa's largest trading partner for 12 consecutive years. The total bilateral trade volume between China and Africa reached US $254.3 billion, an increase of 35.3% year-on-year. Among them, exports from Africa to China reached US $105.9 billion, an increase of 43.7% year-on-year. The rising trade between China and Africa bucked the trend to promote Africa's economic growth and enhanced the resilience of Africa's economy to withstand the challenge under the pandemic. According to the report released by The Swiss African business circle Association in February 2022, China has been the largest investor in Africa in the past 10 years. During the past 10 years, China has created 18,562 new jobs for Africa on average every year.[2]

The Forum on China Africa Cooperation—Beijing action plan (2019–2021) proposed, China and Africa share their experience and jointly grasp the opportunities in digital technology development and encourage enterprises to collaborate in the fields of information and communication infrastructure, the Internet, digital technology and so on.[3] Digital technology collaboration between China and Africa not only provides new opportunities for African countries to equally integrate into the global industrial and value chain, but also creates convenient conditions for China to share the benefits of digital transformation. The collaboration between the two sides faces a rare historical opportunity encounter. Vera Songwe, executive secretary of the United Nations Economic Commission for Africa, pointed out digital technology is the highlight of future cooperation between Africa and China and Africa needs to bridge this digital divide and is eager to learn from China in this regard. The 'One

---

[1] Forum on China–Africa Cooperation.2018.10. http://www.china.org.cn/english/china_key_words/2018-10/29/content_68888913.htm.

[2] China Africa trade has reached a new high under the epidemic, highlighting the resilience of China Africa economic and trade cooperation. Xinhua News Agency. 2022. 3. http://m.news.cn/2022-03/01/c_1128425904.htm.

[3] Forum on China Africa Cooperation—Beijing action plan (2019–2021). Ministry of foreign affairs of China. 2018. 9.

Belt and One Road' initiative is consistent with the African Union's agenda 2063 in the aspect of digital technology. The agenda 2063 formulated by the African Union is a common strategic framework for inclusive growth and sustainable development in Africa, which defines the vision of developing African digital technology and aims to build African countries into integrated digital economies and enable African governments, commercial economies, self-employed households enjoy the safe and reliable information and communication technology services.[4] Under the favourable policy environment, China–Africa digital technology collaboration has been in the ascendant in recent years. South Africa, Nigeria, Kenya, Uganda, Rwanda and other countries have strengthened their digital technology Cooperation with China. China has promoted the development of digital technology in Africa, and African countries also play an irreplaceable role in China's construction of a new 'double cycle' development pattern.

20.2   Digital Technology

Digital technology refers to the products or services embedded in or supported by information and communication technology, including digital artifacts, digital platforms and digital infrastructure (Briel et al., 2018; Nambisan, 2017). The feature of digital technology is editable and scalable. Editability refers to the ability to access and modify objects beyond their behaviour. Scalability refers to the ability to enhance functional performance at a low cost and high speed. With less modification or even just the addition of hardware or software, higher performance can be achieved to handle large-scale businesses (Nambisan, 2017). The structural characteristics of digital technology include openness and relevance. Openness is conducive to sharing technological infrastructure, providing support for the opportunity of diversified users and realizing value creation. Meanwhile, this openness also realizes the visualization of numerous data and reduces the asymmetry of market information to a certain extent (Nambisan et al., 2018; Smith et al., 2017). Relevance can promote the connectivity and interaction between multiple actors which provides a channel for obtaining resources, and also promotes the direct connectivity between global customers (Liu et al., 2018).

20.3   Inclusive Development Theory

In recent years, inclusive development or inclusive growth has become a research focus of the world bank, the Asian Development Bank, the United Nations Economic and Social Council, the United Nations and other international organizations. International organizations such as the Asian Development Bank and the world bank have

---

[4] African Union Commission, Agenda 2063: The Africa We Want. September 2015. 5.

grafted the inclusive feature of Indian innovation on growth and development, and formally used the terms of inclusive growth and inclusive development (Ali, 2019; Bolt, 2004; Sarah, 2006; World Bank, 2006).

From an international perspective, uneven development between the North and the South keeps widening. According to World Population Review, 696 million people still live in extreme poverty, surviving on less than $1.90 (INT) per day, particularly in Sub-Saharan Africa. Thus, the task of combating poverty worldwide is still extremely severe. As far as less developed countries are concerned, they not only fail to fully enjoy the benefits of globalization, but also suffer from political repression and economic exploitation by developed countries. In the industrial division pattern based on the global value chain, developed countries take their leading advantages in science and technology, finance, organization and management to pursue their own interests, while they lack sufficient support for meeting the development and interests of less developed countries. The less developed countries face the challenge for breaking the bottleneck of development and have to subject to part of a vertical or horizontal integration of transnational corporations from developed countries, which leads to enlarge the gap between countries and regions. Therefore, it makes sense for the United Nations to call for world sustainable development. China suffered civil wars and imperialist aggression in history. The painful experience makes China have a deep understanding of the challenges faced by developing and less developed countries, such as poverty, starvation, and unbalanced regional development. That's why China actively responds to the initiative of inclusive development, proposes and practices a new development concept as "Innovation, Coordination, Openness, Greenness and Sharing".

In the digital era, with the wide use of cloud computing, big data, artificial intelligence and other technologies, digital technology is able to achieve leverage effect (e.g., less resource input, more output) (Huang et al., 2017) so as to help developing countries catch-up with the wealthy economies at the development curve. On the one hand, the openness of digital technologies reduces the asymmetry of market information to a certain extent and promotes the connectivity and interaction between multiple actors. This relevance provides a channel for the acquisition of resources, and also enhances the direct connectivity of global customers (Liu et al., 2018); On the other hand, digital technology is conducive to promoting inclusive development, easing international contradictions, narrowing the development gap, solving social problems as well as promoting the comprehensive and coordinated development of social economy and enhancing the sustainability of development.

20.4 Digital Technology Development in China and Africa

20.4.1 Current Digital Technology Development in China

The 2021 digital economy report[5] released by the United Nations Conference on Trade and development pointed out that China is becoming one of the major leaders in digital technology field. In terms of the ability to participate and benefit from data-driven digital technology, the United States and China set themselves apart from other competitors. Half of the world's large data centres are in these two countries. In the past 5 years, China and the United States have obtained 94% of the total financing of AI start-ups, 70% of the world's top AI researchers, and nearly 90% of the market value of the world's largest digital platform. For example, Chinese companies such as Tencent and Alibaba are increasingly investing in every link of the global data value chain, and these companies have data advantages because of their platform businesses. They are no longer just digital platforms, instead, these companies have become global digital leaders with a large amount of user data, strong financial, market and technical forces worldwide.[6] The year of 2021 witnessed 5G base stations reached 1.425 million in China, with 10.1 5G base stations per 10 thousand of people. The world's largest 5G network has been built in China. At present, the 5G coverage rate in all cities has reached 100%, and the rural coverage rate is nearly 90%.[7] According to the statistics in the white paper on China's digital economy issued by the China Academy of information technology, the scale of China's digital economy in 2020 was nearly $5.4 trillion, ranking second in the world.[8]

20.4.2 Challenges Faced by Africa in Digital Technology Development

20.4.2.1 Digital Divide in Infrastructure Construction

The level of digital technology in African countries is closely related to their industrialization development level, economic development stage and global industrial

---

[5] The United Nations Conference on Trade and Development. UNCTAD (2021). Digital Economy Report: Cross-border data flows and development: For whom the data flow. https://unctad.org/web flyer/digital-economy-report-2021.

[6] The United Nations Conference on Trade and Development. UNCTAD (2021). Digital Economy Report: Cross-border data flows and development: For whom the data flow. https://unctad.org/web flyer/digital-economy-report-2021.

[7] China has built the world's largest 5g network. Global network 2022.06. https://3w.huanqiu.com/a/9b216e/48KtCiLCcO6.

[8] White paper on the development of China's digital economy, China Institute of Information and Communications, 2020.

division. Since the beginning of the twenty-first century, although Africa's international status has been raising, its endogenous economic development motivation has strengthened, and inflow investments to Africa has keeping upward, the overall industrialization development level of Africa is still comparatively low, the scale of African economy does not account for a proportion matching its size and population in the world economy, and the contribution of digital technology to African economic development is very limited for the moment.

At present, African digital technology is still at the early stage of development. Most African countries have relatively underdeveloped digital infrastructure, low intercontinental optical fibre coverage, poor submarine optical cable connection, imperfect transnational and cross regional communication networks, limited broadband network service coverage and high access costs. In addition, the weak foundation of logistics supply chain, the lack of mature e-commerce platform system, and the relatively weak development of electricity and supporting facilities are also significant factors that cause the overall low level of digital technology development in African countries. In 2018, 'One Belt and One Road' big data centre of China's National Information Centre evaluated the digital silk road smoothness index of 71 countries along the 'One Belt and One Road' zone, of which South Africa (11th), Egypt (16th), Morocco (50th) and Ethiopia (60th) were shortlisted. However, other African countries are not included in the list due to weaker construction and development of digital infrastructure. This reflects the regional imbalance and fragmented development of digital technology in Africa, which brings many difficulties and challenges to eliminate the digital divide between China and African countries and share the benefits rising from digital technology development.

20.4.2.2   Unbalanced Digital Technology Development

The biggest contradiction brought about by the digital divide is the unbalanced development of digital technology in various countries. Only 20% of the population in the least developed countries get access to the Internet. Moreover, Internet users complain slow download speed, high online tariff and their purpose of internet surfing is diverse. For example, in some developed countries, 8 out of every 10 Internet users shop online, while in many least developed countries, the proportion of online shopping is less than one tenth (United Nations, 2021). In addition, within the country, there is a clear gap in the use of the Internet between rural and urban areas and between men and women. The least developed countries in Africa have the largest gender gap in Internet use. Africa is an ancient and young continent with a population about 1.3 billion. The median age of many countries is less than 20 years old, which means that a large part of the African population is teenagers and children who are in urgent need of educational resources. In recent years, the Internet penetration rate in African countries is slowly increasing. According to incomplete statistics, by 2020, about 39.3% of the population in Africa use the Internet with a total population of about 510 million, despite this percentage still lower than 60% of the world's average internet access level. Based on the China Internet Development Report, by December

2021, the number of Internet users in China reached 1.032 billion with the Internet penetration rate of 73.0%; the number of mobile Internet users attained 1.029 billion, accounting for 99.7% of the total mobile users; and the number of online payment users achieved 904 million, occupying 87.6% of the total.

20.5   China–Africa Digital Technology Collaboration

20.5.1   Deepening Collaboration in Digital Technology Infrastructure Construction

In response to the call of African countries for developing digital technology, Chinese companies actively assist the African continent to build digital infrastructure and improve its Internet penetration. In 2020, China Mobile and eight other foreign companies around the world undertook 2Africa submarine cable project in Africa. The submarine cable serves the African continent and the Middle East. It is the widest coverage submarine cable in the African continent. The cable connects with other submarine cables through East Africa to Asia. After completion, it will greatly enhance the connectivity of the whole Africa and the Middle East and is expected to form a digital ecology in the Asian-African continent. 2Africa submarine cable, with a total length of 37,000 kms and 21 landing points in 16 African countries, will seamlessly interconnect Africa with Asia (eastward through Egypt) and the Middle East (through Saudi Arabia), meet the urgent need of Africa for high-capacity and reliability of the Internet, further support the rapidly growing capacity demand, and lay a solid foundation for satisfying the future needs of hundreds of millions 4G, 5G users and fixed broadband access.[9]

Due to the lack of communication facilities, a large number of remote areas in Africa are still unable to make phone calls and get access to the Internet, just like an 'Information Isolated Island'. These areas usually have a scattered population with no stable electricity supply and inconvenient traffic conditions. Traditional iron tower stations are not suitable for deployment in these remote areas due to high cost, poor transportation and other difficulties. To solve this communication problem, Huawei has creatively designed a base station that can be built on a wooden pole after a long time of research and development. This base station is named as Rural Star which has its own power supply and low power consumption. It keeps simple and small but fully meets the customers' needs of fast and low-cost station building. Thanks to its unique advantages, Rural Star has been widely used, helping countless people in remote areas realize the hope of connecting the external world. For example, in Ghana, a Rural Star base station was built in only 3 days, reducing the time and cost by 70%. Great changes have taken place in the lives of villagers since

---

[9] Nine institutions around the world jointly build submarine cables to upgrade Africa's digital infrastructure. Sina Finance.2020.5. https://baijiahao.baidu.com/s?id=1667129689681073798&wfr=spider&for=pc.

then; In Nigeria, Rural Star has brought convenient communication and business opportunities to the villagers of Tobolo, making their lives better and opening the door to the outside world; Regardless of plains, hills, deserts, islands, or rural areas, urban villages, highways, tunnels, wherever Rural Star goes, network connectivity has changed from 'impossible' to 'possible'. More and more African regions have achieved network connectivity and begun to enjoy the benefits rising from the digital technology. In 2020, Huawei launched the Rural Star Pro innovative solution, which can provide high-quality mobile broadband services for more distant villages. This very innovative solution adopts the integrated design of access and return, which cuts down the power consumption of the whole station to 100 watts, largely decreases the end-to-end cost, and greatly promotes the digitalization process of African villages. At present, a series of Rural Star solutions have provided mobile Internet services to more than 60 countries and regions, covering more than 50 million people in African remote areas (Huawei Sustainability Report, 2020).

Secondly, according to the report of the International Telecommunication Union (ITU), about 61% of households in Africa still have no access to the Internet. The high cost of fixed cable installation, the difficulty in acquiring station building sites due to private land ownership, and the long period of project deployment have caused enormous challenges to the last mile connection. Huawei provides the WTTX (wireless broadband to home) solution, which is able to address the problem of home broadband access efficiently and effectively. The scheme successfully reduces the deployment cost by up to 75% by using the existing network and base station site structure. Besides, Huawei Marine,[10] a joint venture of Huawei and Global Marine (UK company) has contracted to build the Cameroon Brazil transatlantic submarine cable system to connect Africa and Latin America. The CBCS system directly connects Kribi in Cameroon and Fortaleza in Brazil, with a total length of about 6000 km. It adopts the leading 100 g transmission technology and designs 4 pairs of optical fibres with an initial system capacity of 32tbps. The system was put into use at the end of 2017. Huawei is devoted to improving communication connectivity in the Africa. In addition to put network connectivity in priority, Huawei also pays attention to user experience. It provides users valuable services and makes great efforts for ameliorating the well-being and security of African people and local productivity. The construction of basic communication facilities has significantly narrowed the digital divide (Huawei news, 2015).

20.5.2 Innovative Collaboration in Inclusive Development of Digital Business Ecology in Africa

After China initiated the 'digital silk road', China–Africa cooperation in the field of digital commerce moves to a new stage. Since 2016, Alibaba Group has put forward

---

[10] Huawei Marine: It is a joint venture of Huawei and Global Marine (a British company). Huawei sold out its stakes of the joint venture to a British company called Hengtong Group after 2018.

the World Electronic Trade Platform (eWTP) initiative, calling for conforming to the trend of rapid digital technology development, better helping the development of small, medium and micro enterprises, promoting the growth of global inclusive trade and incubating new rules of global trade in the Internet era. According to Alibaba platform data, from April 2018 to March 2019, the import of goods from Africa increased by 98% year-on-year, the number of consumers buying African goods increased by 64%, and the transaction volume of export goods to Africa increased by 69%. Under the framework of the digital silk road, China, Egypt and Rwanda signed memorandums of understanding on e-commerce cooperation in 2017 and 2018 respectively, providing policy and platform support for African small business suppliers to participate in cross-border e-trade. On October 18, 2018, the 'gather the world project' of Alibaba's 'gather cost-effective' platform launched several types of coffee from Rwanda. The turnover of Rwanda coffee in 1 day is equivalent to the total sales volume of past year. President Kagame of Rwanda said that Rwanda joined eWTP and sold local high-quality coffee to Chinese consumers through e-commerce platforms, which remarkably increased farmers' income (Ali Research Institute, 2019). In the aspect of digital technology boosting Africa's logistics, China National Building Materials Group Co., Ltd. proposes a new business model such as cross-border digital trade + shared overseas warehouse. It is a kind of foreign trade comprehensive logistics service operation mode for Africa, integrating e-commerce elements into traditional international trade mode, and providing one-stop comprehensive foreign trade services for African countries such as South Africa, Kenya and Sudan. Some African countries have begun to explore the feasible path of demand docking with China in the field of digital technology. Meanwhile, Chinese companies are also further exploring the needs of African market, making efforts for achieving the comprehensive and coordinated development of China–Africa digital technology and enhancing the bilateral sustainability of development.

20.5.3 Advanced Collaboration the Development of Mobile Payment and Inclusive Finance in Africa

Developed countries have established a credit card payment system based on bank accounts for a long time, while Africa has a low credit card holding rate due to underdeveloped credit economy. However, the availability of mobile phones would make Africa possible for directly transforming into a cashless, paperless, virtual, digital and not necessarily bank account binding required payment method. To meet the need of African consumers for mobile phones, some world-famous brands of mobile phone like Huawei, Transsion, Oppo and Xiaomi offer best-cost cell phones to local buyers. China has become one of the main mobile phone suppliers in Africa, sharing more than 50% of the market share. While satisfying the mobile phone need of African people, Chinese companies are dedicated to meeting continuously increasing new payment needs in Africa through launching new mobile payment products, such as adding digital wallet, mobile payment, digital ID card and other application functions to mobile devices. These functions have been successfully practiced in China. Most Chinese people embrace mobile payment. The highest rate of mobile phone penetration in the world and the high acceptability of consumers in China make mobile payment become very popular for Chinese people in daily life.[11] Chinese companies lead innovation in the digital field and are open to share the innovation with African companies. On the one hand, it has accelerated the mobile payment process in Africa. On the other hand, it has also promoted the internationalization of RMB. China and Africa both sides benefit from the development of mobile payment.

South Africa is the first African country to access China's mobile payment. Alibaba's Alipay lands in Africa through cooperation with Zapper, a South African payment company. Nowadays mobile payment is becoming more and more popular in Africa. China's mobile payment has also brought new possibilities for inclusive finance in Africa. Many people in Africa do not have bank accounts. Even if they own bank accounts, they ought to pay high fees. People still rely on cash for transactions in their daily life. Most people lack access to high-quality financial products and services, which restricts the release of Africa's huge financial market potential. Developing inclusive finance has become extremely urgent. At the same time, in just a few years, the development of mobile digital technology has changed the communication situation in Africa. Africa has directly jumped into the digital era. Africa has become a "Mobile Goes First" market. Most Africans rely on mobile phones for their first contact with the Internet. The popularization of mobile payment brings very convenient financial services to African people.

In Kenya, within 10 years, the financial penetration rate has soaring from 26 to 75%, and the key factor driving this trend is digital technology innovation. Standard bank, the largest bank in Africa, has chosen a more direct path: join hands with Wechat of Tencent to share the growing mobile payment market in Africa. The Wechat wallet supported by the bank was launched in South Africa, the most industrialized country in Africa, which enables users to use various services as same as Chinese Wechat version, including account transfer, taxi payment and other service fees (Global network, 2016).

20.5.4 Digital Collaboration in the Medical and Health Field

China–Africa digital technology collaboration has played a key role in Africa's fight against the COVID-19. Through online video teleconferences, experts from China and Africa communicated their experiences in combating the pandemic, discussed the intelligent and digital medical treatment and other related new projects at the post-pandemic era. The global combating COVID-19 platform established by Ali Health launched training programs for African medical staff. Ali Health offered

---

[11] See the web link: http://www.chinadaily.com.cn/a/202101/15/WS60010cd3a31024ad0baa2df7.html.

online trainings for more than 3000 African medical staff together with several assistance measures such as digital communication technology and digital administration affairs. The series of actions improved Africa's ability to fight against the pandemic, promoted the sustainable development of Africa's health care system, and catalysed Africa's digital transformation. Thanks to the increasing popularization of communication technology in Africa and the in-depth collaboration of China–Africa in digital technology, China–Africa medical cooperation will usher in a new trend of digital transformation. Taking the opportunity of China–Africa digital technology collaboration will strengthen cooperation with African countries on high-tech health products. Through digital health platform, African people can not only be provided with digital health solutions, but also enjoy higher quality health care products and services (Hu, 2020). In the middle of 2020, the Chinese Embassy in South Africa organized and coordinated experts in the digital field of the two countries to exchange 'big data' combating pandemic experience through video connectivity. Experts from China shared and exchanged China's big data combating pandemic experience at the meeting, discussed with South African experts to how to jointly promote smart city through cooperation and help South Africa win the war of pandemic prevention as soon as possible (Du, 2021).

Furthermore, the trade volume of China–Africa cross-border e-commerce platforms witnessed rapid growth after the outbreak of the COVID-19. China International Import Exposition and Canton Fair take a form of 'Cloud Promotion' to present products and services from Africa. The events attract the participation of numerous African businesses and entrepreneur representatives from Egypt, Kenya and Ethiopia. Since then, more and more African commodities have entered the Chinese market through cross-border e-commerce. On 29 June 2020, representatives of relevant government departments and industry associations from many African countries attended the 'cloud opening' of the first China Africa Digital Trade Week. New communication models, such as cloud conference, cloud contract signing, cloud exhibition and sales, and cloud docking, were used to organize dialogues and conduct idea exchanges during the week.[12] On 26 October 2020, the China–Middle East Africa (Kenya) International Trade Digital Exhibition was held visually for showing products in health care, agriculture, manufacturing and other fields from both sides through the 'cloud trade promotion exhibition', and achieved fruitful results.[13] Evidence proves that China–Africa economic and trade cooperation is not constrained by the COVID-19 and keeps continuously upward. One of the mystery lies in the China–Africa's successful collaboration in digital technology development. Digital economy has become an important force in upgrading China Africa relations.

---

[12] Cloud opening of China Africa Digital Trade Week. XinHua News Agency. 2021. 03. 29. https://baijiahao.baidu.com/s?id=1670892613609435187&wfr=spider&for=pc.

[13] China Middle East Africa (Kenya) International Trade digital exhibition opens in Beijing.2020.10.27. https://baijiahao.baidu.com/s?id=1681672286401763217&wfr=spider&for=pc.

20.5.5 Digital Technology Collaboration in Promoting African Education Development

According to the report of Progress in International Reading Literacy Study (PIRLS), 78% of grade 4 students in South Africa lack basic English reading comprehension ability. In response to the call of the South African government for ensuring that all children can read fluently and understand the relevant course content by the end of the third grade, in July 2020, Huawei cooperated with the South African operator Rain and the non-profit educational organization Click Foundation to carry out the Digi school project. The project plans to connect 100 local primary schools within 1 year. Huawei provides connectivity equipment and funds to Click Foundation to make high-quality learning resources available for schools, while Rain provides 4G and 5G networks to enable schools to get access to the Internet. By the end of 2020, the Digi school project has completed the connectivity of 29 schools, and more than 22,000 students have benefited from it (Huawei Sustainability Report, 2020).

It is estimated that the COVID-19 has led to the suspension of nearly one billion students around the world. Like many countries, schools in Senegal have been forced to stop classroom teaching due to the pandemic. To ensure the continuity of education, UNESCO established the Global Education Alliance in March 2020. As a member of the Global Education Alliance, Huawei, together with the Ministry of Education of Senegal, the UNESCO West Africa Office and the local operator Sonatel, launched the Digi school project in August 2020 to provide distance teaching empowerment for local teachers and help them stop classroom teaching but without stopping learning. Before joining the alliance, Huawei has cooperated with the Ministry of Education of Senegal to build smart classrooms in some schools in Dakar to make local students experience interactive multimedia learning. The Digi school project obviously further deepens the collaboration between Huawei and the Ministry of Education of Senegal to promote fair and high-quality education in the region. The project aims to cover 200 schools, benefiting 20 thousand teachers and 100 thousand students. By December 2020, more than 200 teachers had received the digital skills training required for distance teaching, and more than 60 schools and 15 thousand students had benefited from it.

Another example is Huawei's 'Future Seed' project which was created in 2008. It is a flagship Corporate Social Responsibility (CSR) project with a long history and receiving the most funding from Huawei. The project is committed to cultivating high-end ICT talents and promoting the development of the local ICT industry. Through the 'Future Seed' project, outstanding students from all over the world get together. They are able to not only have close contact with advanced ICT technologies and products, exchange and interact with global elites and industry experts, but also broaden their horizons and increase their knowledge in cross-cultural exchanges, so as to sow seeds for cultivating future ICT industry elites (Huawei Sustainability Report, 2020).

The next example is the Panda Pack project of Alibaba, an international public welfare program for children. It distributes love packages to poor primary school students in developing countries along the 'One Belt and One Road' zones to improve their basic learning and living conditions. The total amount of funding raised for the project is expected to exceed 100 million RMB, which will benefit 1 million children. The first batch of international love package partners include several African countries: Ethiopia, Namibia, Sudan and Uganda. In 2019, Alibaba Africa youth entrepreneurship fund was officially launched. The fund will provide 10 million US dollars in the next 10 years to encourage the development of young entrepreneurs in Africa. Entrepreneurs from 54 countries in Africa can apply for it and every year a competition will be held in Africa to provide them with a reward totalling 1 million USD (Ali Research Institute, 2019).

20.6   Conclusion and Future Prospects of China–Africa Digital Technology Collaboration

Digitalization is one of the key components in the fourth industrial revolution. As the world's largest developing country, China has obtained remarkable achievements in artificial intelligence, mobile payment, digital commerce and other digital fields. How China shares its experience in advancing digital technologies with other developing countries is worth of studying. Based on inclusive development theory, our study took two Chinese high-tech giants like Huawei and Alibaba as examples together with other secondary data to discuss how China collaborates with Africa in digital technologies and helps Africa to solve digital technology-based problems. Chinese companies like Huawei and Alibaba have assisted Africa in building digital infrastructure, launching digital education, developing digital commerce and digital finance. As many developing and less developing countries have not prepared well for embracing digital knowledge economy, the cooperation between China and Africa in digital technology is not an easy journey. Poor digital infrastructure, lack of digital technology talents, huge market but without efficient e-commerce platforms, digital finance at the infant stage are the present situation and challenges which many African countries are facing. However, China and Africa are willing to cooperate with each other for handling the challenges based on mutual trust and benefits. There are plenty of cooperation potential for China and Africa in digital technologies. Daily progress makes great advancement in the future. The successful cooperation between China and Africa provides an excellent example for building common destiny community for all humanity in the history.

In May 2020 the African Union announced the African digital transformation strategy, which puts forward the overall goals and sub goals of the African Union's digital transformation and guides African countries to develop digital technology construction. Agenda 2063 of the African Union has formulated the strategy of digital networks and services in Africa and proposed corresponding policy recommendations, hoping to turn Africa into a digital society from the aspects of network infrastructure and narrowing the digital divide (African Union Commission, Agenda

2063). African countries have taken positive actions and issued policies suitable for the development of their own digital economy, such as supervising mobile sector, providing financial support, technically improving network flexibility, and ensuring the affordability of the network and keeping accessible to important connectivity. African countries pay much attention to the development of digital technology and strongly support it.

China's 'One Belt One Road' initiative matches well the African Union's vision of strengthening digital technology transformation in the Agenda 2063. Many countries including African countries have actively participated in it and benefited from it. Generally speaking, 'One Belt and One Road' initiative received rational, positive, and constructive comments from the international community. However, there is still some one-sided, subjective, and negative criticism, which views it as a debt trap instead of a development opportunity and the Chinese version of the 'Marshall Plan' to transfer China's domestic excess capacity and deprive local employment opportunities. As a matter of fact, China has waived a huge sum of debts of African countries and never colonized the African continent. Bilateral cooperation agreements are reached based on a win–win model. China is historically and constantly committed to multilateral comprehensive cooperation with the African continent. In the context of One Belt and One Road initiative, China–Africa digital technology collaboration will be further deepened, yield more fruits and better benefit Africa and local people in the future.

The United Nations Economic Commission for Africa points out that digital transformation is crucial for Africa to enhance its global competitiveness in the twenty-first century. It not only contributes to African integration, promotes inclusive growth, and creates employment opportunities, but also eliminates the widening digital divide, helps Africa eradicate poverty and benefits African people. China has the world's leading digital productivity, which can effectively enable and support the development of digital infrastructure in Africa; China has cutting-edge digital technology represented by various new business forms, which is able to provide resources support to African countries for promoting the integrated development of digital technology and substantial economy. In August 2021, the China–Africa Digital Innovation Partnership Program was officially launched.[14] In the future, China and Africa will further promote the China–Africa Digital Innovation Partnership Program and carry out in-depth cooperation in six aspects: digital technology infrastructure, digital economy, digital education, digital inclusiveness, digital security, and building a digital cooperation platform, directly addressing Africa's most urgent needs. China–Africa digital technology collaboration will undoubtedly inject a booster into the depressed world economy under the pandemic and help the recovery of the world economy.

---

[14] China will work with Africa to formulate and implement the China Africa digital innovation Partnership Plan. Ministry of Foreign Affairs. 2021. 08. https://www.fmprc.gov.cn/wjbxw_673019/202108/t20210824_9138400.shtml.

References

Adonu, G. (2021). Catalysing digital economy in Africa: The role of sovereign. *UNILAG Law Review, 4*(2), 23–53.

Ali Research Institute. (2019). *Building the 21st century digital silk road—The practice of Alibaba economy.*

Bolt, R. (2004). Accelerating Agriculture and Rural Development for Inclusive Growth: Policy Implementation, *ERD Policy Brief Series*, 29, Asian Development Bank.

Briel, F. V., Davidsson, P., & Recker, J. (2018). Digital technologies as external enablers of new venture creation in the IT hardware sect. *Entrepreneurship Theory and Practice, 42*(1), 47–69.

China Institute of Information and Communications. (2020). *White paper on the development of China's digital economy.*

Global Network. (2016). *Tencent cooperates with Naspers to promote WeChat and accelerate the pace of market development in Africa.* 2016.4. https://tech.huanqiu.com/article/9CaKrnJTy5F

Global Network. (2022). *China has built the world's largest 5g network.* 2022. 06. https://3w.huanqiu.com/a/9b216e/48KtCiLCcO6. https://baijiahao.baidu.com/s?id=167089261 3609435187&wfr=spider&for=pc

Huang, J., Henfridsson, O., Liu, M. J., et al. (2017). Growing on steroids: Rapidly scaling the user base of digital ventures through digital innovation. *Mis Quarterly, 41*(1), 301–314.

Huawei News. (2015). *Huawei ocean undertook the Cameroon Brazil transatlantic submarine cable system to connect Africa and Latin America.* 2015.10 https://www.huawei.com/cn/news/2015/10/huaweihaiyangchengjian

Huawei Investment & Holding Co. Ltd. (2020). *Sustainability report.*

Liu, G., Chen, P., & Guo, A. (2018). Is the demand-side perspective a useful perspective for entrepreneurship research? *Academy of Management Proceedings.* Briarcliff Manor, NY 10510: Academy of Management.

Liu, H., & Luo, J. (2021). The establishment of Forum on China–Africa Cooperation and a new chapter in China–Africa cooperation. In: *Sino-African Development Cooperation.* Research Series on the Chinese Dream and China's Development Path. Springer. https://doi.org/10.1007/978-981-16-5481-7_2

Ministry of Foreign Affairs. (2021). *China will work with Africa to formulate and implement the China Africa digital innovation Partnership Plan.* 2021. 08. https://www.fmprc.gov.cn/wjbxw_673019/202108/t20210824_9138400.shtml

Ministry of Foreign Affairs of China. (2018). *Forum on China Africa Cooperation—Beijing action plan (2019–2021).* 2018.9. https://www.mfa.gov.cn/web/ziliao_674904/tytj_674911/201809/t20180905_7948514.shtml

Nambisan, S. (2017). Digital entrepreneurship: Toward a digital technology perspective of entrepreneurship. *Entrepreneurship Theory and Practice, 41*(6), 1029–1055.

Nambisan, S., Siegel, D., & Kenney, M. (2018). On open innovation, platform, and entrepreneurship. *Strategic Entrepreneurship Journal, 12*(3), 354–368.

Sarah, C. (2006). Structural change, growth and poverty reduction in Asia: Pathways to inclusive development. *Development Policy Review, 24*(s1), 51–80.

Science and Technology Daily. (2021). *China Africa cooperation helps African digital economy to usher in new development.* Du Huabin. 2021.12.06 (002).

Sina Finance. (2020). *Nine institutions around the world jointly build submarine cables to upgrade Africa's digital infrastructure.* 2020.5 https://baijiahao.baidu.com/s?id=166712968968 1073798&wfr=spider&for=p

Smith, C., Smith, J. B., & Shaw, E. (2017). Embracing digital networks: Entrepreneurs' social capital online. *Journal of Business Venturing, 32*(1), 18–34.

United Nations. (2021). *Digital Economy Report 2021: Cross-border data flows and development: For whom the data flow.*

World Bank. (2006). India inclusive growth and service delivery: Building on India's Success, *Development Policy Review Report*, No. 34580-IN, Washington, DC.

Xinhua News Agency. (2021). *Cloud opening of China Africa digital trade week*. 2021.03.29.
Xinhua News Agency. (2022). *China Africa trade has reached a new high under the epidemic, highlighting the resilience of China Africa economic and trade cooperation*. 2022. 3. http://m. news.cn/2022-03/01/c_1128425904.htm

Chapter 21
Analysis of Africa's Business to Consumer E-commerce Development and Outlook on China's Investment in Africa

Zhan Wang, Ruoyun Liu, and Siqian Wang⊙

21.1 Introduction

Over the past few years, e-commerce has become an integral part of the global retail framework. Like many other industries, retailing has undergone massive transformation with the advent of the internet. Due to the continuous digitization of modern life, consumers in almost every country can benefit from online transactions.

The huge consumer base attracted us to focus our research on Business to Consumer (B2C). Business to Consumer is the consumer-facing portion of e-commerce. It refers to Internet transactions of goods and services between businesses and private consumers. Globally, Asia has the largest B2C e-commerce market (47.7% share in 2017). Asia's prominence in e-commerce is reflected in the strength of players such as China's Alibaba Group. While China has become the largest B2C e-commerce market in the world, we are curious about what is going on in Africa and set our sights on Africa. Studying the development of e-commerce in Africa will help us better understand Africa and promote cooperation between China and Africa in more aspects.

Since ancient times, Africa has been the centre and hub of trade between Europe and Asia. Its important seaports and the Suez Canal have to date played an important role in international markets. With the rapid development of electronic information

Z. Wang
College of Foreign Languages of Huazhong Agricultural University, Wuhan, China
e-mail: wangz13@aliyun.com

R. Liu (✉)
African Research Center of Wuhan University, Wuhan, China
e-mail: 279440580@qq.com

S. Wang
China-Africa Innovation Cooperation Center, Wuhan, China

© The Author(s) 2024
M. Muchie et al. (eds.), *China-Africa Science, Technology and Innovation Collaboration*, https://doi.org/10.1007/978-981-97-4576-0_21

technology in today's world, the importance of information consumption is beyond doubt. The age-old African continent also needs to establish new channels for trade and consumption through e-commerce platforms to enjoy the benefits of economic globalization. The growing consumer demand of African people will provide more opportunities for e-commerce development. Meanwhile, the rooting and maturing of e-commerce in Africa will also contribute to its growth in international trade. As the continent with the most concentrated developing countries in the world today, Africa is taking on a new look in front of the world's people. Africa's economy is in dire need of transformation, its online trade is increasing, Africa is increasingly cooperating internationally, and its e-commerce is coming to the fore. Africa abounds with investment opportunities, and now is the perfect time for China's enterprises to develop e-commerce and related businesses in Africa.

21.2   Literature Review

The development of e-commerce in Africa cannot be achieved without the innovation of Internet information technology, the improvement of Africa's infrastructure, the rise of global mobile payment and the development of cross-border logistics. Some foreign scholars have analysed the development of e-commerce in African countries from a country perspective, and some have analysed the environment and potential of African e-commerce market as a whole and put forward the opportunities and challenges that African e-commerce development will face. With Senegal as an example, Abdoulaye Ndiaye analysed the political, economic, judicial, and infrastructural environment for e-commerce development in the country, and analysed the national and local level barriers to e-commerce development (Abdoulaye Ndiaye, 2000). Ducass Alain and Jean-Marc Kwadjane analysed the basic environment for e-commerce in Morocco, Tunisia, Senegal and Côte d'Ivoire, and summarized the advantages and disadvantages of e-commerce development in the African region (Ducass Alain et al., 2015). Yang Jianshi and Wang Qiuhua explored the role of e-commerce in developing tourism market in African countries. Wang Peixin examined the overall development of the e-commerce in Africa and believed that the e-commerce in Africa is weak but growing fast. He held the view that the e-commerce will bring new opportunities such as economic growth, technological innovation and industrial chain upgrading to Africa and that the e-commerce development in Africa is also restricted by factors such as policy environment, infrastructure, financial service, information and network security, education and human resources. Therefore, he suggested that the e-commerce development in Africa should be tailored to local conditions and on its own path. Tang Xiaoman analysed the development model of e-commerce in China-Africa trade from three aspects: construction, popularization and implementation strategies of China-Africa e-commerce trade platform. Wu Jiang also believed that e-commerce is necessary for China-Africa trade and that an e-commerce platform can be built to promote the development of China-Africa trade. Zhang Xiaheng investigated the overall cross-border e-commerce development in Africa and concluded that there are obstacles from the aspects such as safety, logistics, e-payment, language and religion, cultural conflicts and mobile terminals for cross-border e-commerce development. He put forward that attention should be paid to the cross-border e-commerce market in Africa, cross-border e-commerce business can be launched by taking advantage of the mobility characteristics, and local e-commerce resources should also be combined to develop cross-border e-commerce. He also emphasized the strengthening of cooperation with local government, communications operators and financial institutions. With Ghana, an African country, as the research object, Yang Jianzheng and Xia Yunchao analysed the overall e-commerce development in Ghana and believed that the certain development has been achieved in communications technology of Ghana while some issues such as payment methods and e-commerce legislation still need to be addressed for the e-commerce development.

The development of e-commerce in Africa cannot be achieved without the innovation of Internet information technology XE "information technology (IT)", the improvement of Africa's infrastructure, the rise of global mobile payment and the development of cross-border logistics. Some foreign scholars have analysed the development of e-commerce XE "e-commerce" in African countries XE "African countries" from a country perspective, and some have analysed the environment and potential of African e-commerce market as a whole and put forward the opportunities and challenges that African e-commerce development will face. With Senegal as an example, Abdoulaye Ndiaye analysed the political, economic, judicial, and infrastructural environment for e-commerce development in the country, and analysed the national and local level barriers to e-commerce development (Abdoulaye Ndiaye, 2000). Ducass Alain and Jean-Marc Kwadjane analysed the basic environment for e-commerce XE "e-commerce" in Morocco, Tunisia, Senegal and Côte d'Ivoire, and summarized the advantages and disadvantages of e-commerce development in the African region (Ducass Alain et al., 2015). Yang Jianshi and Wang Qiuhua explored the role of e-commerce in developing tourism market in African countries. Wang Peixin examined the overall development of the e-commerce in Africa and believed that the e-commerce in Africa is weak but growing fast. He held the view that the e-commerce will bring new opportunities such as economic growth, technological innovation XE "Innovation" and industrial chain upgrading to Africa and that the e-commerce development in Africa is also restricted by factors such as policy environment, infrastructure, financial service, information XE "information" and network security, education and human resources. Therefore, he suggested that the e-commerce XE "e-commerce" development in Africa should be tailored to local conditions and on its own path. Tang Xiaoman analysed the development model of e-commerce in China-Africa trade XE "China-Africa trade" from three aspects: construction, popularization and implementation strategies of China-Africa e-commerce trade XE "trade" platform. Wu Jiang also believed that e-commerce is necessary for China-Africa trade and that an e-commerce platform can be built to promote the development of China-Africa trade XE "China-Africa trade". Zhang Xiaheng investigated the overall cross-border e-commerce development in Africa and concluded that there are obstacles from the aspects such as safety, logistics, e-payment, language and religion, cultural conflicts and mobile terminals for cross-border e-commerce XE "e-commerce" development. He put forward that attention should be paid to the cross-border e-commerce market in Africa, cross-border e-commerce business can be launched by taking advantage of the mobility characteristics, and local e-commerce resources should also be combined to develop cross-border e-commerce. He also emphasized the strengthening of cooperation with local government, communications operators and financial institutions. With Ghana, an African country, as the research object, Yang Jianzheng and Xia Yunchao analysed the overall e-commerce development in Ghana XE "Ghana" and believed that the certain development has been achieved in communications technology of Ghana while some issues such as payment methods and e-commerce legislation still need to be addressed for the e-commerce development.

Overall, both the domestic and foreign scholars attach great importance to the e-commerce development in Africa and mainly concentrate on the analysis of e-commerce environment in African countries and the studies on China-Africa trade via e-commerce.

21.3   Overview of the Global B2C E-commerce Development

Global B2C e-commerce showed an upward trend in transaction value from 2010 to 2017, with an average annual growth rate of 19%. The global B2C transaction volume reached $1.53 trillion in 2016, up 12% against 2015. As predicted based on the 2017 Global E-Commerce Development Report, the total global B2C e-commerce transactions will reach $1.843 trillion in 2017. The biggest transaction volume of B2C e-commerce was achieved in the Asia–Pacific region, which accounts for 47.78%, followed by 25.64% in Europe and 24.86% in North America, and finally 1.53% and 0.49% in South America and the Middle East and Africa, respectively. The development of global B2C e-commerce presents the following distinctive features:

(i). *It complements the traditional retail and promotes the development of the retail industry*

While traditional retail is growing globally, e-commerce is also undergoing a new round of expansion. E-commerce and retail are not two factors that inhibit each other; on the contrary, e-commerce is one of the main drivers of retail growth. In the early stage of Internet development, e-commerce showed strong growth, compared with retail. As the Internet economy grows more sophisticated, retail giants are finding ways to sell online, retail brands are realizing their global business growth through online sales, and e-commerce becomes a new channel for retail development.

(ii). *E-commerce development shows an internalization tendency*

Many e-commerce brands became dedicated to business development outside of their home countries and made transactions through a cloud-based platform that provides an interface to partners around the world.

(iii). *Mobile e-commerce becomes more popular*

With the increase of mobile phone users and the more convenient mobile phone network, many users use less and less the computer browsers but spend more and more time on mobile phone. Among the visits to many e-commerce sites, more than 50% are from mobile users. The cooperation between e-commerce platform and the social media and social network opens the gate of mobile consumption.

(iv). *The emerging technology fields become more closely connected with the e-commerce*

With the development of artificial technology, unmanned aerial vehicle (UAV) and Augmented Reality (AR)\Virtual Reality (VR) technology, it is also hoped to take advantage of the emerging technology to promote the e-commerce development. Artificial intelligence can help understand the customers' needs and thus provide personalized experiences to the customers, such as 24-h services all the year around. Artificial intelligent can help analyse from the numerous data modules which customers may be the ultimate buyers of the products, greatly improving the conversion rate. The research and application of UAV technology in the field of e-commerce logistics are emerging endlessly. With help of the AR\VR technology, customers will gain more real and colourful experiences in e-commerce transactions.

China is the country with the most rapid B2C e-commerce development and the largest market of B2C e-commerce. With the B2C e-commerce development, it is expected that the total of B2C e-commerce transaction volume in China will reach $680 billion in 2017. Africa is the region with weakest B2C e-commerce development. South Africa, a country with relatively good e-commerce development, is expected to reach a transaction volume of $833 million.

21.4 B2C E-commerce Development in Africa

21.4.1 Overview of B2C E-commerce Development in Africa

The B2C e-commerce development in African countries overall is weak. However, some countries have better e-commerce development due to their good economic environment, ICT environment, infrastructure, and financial environment. South Africa, Nigeria, and Egypt are among the well-established countries in e-commerce in Africa. Kenya, Morocco, and Côte d'Ivoire are the emerging countries of e-commerce development.

E-commerce in South Africa is growing steadily in the retail environment. South Africa's online sales are expected to grow to R53 billion by the end of 2018, with an annual growth rate of 15% until 2021, and online sales value of 1% of total retail value.[1] Credit or savings cards are still the main payment method for e-commerce in South Africa. Due to the growth of credit card fraud, the Payments Association of South Africa mandated the use of the 3D security system in 2014. Meanwhile, with the popularity of mobile Internet and cell phones, mobile payment platforms and e-wallets launched jointly by banks, communication operators, retailers and others begin to become the new payment method, playing a particularly obvious role in e-commerce.

Nigeria is one of the countries with more established e-commerce on the African continent, and the country has the highest number of start-up e-commerce businesses. Nigeria has a very favourable financial environment and successfully introduced the payment service providers, Visa and Master, allowing Nigerians to use electronic means of payment and gradually apply it in the field of e-commerce. Since 2012, e-banking and mobile payments have been piloted in Lagos, its largest port city, and have since spread to 27 states and territories. Cashless payments have facilitated e-commerce and financial services in the country. The increased online transaction volume has also attracted payment service providers from Europe and Asia to invest in digital infrastructure projects in Nigeria. Meantime, the country's rapidly growing youth population, spending power and the popularity of smartphones have led to a rapid rise in the volume of online transactions and financial services in Nigeria. E-commerce spending in Nigeria is currently estimated at $12 billion, with McKinsey forecasting a projected $75 billion by 2025. The better developed e-commerce platforms in the country are Jumia and Konga respectively. In 2015, the Federal Government of Nigeria signed the Cybercrime Act to prevent fraud in e-commerce. The Cybercrime Act 2015 also provides for penalties and establishes the institutional framework for enforcement. It aims to protect e-commerce transactions, corporate copyrights, domain names, etc. in Nigeria. The difficulties facing e-commerce in Nigeria now mainly lie in poor infrastructure, inadequate road transport and electricity supply, and poor telecommunication facilities, which inhibit the potential of the e-commerce market.

With the development of mobile internet and smart phones, the e-commerce industry in Kenya has grown by leaps and bounds. According to the Communications Authority of Kenya, the e-commerce market in Kenya was valued at approximately KR4.3 billion in 2014. The e-commerce has been significantly used by more and more people across Kenya, especially among SMEs. Kenya published the Communications Act (Amendment) in 2008. This Act seeks to promote e-commerce by increasing public confidence in electronic transactions, recognizing the legal nature of electronic records and electronic (digital) signatures, creating new offence for cyber-crimes and crimes involving electronic records and transactions and the use of computers and telecommunications equipment, and recognizing electronic transaction records as evidence in court proceedings. According to the Communications Authority's (CA) quarterly report, broadband sign-ups increased by 6.7 percent to 12.7 million in the fourth quarter of 2016, up from 11.9 million sign-ups in the

---

[1] https://www.export.gov/article?id=South-Africa-ecommerce.

previous quarter. Mobile network usage in the country accounts for 99% of that of all broadband.[2] Mobile payment in Kenya is a very favourable factor for the development of e-commerce in the country, and mobile applications (apps) have rapidly become widely used in the e-commerce market and play an important role in facilitating e-commerce.

Côte d'Ivoire has good telecommunications infrastructure and good internet market has also laid good foundation for the e-commerce development in the country. According to statistics of the Ministry of Telecommunications of Côte d'Ivoire, 37.45% of the population of Côte d'Ivoire surfed on the internet in 2015, a remarkable increase compared to 5% in 2013. And the growth of mobile network users is an important reason. Banks began to cooperate with the internet and telecommunications operators. Orange and MTN provide small-amount mobile payment service for the users through cooperation with the banks. The cell phone company Moov cooperates with the postal service agency to realize the online transfer through postal service. In 2012, the government issued 3G license and the African Coast Europe-ACE project made the access to the broadband network even more convenient in Côte d'Ivoire and provides more services to the consumers through integration of land telecommunications lines, the internet and mobile network. The government plans to install 7000 kg-long optical fibre lines to improve the coverage of telecommunications service in the country. According to statistics of ARTCI, the popularity rate of mobile phones in the country was 97.5% in 2014 and 99.5% in 2015.[3] In August 2006, the West African Central Bank has established an automated interbank payment system to reduce delays in bank settlement operations. Although credit cards are rarely used in Côte d'Ivoire, the relative novelty of electronic payment systems has helped facilitate the rapid growth of e-commerce. In 2014, MTN partnered with online retailer Jumia (www.jumia.ci) to provide users with a full range of online and mobile services. Thus, it seems that e-commerce in Côte d'Ivoire is very promising.

21.4.2 Characteristics of E-commerce Development in Africa

(i) *African countries rely on imports and the e-commerce accelerates the supply and demand*

The development of import and export trade in Africa is unbalanced. What is exported are pure traditional agricultural products and low value-added products. The countries are relatively backward in industry and mainly focus on a single agricultural economy. Daily necessities, industrial products and high-tech products are all imported. E-commerce has now sprouted in Africa and is rapidly growing to cover major African cities. Benefiting from the e-commerce platform, the imported products can reach more and more people, the supply and demand is promoted, the sales

---

[2] https://www.export.gov/article?id=Kenya-eCommerce.

[3] https://www.export.gov/article?id=Cote-D-Ivoire-ECommerce.

channels are expanded, the turnover links are reduced, and a new investment boom is sparked. E-commerce is gradually accepted by more consumers as time is saved and security is improved by shopping online. Online sales and offline stores complement each other. In this way, consumers have more shopping options, the goods are cost-effective, the competition is fair, and there is a wide range of consumers. Hence, e-commerce can be considered as a booster for future economic development.

(ii) *The Internet develops rapidly and the enterprise e-commerce channels are gradually improved*

Africa's network infrastructure is gradually improving, and the Internet service popularization is being stepped up in many coastal areas and major port cities. The society-wide informatization is developing rapidly, and African governments are encouraging e-commerce development and increasing investment in e-commerce to lay the foundation for the popularization of e-commerce. Regulators have introduced relevant policies and regulations to address the fearfulness of the population towards online shopping. Meanwhile, enterprises are also investing more in e-commerce and gradually improving their online shopping platforms so that they can reach major cities and port cities. The e-commerce channel of the pillar enterprises in each country has been gradually established and has become an indispensable means for business transactions.

(iii) *The e-commerce development region is concentrated, and second-tier cities have great potential*

The transportation network between the capital cities and coastal cities in Africa is getting better and better, and the cable coverage and smartphone coverage are increasing year by year. No monopoly e-commerce platform has emerged, together with the rising middle class and increasing consumer demand, it is evident that the development opportunity of e-commerce has arrived. The major cities in Africa are located on the coast. The logistics and transportation network are well developed, the population is concentrated, the economy is prosperous, and e-commerce is concentrated in the capital and coastal port cities. Although the development of second-tier cities is restricted, but the demand for daily necessities and technology products such as cell phones is also considerably large, so there is great potential for the development of e-commerce platforms in second-tier cities.

(iv) *The infrastructure of inland cities lags behind, and logistics and distribution become a bottleneck for development*

Africa's inland cities have a lack of road and rail transportation resources, slow waterway transportation, and expensive air transportation, so logistics and transportation issues are the condition limiting the development of e-commerce. The population in Africa is very scattered and the streets are not equipped with door signs and mailboxes, resulting in low accuracy and long delivery time of logistics and low satisfaction of online shopping experience. Logistics companies operate with high efficiency compared to postal transportation, but the costs are also high.

Therefore, large e-commerce platforms build their own logistics systems and establish warehouses and distribution centres in African countries to expand their vertical value chains and support industrial development. In addition to the impact of transportation facilities, the import policies and terms and conditions of African countries that rely on imported products vary greatly, tariff issues constrain the development, and many goods are backlogged in the port. Therefore, the establishment of a unified trade development policy and terms and conditions as well as taxation is one of the urgent issues to be resolved.

(v) *Banking service is more backward and there are multiple electronic payment means*

Payment in Africa is still predominantly in cash. Due to the public's shopping habits and distrust of online transactions, online purchases are also mostly cash-on-delivery transactions. As for bank cards and credit cards, the development of bank card and credit card business is slow because of high service fees, few outlets, single service targets and a wide range of materials. As a result, African countries are embarking on e-banking promotion, reducing procedures, facilitating payment methods, improving e-wallets, e-money, and mobile banking services, and introducing diversified online payment methods.

(vi) *Offline trade is still the mainstream and the concept of network trading needs to be changed*

E-commerce accounts for a low percentage of total trade in Africa. In particular, trade in small and medium-sized cities is still confined to the form of bazaars and convenience stores, and large sellers and hypermarkets have not reached the hinterland. Meanwhile, the offline trade has few products, an imbalanced structure and small point-of-sale coverage. So online trade is urgently needed as a supplement. Due to the low education level of most African people, their low awareness and acceptance of the Internet economy, and their concern about the security of online payment, as well as the deep-rooted traditional cash-and-goods transaction model and inconvenient logistics and distribution, the task of e-commerce inlandization is very difficult. Consumers must change their consumer mindset and recognize the positive impact of online transactions.

21.5 Outlook on China's Investment in E-commerce (B2C) in Africa

Overall, the prospect of B2C e-commerce development in Africa is optimistic, but there are still shortcomings and deficiencies in economic infrastructure and supporting facilities. Only through continuous improvement and learning can the e-commerce development in Africa bring practical and reliable benefits to African people and bring new vitality to Africa's economic and information technology development.

21.5.1 Communications and Transport Infrastructure

China-Africa relations have a long history. In recent years, China has proposed the "Belt and Road" development strategy, which is a driving force for bi-lateral economic trade between China and Africa, and also provides favourable policies for infrastructure construction in the Belt and Road related countries. In view of the imperfection of communication infrastructure and the backwardness of transportation infrastructure in the relevant African countries at present, plus the urgent desire of African countries to improve the current situation, but the lack of technology and experience in information and road traffic facilities construction, in terms of investment in Africa, China can first focus on the construction of infrastructure areas: fibre optics, cables, railroad, highway, airport and port.

21.5.2 Logistics Industry and Delivery System

Boosted by e-commerce, China has built a diversified logistics system and has become one of the world's largest countries in terms of express business. While maintaining years of rapid growth, China's express giants should accelerate the process of globalization. Considering the competitive pressure of the logistics industry in the developed global economy and responding to the national "going out" and "Belt and Road" strategies, Chinese enterprises can take Africa as the first choice for investment in the logistics industry. At present, the express logistics system in most African countries is still immature, with a single form of logistics, small scale of logistics companies and poor informationization level, etc. Stronger logistics companies are needed to enter and help them establish a more complete and efficient express logistics system. Therefore, it is suggested that Chinese express giants may consider investing in the African market by acquiring smaller local logistics companies and solving problems such as cultural barriers and policy constraints with their localized resources. In the early and middle stages, investment should be focused on large and medium-sized cities, and the "last mile" distribution system covering communities and streets should be established as soon as possible. When the time is right, the distribution system in populated villages should be established gradually. Meanwhile, cooperation with local e-commerce platforms, cross-border e-commerce platforms, trade platforms, etc. should be strengthened to broaden business channels and use the large-scale express business volume to reduce logistics costs in Africa. If possible, the construction of an intelligent delivery system should be explored, and new delivery modes such as delivery by drones should be actively explored to reduce the negative impact caused by the lack of couriers and poor transportation facilities. With a series of investment and business expansion, we will gradually form a perfect distribution system, so as to develop synergistically and win–win with the African B2C e-commerce market.

21.5.3 Multi-level E-commerce Development Platform and Cross-Border E-commerce Trade

Driven by the process of globalization and informatization, China's cross-border e-commerce trade is booming in the Asia–Pacific, European and American markets, but has not yet ventured into the African region. Chinese e-commerce giants should see the dependence of African countries on imported products, and high-tech products, and focus on breaking through logistics and tariff barriers to develop cross-border e-commerce trading platforms in Africa. In the meantime, with further development of information and communication technology in Africa, the social environment is gradually stabilized, the economic environment is gradually improved, and the consumption level in Africa remains high. While African regions rarely see strong B2C e-commerce platforms. Therefore, it is recommended that Chinese B2C e-commerce giants seize the opportunity to establish influential large-scale comprehensive e-commerce platforms in Africa. To avoid obstacles caused by culture, language, currency, laws and regulations, enterprises can adopt acquisition and joint venture strategies to introduce advanced technologies and concepts of domestic e-commerce development into Africa based on small local e-commerce platforms in African countries. Finally, considering the differences in the level of economic development in Africa and the regional characteristics of online shoppers, it is suggested that Chinese investing enterprises can first frame the scope of e-commerce services in large cities and port cities with good economic footing, while locking online consumers in high-income groups above the middle class. After the B2C business in Africa gradually matures, then they can expand from large cities to small and medium-sized cities, and gradually develop e-commerce trade.

21.5.4 Bank Cooperation and the Third-Party Payment Platform

E-commerce has sprouted in Africa, and countries have launched e-money and third-party payment means one after another. However, the number of users of third-party payment platforms in Africa is still relatively small due to the limited scope of services and insufficient security. Chinese smart terminal manufacturers, telecom operators, Internet companies and others can launch third-party payment services in Africa based on their own platforms and advantages. Considering the access, reliability, security, and user volume of payment platforms, in terms of investment in payment platforms, Chinese enterprises should first pay attention to communication with local governments to support and encourage the entry of Chinese third-party payment platforms; then they need to consider cooperation with banks in African countries to dock the banking system with third-party payments; finally, they need to consider the acceptance of users, and can carry out application promotion through new media publicity on the Internet and cooperation with offline physical stores to improve the credibility of the platform and develop the user population.

21.6 Conclusion

Through a large amount of literature research, we found that the e-commerce in Africa already has some characteristics. Compared with the traditional retail industry in Africa, the share of e-commerce is still small, but it is in a period of rapid growth, and the contribution of e-commerce cannot be ignored; e-commerce can be of great help to most African countries that rely on imported industrial products. At the same time, with the help of various favourable factors, enterprises have also increased investment in e-commerce, gradually improving the online shopping mall platform, and making it enter major cities and port cities.

We also found that African E-commerce develops intensively relying on the capital and coastal port cities with well-developed logistics and transportation networks, concentrated population, and prosperous economy. African countries attach great importance to the development of communication digital technology and actively improve communication infrastructure. The transportation infrastructure in most African countries is weak, and logistics and transportation problems are the constraints for the development of e-commerce. Finally, E-commerce platforms and mobile payments have entered people's lives and shopping.

In view of the actual situation of the development of e-commerce in Africa, we put forward the prospect of China's investment in this field, including four aspects in total. China's investment can first focus on infrastructure, such as investment in African optical fibre, cable construction, railway transportation, road transportation, airports, and port construction. Secondly, it is suggested that China's investment in e-commerce platforms should be framed in large cities and port cities with good economic foundations. Thirdly, it is recommended that Chinese logistics giants consider investing in the African market by acquiring smaller local logistics companies and use their localized resources to solve problems such as cultural barriers and policy constraints. Finally, China's smart terminal manufacturers, telecom operators, Internet companies, etc. can launch mobile payment services in Africa based on their own platforms and advantages.

References

Alain, D., & Kwadjane, J. -M. (2015). *Le commerce électronique en Afrique, Institut de Prospective économique du Monde méditerranéen (IPEMED)*. https://www.ipemed.coop/fr/publications-r17/collection-construire-la-mediterranee-c49/le-commerce-electronique-en-afrique-maroc-tunisie-senegal-et-cote-divoire-a2751.html

Ndiaye, A. (2000). *Développement du commerce électronique en Afrique: le cas du Sénégal.* https://fr.slideshare.net/siliconvillage/dveloppement-du-commerce-electronique-en-afrique-cas-du-senegal

Part VII
China-Africa Collaboration in Finance and Renewable Energy

Chapter 22
Sino-Ghana Collaboration on Cryptocurrency: Practices and Implications

Yan Zhao◉ and Gideon Kinnah◉

22.1 Introduction

A story is told of a young Laszlo Hanyecz buying pizza with 10,000 btc on 22nd May 2010 (Forbes, 2022). A decade after, the same 10,000 btc is worth more than $300 m. This rapid appreciation in value signifies and ushers in yet another evolution phase of the financial system. It has proven financial commodities can be built outside the traditional system with exponential value growth and possibilities. The era of cryptocurrencies has made possible faster, cheaper, peer-to-peer global transactions. The innovation is built on a decentralized system which makes crypto transactions possible without going through third parties/intermediaries; banks, governments, and financial organizations. This makes up the core attractive feature of this new system built on a blockchain of immutable records. This brings in the concept of a decentralized system and anonymity. Contrary to the traditional way of exchanging value, cryptocurrencies allow engaging in e-commerce without the well-known KYC principle. Against the backdrop of rapid growth and efficiencies, the decentralized anonymous system poses challenges and risks of its own. Talk of fleecing activities cooked in fraud, Ponzi schemes, nefarious activities that fuel cybercrime and other terrorist activities.[1] Cryptocurrencies began on a quiet note, diffused into tech nerds, academia and into mainstream fintech garnering the attention of the global

---

[1] Not to say the decentralized anonymous feature is all evil. Same feature made it possible for funds to reach people fighting just course like the End-Sars protest in Nigeria, resistance forces in war-torn countries among others.

Y. Zhao (✉)
School of Management, Shanghai University and Director, Center for Innovation and Knowledge Management, Shanghai University, Shanghai, China
e-mail: zhaoyan87@shu.edu.cn

G. Kinnah
School of Management, Shanghai University, Shanghai, China

M. Muchie et al. (eds.), *China-Africa Science, Technology and Innovation Collaboration*, https://doi.org/10.1007/978-981-97-4576-0_22

405

finance industry. Following the innovation curve, birthed many altcoins and stable-coins among other use cases. Its evolution has given way to yet another disruptive wave in the global financial structure. Defi: decentralized finance, an open and alternative global financial platform running autonomously on cryptocurrency without a single person or central authority. Now, one could ask what this means to traditional finance or cryptocurrency as we knew it. Defi is built on smart contracts primarily the Ethereum blockchain completing specific functions determined by underlying smart contract codes. Defi cuts across global remittance, access to stable currencies, trade-in tokens, crypto savings, and even traditional banking functions like insurance, collaterals, and building your portfolio. The booming crypto economy has given people the opportunity to borrow or lend in both short and long terms, earning interest and financial liberty. The growing development in the crypto economy has sparked conversations on control, stability, legality, and regulation of cryptocurrencies and its related innovations in various economies, and regions around the world. Amidst these concerns, Meta's Libra, state-issued virtual currencies (mostly in development) have surfaced in recent times. Cryptocurrencies have different adoption levels across the globe among local practices. The 2021 Chainanalysis report puts worldwide cryptocurrency adoption at 880% with P2P platforms leading this expansion. Likewise, various governments and regulating bodies have diverse treatment and positions on cryptocurrencies. For instance, China prior to 2020 processed 90% of global cryptocurrency transactions due to cheaper labour and power supply. Today, the mining, promoting, and processing of cryptocurrency-related bank transactions are banned in the mainland owing to changes in policy and an attempt to address challenges posed by the decentralized system (Arjun, 2022). USA stands on the flip side, decriminalizing innovation, and classifying virtual currencies as commodities under the Commodity Exchange Act while putting efforts into its future development. This enabling environment has led to the term 'crypto-immigration'—companies moving their mining operations from China to the US. China and the US by far represent 2 ends of cryptocurrency treatment, with a majority of countries of the developed and developing world spread between these ends.

Sub-Saharan African countries are also distributed among this divide. 20% of sub-Saharan African countries have banned crypto assets, with others implementing some form of restrictions. Ghana, Kenya, and South Africa have enabling environments for the adoption and use of cryptocurrencies, despite not being accepted as legal tender (Chainanalysis, 2021). Nigeria in 2021, took an entrenched position to ban cryptocurrency transactions and its development within its borders. According to Chainanalysis (2021), Africa comparatively has the smallest cryptocurrency economy, yet it's the most dynamic and exciting region in the crypto space. In 2021, Africa recorded a 1200% growth in cryptocurrency market value. Its adoption continues to deepen owing to the growth of P2P platforms, easier means for P2P remittances as well as international transactions both within and outside the continent (Chainanalysis, 2021). This shows cryptocurrency has huge adaptation potential in the region with a 3% share of global value received. Users out of necessity turn to the asset class, preserve savings against unstable local currencies, and remittances previously cumbersome to make via traditional banks. Ghana, an emerging cryptocurrency market has a global adoption index score of 0.14 with 10 P2P exchange trade volumes. Bank of Ghana in its April 2022 notice, reiterates its position on cryptocurrency not being legally accepted in Ghana, urging investors and users to exercise caution. In the past couple of years, the government through the central bank of Ghana has taunted the intention to adopt and launch a state-backed digital currency. It has also launched a Sandbox initiative tailored to the exploration of blockchain, cryptocurrency adoption and further development to find practical solutions to local problems in the finance and commerce industry (Sandali, 2021).

According to Chainanalysis 2021 global cryptocurrency report, China has a cryptocurrency adoption index of 0.16, with 155 P2P exchange trade volume (Defi 0.62 2 on-chain retail Defi value received). Before the Chinese Communist Party crackdown on cryptocurrency operations in the mainland, it controlled most of the global crypto mining. The government cited concerns about stability and environmental impact. However, experts believe, this change in policy tackles huge capital flight often evading state control, curbing the demand for cryptocurrency assets, consolidating the sovereignty of the Chinese Yuan and to pave way for the roll-out of the Central Bank Digital Currency (CBDC). The Chinese digital yuan takes a cue from Blockchain innovation. Compared to the flagship Bitcoin, the CBDC is programmed to handle 300,000 transactions per second (Forbes, 2019). In contrast to the financial liberty brought by cryptocurrency, the CBDC provides a deeper look into how people spend money, giving the central more control over financial flows. China through this development has carved its path in blockchain and cryptocurrency innovation. Owing to a wealth of experience and research into the underlying technology, its implication to the Chinese economy and their global ambition. China is Africa's biggest development partner (The Economist, 2022). In recent years, China has invested heavily in development programs in various African countries from agriculture, roads and transportation network, health, technology, and infrastructure, among others (Xinhua, 2021). Africa to a fair extent does benefit from Sino-Africa collaboration and a learning opportunity to development programs like the purported poverty alleviation program of President Xi Jinping. Extending the Sino-African collaboration to blockchain and cryptocurrency-related innovations opens various governments and the African economy to a wealth of experience, and technical ability to carve their cryptocurrency future at a much faster pace.

In 2021, Ghana was commissioned as the African Continental Free Trade Area (AfCFTA) secretariat. AfCFTA creates a combined market of 1.2 billion people with a $3 trillion GDP (Boateng, 2019). Before AfCFTA came into effect, intra-continental transactions were heavily dependent on the US dollar. The Ghanaian banking industry went through a restructuring phase in 2018, coupled with development growth in the crypto space sparked social conversations on Ghana's response to these growing new technologies. Others ask how Ghana could take advantage of this strategic position. Practices are peculiar to economic and local settings. We use the narrative approach to build our analysis on cryptocurrency/blockchain from the lens of a Sino-Africa collaboration. The chapter first seeks to inquire about cryptocurrency practice and its implications in Ghana. Building on that, it also discusses how a Sino-Africa relationship in the context of cryptocurrencies. and blockchain technology could be of benefit and or influence Ghana on its digital transformation journey carving its crypto future.

22.2 Cryptocurrency Practices and Implications

22.2.1 Cryptocurrency, Blockchain Fintech Ecosystem

The crypto economy disrupts the financial industry, redefining how financial solutions are built. Ghana is the second-largest economy in West Africa. It's a hub for mobile payment systems in Africa. The fintech industry in Ghana has undergone several evolution and diversification phases. The government of Ghana through its Digital elite (Digital Ghana Agenda) policy through the Bank of Ghana launched a Sandbox and Fintech and Innovation Office (FIO). This strategic step creates a conducive atmosphere for innovations within the financial sector prioritizing blockchain. Before this initiative, the Ghana fintech services were primarily underpinned by foreign telcos operating in Ghana. The adoption of entwined cryptocurrency and blockchain created an opening for young fintech startups to leverage the innovation to provide blockchain and cryptocurrency-backed services. The African market is worth over $3 trillion. However, it's riddled with trade barriers that limit holistic access to this large market by both SMEs and individual consumers in the continent. Through the practices of blockchain and cryptocurrencies, fintech startups like JangoSend (formerly Sesacash) provide a digital wallet that gives access to about 20 traditional and global digital currencies. This aims to curb the 50%-60% of individuals resorting to carrying cash for cross-border payments within the continent. The platform has a crypto cashback as part of its growth strategy. Thus, instead of a point-based system, users are rewarded with cashback in cryptocurrencies for transactions and the community involvement-induced affiliate programme. Ghana's Nestcoin in 2022 first quarter, raised $6.45 m pre-seed to accelerate crypto and web3 adoption in the African market. The project purports to be powered by the decentralized financial system—Defi. The IMF encourages adapting to the blockchain/cryptocurrency technology is significant to reap its benefits. Building local infrastructure and services will accelerate its adoption beyond crypto trading (Kinnah, 2022). Mazzuma also leverages the cryptocurrency Defi distributed ledger to develop its token system that facilitates seamless cross-border transactions. Their architecture breaks down the cryptocurrency complexities into easy user-friendly modes. Accessibility is core to digital and financial inclusion, crypto backed practices such as these fintech consolidates to foster and achieve the 85% financial inclusion goal by 2023.

22.2.2   None-Fungible Tokens (NFT) Markets in Ghana

The disruptive nature of cryptocurrencies has changed the paradigms in the financial system. Cryptocurrency also promises financial liberty and digital inclusion. Its development has gone beyond mere remittances between peers. Talk of Initial Coin Offering (ICO) issued in cryptocurrencies, altcoins and stablecoins and Defi development. None-Fungible Tokens (NFT) encompasses notable development in the cryptocurrency ecosystem. NFT in recent times have made massive waves in the crypto economy (Chainanalysis, 2022). One will ask how NFT is related to cryptocurrency. NFT is also built on the decentralized blockchain system of cryptocurrency, popularly on the Ethereum and Bitcoin blockchains. In practical terms, NFT is a sub-product of cryptocurrency. They are purchased, used and stored in a decentralized ledger. Like cash, Bitcoins, Ethereum and other cryptocurrencies are fungible with economic value. Thus 1 Btc = 1 Btc, 1ETH = 1ETH in value trade across the chain. However, None-Fungile-Tokens are especial and different from each other. It envelopes a set of art, GIFs, videos, collectables, Music, tweets, etc. Basically, it can make an NFT out of anything. It is like the traditional auction system done in a digital form. NFT also divides between sceptics and supporters. Many people doubt its authenticity and sustainability. Some experts view it as a practical key to Metaverse or web3. Chainanalysis puts the NFT market at a whopping $41 billion market as of 2021 (Dailey, N. 2022, Chainanalysis, 2022).

Studies into cryptocurrency show a youthful and dense population in the relatively small Ghana crypto space (Kinnah, 2022; Owusu, 2020). The youth continue to spearhead the adoption and adaptation of cryptocurrency and related innovations to solve local problems and as a lever to the global market. The Ghana- Africa crypto economy has caught on to the token system and NFT seeking to have its share of the $ 41 billion market. A video of Ghanaian pallbearers dancing with coffins on their shoulders went viral during the 2020 pandemic lockdown. It later became an internet meme.[2] In April 2022, a NFT of the dancing pallbearers was sold for 372 ETH ($1.046 million). It's currently the highest-sold single NFT in the African crypto economy. Why pay for expensive art or video when it can be screen crabbed, or even found freely everywhere on the internet? NFT gives ownership of an original item. It contains built-in authentication, serving as proof of ownership verifiable in the blockchain. The Royals, AzeAppSociety, and Fufu Sapiens, among other fintech startups and communities in Ghana, are tapping into the rich African cultural heritage, showcased in diverse art, and sold as NFT. The system allows creators to earn a greater portion of their works and royalties in perpetuity whenever a NFT is sold to a new buyer. Ghana has oceans of untapped creativity and innovations. NFT is seen as an innovative solution to the lack of proper intellectual property rights and royalty systems in the Ghanaian creative industry. Token systems and NFT are relatively new with unknown variables and future prospects. Africa is a major hub for digital transformation and development in the crypto economy. Cryptocurrencies not only

---

[2] Urging people to stay safe during covid or be ready to be carried away in a coffin. The meme also relates to daring or risky activities of all nature, indicating danger or seemingly bad after fact.

offer digital inclusion to the emerging African market or reducing migrant remittance cost, but it also offers a unique opportunity for generating value and access to global markets with minimal barriers.

22.2.3   Gaming and Virtual Token

Cryptocurrencies and its entwined blockchain technology were the missing item in the push for global digital transformation in this IoT era. The decentralized system is the soul of derivative innovations in the crypto space. The dynamics of digital inclusion and financial inclusion work in datum. In the advanced worlds, gaming goes beyond hobbies, it has added financial value to the gamers themselves. It's evident in the sale of virtual game tokens, rewards/prices from gaming tournaments and championships to the extent people have made profitable careers in gaming (Bányai et al., 2020). The gaming industry started decades ago. However, the African market often lags in access to the latest innovations in the gaming industry. It's a combination of factors such as developing the gaming culture, the high cost of internet services, and purchasing power as gaming is an expensive hobby. Gamers analytics and Carry1st 2021 report 24 million gamer population in Sub-Saharan Africa. Ghana accounts for 27% of the gamer population (Kpilaakaa, 2022). A greater majority of them are mobile gamers, predominately for the sheer fun of gaming. The launch of Metaverse sparked some excitement in the gaming industry, showing a glimpse of what future gaming could be like in the virtual world. Combining cryptocurrency and blockchain innovation with virtual gaming breaks the financial inclusion barrier to the emerging African market offering gamers an opportunity to generate monetary value in the real world through cryptocurrency while playing their favourite games. The Ghana gaming market is picking up on this new gateway to generate active and passive incomes via NFT, cryptocurrencies adding to the practices in the crypto ecosystem. Virtual gaming companies like OpiPets have expanded their non-fungible token (NFT) play-to-earn (P2E) games to the Ghanaian market. Metaverse Magna (MVM), a Ghanaian company also provides a fun and safe space for gamers to compete and earn rewards in cryptocurrencies. It has an embedded community style that connects gamers. The MVM platform currently has over 10,000 gamers across Africa. This initiative further lowers the entry barrier to the adoption of decentralized gaming opening game lovers to the broader global gaming world. It's leveraging the crypto economy to access untapped crypto-gaming markets across Africa. This further narrows the digital inclusion and financial inclusion gap between advanced markets and emerging markets in Africa.

22.3 Implications of Cryptocurrency Practices in Ghana

22.3.1 Access to Global Market and Financial Inclusion

We can draw a nexus relationship between the advancement of blockchain and digital inclusion and financial inclusion. The blockchain centralized system affords the necessary tools to build other value-creation financial solutions. These solutions contribute to the furtherance of the digitalization of the economy. Sub-Saharan Africa has the largest unbanked population. Cryptocurrency development affords individuals to explore other financial solutions. The average person has more access to be part of the global financial system in the crypto economy by engaging in crypto trading, and access to other useful and affordable financial solutions and products. Ghana has the largest mobile money payment system in West Africa. The success of mobile money payment gateway corresponds to opening access to the larger unbanked population. This encompasses low- middle-income earners not captured in the formal financial industry. The poor and low-income earners in the informal sector are often trapped in a cash economy. Mobile money platform builds a foundation important to the advancement of digital transformation in the Ghanaian economy. Cryptocurrency adoption in emerging markets benefits from easing the transition into a decentralized system as market stakeholders become more conversant with mobile payment. The interoperability development on the Ghana interbank payment services (GHIPS) platform amplified the adoption of mobile payment. GHIPS made it possible for remittances across all telco networks, across and within banks. Through this development, Ghana made grounds for financial inclusion. Until May 2022, the mobile money payment growth began to stall due to panics brought on by the 1.5% electronic tax (E-Levy). However, regardless of the effect of e-levy, the mobile payment system has limitations to the extent of financial inclusion.

Practical adaptation of cryptocurrencies and its backing blockchain further opens the individual and businesses in the Ghanaian economy to a much larger global market. Cryptocurrency extends inclusion beyond the local bothers into the bigger African and world markets. Through the efforts of fintech providing solutions backed by the decentralized finance system and accessibility to global virtual currencies, businesses and people can engage in business and other economic activities over the internet with minimal barriers. For instance, the NFT market built on the blockchain system showcases Ghanaian NFT to global buyers—dancing pallbearers sold for 372 ETH ($1.046 million). Cryptocurrency reduces the remittances cost compared to the traditional means of exchanging value across borders. Businesses are encouraged to expand markets beyond their geographical operations as business transactions now could be done much faster and easier through decentralized fintech platforms with minimal interference from the financial institutions.

22.3.2 Cryptocurrency Implication on Taxes in Emerging Markets

Cryptocurrency and blockchain are hoped to be the calvary to emerging markets of Africa. A calvary to provide alternatives to consistently dwindling local currencies, further deepen digital inclusion to access global e-commerce markets, etc. Many people also hoped for the decentralized system to end the imbalance of huge remittance costs in developing economies deemed as high-risk regions (Aysan and Kayani, 2021) Yes, the innovation has created alternative solutions and global financial inclusion for many across Sub-Sahara Africa. It revolutionized the dynamics in the financial industry on how value is created and exchanged. The decentralized and autonomous feature rails in concerns on regulation among other treatments of cryptocurrencies. Ghana an emerging market proves to be of high potential benefit to the development and adoption of cryptocurrency and its entwined blockchain technology. Various financial products and services backed by decentralized technology are built on existing local financial practices while cunning innovative solutions. Relevant regulatory bodies in Ghana have a lazy-fare attitude in effectively reacting to the full extent of the disruption brought by cryptocurrencies.

With cryptocurrencies, their derived products, solutions, and services bring in new and alternative ways of generating value and exchanging value. It is an existing practice in traditional finance to tax gains, assets and even expenses (such as tax of remittances). The Bank of Ghana, responsible for the creation, and distribution of legal tender and regulates the financial markets, has reiterated its position and treatment of cryptocurrency, its derived products, and solutions as illegal within the jurisdiction of Ghana (Bank of Ghana, 2018). The Ghana Revenue Authority likewise have no stipulated tax guidelines on digital currencies and their derived products. Exchanges and platforms in the crypto ecosystem are mostly registered under the fintech provision by the Bank of Ghana. These entities are liable to pay tax on business gains. However, individuals earning fortunes and incomes in the cryptocurrency space have no legally stipulated guidelines to pay taxes. Bear in mind, most of the youth in Ghana engage in Bitcoin for its value gain opportunity and the alternative earning potentials in the cryptocurrency ecosystem. Neither NFT, crypto trades nor tokens are captured as asserts, therefore they are not expressly subjected to any assert gain taxes or personal gain tax. The digital economy is a trillion-dollar industry and keeps expanding. The government of Ghana stipulated e-levy in May 2022, to tap into the digital market previously not captured under the tax net. However, this new tax policy does not effectively tackle the crypto ecosystem. The interoperability system allows all digital transactions to be tax 1.5% subject to certain exceptions. However, the cryptocurrency ecosystem goes beyond the Ghana payment platform interoperability. The Ghana cryptocurrency market continues to grow amounting to billions of dollars in value. Therefore, a larger amount of gains realized from the crypto market go untaxed compared to other economies like America, Canada, Portugal, Spain, and Turkey, with clear tax guidelines on treating cryptocurrency markets (Forbes, 2019). Such treatment creates a conducive space to evade taxes.

Extending the tax net to the crypto ecosystem could curb such practices, broaden the tax net, and ease heavy taxes imposed on other core industries like ports and other commerce.

22.4 Sino-Ghana Collaboration on Cryptoization Conundrum

In recent times, China has deepened its engagement with Africa through various economic and development initiatives. The Sino-Africa relationship can be categorized into political inter-trust, cultural exchanges, economic cooperation and peace and security. At the 8th Ministerial Conference of the Forum on China-Africa Cooperation (FOCAC), President Xi expressed a key commitment to the Sino-Africa cooperation; 10 billion US dollars of trade finance to support African export, 10 connectivity projects for Africa, form an expert group on economic cooperation with the AfCFTA secretariat, establish a China-Africa cross-border RMB centre, etc. The Chinese provided an alternative support (in lieu of the west) wrapped in a supposed win–win cooperation agenda (Xinhua, 2021). There are diverse interpretations of the win–win or the actual cost and benefits of the Sino-Africa relationship (Wang, 2018). Some may view it as low interest (loan) development partnership with minimal governmental interference.

China has long deep ties with Ghana. History puts the start of Sino-Ghana relationship to the Osagyefo Dr Kwame Nkrumah reign, where Ghana established trade relations with China. China from then has been an important trading and development partner. Ghana was a significant ally in China's readmission into the WTO. The Sino-Ghana relationship has predominately existed at the diplomatic and institutional levels with little reflection on people-to-people ties. Throughout the 4th republic, the Peoples' Republic of China continues to have bilateral relations, a significant trade partner for Ghana's natural resources, infrastructure development, research, and academic exchanges. The political stability and economic growth have attracted Chinese companies in the automobile, and manufacturing industries to set up permanent operations in Ghana, serving Ghana and West African markets. There also exist periodic cultural exchanges. Some experts, researchers, opinion leaders and the media express the real cost of the Sino-Africa relationship far exceeds the face value to the detriment of Africa as a whole. Such positions call the objective research into the Sino-Africa relationship rather than going alone with popular opinions in the media. This chapter takes an insight into how extending Sino-Ghana relationship into the Cryptocurrency context could be of benefit to Ghana in carving its own crypto future. We first investigate CBDC status and development by the central banks of China and Ghana.

22.4.1 Central Bank Digital Currency (CBDC) Development in Digital Economy

The development of digital currencies or cryptocurrency was mainly fuelled by the 2008 global financial market collapse. It revealed a centrally controlled financial system is prone to failure. As a solution, a decentralized financial system was developed whereby control is distributed across the system with no single point of failure. Cryptocurrency is the missing link to an efficient IoT system under digital transformation. Cryptocurrency is built on an anonymous and distributed ledger and its value subject to the invincible hand of market demand and supply. Thus, transactions are not under the control of intermediaries and cannot be tracked in the system and their value is highly volatile. Cryptocurrency development requires technological expertise, and monetary investments in the forms of infrastructure and systems, just to mention a few. Emerging markets are incapacitated in developing a CBDC by a lack of proper infrastructure, lack of needed investment into research and development and advancement in technology. Sino-Africa relationship is a major source of trade and development. In moving forward from a cash economy into a digital economy, tapping into cryptocurrency, how best can the Sino-Africa partnership be leveraged to this digital transformation process in win–win cooperation. China is an important global player with advanced technological development and financial structure. It's the biggest cashless economy of the world. It has a developed financial infrastructure, primarily spearheaded by private fintech. Alipay and WeChat Pay dominate currency circulation in the digital economy including B2C, C2B, C2C, G2C C2G. China checks all the needed resources and capacity to develop a state-backed digital currency. CBDC works in contrast to financial liberty, giving regulatory bodies more access and control in the financial economy. Major concerns of stakeholders in the finance world were the use of the system for criminal activities, regulation, and stability. The disruptive nature of cryptocurrency challenges the fundamental role of central banks: creation, distribution of and destruction legal tender, monetary, and financial regulation. In the last few years have seen major banks, fintech and financial institutions adapting cryptocurrencies as means of payment, investment assets or building on its decentralized system, blockchain. The rapid development of cryptocurrency prompted research into its implications to the financial market, adaptability into the mainstream financial market, etc. Out of these efforts came the idea of developing a centrally controlled digital currency that ushers central banks and financial institutions into the digital economy and also solves concerns brought by cryptocurrency innovation. With time various governments of the world invested in developing state-backed digital currencies. Developing a central bank digital currency (CBDC) is heavily dependent on the technological capability and infrastructure, financial investment, and political will.

22.4.2   Peoples Bank of China (PBOC) CBDC Design and Implementation

Cryptocurrency disrupted currency distribution and regulation. To some extent, it impeded the effectiveness of monetary regulation and policies. In recent times, financial sovereignty has become an important discussion than ever before. USA imposed taxes on Chinese goods during the US-China trade war, and weaponized the dollar to sanction Russia. Although China cited environmental impact, and stability concerns, these above factors hastened the investment to develop a CBDC. China played a major role in the initial adoption and growth of cryptocurrencies. This facilitated hundreds of new private currencies coming into the system. Coupled with the already privately dominated digital circulation threatened the tightly state-controlled economy. The People's Bank of China (PBOC) announcing the launch of a state-backed digital currency had embedded motives; cut off demand for private currencies that threatens the sovereignty of the Chinese Yuen, to minimize private control in the digital economy and deepen control on public financial spending (Kharpal, 2021; Forbes, 2019).

Cryptocurrency innovation further enabled the development of decentralized finance—Defi on which diverse financial solutions and products can be built. The People's Bank of China (PBOC) took a different route. The launched CBDC is designed to leverage its first-mover advantage to promote the internationalization of the eCNY. It's an effort to curb its dependence on the US dollar.[3] Cryptocurrencies are anonymous in design, the eCNY design gives the government through its central bank absolute control. The system leverages transparency and security. Thus, all transactions are tracked. China already has a sophisticated surveillance system even through fintech companies. Taking back circulation control in the digital spaces gives direct and further surveillance on how people earn and spend. The Chinese digital currency has the full economic functions of fiat currency. It has a 1:1 conversion ratio. It doubles as an electronic payment (del Castillo, 2019). Therefore, it can be used for transactions with or without the use of an internet connection. The Chinese CBDC is programmable for monetary policy measures and emergency cases. Thus, it can be programmed to carry out various quantitative easing programs, taxation and tax refunds, and emergency state support among other state-backed activities.

China has a robust financial infrastructure and fintech development. The eCNY implementation adopts a two-tier system approach. The CBDC is under the singular control of the PBOC. The central authority within its monetary policy controls and issue the digital currency to commercial banks, authorized fintech[4] and distribution entities which form the second tier. The second tier are responsible for user wallet registration, exercise bank and finance regulation activities like KYC, anti-money

---

[3] According to SWIFT February 2022 report, RMB has 2.29% share of international payments currency compared to 43.08% US dollar share. In the global trade finance market, RMB is 3rd with 1.92% share compared to 87.38% USA dollar share.

[4] During the 2019 singles day, the central bank partnered with Tencent and Alipay to distribute the digital currency.

laundry controls, customer due diligence, transaction records and data. This generated information is directly fed to the central authority. End users comprising of business entities and individuals access the eCNY by registering through the authorized second-tier operators (del Castillo, Forbes, 2019, Digital Yuan e-CNY white paper, 2021). The system is designed to ease the smooth acceptance of the digital currency. It gives the central authority total control on monetary policies and factors that affect the economic stability, consolidating national currency sovereignty.

22.4.3   Bank of Ghana (BoG) CBDC Design and Implementation

This section investigates how extending the Sino-Ghana relationship into the cryptocurrency context could benefit Ghana in carving its own crypto future. After learning of China's CBDC, we explore the status and characteristics of CBDC in Ghana. Development in the financial industry prompted discussions and investment into the possibilities of a Central Bank Digital Currency (CBDC). Various states across the world have research and developments in various stages. Ghana is a hub for mobile payment services in Africa. The uptrend development in the mobile payment system has furthered access to the unbanked population. Owing to global development in the financial industry and government digitalization policy, the Bank of Ghana in 2019 announced its intention to develop and pilot a state backed digital currency—eCedi. It's an integral product out of the Sandbox initiative focused on blockchain and cryptocurrency adoption and development. The Ghana payment landscape includes among others E-zwich and Gh-link, GhiPS Instant Pay and Mobile Instant Pay (operated by Telcos). According to Bank of Ghana 2020 report, it experienced uptrend growth in transaction volumes and values. The digital Ghana agenda interoperability made it easier for financial activities to be done across above payment gateways.[5] The Bank of Ghana's strategic goals aims to promote financial inclusion, promote digital payment adoption and to develop an integrated electronic payment infrastructure that will enhance the interoperability of payment for efficient, reliable, and safe and secured digital finance infrastructures, products, and services. The growth in mobile money payment fulfils the financial inclusion bid as it has become the predominate means of daily transaction, laying the fundamental blocks for future digital financial innovation (Bank of Ghana, 2019). The design of the Ghana CBDC aims at leveraging its existing goals to achieve digital finance inclusion in partnership with established fintech.

Ghana leads initiatives to develop and adopt a centrally backed digital currency in the Sub-Saharan region. The proposed eCedi is said to be in advanced development stages. The digital currency seeks to be made widely accessible to the public and

---

[5] Mobile Money had GH¢564.16 billion worth of transactions, GH¢ 6.98 billion total float balance (92.1% growth) with 38,473,734 registered mobile money account holders (18.5% increase) as of December 2020.

business entities. According to the Bank of Ghana eCEDI press release in 2019, it will be designed to be integrated into the existing payment infrastructure to minimize potential risks and consolidate its position as an intermediary. The eCedi is built as a Token-based CBDC. A token-based digital currency takes on the full functional characteristics of cash set at a 1:1 ratio. Although built on the blockchain system, its issued and backed by the central bank, acceptable for retail payments. eCedi CBDC is designed to be stored on wallet. Thus, users with the use of public wallet address can exchange tokens much like the traditional cash system. One of the goals is to ensure accessibility to the public in an effort to bank the unbanked with minimal technological literacy or access to internet connection. Based on these needs, the eCedi is designed to be compatible with hosted wallets used on web-based platforms, mobile apps and hardware wallets using secure and portable smart devices (smart cards, wearables) used by individuals. Ghana has a limited internet coverage and accessibility. Therefore, it's also designed to support offline transactions. The Ghana CBDC like any technology has its success contingent on end-user acceptance and adoption. Therefore, its design, distribution, and implication consider the current local user behaviour and integrated into existing financial infrastructure (GhiPS, e-zwich, Gh-Link, etc.). The overall user experience is expected to be simple intuitive with fewer steps required to register and complete a transaction with 24/7 tech support system.

According to Bank of Ghana, the implementation of the eCedi is built on 4 converging principles: Governance, Accessibility, Interoperability, and Infrastructure. Similar to the Chinese CBDC, eCedi infrastructure is tailored to consolidate the existing role of Bank of Ghana as the singular central authority while serving operational needs and roles of banks, fintech, consumers and special deposit-taking institutions. Based on this, a two-tier distribution architecture is adopted. It's the optimal and efficient approach solidifying the BoG role as the sole issuer, distributor, and regulator. The second-tier consist of banks, special deposit-taking institutions (SDIs), authorized fintech and added value services. These entities are tasked with user wallet registration, KYC compliance, Anti-money laundry controls, customer due diligence. All activities in the eCedi ecosystem are subject to stipulated banking regulation and monetary policies. The BoG is yet to announce it roll out strategy. However, it garners for the support of relevant stakeholders. Like cash has, eCedi zero interest rate. Its scalable and meets international security standard with no single-point failure.

Ghana is an important trade partner and economic hub in Sub-Saharan Africa. It's the gateway connecting its region and the rest of the world. Ghana currently holds the seat for African Continental Free Trade Area AfCFTA Secretariat and represents the region at the global security council. It has played an important role in cryptocurrency adoption in the sub-region. The Ghanaian economy is an epicentre for mobile payment and digital transformation. Embarking on an economic migration into a digital elite system requires an adjustment of government monetary practices. With that background, the eCedi is set to be programmable, adjustable into diverse government monetary programs, including government to people (G2C), people to government (C2G) payments, welfare, and other monetary policies. Positioning Ghana at the center of socio-economic development and security partner, the Ghana CBDC is expected to be adaptable for future global single multi-CBDC systems, regional commerce, possibly adopted as a means of payment within the AfCFTA for cross border transactions.

22.5   Areas of Sino Ghana Collaboration—How

The practical use of cryptocurrencies and its underlying platform goes beyond the financial industry. Talk of applications in energy, education, logistics, food and agriculture, data, just to mention a few. The advancement of cryptocurrency technology encapsulates the existence of supporting advanced technologies, technical knowhow and financial investment capacity. According to IMF, the real benefit of the technology comes from adapting to the technology beyond its face adoption.

22.5.1   Create Knowledge Base for Sino-Ghana Cryptocurrency Technology Transfer

Cryptocurrency and its entwined blockchain are in enhance advanced technologies, making it possible for collaborative efforts to take place. Like any other technology, it can be transferred. Transfer of technology is a global practice where establishments, business and countries import advanced technologies of which they do not have the capacity for home-grown technology. China initially supported the financial application of cryptocurrency in respect to mining and trading digital coins. Its policy changed to aggressively crackdown on such operation in 2020. However, China still sees the immerse relevance and use of the technology beyond just mining digital coins. Beyond developing a state backed CBDC, the China has begun pilot programmes involving 15 pilot zones (Helms, 2022; Kharpal, 2022). These programs are tasked to researching and testing cryptocurrency/blockchain application across education, health, trade, manufacturing, energy, government and tax services, law, and finance, and cross-border finance. This collaborative advancement approach targets optimizing the technology to cost efficiency, data sharing, and building credible systems. Results from this initiative adds to China's already advanced knowledge and expertise as a leader in cryptocurrency and blockchain space. China compared to Ghana is far advanced in technology. Historically, during its global dominance journey as both manufacturing and technology hub adopted the technology transfer approach to better its home-grown solutions and introduced new technologies. It used access to its large market to porch and attract advance technology from the west via foreign direct investments, joint ventures, etc. China also strategically sent its people to learn advance technologies from Europe and American universities and establishments (Fedasiuk & Weinstein, 2020).

Ghana envisions to have a fully digitized economy in 2030. The government of Ghana through relevant bodies have ongoing digital transformation in Banking and finance sector, port and harbour, government agencies, etc. Ghana doesn't have market the size of China nor does it have effective execution of the study abroad technological transfer programmes leading to brain drain. China has a well of experience and technological expertise in the crypto and blockchain ecosystem. China through its Sino-Africa is not an alien to deliberate technology transfer to Africa. Talk of joint research, technical cooperation, direct foreign investments, joint ventures. Access to advanced technology is pivotal in adapting cryptocurrency and entwine blockchain to advance digital inclusion. Sino-Ghana relationship can be leveraged through Knowledge base, organizational learning, joint-venture, technical cooperation, and strategic alliance to tap into the well of expertise of China in the cryptocurrency ecosystem while addressing stipulated concerns. Creating access to this technology base provide Ghana short-term and medium-term capability to further its digitization agenda applying cryptocurrency and blockchain technologies across industries.

22.5.2 Development of Supportive Infrastructure Systems

Digital transformation involves changing from current analogue systems to digital support equipment and infrastructure. Digitization also implies change from paper system to paperless systems. Cryptocurrency and the blockchain platform require smart and modern technological infrastructure support. One important element of cryptocurrency adoption is access to advanced internet connection and service. According to data from Statista, as of January 2022 Ghana has 53% internet coverage with an average cost of $0.66/1gb. Compared to advanced world, Ghana has an expensive and poor internet coverage and service. This adds to the digital divide that exist between Ghana and the global world. Living in the Accra—the capital city, one is not guaranteed access to fast, affordable, and reliable internet access. Internet access today is fundamental to access global markets, vital information, education, and a catalyst for economic development. On the contrary, China with state companies: China Mobile, China Unicom, China Telecom, have advanced coverage, fast, reliable and access to affordable internet services. China currently have 5G for both business, domestic and industrial use at very affordable fees. This has fuelled the establishment of cryptocurrency ecosystem, e-commerce, virtual learning, ever-expanding gaming industry among other technological and socio-economic developments. Taking out the recent government reacquisition of Airtel-Tigo Ghana, the telecom industry is dominated by foreign companies providing bad services at expensive rates compared to their home country. Upgrading Ghana to reliable, fast, affordable, and country-wide access to internet service will play an integral part in the digitalization agenda and financial inclusion. The mere means of breaking the digital divide and upgrading internet connectivity in Ghana, opens the portal for business development, and other positive ripple effects. Development partnership forms a core part of the Sino-Ghana relationship. Therefore, it's not far reaching to extend such relationship to infrastructure development that improves Ghana's ability to take advantage of innovation brought by cryptocurrency and blockchain. Such a collaboration could give access to technological systems and machinery. Speaking of internet—the building blocks for a digital economy, the government of Ghana through bilateral relationship with China, can partner with China mobile to develop a technology to further coverage and access to fast internet connection like 5G at a much affordable rate. Such government backed internet action plan takes back national control to lead digital transformation in the telecommunication industry.

22.5.3   *Expand Access to Renewable Energy in Ghana for Industrial Cryptocurrency and Blockchain Use*

With global digitalization and technological advancement, it becomes more important to look at the effect of energy consumption. Various environment-focused organizations and specialists have stipulated guidelines, goals, and commitment to this course. Recently, talks on sustainability, Go Green, net zero carbon, etc. have become trendy and a basic corporate target and requirement. Cryptocurrency, a derivative of digital transformation is an energy[6] intensive innovation. Therefore, the nexus between cryptocurrency, digital transformation, financial inclusion, and energy needs to be properly considered. Both proof of work and proof of stake of blockchain are energy intensive although the latter is relatively less. Energy is a cost of digital transformation: electrical systems and machinery. China is both a global leader in carbon emission from coal and development of alternative energy sources. Much of China's renewable energy was used in crypto mining adding to 42% in 2020 renewable energy powering crypto (Forbes, 2022). According to the Renewable Energy Act, 2011 (Act 832) of Ghana, renewable energy sources include non-depleting such as hydro, solar, wind, biomass, landfill gas, biofuel, sewage gas, ocean energy, geothermal energy, and any other energy source (Aboagye et al., 2021). Ghana has huge untapped sources of renewable energy. In the past decade Ghana has been plagued with recurring power outages. Several short-term solutions such as importing energy badge were put in place. However, Ghana's renewable energy continues to be untapped. To further develop cryptocurrency, CBDC and blockchain in Ghana entail more demand for energy. Energy for crypto mining firms, fintech, and digital systems, and to meet demand from businesses that might spring up as the country achieve financial inclusion and digital inclusion. Therefore, a Sino-Ghana relationship to advance the development of cryptocurrency ecosystem in Ghana could cover bilateral effort to expand the available energy capacity of Ghana. This could be done by investment in technologies and infrastructure that can tap into these sources. Such as large-scale solar farms, windmill farms, localized production, or assembly of clean

---

[6] Bitcoin consumes electricity at an annualized rate of 127 terawatt-hours (TWh): 707 kilowatt-hours (kWh) per transaction (11 × Ethereum) (Schmidt, 2022).

power companies like Sungrow, among other renewable energy programs in China to efficiently harvest, store and use renewable energy and prevent energy wastage.

22.6 Benefits of Sino-Ghana Collaboration on Cryptoization Conundrum

22.6.1 Financial Inclusion and Digital Inclusion in Light of Modern Crypto Economy

China is a long-standing development partner to Ghana. Extending the Sino-Ghana relationship to cryptocurrency development in Ghana has inherent digital inclusion and financial inclusion benefits. Tapping into benefits in the crypto ecosystem goes hand in hand with digital transformation. This is in the sense that, current financial and technological infrastructure needs to be updated to digital systems that support the rollout and adoption of cryptocurrency, digital currencies and their related products and services. Sino-Ghana collaboration points in a bid to advance cryptocurrency future could also benefit from socio-economic developments such as expanding coverage and access to high speed, affordable internet, digital infrastructure, and research base. The effect of such collaboration breaks the digital divide, opening the economy to global markets, leading to access to financial opportunities both within the country and across the world. This further deepens the Sino-Ghana relationship and provides China with an advanced market for advanced technological products.

22.6.2 Gain Strategic First Mover Advantage

Since 2018, African governments have invested into developing central bank-issued digital currencies and cryptocurrencies. Tunisia, Kenya, Nigeria, Senegal, Ghana have CDBC either in the research state or have piloted. African CBDCs such as Tunisia's e-Dinar, and e-CFA have failed while e-Naira continues to struggle. The success of an innovation depends on users' behavioural adoption reaction and acceptance. Ghana's proposed e-Cedi is still at the development state. Therefore, the country has the opportunity to learn from failures of CBDC by other African countries. With a Sino-Ghana collaboration mainly affords Ghana access to a worth of experience and technological know-how placing Ghana at an advantage to launch a CBDC that will meet end-user needs and expectations. Intra-Africa trade is heavily dependent on the American dollar for remittances. Holding the seat of the African Continental Free Trade Area (AfCFTA) secretariat coupled with proven political peace, stability, and security, with a vibrant economy, positions Ghana at a strategic advantage to enjoy a first mover advantage. Ghana isn't the first to launch or pilot a CBDC. However, with the success of e-Cedi (while tapping into technology capacity from the Sino-Ghana relationship) with its international use capability, Ghana could make a good case for it to be adopted as an acceptable mode of payment for remittances and settlement of goods and services within the continent. Such functionality will rid of the continent's heavy dependence on western currencies which go a long way to minimising debt and stabilising African local currencies.

22.7 Challenges of Sino-Ghana Collaboration on Cryptoization Conundrum

22.7.1 Ineffective Use of the Technology Transfer

Sino-Ghana collaboration even in the scope of advancing cryptocurrency development should be based on a win–win situation. Sino-Africa relationship is divided into 4 sessions: cultural exchanges, political inter-trust, economic cooperation and peace and security. Historically, economic cooperation has come in the form of technical cooperation, strategic alliance, and foreign direct investments (Xinghua, 2021). Sino foreign direct investments and joint ventures have come under criticism for their benefit to Africa (Wang, 2018). Some opinion is about ineffective technological transfer on technical cooperation in infrastructure projects and manufacturing foreign direct investments as most of the advanced technological applications are mostly rather done by Chinese nationals undermining the technology transfer motives. Collaborating with China on importing technology and expertise that will help Ghana to advance cryptocurrency/blockchain might not lead to the desired long-term benefit of technology transfer via knowledge base and organizational learning. When such happens, it brings imbalances to an ideal win–win Sino-Ghana collaboration. Thus, Ghana needs to improve its absorption capability to acquire the advanced technology transferred from China and enhance its innovation capability.

22.7.2 Holistic Trust in Chinese Technology

An underpinning value of cryptocurrency is the decentralized system that puts trust in its user base. Financial liberty plays an incentive role in the adoption of cryptocurrencies. The adoption and development of cryptocurrency bother regulation. Ghana and China have different stances on regulating cryptocurrency and related products and services. E-Cedi and e-CNY are similar in design characteristics and implementation structure. Both have the central bank as the singular authority. Ghana CBDC is yet to detail the extent of transparency embedded in the e-cedi. However, the e-CNY affords the Chinese government total access to financial data which can regulate how people earn and spend. Sino-Ghana collaboration to further develop a CBDC or cryptocurrency development is likely to send a positive signal across the ecosystem. It might amplify questions on government intentions as a digital banker to the public, leading to adoption decline and complete failure of e-Cedi.

22.8   Summary and Conclusion

Cryptocurrency and its entwined blockchain, have advanced digital transformation in the present day. The practical use of cryptocurrency goes beyond regulation. It has implications across industries and forms a catalyst for the development of derived innovations such as decentralized finance—Defi. Defi makes it possible for financial solutions and products to be built on a decentralized platform. Efforts to adopt cryptocurrency lead to the research and development of central bank digital currencies CBDC with the central bank/government as the singular authority. This chapter examined cryptocurrency practices and implications in Ghana. Blockchain also created an opening for young fintech startups to develop an ecosystem of blockchain and cryptocurrency services and products on the decentralized system and Defi. Crypto practices in Ghana tap into the $ 41 billion NFT market often seen as an innovative solution to the lack of proper intellectual property rights and royalty system in the Ghanaian creative industry. Gamers also leverage the decentralized crypto economy to access untapped crypto-gaming markets across Africa and the broader global gaming world through non-fungible token (NFT) and play-to-earn (P2E) models. These activities minimize adoption barriers to achieve both digital inclusion and financial inclusion. Ghana currently has no legally stipulated guidelines to effectively tackle taxes in the crypto ecosystem creating a conducive space to evade taxes. Implications of such crypto practices lead to access to global markets and financial inclusion enabling practical adaptations that open up the Ghanaian economy to a larger global market. Crypto-based solutions extend inclusion beyond local borders and offer access to the bigger African and world markets. For instance, the NFT market built on blockchain technology has enabled the sale of Ghanaian NFTs like the dancing pallbearers for 372 ETH ($1.046 million). Cryptocurrency also reduces remittance costs compared to traditional methods of exchanging value across borders. Businesses are encouraged to extend their market reach beyond geographical boundaries as transactions can now be done faster and easier through decentralized fintech platforms with minimal interference from financial institutions.

The practice of cryptocurrency also has tax implications in emerging markets such as Ghana. The Ghana Revenue Authority and the Bank of Ghana have not yet issued any guidelines on the taxation of digital currencies, derived products, and services related to cryptocurrency within their jurisdiction (Bank of Ghana, 2018). As a result, more profits made in the crypto market remain untaxed, in contrast to countries such as the United States, Canada, Portugal, Spain, and Turkey, which have established tax guidelines for the treatment of cryptocurrency markets (Forbes, 2019). This lack of taxation may provide an opportunity to evade taxes. Enhancing the taxation system to encompass the crypto ecosystem could help to reduce such practices, broaden the tax base, and reduce the high taxation rates imposed on other main sectors such as ports and other industries.

We also looked at how a Sino-Ghana collaboration could be of benefit in Ghana carving its cryptocurrency future. Ghana and China have a deep historical relationship. Both countries have active plans to further develop cryptocurrency and a digital economy. This could be looked at from the lens of development partnership in the digital transformation process. China eclipse Ghana in technological capacity and expertise to develop and advance cryptocurrency/blockchain in a new digital elite economy. The Sino-Ghana relationship comes in handy in creating a knowledge base. This provides Ghana with short-midterm capabilities to further cryptocurrency, and blockchain development across industries optimizing the technology to cost efficiency, data sharing, and building credible systems in a new digital economy. Cryptocurrency development and a digital economy are both energy intensive, they could further burden the already insufficient energy resources. Hence, it's imperative to leverage the Sino-Ghana bilateral relationships to invest in investment in technologies and infrastructure and expand the available renewable energy capacity of Ghana to meet new energy demands brought by the digital economy. The development of these supportive infrastructure systems adds to breaking the digital divide that exists between Ghana and the global world.

In conclusion, cryptocurrency advancement and practices have a bearing on digital inclusion and financial inclusion. The chapter draws attention to this nexus as Ghana channels efforts and investment to further develop cryptocurrency in its new digital economy. The chapter found an interesting focal point to which the Sino-Ghana relationship could be extended. Provisions under such collaborative efforts might provide quick fixes and temporal capacity. Recent technological developments such as Defi, Metaverse, blockchain & cryptocurrency, make Ghana look at long-term solutions and capacity building. Starting modern technology education at the early education, technological transfer through better and effective study abroad programs as well as strategic use and inclusion of Ghanaian experts in the diaspora could be sources of technological capacity building in anticipation of future disruptive innovations.

References

Aboagye, B., Gyamfi, S., Ofosu, E., & Djordjevic, S. (2021). Status of renewable energy resources for electricity supply in Ghana. *Scientific African.* https://doi.org/10.1016/j.aglobe.2021.100023

Arjun, K. (2022). *China names blockchain trial zones after its crackdown on cryptocurrencies.* https://www.cnbc.com/2022/01/31/china-names-blockchain-trial-zones-after-crackdown-on-cryptocurrencies.html

Aysan, A. F., & Kayani, F. N. (2021). China's transition to a digital currency does it threaten dollarization? *Asia and the Global Economy, 2*(1), 100023. https://doi.org/10.1016/j.aglobe.2021.100023

Bank of Ghana. (2018). *Digital and virtual currencies operations in Ghana.* Notice No. Bg/Gov/Sec/2018/02

Bank of Ghana. (2019). *Design paper for the digital cedi (eCEDI).* https://www.bog.gov.gh/news/design-paper-of-the-digital-cedi-ecedi/

Bányai, F., Zsila, Á., Griffiths, M. D., Demetrovics, Z., & Király, O. (2020). Career as a professional gamer: Gaming motives as predictors of career plans to become a professional esport player. *Frontiers in Psychology.* https://doi.org/10.3389/fpsyg.2020.01866

Boateng, G. (2019). *The AfCFTA may be the last opportunity for Africa's economic transformation.* https://acetforafrica.org/research-and-analysis/insights-ideas/commentary/the-afcfta-may-be-the-last-opportunity-for-africas-economic-transformation/

Chainanalysis. (2021). *The 2021 Geography of Cryptocurrency Report.* https://go.chainalysis.com/2021- geography-of-crypto.html

Chainanalysis. (2022). *The 2021 NFT Market Report.* https://go.chainalysis.com/nft-market-report.html

Dailey, N. (2022). *NFTs ballooned to a $41 billion market in 2021 and are catching up to the total size of the global fine art market.* https://markets.businessinsider.com/news/currencies/nft-market-41-billion-nearing-fine-art-market-size-2022-1

del Castillo, M. (2019). Alibaba, tencent, five others to receive first Chinese government cryptocurrency. *Forbes* (August 27). https://www.forbes.com/sites/michaeldelcastillo/2019/08/27/alibaba-tencent-five-others-to-recieve-first-chinese-government-cryptocurrency/?sh=1fe27e251a51

Fedasiuk, R., & Weinstein, E. (2020). Overseas professionals and technology transfer to China. *CSET Issue Brief.* https://cset.georgetown.edu/wp-content/uploads/CSET-Overseas-Professionals-and-Technology-Transfer-to-China.pdf

Forbes. (2019). *Alibaba, tencent, five others to receive first chinese government cryptocurrency.* https://www.forbes.com/sites/michaeldelcastillo/2019/08/27/alibaba-tencent-five-others-to-recieve-first-chinese-government-cryptocurrency/?sh=1fe27e251a51

Forbes. (2022). *What is bitcoin pizza day, and why does the community celebrate on May 22?* https://www.forbes.com/sites/rufaskamau/2022/05/09/what-is-bitcoin-pizza-day-and-why-does-the-community-celebrate-on-may-22/?sh=32c163affd68

Helms, K. (2022). *China designates 15 national pilot zones and 164 entities for blockchain projects.* https://news.bitcoin.com/china-designates-15-national-pilot-zones-164-entities-for-blockchain-projects/

Kharpal, A. (2022). *China names blockchain trial zones after its crackdown on cryptocurrencies.* https://www.cnbc.com/2022/01/31/china-names-blockchain-trial-zones-after-crackdown-on-cryptocurrencies.html

Kinnah, G. (2022). Bitcoin adoption: An empirical study on user behaviour in Ghana. *International Journal of Novel Research in Humanity and Social Sciences, 9*(3), 29–45. https://doi.org/10.5281/zenodo.6631557

Kpilaakaa, J. (2022). *Opis group expands free-to-play, play-to-earn NFT game "OpiPets" to Africa, announces $500 giveaway contest.* https://www.benjamindada.com/opipets-nft-game-expands-to-africa/

Owusu, A. (2020). Preliminary insights into the adoption of bitcoin in a developing economy: The case of Ghana. *IGI Global.* https://doi.org/10.4018/978-1-5225-9715-5.ch064

Sandali, H. (2021). *Ghana to prioritize blockchain projects in new regulatory sandbox.* https://www.coindesk.com/markets/2021/02/26/ghana-to-prioritize-blockchain-projects-in-new-regulatory-sandbox/

The Economist. (2022). *The Chinese-African relationship is important to both sides, but also unbalanced.* https://www.economist.com/special-report/2022/05/20/the-chinese-african-relationship-is-important-to-both-sides-but-also-unbalanced

Wang, J.-A. (Juexuan) (2018). *Perception and prejudice: Sino-Ghanaian relations within the service sector and the wavering perception of China on the global stage. Independent Study Project (ISP) Collection.* https://digitalcollections.sit.edu/isp_collection/2903

Xinhua. (2021). *Keynote speech by Chinese President Xi Jinping at opening ceremony of 8th FOCAC ministerial conference.* http://www.focac.org/eng/ttxxsy/202112/t20211202_10461079.htm

Chapter 23
Innovative Technology of the Smart Transport and Smart Cities for Sustainable Infrastructure to Support Resilient Economic System in China and Africa

Yuchen Wang, Mi Zhou, and Baorong Yang

23.1 Background

On 14–16 October, 2021, the United Nations hosted the 2nd Global Sustainable Transport Conference (2nd GSTC) in Beijing, China (UN Secretary-General, July 2021, Tentative Programme, Oct., 2021). Transport enables the movement of people and goods, supporting livelihoods and jobs and thus contributing to poverty eradication, food security and inequality reduction. Through improving access to quality services, such as health, education, and finance, transport can empower women and other vulnerable groups by enabling inclusive economic growth. The Beijing Statement of the (2nd GSTC) mentions that transport enables connectivity at all levels (Concept Note Forum, Sept., 2021) among all countries, including land, sea, tube and air transport modes, which links community interaction, integrates markets and economies, enhances rural and urban interlinkages, facilitates trade, underpins supply chains and can boost resilience and contribute to the attainment of the Sustainable Development Goals (SDGs)(Resolution Adopted by the General Assembly, Dec., 2017).

Y. Wang (✉)
University of Venda, South Africa-China Transport Co-Operation Centre (SACTCC) and China-Africa Transport Strategy Research Institute (CATSRI), Thohoyandou, Pretoria, South Africa
e-mail: sactccwangyc@126.com

M. Zhou
Chang'an University (CAU), Xi'an, China

B. Yang
China-Africa Institute (CAI), Beijing, China

© The Author(s) 2024
M. Muchie et al. (eds.), *China-Africa Science, Technology and Innovation Collaboration*, https://doi.org/10.1007/978-981-97-4576-0_23

427

Transport is a basic, leading, strategic industry and an important service industry in the national economy. China has opened roads in the mountains and built bridges in encounters with rivers and oceans. There are huge transport facilities developed in China, which are accelerating the construction of a transport power country. China built the world's largest high-speed railway network, highway network, and world-class port cluster. Aviation and navigation have reached the world, and the comprehensive transport network has exceeded 6 million kilometres. It should be noted that as the demand for transport continues to grow, related issues such as resources, energy, and the environment have also become prominent. Realizing the sustainable development of transport is of great significance to both the industry and the high-quality development of the economy and society (see Fig. 23.1).

From online car-renting to unmanned distribution, from unmanned driving to automated terminals, whether it is to serve people or facilitate the flow of goods, intelligence, and digitization are becoming a new bright spot in China's transport development. New technologies such as smart road networks, smart rails, smart ports, and smart shipping are widely used. New online and offline consumption models such as online car-hailing, shared bicycles, and shared cars are becoming more popular. Scanning face to entry, "paperless" boarding, and drones for delivery, contactless delivery, smart parking, and road passenger transport customized services have developed rapidly. In recent years, China has accelerated the promotion of the integration of emerging technologies such as 5G, big data, and artificial intelligence with transport, so that people can enjoy their travels and goods. The flow continues to reach new realms, injecting strong momentum into the sustainable development of

Fig. 23.1 Overview of sustainable transport

the transport industry. Facts have proved that facing the needs of economic and social development, seizing the opportunities of the new round of scientific and technological revolution and industrial transformation, and vigorously developing intelligent transport and intelligent logistics can better contribute to sustainable development.

New and emerging technologies, when properly applied, are key to solving many of the challenges to sustainable transport. The deployment of existing solutions, such as zero-carbon vehicles, automated safety, and intelligent transport systems, has accelerated, accompanied by the creation of the necessary fuel, power, and digital infrastructures.

A case study in Shenzhen, China serves as an example (Website of Shenzhen Municipal Government, 2022).

23.2   Smart Transport

23.2.1   Definition and Scope

Smart transport refers to an efficient and safe transport service system that is finally formed after real-time information analysis and processing by relying on various information technologies such as the Internet, big data, the Internet of Things (IoT), and artificial intelligence. Smart transport mainly covers smart design. smart equipment, smart logistics, smart management, and smart road and rail networks. Smart transport is established on the basis of intelligent transport, and both are the products of the application of information technology, sensor technology, communication technology, and other technologies in the transport field. The essence of intelligent transport is to apply advanced technologies such as computers, control, communication, sensing, and networks to the entire transport system to improve the traditional transport information transport system. Intelligent transport mainly focuses on the informatization of various transport applications. In addition to collecting and transmitting traffic information, intelligent transport pays more attention to the analysis and decision-making response of traffic information. This is the "smart" aspect of intelligent transport. Smart transport not only effectively integrates various information technologies in traffic operation and management, but also emphasizes systematic, real-time, and information interaction, which can realize the automation of transport system functions and intelligent decision-making. In terms of system construction, smart transport focuses on the intelligence of system integration and the flexibility of coordination; in terms of public services, smart transport is biased towards the personalization of service content and the humanization and intelligence of service models (see Fig. 23.2).

Fig. 23.2 Bird's eye view and side view of smart transport

23.2.2 Main Business Fields of Smart Transport and Applications in China

The upstream of the smart transport industry is mainly data providers, chip and circuit integration manufacturers, software developers, and hardware manufacturing with quotient and algorithm providers. Hardware equipment is the foundation of smart transport, including storage and processing equipment, sensing and communication equipment, etc., such as sensors, servers, and network equipment.

In the software industry, the development of China's software technology is changing with effective efforts currently. The structure of the industrial service industry is continuing to adjust. The industry chain continues to improve (World Transport Convention, 2017, 2018, 2019).

The middle reaches of the industry chain are consulting service providers, operation service providers, and system solution providers. The main business is based on the specific needs of downstream users to provide customers with architecture design services, mainly including hardware and software integration and improvement. The rapid development of technology has become the rise of the midstream industry as an important driving force. Based on the rise of emerging technologies such as cloud computing and the IoTs, midstream service providers, and system solutions providers, merchants can continuously improve the system according to downstream needs. For example, in order to improve the convenience of payment and enhance the downstream payment experience, Guangzhou Metro of China has established a multi-channel integrated payment platform for the entire network. The establishment of this platform has realized the Guangzhou Metro mobile payment and financial IC multi-channel payment such as cards, which can support multiple payment methods across the entire network line to achieve a real-time grasp of material and material consumption information. The Wuhan Metro of China has added ZTE's IPSA system to its Line 1 to improve the information processing process.

The downstream of the industry chain mainly links traffic construction operators, who can come from the city management offices, municipalities, traffic, safety, etc.

23.2.3   Market of China's Urban Rail Transport Industry

With the continuous advancement of urbanization, the pressure on urban traffic has increased dramatically. To solve the urban traffic pressure, China vigorously develops urban rail transit. The construction of an urban rail network is the basis for the development of smart city rail transit. According to the data of the National Bureau of Statistics, the length of China's urban rail transit network reached 9206.8 km on Dec. 31, 2021, indicating that urban rail informatization has further increased. The construction of the transport information system continues to expand, and the construction of China's smart cities is deepened.

In 2020 investment had been increased from 547.0 billion yuan in 2018 to 628.6 billion yuan, which was maintaining a speedy growth. In 2021 it reached 586.0 billion yuan due to the pandemic. It is expected that the scale of China's urban rail smart transport market will increase by 5% from 2021 to 2023. The rapid development of urban rail transit will also drive the construction of its supporting information system, thus it makes the market scale of the entire industry different.

23.2.4   Links of Smart Transport and Sustainable Development

According to the Beijing Statement of the 2nd GSTC (Fig. 23.3), attaining sustainable transport would mean fully delivering on the benefits while avoiding or alleviating the associated costs of mobility. But the world is falling short in making progress towards this objective. At the same time, the world is also presently not on track for eradicating poverty, reducing emissions, and achieving other substantial portions of the 2030 Agenda as well as the Paris Agreement, including the objective of limiting global average temperature rise. Accelerating the transformation towards sustainable transport will be central to creating a community of shared future for humankind. It will also be crucial for achieving the world's common objectives for people, the planet, and prosperity.

Fig. 23.3 Beijing statement of the 2nd GSTC

23.2.5   COVID-19 and Global Economic Recovery

From early 2020 to date, COVID-19 has affected the sustainable rebuilding and smooth flow of international supply chains and the cross-border movement of people and goods, including medical supplies. It needs to find the right way for global response to the COVID-19 pandemic and global economic recovery.

The 2nd GSTC has strengthened regional and interregional connectivity and joint action with regard to "hard" transport aspects, like regional and trans-border infrastructure, and "soft" transport elements, such as streamlined customs and border-crossing regulatory frameworks, including through global initiatives, regional and interregional strategies, and plans.

It is urgent to promote international cooperation, capacity-building, and knowledge exchange among countries which should be encouraged with a view to advancing sustainable transport technology and innovation and learning from good practices. Action is also needed to promote the implementation of international transport with related conventions, regulations, and agreements.

The Global Innovation and Knowledge Centre for Sustainable Transport in China plays an active role in the global economic recovery after COVID-19. The first action was that it hosted an international implementing platform in Sept. 2023. The second action will aim the successful practices within China to introduce the neighbouring countries and the Belt and Road regions around the world. For African countries, it will be a good opportunity to make use of this platform to serve their own transport infrastructure development.

23.3   Smart Cities

23.3.1   Definition and Scope

Smart Cities refer to the use of various information technologies or innovative ideas to integrate the system and services of the municipality to improve the efficiency of resource utilization, optimize municipal management and services, and improve the quality of life of citizens (Fig. 23.4). Smart cities make full use of the new generation of information technology in all aspects of life in the municipality. The advanced form of urban informatization based on the next-generation innovation of the knowledge society (Innovation 2.0) realizes the in-depth integration of informatization, industrialization, and urbanization, and helps to alleviate "huge city disease", improves the quality of urbanization, realizes refined and dynamic management, and enhance the effectiveness of urban management by improving the quality of life of citizens. The specific definition of smart cities is relatively broad. The definition widely recognized internationally is that smart cities are a municipal form supported by a new generation of information technology and a next-generation innovation in a knowledge society (Innovation 2.0), emphasizing that smart cities are more than just

Fig. 23.4 Smart cities

the application of a new generation of information technology such as the Internet of Things and cloud computing, more importantly, they are to build a sustainable urban innovation ecosystem characterized by user innovation, open innovation, mass innovation, and collaborative innovation through the application of the methodology of Innovation 2.0 for the knowledge society (Steyn et al., 2018, 2019, 2020).

Smart cities promote sustainable economic growth and high quality of life through investment in human and social capital, transport and information and communication infrastructure, and scientific management of the above resources and natural resources through participatory management. Some pioneer cities for smart city construction are also increasingly highlighting people-oriented sustainable innovation. For example, the European Union has launched the Living Lab program for Knowledge Society Innovation 2.0, which is committed to building cities into open innovation spaces where all parties participate, based on the needs of citizens. For example, the six indicators of the University of Vienna's evaluation of the municipal system are smart economy, smart transport, smart environment, smart residents, smart life, and smart management. At present, the construction of smart cities is a hot issue in the fields of information technology and urban planning.

23.3.2 History of Smart Cities in the World

The concept of smart cities began as far back as the 1960s and 1970s when the US Community Analysis Bureau began using databases, aerial photography, and cluster analysis to collect data, direct resources, and issue reports in order to direct services, mitigate against disasters and reduce poverty. This led to the creation of the first generation of smart cities.

The first generation of smart cities was delivered by technology providers to understand the implications of technology on daily life. This led to the second generation of smart cities, which looked at how smart technologies and other innovations could create joined-up municipal solutions. The third generation of smart cities took the control away from technology providers and municipal leaders, instead creating a model that involved the public and enabled social inclusion and community engagement.

The third-generation model was adopted by Vienna, which created a partnership with the local Wien Energy company, allowing citizens to invest in local solar plants as well as working with the public to resolve gender equality and affordable housing issues. Such adoption has continued around the world, including in Vancouver, where 30,000 citizens co-created the Vancouver Greenest City 2020 Action Plan.

23.3.3 Key Elements of Smart Transport for Smart Cities

According to application scenarios, smart transport is a core for smart cities. Key elements can be roughly classified into three categories: urban rail transport, urban road transport, and urban expressways. Urban rail transit refers to a vehicle transport system that adopts a track structure for load-bearing and guidance, mainly in the form of trains and bicycles. The construction of smart transport in rail transport mainly focuses on the comprehensive monitoring of urban roads and the creation of an integrated urban rail transport management and service system; urban road transport refers to the general term for roads that are responsible for access to various regions and connect to the city's out-of-city traffic as mainly motorized vehicle lanes, non-motorized vehicle lanes and sidewalks. The construction of smart transport in the field of urban road subdivision mainly focuses on alleviating traffic congestion and improving traffic conditions, so that the four factors of people, vehicles, roads, and the environment can be coordinated to maximize the effectiveness of urban traffic; urban expressways are defined by the government. The fully enclosed or semi-enclosed high-grade highways identified by the uniform number are divided into two categories: national highways and provincial highways according to management and strategic significance. Smart Expressway is to gradually establish a complete infrastructure monitoring system, an intelligent road network operation perception system, a real-time forecast and early warning system, and an efficient emergency support system through technologies such as the Internet of Things, cloud computing, and big data analyses.

23.3.4 Interactions of Smart Transport and Smart Cities

Smart transport is an important core of smart city construction, which is based on the real-time management and traffic guidance of the geographic information system. With dispatching and other functions, the efficiency of the transport infrastructure is exerted, thereby improving the operation efficiency and management level of the transport system and realizing intelligent services have been well implemented. The system, real-time, and interactivity of information traffic in smart city traffic by Sullivan's data, China's total expenditure on smart transport was US$2.74 billion in

2017, accounting for 39.3% of the investment in key areas of China's smart cities. Currently, its investment in smart cities is increasing rapidly.

The construction of smart cities has promoted the expansion and integration of smart transport products in various application fields. Designed to expand and advance the construction of the concept of smart transport, industry applications are gradually becoming deeper and more integrated. In the field of traditional transport, depending on frequency monitoring system, as an important application in intelligent transport products, it is mainly used for the safety monitoring needs of traffic status.

The urban infrastructural construction of the municipality has promoted the development of smart transport, and the role of video surveillance systems has gradually shifted to road traffic monitoring, electronic police, and expansion of functions such as intelligent recognition. At the same time, video surveillance products are integrated with technologies such as the Internet of Things and big data, they are integrated with urban traffic command.

The dispatching application platform is docked, and the collection and analysis tools of transport information technology are used to provide a guarantee for the safety of smart transport.

The construction of smart cities provides conditions for the expansion and integration of smart transport products in the application field.

23.3.5 Sustainable Development of Smart Transport with Smart Cities in China

Encouraged by a series of favourable policies in China, its smart transport has built high-definition video surveillance for urban traffic flows. A relatively complete system has been formed in terms of traffic volume, traffic analysis system, traffic guidance system, GPS monitoring system, etc.

According to data released by the Ministry of Transport of China in 2018, 280 cities had adopted smart transport technology to improve urban road and rail transit management systems. The annual rate of traffic accidents is reduced by more than 20.0%, and the rate of transport efficiency increased by more than 50% by the public reports. It can be seen that with the strong support of the government, dividend policies have been continuously introduced, providing a good development for China's smart transport industry. The environment has effectively promoted the healthy development of the industry.

In recent years, with the gradual popularization of the concept of smart cities, smart transport has become an important support for the development of smart cities. As the new generation of information technologies, such as artificial intelligence and cloud computing technologies, continue to penetrate, the industry boundaries of traditional transport are constantly being broken.

At this stage, the operation and service capabilities of intelligent transport have been upgraded in both directions through a new generation of information technology, and the industry as a whole has become intelligent. Currently, the world is gradually entering the era of smart transport. Due to the wide coverage of the smart transport industry, it became more market segments and more supply side entities.

In addition to diversification, various participants use their respective advantages to cut into the subdivision of the smart transport industry chain, relying on their respective software and hardware technologies. Strength is empowered by intelligent transport. However, companies in the industry mainly focus on cultivating a certain field, and they have not yet formed a company with strong comprehensive strength.

23.4 Shenzhen Smart Cities Development

23.4.1 Overview of Shenzhen

Shenzhen is the first special economic zone located in Southern China, close to Hong Kong, which was started in 1980 (Fig. 23.5).

Shenzhen is a municipality under the jurisdiction of Guangdong Province, a sub-provincial city, a global metropolis, and a national comprehensive supporting reform pilot zone. It is one of the central cities of the Guangdong-Hong Kong-Macau Greater Bay Area. It is located on the east bank of the Pearl River Estuary and eastern Daya Bay and Dapeng Bay bordered by Lingding Ocean to the west, Hong Kong to the south, and Huizhou and Dongguan to the north. The municipality covers an area of 1997.47 square kilometers and a sea area of.

One thousand, one hundred forty-five square kilometers. It has 9 administrative districts and 1 new district under its jurisdiction. In 2020, the municipality had a permanent population of 17.56 million, making it one of the most populous cities in China (Website of Shenzhen Municipal Government, 2022).

Fig. 23.5 Location of shenzhen and its night view

Shenzhen was established on the basis of the original Baoan County in January 1979. In August 1980, the Standing Committee of the National People's Congress approved the establishment of China's first special economic zone—Shenzhen Special Economic Zone in Shenzhen, Guangdong Province. Due to the reform and opening policy, Shenzhen has developed rapidly, and has risen from a border agricultural county to a first-tier metropolis in the country. In 2016, Shenzhen became the third largest municipality in mainland China in terms of economic aggregates. In the list of global metropolises announced by GaWC in 2020, Shenzhen once again entered the Alpha-level, ranking 46th, and it was also one of the six first-tier cities in the world (Alpha-level) selected by Greater China.

Shenzhen is one of the four pillar industries with high-tech industry, financial industry, modern logistics industry, and cultural and creative industry. Shenzhen forms an important high-tech R&D and manufacturing base in southern China and is often referred to as the "Silicon Valley of China". The container throughput of Shenzhen Port has ranked third in the world for many consecutive years, and the total export volume has ranked first in mainland China for more than 20 consecutive years. Shenzhen Bao'an International Airport is the fifth largest civil aviation airport in China. The number of IPOs on the Shenzhen Stock Exchange was ranked first in the world from 2009 to 2015, and it became an important financing platform for Chinese companies. In 2019, the Central Committee of the Communist Party of China and the State Council supported Shenzhen to build a pilot demonstration zone for socialism with Chinese characteristics and assumed the important mission of experimentation and demonstration in China's system innovation and opening up.

23.4.2 Introduction of Smart Cities at Shenzhen

Smart cities refer to the use of various information technologies or innovative concepts to open up and integrate the city's systems and services to improve the efficiency of resource utilization, optimize municipal management and services, and improve the quality of life of citizens.

Across the country, Shenzhen proposed the concept of "Building a Smart Shenzhen" as early as 2010, and then the Shenzhen Smart Industrial Park landed in Shenzhen. Leading companies such as Huawei have successively carried out smart cities' construction cooperation with Shenzhen, and finally, Shenzhen became a smart metropolis in 2018. It has made efforts for the metropolis with the highest degree of smart cities in China and has been maintained to date. Shenzhen's achievements in building smart cities are reflected in all aspects of e-government, smart subways, and smart airports. It was pointed out that in 2020, the overall index of national key cities' online government service ability ranked Shenzhen first in the whole country (Fig. 23.6).

Fig. 23.6 Lasers and LED light-up skyscrapers of Shenzhen's Futian CBD

23.4.3 Key Elements of Shenzhen Smart City

23.4.3.1 Internet of Things (IoTs)

IoT is an important part of a new generation of information technology in Shenzhen, and it is also an important development stage of informatization. The essence of the Internet of Things is to realize the connection of things, and its core foundation is still the Internet. Users at both ends of the Internet of Things can extend to any item to complete the exchange of information. In the construction of smart transport, the Internet of Things technology can be used to comprehensively perceive the construction of transport infrastructure and transport vehicles and provide a basis for smart perception. The integrated monitoring system based on the Internet of Things accelerates the development of the smart transport industry, which enables the circulation of real-time traffic data and information and realizes the interconnection of data collection and traffic information, and monitors the entire traffic operation.

23.4.3.2 Big Data (BD)

BD refers to a collection of data that cannot be captured, managed, and processed with conventional software tools within a certain time frame due to the rapid growth of data volume. It is a massive amount of data that requires new processing modes to have stronger decision-making power and process optimization capabilities with a high growth rate and diversified information assets. The essence of big data is not to study how to process data but to better discover the value contained in massive data. Driven by the construction of smart transport, big data on driving and road segments can realize data sharing and information sharing among vehicles on the same road, between roads, and vehicles on different roads. For the traffic management department, the use of traffic big data can mine and utilize the deep value of information data. After analyzing the data, the existing real-time data can be fully utilized, which is beneficial to the management and decision-making of the transport department. The application of big data technology is a powerful driving force for intelligent transport. Compared with the location information of mobile phone map navigation, the location information of communication big data has unique advantages. It can not only actively locate, where the location information is more comprehensive, but also can identify user value. For example, merchants can use customer location data and e-maps to better understand and improve the flow of people and can adjust the store layout and merchandise.

23.4.3.3 Cloud Computing (CC)

CC is a pay-per-use model that provides usable, convenient, and on-demand network access and enters a configurable computing resource-sharing pool (resources include networks, servers, storage, application software, and services). The resources can be provided quickly with little management effort or little interaction with service providers. According to the service target customers, cloud computing can be divided into three categories: Infrastructure as a Service IaaS, Platform as a Service PaaS, and Software as a Service SaaS. IaaS service providers mainly provide IT infrastructure services to enterprises, government departments, and other institutional customers. PaaS service providers mainly provide development environment services to developers and independent software company ISVs, while SaaS mainly provides application software services to enterprises and individual customers. There are many problems in the actual application of big data. As an ideal carrier for big data analysis, the cloud computing storage structure is suitable for handling complex and changeable data sources. In addition, cloud computing service providers can also provide powerful data analysis software services, and big data analysis services may also enhance the market competitiveness of cloud computing platforms and enhance the added value of cloud computing. Therefore, in the construction of smart transport, cloud computing can provide a new mode for the storage of various types of transport data, and it can also realize the sharing of information resources and the interconnection of systems.

23.4.4 Current Development and Prospects

As a full economic central metropolis, Shenzhen is once again at the forefront of the times in terms of smart city construction. From digital cities to smart cities, to today's smart cities, this benefits from high-quality geographical location and policy support. Now Shenzhen has become one of the first pilot cities for smart cities in the country and has quickly established a leading edge in the popularization of the Internet and the digital economy. According to Deloitte's latest "Super Smart Cities Report", Shenzhen ranks first in the first echelon of China's super smart cities.

Mayor of Shenzhen, Mr. Chen Rugui indicated at a special press conference held in June 2020 after the conclusion of the sixth session of the sixth Shenzhen Congress that "Smart cities" involve all aspects. During the construction process, Shenzhen should focus on the security management of big data and the protection of personal privacy. Here the protection means to ensure data security and cyberspace security. Data is a resource. Under the premise of ensuring safety, a series of analyses, application and development should surely bring innovation in many aspects and promote a new round of high-quality economic development (Fig. 23.7).

Big data definitely brings changes to organizational models, innovation models, and marketing models, and they can provide a lot of vitality and motivation for innovation and development. They promote a new round of economic innovation and development through the construction of smart cities. Shenzhen's electronic information industry is very developed, mobile payment and the Internet are relatively mature, and there is a solid foundation for building a smart metropolis. Moreover, promoting the overall transformation and upgrading of the industry through technological innovation may also drive several industries to adjust and optimize the structure. Smart cities involve all aspects. In the construction process, Shenzhen has focused on the security management of big data and the protection of personal privacy to ensure data security and cyberspace security. Data are a resource. Under the premise of ensuring safety, a series of analyses, applications and developments

Fig. 23.7 Shenzhen has digitized most government administrative affairs in recent years

will surely bring innovation in many aspects and promote a new round of high-quality economic development.

Currently, Shenzhen is promoting ten major projects including high-speed broadband networks, comprehensive perception systems, urban big data, and smart cities operation and management. The highest goal of building a smart metropolis is to realize the perception of all things, the interconnection of all things, and the intelligence of all things. Shenzhen planned to use three years to realize basically the comprehensive perception of a picture, a number of travels throughout Shenzhen, a key to knowing the overall situation, integrated operation and linkage, and one station on innovation and entrepreneurship, one screen smart life, and strive to achieve "technology which makes urban life better".

Shenzhen has desired to promote scientific, refined, humanized, and high-quality urban management, the key is to rely on informatization and intelligence. Shenzhen has done a good job in transport services, including aviation, ports, ground, underground, fast, and slow systems, built a smart and intelligent integrated transport service system, and promoted the construction of industrial networking and business innovation through the construction of smart cities. In the future, Shenzhen will continue to improve the level of manufacturing, vigorously develop intelligent manufacturing, and increase production capacity and industrial value-added rate.

With the support of smart city services, Shenzhen has built an integrated comprehensive transport service system ranging from airports, ports, and cruise ships to high-speed rail, subways, buses, shared bicycles, to unmanned driving, to smart parking, providing integrated, intelligent, and smarter transport and travel services to make travel more convenient for citizens.

In terms of government services, Shenzhen has strived to achieve "one-time information entry, information sharing, and municipality-wide operation "for all individuals and companies living and doing business in Shenzhen, and it has strived to achieve. Doing business does not need to go out of the street, and individuals do not need to run errands" (Website of Shenzhen Municipal Government, 2022).

In terms of the construction of smart communities, Shenzhen further optimizes the smart services of social insurance, community medical care, employment services, property management, garbage sorting, home care services, etc., which are closely related to the people, so that the citizens have more senses of gain and happiness.

In terms of technical support, Longgang District Government and Huawei have launched a comprehensive strategic cooperation. Huawei will serve as the "chief designer" to incorporate the design, construction, operation, and maintenance of information network systems, information resource systems, and information application systems into the scope of management. Shenzhen is pushing Longgang District to become a global model point for smart city construction. In addition, Huawei has been in-depth and thorough in various fields, and continuously empowered application scenarios through technological innovation and cutting-edge algorithms. The R&D and application of technology are inseparable from the corresponding technical talents, and Huawei attaches great importance to the cultivation of talents in this field. Huawei's talent training system has also brought together a group of partners and industry customers. For example, when some smart cities' projects are implemented,

Huawei jointly builds vocational education parks with local governments to cultivate ICT talents.

23.5 Progress on Smart Transport and Smart Cities in Africa

It is predicted that by 2050, 70% of the world's population will live in smart cities. By 2030, at least six of the world's megacities (metropolitan areas with a population of more than 10 million) will be expected in Africa (Fig. 23.8). South Africa's cosmopolitan cities include Johannesburg, Pretoria, Cape Town, Durban, Port Elizabeth and East London.

South Africa is the most advanced in smart transport and smart cities as smart mobility in Africa (Fig. 23.9). As its economic hub: Gauteng Provincial Government launched its Smart Mobility 2030 Vision in Oct. 2020 (Fig. 23.10). Smart mobility involves creating connected and integrated transport systems, which can offer flexibility and efficiency. This connectivity and integration are cutting across all modes of transport forming a compact, integrated, and efficient transport system.

Fig. 23.8 Map of Africa

Fig. 23.9 Map of South
Africa

This Smart Mobility 2030 plan and its implementation put on three strategic focus areas: infrastructure, operations, and institutions. Gauteng is the freight and logistics hub through establishing an intelligent freight network supported by efficient freight and logistics handling capacity with data-centric mobility.

Expanding the road network in this province will connect new nodes to improve efficiencies in the movement of people and goods. These new links should be equipped with intelligent transport systems, integrating into other systems that form the smart cities. Gauteng Province is working together with all spheres of Government. It is using smart mobility to reduce traffic congestion, increase road safety, improve the environment, and make transportation more accessible and affordable. In South Africa's smart cities' construction, it is at the forefront of Africa. Cape Town has proposed a five-year strategy, Johannesburg is expected to achieve smart city goals by 2040, and around 2030, the administrative capital, Pretoria, will be with a smart management style.

At present, it has reached a certain scale in digital infrastructure, digital inclusion, e-government, and digital economy management. Municipal administrations already use data for emergency matters such as fire, rescue, law enforcement, and disaster risk management. Basic elements of livelihood such as energy and water usage and waste are vital data in smart cities. The City of Cape Town has implemented remote utility meters, enabling the government to analyze electronically recorded consumption data and use these data to plan how and where to invest in new resources. Urban dwellers are also increasingly using the internet. For example, they can now pay all their utility bills, report crimes or emergencies, and apply for permits online. Connectivity is critical to a smart city, and Cape Town is making progress in this regard. The Pretoria Municipal Administration operates on grid electricity, drinking and sewerage management, property tax payments, and public telecommunications with the online mode.

Fig. 23.10 Map of Gauteng

The smart city's development of Nairobi in Kenya is at a relatively fast level. The costs of mobile communication and mobile data are quite low. Nairobi is the leader in the African continent in this field (Fig. 23.11). The Kenyan City of Konza, 60 km from the center of Nairobi, is also expected to become Africa's Silicon Valley. The smart development of this city is not only due to the arrival of technology companies but also because of the place itself becoming smart. Municipal authorities use smart devices and sensors buried under roads and buildings to collect data to manage traffic in a way that optimizes traffic flow. The smart cities will be a popular term that will gradually become a reality.

Twenty participating countries of the Smart Africa Coalition launched by Rwanda and they would make technology part of their national development plans.

China's Huawei is a leader in the construction of mobile communications and mobile data infrastructure in Africa, with a market share of about 70% according to statistics. South Africa's two major telecommunications mobile operators, VodaCom

Fig. 23.11 Nairobi, Kenya

and MTN, dominate the country and Nigeria. France, the United Kingdom, Vietnam and other countries have also developed in the field of mobile networks in Africa recently.

In Nov. 2021, African Union and China hosted the 8th Ministerial Meeting of the Forum of China-Africa Co-operation (FOCAC) in Senegal. Nine focusing projects, including the digitally smart transport and smart cities were announced at the Dakar Action Plan (2022–2024). It is a right period for using innovative technologies on smart transport and smart cities to support resilient economic system in both Africa and China via deep co-operations at present.

23.6   Conclusions

The fourth industrial revolution has resulted in smart transport and smart cities which have achieved remarkable economic development in the world. IoTs, Big Data and Cloud Computing are the key elements for the smart transport and smart cities. Sustainable development of transport is of great significance to both the industry and the high-quality development of the economy and society. It links to the climate change and global SDGs promoted by the UN. The smart transport and smart cities have been deeply interacted from all levels for innovative and commercial aspects with new and emerged technologies. The developed and developing countries should pay attention to working together for a community with a shared future for China and Africa.

South Africa is leading in smart transport in Africa. Gauteng Province implements the Smart Mobility 2030 which aims to reduce traffic congestion, increase road safety, improve the environment and make transportation more accessible and affordable. The main challenges faced by smart cities in Africa are the low level of infrastructure, backward urban management, low number of communication network connections, many traffic restrictions, and difficulty in large-scale use. Compared with developed countries and new market economy countries, there are still many gaps. Most cities also rely on international investment and private enterprises and their influence on urban development in Africa.

China and Africa should work together to innovate the technologies of the smart transport and smart cities which are under the Dakar Action Plan (2022–2024) of the Forum of China-Africa Co-operation (FOCAC).

References

2030 Global Sustainable Development Goals (United Nations 2015).

African smart cities' innovative development. (2020). Available at: https://zhuanlan.zhihu.com/p/111366768.

Baolin, S., & Jie, G. (2021). *Presentations of the 2nd UNGSTC, China.*

Beijing Statement. (2021). *The 2nd UN Sustainable Transport Conference, China.*

China-Africa Co-operation on Combating Climate Change. (2021). *Declaration.*

Concept Note Forum of the 2nd UNGSTC. (2021).

Dakar Action Plan. (2022–2024). *Forum of China-Africa Co-operation (FOCAC).*

Dakar Declaration. (2021). *Eighth Ministerial Conference of the FOCAC.*

Dozen Reports. (2020–2023). *Smart transport and smart cities development both China and Africa.*

General Sessions of the 2nd UNGSTC. (2021).

Growing Gauteng Together Through Smart Mobility. (2020). South Africa.

GSTC Programme of the 2nd UNGSTC. (2021).

Invitation of Secretary-General of the UN. (2021).

List of Speakers of the 2nd UNGSTC. (2021).

Overview of the 2nd UNGSTC. (2021). China

Steyn, W., Maina, J., & Wang, Y. (2018, 2019, 2020). Presentations of the smart transport and big data, South Africa.

Second Global Sustainable Transport Conference, United Nations' website (2021)

UN General Assembly Resolution Adopted. (2017).

Website of Shenzhen Municipal Government. (2022). China.

White Paper of the Chinese Transport Sustainable Development. (2020). China

World Transport Convention. (2017, 2018, 2019), China.

Chapter 24
China's Green Energy Investment in Kenya, Ethiopia and Tanzania: Implications for Aid Policy

Joseph Onjala

24.1 China's Green Finance in Africa

In the past two and half decades, China has provided sustained development financial assistance to African states. The Forum on China-Africa Cooperation (FOCAC), which was established in 2000 to streamline Beijing's partnerships, has spelt out action plans for the sectors to benefit from financing by Beijing. Commitments in the action plans emphasize the goal of supporting African flagship projects. Many African countries rank highly the infrastructural projects needed to unlock development potential. Such projects comprise energy sector development such as power generation, railroads, highways, airports, seaports, water supply, information and communication technology infrastructure projects. The composition of infrastructure projects also reveals Beijing's twin goals of connecting Africa. The sectors reflect areas in which Chinese firms have particular experience and successfully compete for contracts under multilateral financing. They are also sub-sectors that have received less interest from private investment in Sub-Saharan Africa (Guttman et al., 2015).Without being explicit on China's gains in development financing to Africa, Wang observes:

> "China's first African Policy Paper, issued in January 2006, reiterates the intention to build a new type of strategic partnership, of "mutual benefit," and "win-win" results in development flows. Key actions are to promote trade, investment, agriculture, infrastructure, resources development, and tourism. The strategy includes support by technical assistance in such noncommercial areas as health and education. The thrust of this policy is vividly demonstrated in all China's FOCAC commitments to Africa."(Wang, 2011).

It is implicit in China's policy statement that the nation has as much interest to gain from its "win–win" and "mutual benefit" partnership in supplying development finance to Africa. What is not clear is the magnitude of such gains and at what cost to

J. Onjala (✉)
Department of Economics and Development Studies, University of Nairobi, Nairobi, Kenya
e-mail: jonjala@uonbi.ac.ke

© The Author(s) 2024
M. Muchie et al. (eds.), *China-Africa Science, Technology and Innovation Collaboration*, https://doi.org/10.1007/978-981-97-4576-0_24

recipients in Africa. In this paper, we analyse the nature and implications of China's financial flows towards greening the energy sector in Eastern Africa countries, mainly Kenya, Ethiopia and Tanzania. This is examined within the broader context of overall financial flows to the countries. We discuss the approach and limitations of the financial flows to have beneficial impacts in the African countries and its implications for aid policy.

24.2   Infrastructure Development Resource Gaps in Eastern Africa

24.2.1   The State of Energy Services in Kenya, Ethiopia and Tanzania: Development Gaps

There are large infrastructural deficits in East African countries of Kenya, Ethiopia and Tanzania. Because of volatile aid flows, African countries have embraced the Chinese aid and will continue to depend on it to meet the infrastructure development gap. Compared to levels of infrastructure development in other regions across Africa, the East African countries perform relatively well (AfDB, 2013).

By the year 2000, it was estimated that in order to catch up with developing world, Tanzania required an infrastructure investment of $2.4 billion per year (Shkaratan, 2010). On the other hand, Ethiopia needed $5.1 billion (Foster & Elvira, 2010) and Kenya required $4 billion per year (Briceno-Garmendia and Shkaratan, 2010). Rural accessibility of modern energy remained an important issue across the regional countries.

Trends in access to electricity and production from non-green sources are in shown in Figs. 24.1a and b respectively. Electricity access in Ethiopia and Tanzania is consistently lower than in Kenya and below the average for Sub-Saharan Africa. Ethiopia currently has an electricity access rate of 45%, with 11% of its population accessing it through decentralized solutions. In terms of total primary energy consumption, biomass represents 90% of the energy consumed in Tanzania. Electricity represents 1.5% and petroleum products represent 8% of the energy consumption in the country. Solar, coal, wind and other sources represent around 0.5% of the total energy consumed in the country.

Electricity access is 37% in Tanzania while gas accounts for more than half of current power generation, with the remainder coming from hydropower and oil, the latter used mostly for back-up generators.

Kenya has experienced one of the fastest increases in electrification rates within sub-Saharan Africa since 2013, reaching 75% of the population with access by 2018 (IEA, 2019a, 2019b). In terms of electricity production, Ethiopia relies more heavily on hydro-energy and less on polluting sources such as oil, gas, or coal. On the other hand, Kenya relies more heavily on the polluting sources to complement the renewable sources. Oil remains by far the dominant fuel in end-use sectors in Kenya.

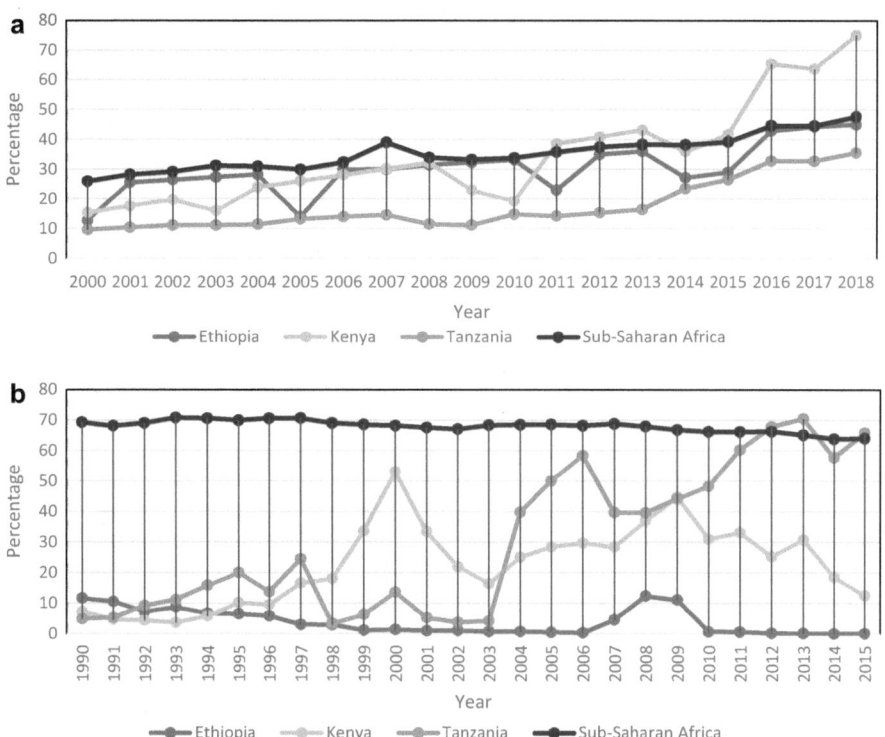

Fig. 24.1 a Access to electricity (% of population), **b** Electricity production from oil, gas, and coal sources (% of total)

In terms of the overall economic and environmental performance, the Eastern African countries show major variations. Figure 24.2a shows that in the last two decades, GDP per capita has generally been below average of Sub-Saharan. GDP per capita has grown much faster in Kenya than in the rest of the regional countries. The GDP per capita remains the lowest in Ethiopia. Access to clean fuels and technologies for cooking (see Fig. 24.2b) has also changed much more rapidly in Kenya compared to other countries. The change in clean fuel access has been marginal in Ethiopia and Tanzania. There is convergence in the overall CO_2 emissions across the regional countries (see Fig. 24.2c), compared to the average for SSA which has been on a downward trend. Finally, as a percentage of total fuel combustion, emissions from electricity and heat production remained lowest in Ethiopia as shown in Fig. 24.2d, reflecting the structure of electricity production that has relied on hydro almost entirely. These emissions have been relatively high in Tanzania and Kenya.

Over 80% of Kenyans rely on the traditional use of biomass as the primary source of energy for cooking and heating—about 90% of rural households use firewood for cooking while 80% of urban households depend on charcoal as the primary source of fuel for cooking (Kenya, 2016a, 2016b). About half of the charcoal in Kenya comes

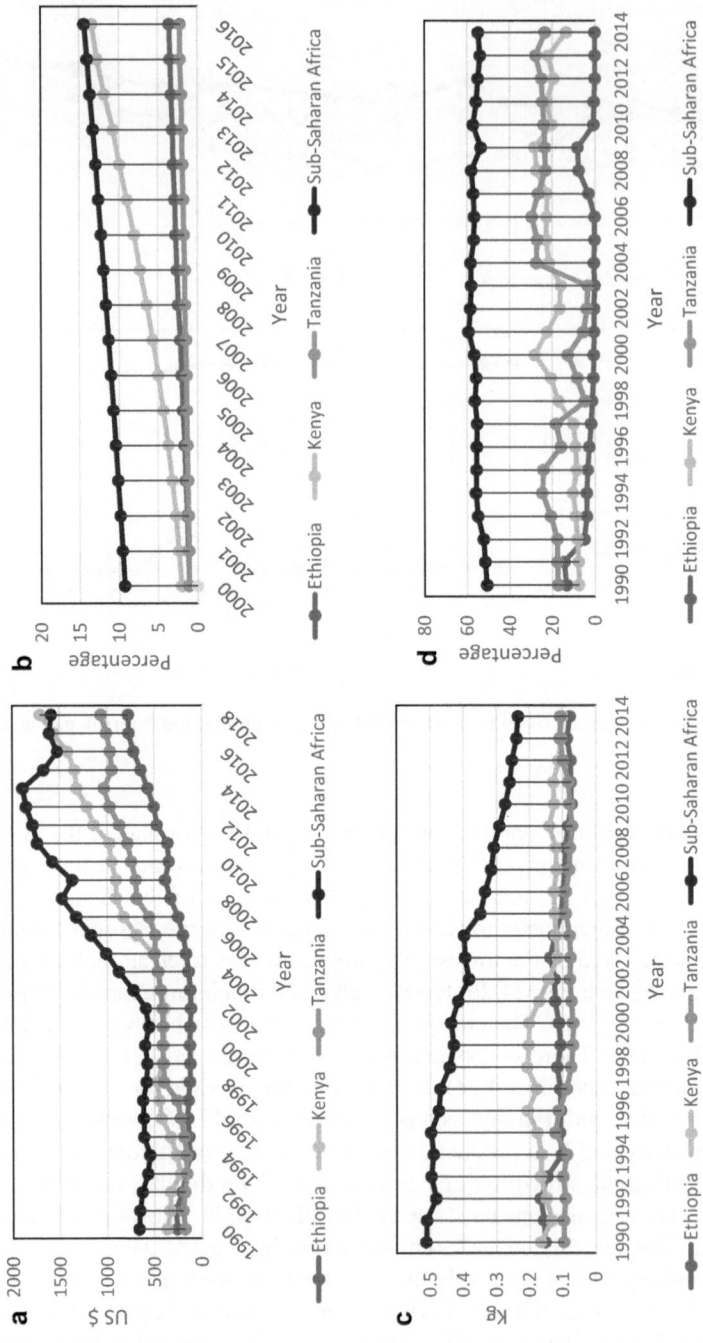

Fig. 24.2 **a** GDP per capita (current US$), **b** Access to clean fuels and technologies for cooking (% of population), **c** CO_2 emissions (kg per PPP $ of GDP), **d** CO_2 emissions from electricity and heat production, total (% of total fuel combustion)

from the Arid and Semi-Arid Lands (ASAL) areas. Charcoal production technologies are inefficient, resulting in massive wastages during wood conversion to charcoal (Kenya, 2016a, 2016b).

24.2.2 Efforts Towards Modernization and Greening of Energy Services in Eastern Africa

The former UN Secretary-General Ban Ki-Moon launched the Sustainable Energy for All Initiative (SE4All) in 2011. This initiative catalysed major new investments to accelerate the transformation of the world's energy systems, pursue the elimination of energy poverty, and enhance prosperity. The goal of the initiative is to: mobilize all stakeholders to action towards universal access to modern energy services; double the global rate of improvement in energy efficiency and; double the share of renewable energy in the global energy mix, within the UN timeframe of 2030. The African governments fully embraced the SE4All Initiative. The considerations for SE4ALL are consistent with green energy development since they are pronounced for bioenergy and hydropower, and are also relevant to the widespread deployment of other clean energy technologies:

> Kenya aimed to reach full electricity access by 2022; the grid would be the principal least-cost solution for the majority of the population (mainly in the south) still lacking access. Today, three-stone fires are still used for most cooking, fuelled mostly by charcoal in urban areas and by wood in rural areas. The Stated Policies Scenarios (STEPS) reflects our measured assessment of today's policy frameworks and plans, taking into account the regulatory, institutional, infrastructure and financial circumstances that shape the prospects for their implementation. Government initiatives lead to 26% of the population having access to clean cooking by 2030 (IEA, 2019a, 2019b). In the Africa Case (AC), everybody gains access to clean cooking by 2030. Most of the 25 million people otherwise without access in rural areas gain access primarily through improved and advanced cook stoves. LPG is the least-cost fuel for most of the urban population. Energy investment amounts to around $60 billion through to 2040 in the STEPS, with renewables and electricity networks accounting for half of this (IEA, 2019a, 2019b).

Projections by the International Energy Agency (2019a, 2019b) on the evolution of green energy developments in the foreseeable future is provided in Table 24.1. In the AC, Kenya could supply an economy six-and-half times larger than today using little more than twice its current energy consumption, if it were to move away from bioenergy and improve energy efficiency. Two-thirds of Kenya's energy currently comes from bioenergy. This share will shrink to 15% by 2040 in the AC due to the increased use of geothermal resources and oil.

In Ethiopia, there is a strong government commitment to reach full electricity access before 2030 in the STEPS. In both scenarios, around 80% of new connections are cost-effectively delivered by grid densification and extension as a large part of the population lives close to the grid. Cumulative energy investment of $100 billion is needed in the STEPS, with electricity access and networks taking the majority. The

Table 24.1 The status of energy services and projections in Eastern Africa Countries

	Stated policies				CAAGR 2018–40	
	2000	2018	2030	2040	STEPS	AC
Kenya						
GDP ($2018 billion, PPP)	76	177	358	627	5.9%	9.0%
Population (million)	31	51	66	79	2.0%	2.0%
With electricity access	8%	75%	100%	100%	1.3%	1.3%
With access to clean cooking	3%	15%	46%	70%	7.2%	9.0%
CO_2 emissions (Mt CO_2)	8	16	27	40	4.3%	6.2%
Ethiopia						
GDP ($2018 billion, PPP)	47	220	493	870	6.5%	8.9%
Population (million)	67	108	143	173	2.2%	2.2%
With electricity access	5%	45%	100%	100%	3.7%	3.7%
With access to clean cooking	1%	7%	34%	56%	9.7%	12.6%
CO_2 emissions (Mt CO_2)	3	14	29	46	5.5%	6.2%
Tanzania						
GDP ($2018 billion, PPP)	57	176	314	585	5.6%	9.3%
Population (million)	34	59	83	108	2.8%	2.8%
With electricity access	11%	37%	70%	80%	3.6%	4.7%
With access to clean cooking	2%	6%	46%	76%	12.2%	13.7%
CO_2 emissions (Mt CO_2)	3	12	24	41	5.9%	8.8%

Source: International Energy Agency (2019a, 2019b). Africa Energy Outlook 2019. World Energy Outlook Special Report. www.iea.org/africa2019

AC needs around 80% more capital, including doubling of investments in renewables and electricity networks compared with the STEP. Ethiopia aims to increase generating capacity by 25,000 MW by 2030: 22,000 MW of hydro; 1000 MW of geothermal; and 2000 MW of wind by 2030. Ethiopia is currently heavily reliant on hydropower. Plans to increase capacity to 13.5 GW by 2040 would make Ethiopia the second-largest hydro producer in Africa. Providing electricity access to all and electrifying productive uses will lead to a fivefold increase in generation in the STEPS, and an even bigger increase in the AC. Solar PV and geothermal account for almost 45% of the power mix by 2040 in the AC (IEA, 2019a, 2019b).

In Tanzania, providing access for all and growth in productive uses lead to a 13-fold increase of electricity demand by 2040 in the AC. This is met with an expansion of gas, hydropower and solar PV. Gas and electricity use in industry is growing strongly, especially in manufacturing industries. However, in the AC, energy efficiency (EE) measures could prevent consumption from being 20% higher than current levels. Despite the low access rate (37%) today, the grid represents more than half of new connections by 2030 in the AC given its existing and planned coverage. Despite the projected policies to promote clean cooking solutions, the number of people relying on traditional use of biomass for cooking is expected to decline from 55 million people today to 44 million in 2030 as efforts to improve access are outrun by high population growth in STEPS.

In the AC, LPG and biogas are the least-cost options for almost half of the population, with improved cookstoves the main way to extend access in rural areas. Recent large discoveries push up gas production to almost 30 billion cubic metres (bcm) by 2040 in the STEPS. Existing infrastructure helps Tanzania to increase domestic gas consumption. Gas demand in 2040 is twice as high in the AC, helped by efforts to promote the use of gas to displace traditional biomass and by support for gas-based industries. Almost $80 billion of cumulative energy supply investment is needed in the STEPS, with most of it being used to widen access to gas and electricity. Tanzania hopes to reduce GHG emissions by 10–20% by 2030 compared to the business-as-usual scenario (138-153 Mt CO_2-equivalent gross emissions). It hopes to increase electricity generation capacity from 1500 MW in 2015 to 4910 MW and achieve 50% energy from renewable energy sources by 2020.

24.2.3 Existing Gaps for Green Energy Development

Kenya is endowed with large renewable energy resources such as wind, geothermal and hydropower but developing them poses a number of challenges. Biomass fuels are the largest source of primary energy in Kenya with wood-fuel (firewood and charcoal) accounting for about 69% of the total primary energy consumption. About 55% of this is derived from farmlands in the form of woody biomass as well as crop residue and animal waste and the remaining 45% is derived from forests. There is a gap between the existing tree cover vis-à-vis the minimum constitutional requirement of 10%. The continuous overreliance on biomass as a primary source of energy is a hindrance to the achievement of 10% forest cover. The Government developed a strategy for the introduction of biofuel blends in the market in 2010. However, there are not enough bio-ethanol feed stocks. Kenya has huge renewable energy resources, but their rate of development has been very low. The gaps include limited:

- Publicly available information on Renewable Energy (RE) resource assessment and mapping to support investment promotion, decision making and energy planning.
- Diversification on policies to address specific issues and challenges associated with the different Renewable Energy Technologies (RETs),
- Incentives for the private sector in the development of the renewable energy technologies,
- Grid access in renewable energy resource endowed areas such as Northern Kenya (Kenya, 2016a, 2016b).

Tanzania has renewable energy potential capacity in, mainly hydro, geothermal, solar, wind and biomass. Geothermal resource potential is being assessed. Tanzania could benefit from a comprehensive Renewable Energy Strategy, and an accompanying resource assessment and investment prospectuses (Tanzania, 2015). There is uncertainty regarding the future direction of the inclusion of non-large-hydro renewables in power generation investment planning. There is also a need to undertake technical studies (including drilling) to understand and identify the precise potential of geothermal resources. Moreover, Tanzania faces constraints in the involvement of the private sector in the development of renewable energy due to limited information and options to access more efficient fuels and technologies; limited understanding of the potential savings from the different economic subsectors and the impact on the projected energy demand if EE measures are implemented. There is a need to follow up on implemented programmes and initiatives in order to understand their impact and identify lessons learned for future initiatives. The main challenges/gaps that Tanzania face to SE4ALL universal access to electricity by 2030 are: The need to strengthen governance in the energy sector with the objective to increase investors' confidence by diminishing perceived and actual risks during project development and operation; the financial capacity of key institutions, The Rural Energy Fund (REF) low volume of investment to support more connections a year; and connection fees that are still a deterrent for rural households and the poor (Tanzania SE4ALL, 2015).

Electricity supply in Ethiopia is dominated by the public sector. This is despite the existing regulations that allow private investment in power production both on and off-grid. The main challenges are: Governance issues—potential private investors existing grid tariff does not allow sufficient margins for return on investment; existing regulations also do not provide sufficient guarantees for investors in case of breach of contract by contracting parties (in particular EEPCO as the off- taker); In the off-grid market, geographic delineation is not clear between grid and off-grid areas to ensure sufficient time for investors to recoup their expenses. (Ethiopia SE4ALL, 2013).

24.3   The Approach and China's Overall Financial Flows in Kenya, Ethiopia, and Tanzania

Like in many other African countries, major infrastructure projects have led to high visibility of China's presence in Kenya, Ethiopia and Tanzania. The major sectors of Chinese activity are roads, railways, energy, and communications. The building of a modern rail network is a recent addition in a number of countries. Many of the sectors targeted are deemed a priority in the Eastern Africa government's national development strategy. The projects are almost exclusively carried out by Chinese firms, which get the financing directly in whole or in part (Jean-Pierre, 2015).Mostly, equipment installations, skilled manpower and technologies in these projects are sourced from China.

China's commitment to support the energy sector in Africa as espoused in the FOCAC action plans is to encourage cooperation in the exploitation of resources and support joint development and proper use of the energy and natural resources, including beneficiation at the source. In a number of African countries, China has established joint ventures with local state-owned companies in order to remain strategically close to political decision-makers in the energy arena. China is a major financier of hydro schemes, a trend of great strategic importance for the African power sector (Hwang et al., 2015).

At the 2009 FOCAC Summit, the then Chinese Premier Wen Jiabao proposed to build 100 clean energy projects in Africa. In 2015 the South-South Climate Cooperation Fund was launched and established China as one of the global leaders for supporting clean energy deployment in developing countries. As a major financial agency for international development in Africa, China has an opportunity to create the structural transformation necessary for long-term sustainability in Africa. Due to China's dominance in renewable energy industries and the large amount of financing for energy projects, China has the greatest potential to influence Africa's development objectives for the better.

24.3.1 Chinese State Financial Promotion Tools

The Chinese Government provides finance to Africa primarily under the category of Other Official Assistance. The Chinese use two models of financing: Strategic partnership and resource-backed package financing. There are five major instruments used in development assistance including: lines of credit to China MNCs; export credit; resource-backed loans; China Africa Development Fund; and China overseas SEZs. It is claimed by some authors that Chinese finance is not out of line, and it has interest rates that are found in the global capital markets (Degele & Seshagiri, 2019).Consistently, the Exim Bank and Other Chinese agencies have been the major conduits, with China Development Bank playing a major role since 2011.

24.3.2 Types of Financial Flows

Kenya, Ethiopia and Tanzania have enjoyed diverse levels of China's financial flows in the period 2000–2014. Bilateral economic and trade relations have scored new progress and both sides have made rapid headway in cooperation in the areas of electric power, communications, investment and project contracts. Figure 24.3 shows trends in all the development assistance flows received by Kenya, Ethiopia and Tanzania. Overall, these flows have been on an upward trend throughout the period 2000–2018.

Between 2000 and 2010, the net development flows for Ethiopia and Tanzania were converging and were much higher than Kenya's. While these flows have continued steadily for Ethiopia, over the period 2010–2018 the flows have dropped significantly for Tanzania. Kenya's net development flows have remained lower than

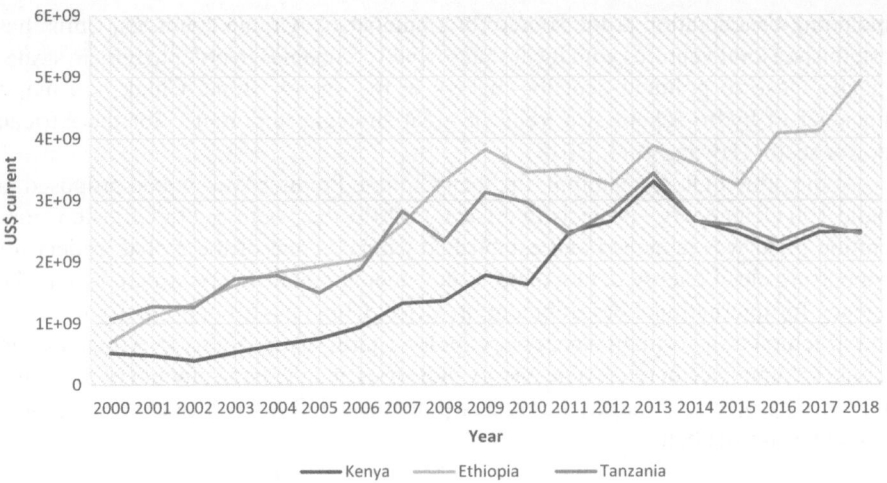

Fig. 24.3 Kenya, Ethiopia and Tanzania Net Development Assistance received 2000–2018 *Source:* Development Assistance Committee of the Organization for Economic Co-operation and Development, Geographical Distribution of Financial Flows to Developing Countries, Development Co-operation Report, and International Development Statistics database[1]

the other countries. Compared to the net development flows from across all the development partners, the flows from China have been erratic over the same period. The remarkable change in development flows for Kenya, Ethiopia and Tanzania coincide with the timings of (FOCAC) action plans.

24.3.3   Bilateral Relations Between China and Kenya, Ethiopia and Tanzania

Since the China-Africa Forum was established in 2000, Kenya has remained an active member of the forum. China's assistance to Kenya is exclusively project-based (Onjala, 2008). The infrastructure support was mainly in road construction projects, modernization of power distribution, rural electrification, water, renovation of the international sports centre, medical and drugs for fighting malaria, and construction of a malaria research centre. Even though China's aid to Kenya has increased substantially in recent years, it still constitutes a very small proportion

---

[1] Data are available online at: https://stats.oecd.org/. Net official development assistance (ODA) consists of disbursements of loans made on concessional terms (net of repayments of principal) and grants by official agencies of the members of the Development Assistance Committee (DAC), by multilateral institutions, and by non-DAC countries to promote economic development and welfare in countries and territories in the DAC list of ODA recipients. It includes loans with a grant element of at least 25% (calculated at a rate of discount of 10%). Data are in current U.S dollars.

of the aid received from bilateral, multilateral and the total aid received in Kenya (Onjala, 2008, 2018).

China and Ethiopia economic relations and the development flows re-emerged following the market-oriented economic system (liberalization) introduced in Ethiopia in 1992 (following erratic period). Since then, China's financial flows to Ethiopia have been rather strategic (geo-political) given the importance of Ethiopia in population, size and militaristic consideration in Eastern Africa as well as symbolic and show cases than real and huge grants as such.

The relationship between Tanzania and China is traced back to the 1960s. Since 1964, China has provided various kinds of assistance to Tanzania. Since 1967, China has also provided technical assistance to a number of sectors, including agriculture, mining, and social services. The most recent project financed by Chinese assistance is the construction of Tanzania National Stadium in 2004, financed through a soft loan from China (Bailey, 1975; Moshi & Mtui, 2008; Yu, 1971).

Over the period 2000–2014, the Chinese development flows for Kenya were dominated by transport at 91.2% ($30,132,435,034), followed by energy generation and supply 3% ($1,001,845,465), and communications 2% ($651,606,048). The figures are shown in Table 24.2. In Ethiopia, the sectors dominating the flows are transport at 84.5% ($66,241,770,000), followed by energy generation—12.2% ($9,530,399,182), and other multisector—2.9% ($2,301,100,574). In Tanzania, the leading sectors in receiving financial flows are business and other services at 62.6% ($14,659,018,580), followed by industry and mining at 27.7% ($6,486,592,040), and transport 4.5% ($1,046,732,871). Agriculture, which is a dominant sector driving economic growth across eastern African countries, has received among the lowest financial flows. Equally, industry and mining have received very low financial resources in Kenya—0.08% ($27,320,844) and Ethiopia—0.0% ($3,400,000).

In Tanzania, the development flows to the sector have been much higher at 27.7% ($6,486,592,040). These patterns of flows point to the view that China may be motivated to support countries with industrial sectors with mineral resources rather than industrial development per se. Energy development projects in Ethiopia and Kenya illuminate the effort to boost connectivity.

Given the position of Ethiopia's capital Addis Ababa as the unofficial capital city of Africa, the Chinese are motivated to make some of their investment and aid directed to Ethiopia as a showcase for other African leaders when they go for official visits to Addis Ababa (Zheng, 2016). For Kenya, transport sector infrastructure supported by the Chinese has been seen largely as a quest to capture markets in Eastern Africa, Kenya being an entry port to most of the countries. For Tanzania, to the extent that the quality of infrastructure is low, the development in the extractive sector and its economy-wide benefits offer value for money, this could be a beneficial engagement mutual for both parties.

Table 24.2 Sector specific development flows 2000–2014

Total amount of aid (US$)	Kenya	Ethiopia	Tanzania
Debt forgiveness	$13,652,177	$141,060,000	$257,813,829
Agriculture, forestry and fishing	$3,063,493	$12,358,741	$16,383,070
Business and other services	$50,820,000		$14,659,018,580
Communications	$651,606,048		
Education	$127,819,138	$28,318,972	$1,898,847
Emergency response	$48,989,664	$24,664,563	$18,138
Energy generation and supply	$1,001,845,465	$9,530,399,182	
General environmental protection	$21,877,810		
Government and civil society	$399,485,774	$113,646,755	$259,461,662
Health	$131,872,410	$431,738	$18,775,709
Industry, mining, construction	$27,320,844	$3,400,000	$6,486,592,040
Other multisector	$59,638,699	$2,301,100,574	$250,695
Other social infrastructure and services	$137,262,098		$131,651,280
Population policies /programmes and reproductive health	$224,221,641		
Trade and tourism	$0		
Transport and Storage	$30,132,435,034	$66,241,770,000	$1,046,732,871
Developmental food aid/food security assistance			$1,807,233
Non-food commodity assistance			$0
Water supply and sanitation			$539,318,350
Women in development			$5082

Source: Dreher, A., Fuchs, A., Parks, B.C., Strange, A. M., & Tierney, M. J. (2017). *Aid, China, and Growth: Evidence from a New Global Development Finance Dataset.* AidData Working Paper #46. Williamsburg, VA: AidData

24.4 China's Green Energy Finance for Development in Eastern Africa

Chinese institutions (notably the powerful State Grid) ruminate about the possibility of a green belt of connectivity which would link renewable energy installations globally through an advanced intelligent grid (cf, *Financial Times*, "China plans super-grid for clean power in Asia", 5 December 2017 (AREI, 2016)). Xinhuanet (Pilling, 2017) quotes State Grid Chairman Shun Yinbiao, speaking at the B20 China Business Council meeting in December 2017. He says that power investment demand would reach $1.5 trillion in Belt and Road countries in the next five years. The demand would come from the 1 billion people without access to electricity in Africa and South

Asia, an upgrade of power facilities in Central and Eastern Europe, as well as energy system transitions in Western Asia.

24.4.1 China and Ethiopia Energy/Power Sector

Ethiopia has vast hydro, wind, solar, and geothermal renewable energy potential. The total exploitable reserves of hydro and wind energy are 45 GW and 10 GW respectively. Only about 5% of Ethiopia's hydro resource and less than 1%of wind resources have been developed thus far (Chen, 2016). Chinese companies' ability to offer low prices based on concessionary finance from Export–Import (EXIM) Bank has contributed to stronger participation and financial flows to Ethiopia. China's energy sector financial flows in Ethiopia have focused on three areas of transmission lines, hydropower plans and wind power plants (Kruger Wikus et al., 2019). China EXIM Bank has financed several hydropower projects, including the controversial Gilgel Gibe III, Finchaa-AmertiNeshe and part of the Grand Ethiopian Renaissance Dam. China Gezhouba Engineering Group financed and was an Engineering, Procurement and Construction (EPC) for Genale Dawa III hydropower project, as well as the EPC for the Finchaa Amerti Neshe project.

In 2006, the Ethiopian Electric Power Corporation (EEPCo) and the China Gezhouba Group Corporation (CGGC) signed an agreement to construct the 197 MW Amerti Neshi Hydroelectric Power plant on the Neshi River. This dam was completed in 2011 for a total cost of $142 million, financed 85% by China Exim Bank. Construction was delayed for more than two years for several reasons, including the 24-month delay of the EXIM Bank of China loan, power interruptions, and logistical problems while transporting construction materials from Djibouti port.

In 2009, HydroChina, financed by a grant from the Chinese government, carried out a survey of solar and wind power potential in Ethiopia. Subsequently, HydroChina and CGCOC signed an EPC contract with EEP to develop a 51 MW wind farm at Adama. Adama Wind Farm is located 95 km southeast of the capital, Addis Ababa. The Export–Import Bank of China provided financing for the project through preferential export buyer's credit. The project cost $117 million, and the China Exim Bank financed 85% of the total cost. In 2010, ICBC provided approximately $235 million and in 2011, China Exim Bank provided $189 million in two separate PEBCs for the project. Dongfang Electric brought Sinohydro into the project in January 2011 construction (Hwang et al., 2015). Adama Wind Farm was the first operational wind farm in Ethiopia. After its inauguration in 2012, EEP signed another contract with HydroChina to add an additional 153 MW of capacity. Adama Phase II has a generating capacity of 153 MW. Again, China Exim Bank financed 85% of the US$345 million project. Phase II was inaugurated in May 2015 after 18 months of construction (Hwang et al., 2015). In 2014, Power China Huadong Engineering Corporation completed the rehabilitation of the Aba Samuel Power Plant. This incurred a cost of $14 million and was financed through a grant from the Government of China (IEA, Africa Energy, 2019a, 2019b).

In 2010, the state-owned Dongfang Electric Machinery Company agreed to supply electrical and mechanical equipment (turbines) funded by a $500 million ICBC loan (Jean-Pierre, 2015). Gibe III drew much criticism: built at the confluence of Gilgel Gibe and Omo rivers, 470 km southwest of Addis Ababa, it led to the forced displacement of 200,000 rural folks, with adverse effects on the water source of Lake Turkana in northern Kenya, thus putting at risk 300,000 Kenyans' access to water (Jean-Pierre, 2015). Ethiopia has since 2011 been building the Millennium Dam to emerge as eastern Africa's powerhouse. The projects are seen as part of the East African Power Pool launched in 2005. Table 24.3 provides a summary of the energy sector projects supported by China in Eastern Africa.

The 254 MW Genale Dawa III hydro project in southeastern Ethiopia is being built by China Gezhouba Group at a cost of around $450 million. The 1045 km Ethiopia-Kenya electricity transmission line is divided between 612 km on the Kenyan side and 433 km on the Ethiopian side. The 104 km electricity transmission project is expected to cost $1.2 billion upon completion, funded by the African Development Bank and the World Bank. The Ethiopian side of the Ethiopia-Kenya electricity transmission line is currently being constructed by China Electric Power Equipment and Technology. The electricity transmission line will have a transmitting capacity of 2000 MW once it is completed. Chinese firms have played an important role in helping Ethiopia achieve its energy sector ambitions. Affirming the government's commitment to continue energy sector partnership with Chinese firms. Ethiopia plans to increase its current 4300 MW electricity generation capacity to 17,300 MW by 2025, with power generation projects in hydro, wind, geothermal and biomass sectors.

The government of Ethiopia has awarded 12 solar energy development projects to three Chinese firms and one Spanish firm. According to the agreement signed by the Ethiopian Electric Utility, the energy projects will be distributed across the country and are estimated to cost $10.3 million. Three of the projects are built in Oromia, two in Amhara and the other two in the southern region. Somali, Tigray, Afar, Benishang, Gumuz and Gambella regions will have one project each. The projects are set to benefit more than 67,700 rural homes upon completion. About 25 projects are under a design phase and the African Development Bank (AfDB) has approved a $15 million loan and grant for the projects. Ethiopia has about 5 MW of off-grid solar. Almost all current solar power is used for telecommunications. Other uses include village well pumps, healthcare and school lighting. About 150,000 households will receive solar power courtesy of government initiative. A solar panels assembly plant was established in Addis Ababa in early 2013 and is capable of making 20 MW of panel per year.

The State Grid Corporation of China has announced the investment of $1.8 billion in Ethiopia's electricity transmission and distribution networks, According *to New Business Ethiopia*, the deal agreement was made following a visit to Beijing by Abiy Ahmed, the Prime Minister of Ethiopia. The Prime Minister held talks with the China Export and Credit Insurance Corporation (Sinosure), which agreed to organise the funds. Although the Ethiopian economy has been growing by more than 10% a year since 2006, it is struggling to service its foreign debts. During Abiy's visit, China agreed to cancel all interest accumulated from its debts, estimated at between $12

Table 24.3 Energy sector financial flows

Year	Kenya	Ethiopia	Tanzania
2005	EXIM Loans for power distribution system modernization and strengthening project $ 13,020,000; Loans for power generation expansion $25 million		
2006	Chemosit-Kisii & Kamburu—Meru Power Distribution System Modernization & Strengthening Project 50.89 millionRMB 7,124,600; In 2006 Nairobi 132 kV &66 kV Network Upgrade and Reinforcement Project Phase 11 382.5 million RMB $53,550,000	EXIM Bank Loans $208 Million USD for Amerti Neshi Hydroelectric Power Plant and Muger Cement Projects $208 million	
2007	Construction of circuit lines and substations Phase I loan 160 million CNY 22,400,000; Completion phase 2 of urban power grid project loan 113 million CNY US$15,820,000	Loans for power company ETB for Beles-Bahir Dar power line extension US$43,400,000; Loan to finance new power generation and expand a cement factory $208 million	
2009	Kenya power distribution system modernization and strengthening Project (Phase 2) 1 loan 161 million RMB 22,540,000	China EXIM Bank Funds Construction of Messabo Harrena Wind Farm $127 million; China Exim Bank loans Ethiopia 1.8 million ETB for expansion of Wolayta Sodo and Gibe III power stations US$1 55,800; EXIM Bank Loans funds for Construction of Adama Wind Farm I 2009 Adama Wind Farm I US$117 million	
2010:	China Loans 7.5 billion KES for construction of geothermal plant 7.5 billion KES $69,750,000; Olkaria IV Geothermal Production Wells Drilling Project loan 38.971 million RMB $545,594	Genale-Dawa III Eximbank Export Buyers Credit EBC2.43%12 (3) $270 million; Gibe III Electric and Hydromechanics ICBC Preferential Export Buyers Credit PEBC $235 million; ICBC loans $420 million for construction of Gibe III Dam	

(continued)

Table 24.3 (continued)

Year	Kenya	Ethiopia	Tanzania
2012	Provision of Drilling Materials for Eighty (80) Geothermal Wells at Olkaria Geothermal Field Project $637 million. Construction of a Nyahururu-Maralal transmission line 1.77 billion KES 16,461,000	Eximbank Adama Wind Farm II US$345 million; Bio-Gas Appliances, Installation, and Training grant 10 million, CNY-$1,400,000 Exim Bank Loans $1.02 Billion Construction of Power Lines	China Loans $1.2 billion to Construct the Mtwara-Dar Es Salaam Gas Pipeline loan $ 1.297 billion
2013	Geothermal power project in Kenya loan 95.4 million$887,220; Nairobi City Center E.H.V and 66 kV Network Upgrade and Reinforcement Project loan 670 million RMB US$93,800,000		

Source: Custer et al (2021), Dreher et al (2017), and Dreher et al (2022)

billion and $20 billion. According to an Addis Ababa newspaper, *The Reporter*, the Ministry of Transport announced that the construction of Ethiopia's electrified railway between Awash and Hara Gebeya has made little progress in the past nine months due to a lack of funds.

In Ethiopia, Sinohydro and Gezhouba Group will work with the Ethiopian SUR Construction on the 385 MW Gibe 1 and 2 hydroelectric complex projects, which are mainly funded by China's EXIM Bank. Another notable project is the 1260 km GDHA 500 kV Power Transmission and Transformation Project. This project, successfully completed by December 2015, constitutes the core part of the East African power grid.

24.4.2 China and Kenya Energy/Power Sector

The power sector in Kenya has also received substantial support under the development cooperation with China. In January 2005, Kenya signed an agreement for a concessional loan with the Chinese Import Export Bank for a $ 1.4 million loan to finance the Kenya Power Distribution System Modernization and Strengthening Project. The project was implemented by China CAMC Engineering Co. Ltd. In September 2005, the Kenya Power and Lighting Co. Ltd (KPLC) received funding of about $20 million concessional loan from the Exim Bank to connect 150,000 new power consumers every year. In 2010, the then Kenyan Finance Minister Uhuru Kenyatta and Chinese Vice-President of the Export–Import Bank of China Mr. Zhu Hongjie signed an agreement for a concessional loan provided by the Exim Bank.

The project aimed to develop a 140 megawatt geothermal plant, specifically through drilling 26 steam production wells at the Olkaria IV geothermal field in Rift Valley (Onjala, 2018). In the financial year ending 2015, China gave Kenya $120 million to the Ministry of Energy and Petroleum. The amount included the large geothermal loans from China EXIM Bank (Sanghi & Johnson, 2016).

In December 2019, Kenyan President Uhuru Kenyatta launched a 50 MW solar power farm located in Garissa, northeast region. The plant is one of the largest photovoltaic electricity stations in Africa. The project was designed and built by the EPC contractor China Jiangxi Corporation for International Economic and Technical Co-operation (CJIC), in conjunction with Kenya's Rural Energy Authority (REA). The project meets the power demand of 70,000 households (some 350,000 people) in Kenya, equivalent to some 50% of the Garissa population.

China's Tianpu Xianxing Enterprises and Kenya's Electrogen Technologies entered into a Sh9 billion ($140 million) partnership to build a solar panel factory in Nairobi —considered to be the first in the Horn of Africa. The project is expected to position solar power as a key source of energy in Kenya by making it more affordable to millions of consumers who do not have access to the national grid and also to those who depend on the unreliable national electricity grid for their energy needs. Tianpu Xianxing Enterprises is a Chinese company that specializes in research, development, design, manufacturing and marketing of solar energy products such as water heaters, semi-finished pipes, vacuum pipes, heat pipes and air heat pumps.

24.4.3 China and Tanzania Energy/Power Sector

Due to its natural resources and geography, Tanzania could have access to a variety of energy sources, including natural gas, biomass, hydropower, geothermal, coal, solar and wind power. Biomass supplies more than 90% of energy consumed. The remaining energy sources are fossil fuels (6.6%), gas (1.5%), hydro (0.6%) and coal and peat (0.2%). Recently discovered gas reserves could aid in alleviating Tanzania's energy constraints. However, so far the country has not benefited from the reserves because of a lack of investment and lack of planning and implementation, including developing pipeline and creating gas processing infrastructure. Major Chinese energy companies are interested in investing in Tanzania's energy sector that has proved to be worthwhile. The investors from China expressed their interest during a meeting with the Deputy Permanent Secretary in the Ministry of Energy and Minerals, Dr. Juliana Pallangyo, officials from the ministry, Tanzania Electric Supply Company (Tanesco), State Mining Company (Stamico), National Development Corporation (NDC) and Energy and Water Utility Regulatory Authority (EWURA). Tanzania, which is currently focusing on exploiting its domestic reserves of natural gas to increase installed generating capacity by 6000 MW, might consider the Chinese interest as part of its 25-year Power Sector Master Plan.

In October 2013, Tanzania government signed $1.7 billion contracts with six Chinese companies for hydropower projects, residential construction and an alternative energy research centre. According to IPP Media, a Tanzanian media group, the contracts were awarded during Tanzanian Prime Minister Mizengo Pinda's official tour to China. The new energy agreements made were:

- Tanzania Electric Supply Company (Tanesco) and Tabiau Electric Apparatus Stock for a $690 million power transmission line project.
- Tanesco and Shanghai Electric Power Company for the construction of the Kinyerezi III power plant.
- Tanesco and China Gezhouba Group Corporation to develop the Rumakali hydropower project.
- Mkonge Energy Systems with Sino Hydro Resources for the $136 million Masigira hydropower project.
- National Development Corporation with Dalian International Economic & Technical Cooperation Group and Hydro China Kunming Engineering Corporation to build a $136 million Centre for Research and Training on Alternative Energy.

In July 2014, The Export–Import Bank of China gave Dar a soft loan of Tsh 226.7 billion (about $136 million) for the wind power project in the Singida region, which has 50 MW generation capacity. The project focuses on mitigating the shortage of power in the country and helping people, especially in rural areas, to get reliable electricity and increase their production and income. Joint Ventures Company has been incorporated as Geo Wind Power Tanzania Ltd to oversee the project under Public Private Partnership arrangement. The project involves three shareholders namely, National Development Corporation, TANESCO and Power Pool East Africa Ltd. located in Singida Municipality.

In January 2020, the Chinese group Sany Heavy Industries embarked on a wind energy project in Tanzania. The company concluded an agreement with the authorities for the development of the project in several stages. It will build a wind farm capable of producing 600 MW, making it the largest facility of its kind in Africa, thus eclipsing the **Lake Turkana wind farm** which supplies 310 MW of electricity to the grid in Kenya. The first stage of the project which will produce 100 MW. Sany Heavy Industries is not new to the wind energy sector in Africa. In 2013, the group won a $95 million contract to supply 102 turbines for the Adama wind project. The turbines, which rotate using wind force, are installed on the rocky hills of the Ethiopian highlands, 100 km southeast of the capital Addis Ababa. Wind is currently a small element of the product mix of Sany, a construction equipment giant that is one of China's largest privately-owned companies but was ranked 14th in its domestic market in 2022. Energy projects constructed by Chinese contractors in Tanzania and Zambia include the 300 MV KIII gas-fired power plant and the 750 MW Kafue Gorge Lower hydroelectric power plant.

24.5   Conclusions and Recommendations

Three features distinguish the types of Chinese financial flows to Kenya, Ethiopia and Tanzania as identified in our analysis for the period 2000–2014.

- First, the bulk of these loans were infrastructure sector-related. Majorly they covered the roads, rail, airport networks, energy and information and telecommunications. The patterns were consistent with previous studies which suggest Chinese infrastructure finances to be invariably directed at facilitating the creation of markets and export of primary commodities from the continent.
- Second, compared to the concessional financing that the continent used to obtain from IFIs such as the IDA window of the World Bank, the financial flows are mainly tied loans that are relatively expensive.
- Thirdly, Chinese firms are also exclusively the ones that were building such infrastructure. A further challenge is the practice of almost exclusive reliance on Chinese workers and supplies in the infrastructural activities as implicit in the financial aid. Apart from the reduced or limited technological development, sustenance of the infrastructural assets in the medium to long term can be quite sub-optimal for the recipient countries.

China is not only experiencing its own green energy transition away from coal and toward renewables, but it can also shape the transition of the rest of the world. China can make significant contributions in shaping energy sector development in Eastern Africa.

24.5.1   Towards Aid Policy in Relation to Chinese Green Finance?

Eastern African countries should leverage China's financial flows to go beyond the traditional scope of building energy infrastructure and begin to agitate the use of modern green technologies in different sectors. As such, the much desired impact of green infrastructure on improving efficiency and competitiveness of domestic production should be the primary policy goal of these countries. The regional countries can emphasize specific aid partnerships with China, which can stimulate modernization of local industrialization, all without mentioning that African labour costs are lower than those of many urban Chinese who are deployed in the infrastructural projects. A clear public debt policy is needed among the African countries. African governments should develop clear borrowing policies and bargain for aid packages that not only fulfil their financial needs but also have deeper impacts in terms of reducing the costs of technology uptake and attracting new investments.

References

African Development Bank (AfDB). (2013). *State of Infrastructure in East Africa*. A Report Prepared by the Statistics Department in the Chief Economist Vice Presidency of the African Development Bank.

African Economic Bulletin. (2017). May 16th–June 15th.

AREI-Action-Plan. (2016). http://www.arei.org/wp-content/uploads/2018/03/AREI-Action-Plan-Nov-2016.pdf. Accessed 13 June 2018.

Bailey, M. (1975). Tanzania and China. *African Affairs, 74*(294), 39–50.

Custer, S., Dreher, A., Elston, T. B., Fuchs, A., Ghose, S., Lin, J., Malik, A., Parks, B. C., Russell, B., Solomon, K., Strange, A., Tierney, M. J., Walsh, K., Zaleski, L., & Zhang, S. (2021). *Tracking Chinese development finance: An application of AidData's TUFF 2.0 methodology*. AidData at William & Mary.

Degele, E., & Seshagiri, R. (2019). Sino-Africa bilateral economic relation: Nature and perspectives. *Insight on Africa, 11*(1), 1–17.

Dreher, A., Fuchs A., Parks, B. C., Strange, A. M., & Tierney, M. J. (2017). *Aid, China, and growth: Evidence from a new global development finance dataset*. AidData Working Paper #46. Williamsburg, VA: AidData.

Dreher, A., Fuchs, A., Parks, B. C., Strange, A., & Tierney, M. J. (2022). *Banking on Beijing: The aims and impacts of China's overseas development program*. Cambridge University Press.

Ethiopia SE4ALL. (2013). *Updated rapid assessment and gap analysis on sustainable energy for all (SE4All): The UN secretary general initiative*. Federal Democratic Republic of Ethiopia Ministry of Water, Irrigation and Energy.

Guttman, J., Amadou, S., & Soumya, C. (2015). Financing African infrastructure: Can the world deliver? Global economy and development, Brookings.

Hwang, J., Brautigam, B., & Wang, N. (2015). Chinese engagement in hydropower infrastructure in Sub-Saharan Africa. *The SAIS China-Africa Research Initiative Working Paper 1*. December 2015.

International Energy Agency (IEA). (2019a). *World Energy Outlook 2019*, IEA, Paris https://www.iea.org/reports/world-energy-outlook-2019. International Energy Agency.

International Energy Agency (IEA). (2019b). *Africa Energy Outlook 2019*. World Energy Outlook Special Report. www.iea.org/africa2019.

Kruger, W., Stuurman, F., & Alao, A. (2019). *Ethiopia Country Report. Report 5: Energy and Economic Growth Research Programme*. Power Futures Lab.

Moshi, H. P., & Mtui, J. M. (2008). *Scoping studies on China–Africa economic relations: The case of Tanzania*. A Revised Final Report submitted to African Economic Research Consortium, Nairobi-Kenya, March 2008.

Onjala, J. (2008). *Scoping studies on China–Africa economic relations: The case of Kenya*. A Revised Final Report submitted to African Economic Research Consortium, Nairobi-Kenya, March 2008.

Onjala, J. (2018). China's development loans and the threat of debt crisis in Kenya. *Development Policy Review*. https://doi.org/10.1111/dpr.12328

Pilling, D. (2017). Chinese investment in Africa: Beijing's testing ground. *Financial Times*. June 13. https://www.ft.com/content/0f534aa4-4549-11e7-8519-9f94ee97d996.

Republic of Kenya (2016a). *Kenya Action Agenda*. Sustainable Energy for All (SE4ALL). www.se4All.org.

Republic of Kenya (2016b). *Kenya National Forest Programme 2016–2030*. Ministry of Environment and Natural Resources.

Shkaratan, M. (2010). Tanzania's infrastructure: A continental perspective. *Policy Research Working Paper 5962*. The International Bank for Reconstruction and Development/The World Bank.

United Republic of Tanzania. (2015). *Tanzania's SE4ALL action agenda*. Ministry of Energy and Resources, December 2015.

Wang, J. (2011). What drives China's growing role in Africa? *IMF Working Paper WP/07/211*. International Monetary Fund.

Yu, G. T. (1971). Working on the Railroad: China and the Tanzania–Zambia Railway. *Asian Survey, 11*(11), 1101–1117.

Zheng, Y, (2016). *China's aid and investment in Africa: A viable solution to international development?* Fundan University, August 2016.

Part VIII
China-Africa Collaboration in Health Sector

Chapter 25
China–Africa Collaboration in the Health Sector Under the Epidemic of COVID-19

Yizhou He

25.1 Introduction

As one of the poorest states of health, both traditional and nontraditional security are intertwined leading to the African continent being a place far away from peace. Various security challenges have unceasingly threatened the African continent for ages. Natural disasters, poor infrastructure, lack of financial resources, a fragile economy, and prevalent poverty leading Africa to lag behind the world in every aspect. Although political situations generally stabilized, it is undeniable that civil wars, military coups, border conflicts, and other security issues are affecting Africa's social development continuously. Meanwhile, nontraditional security problems have risen to prominence, especially in the field of public health. As a result, the African region is not only suffering from extreme poverty but is also being threatened by a severe health emergency. Health is an essential prerequisite for the whole of humanity. A well-functioning health system is a necessity for human well-being and social development. However, Africa is the place that suffers from the highest frequency of public health events, and public health crises have deeply disturbed the security and development of Africa.

Since the coronavirus disease (COVID-19) outbreaks and spreads all over the world, it not only threatens people throughout the world but also seriously ravages social and economic development globally, bringing grave challenges to global public health security. Specifically, because of the poor living conditions with weak infrastructure as well as appalling sanitation, Africa has to take action to struggle with public health to respond to COVID-19 through social and economic aspects. This results in the indispensable help and support of the international community to establish a properly functioning healthcare system in Africa.

Y. He (✉)
Institute of African Studies, Zhejiang Normal University, Jinhua, China
e-mail: heyizhou49@foxmail.com

M. Muchie et al. (eds.), *China-Africa Science, Technology and Innovation Collaboration*, https://doi.org/10.1007/978-981-97-4576-0_25

China and African countries are a community of destiny and have great common interests in maintaining world security and development. The coronavirus pandemic hit the world to undergo profound changes, including China–Africa relations, so it is necessary for China–Africa to cooperate in the health sector. As a result, it is significant to study the current situation and problems of public health in African countries and summarize the effective system between China and Africa in the public health sector to build a better post-pandemic future relationship.

25.2   The Plight of the COVID-19 Epidemic in Africa

Africa is a region that with weaker health systems tend to suffer more in the event of a pandemic. To date, a total of 12,049,613 COVID-19 cases and 255,634 deaths (case fatality rate: 2.1%) have been reported from 55 member states of the Africa Union (AU) as of 5 September 2022. This represents 2% of all cases and 4% of all deaths reported globally, presenting a higher case fatality rate (CFR) than the global CFR.[1]

African countries were prompted to take measures at the beginning of the COVID-19 epidemic to prevent and control the disease on the whole. The government has certain experience and mobilization in epidemic prevention to a certain extent. Most African countries have implemented interventions to prevent large-scale epidemics in the early stage of COVID-19, typically through nonpharmacological measures such as lockdowns of the cities, restrictions on large gatherings, notifying citizens of the curfew, and social distancing to mitigate the spread of the disease. Meanwhile, strategies such as building working groups on the issues, appointing special envoys, establishing funds, launching the rapid testing partnership initiative, striving for debt relief and training the medical personnel was set to fight against the pandemic. The current Chairperson of the African Commission, Mr. Moussa Faki Mahamat considered the response to the COVID-19 pandemic in Africa was effective in the early stage of COVID-19.

Nevertheless, these mitigation efforts are unsustainable due to the severe backwardness of the medical and healthcare system in Africa, and it is difficult to prevent and control the confirmed cases and deaths during the COVID-19 pandemic in the long term. Most African countries lack sufficient preparations for COVID-19. According to the Global Health Security Index (GHS Index) from WHO, only 21 African countries were clinically prepared to deal with epidemics. Since the discovery of COVID-19 in Africa in February 2020, 96% of AU countries have experienced three different waves of COVID-19, 72% have experienced four waves, 17% have experienced five waves, and Kenya and Mauritius have experienced six waves, respectively.

The fact cannot be argued that Africa's public health infrastructure is quite poor, hospitals, medical personnel and medical supplies are seriously lacking throughout

---

[1] Data Source: World Health Organization (WHO).

Africa. At the start of the COVID-19 pandemic, comprehensive and rigorous public health and social measures were introduced in several African countries. However, because of the low level of compliance with control measures and the lag in public health interventions for recovery in many African countries, they have played a relatively small role in reducing the spread of COVID-19. Three waves of the COVID-19 pandemic occurred in 2021, which exposed severe health system imbalances in Africa. The overreliance on external production of vaccines, lack of physiotherapy devices, and even a severe shortage of medical oxygen, intensive care units and beds for patients all make Africa paralyzed. Additionally, according to WHO, there are less than 20 medical doctors per 10,000 population in African countries, and only 2.9% of health workers are distributed in African countries, which is quite low compared to health workers accounting for 16.4% globally. Besides, there are not enough laboratories and testing equipment, and Africa's pharmaceutical production capacity is insufficient, leading to African countries being net importers of pharmaceutical products.

The low testing rate for COVID-19 and the low vaccination rate also hindered Africa's health sector. For the testing for COVID-19, the WHO recommends that the test should be performed at least once a week for every 1,000 people. However, most African countries cannot meet the standard. Countries like Tanzania have only 63 tests per million people, 373 in Niger, 383 in Chad, 467 in the Democratic Republic of Congo, and 563 in Burundi, respectively (Pandey, 2020). Low vaccination coverage is a critical reason for the continued prevalence of COVID-19 in Africa. As of 12 September 2022, a total of 314 million people in the African region were fully vaccinated, representing 22.6% of the region's population. However, twelve countries have not reached 30% of the population vaccinated, and Eritrea is the only country that even has not initiated COVID-19 vaccination.[2] The inequity of the COVID-19 vaccine is particularly evident in Africa, and the number of doses received in the African region represents 40% of the doses required to vaccinate 70% of the population in all countries. 32 of 46 countries reported that the vaccine had expired. The number of expired doses represents 2.3% of the doses received in these countries and 1.5% of the doses received in the African region.[3] Only 18 AU member states have 20% or more of the population fully vaccinated. 16 countries have less than 10%, and 11 countries have less than 5% population coverage.

Besides, public spending in Africa is severely insufficient and Africa is heavily burdened by debt. The WHO calls on all countries to spend at least 9% of public finances on health, but Africa cannot reach the goal. Only two of Africa's 47 countries have made the commitment made by African finance ministers in 2001 to spend 15% of their national budgets on health systems, and only 13 countries spend the global standard of $60 per capita per year, with the rest spending only half that amount (WHO, 2022b).

Poverty and disease are long-standing problems that seriously affect the lives and health of African people. The rising poverty rates may take nearly a decade

---

[2] Data source: WHO.

[3] Data source: WHO.

to return to pre-pandemic levels. Particularly, in earlier surveys, 40% of African countries reported that sexual, reproductive, maternal, new-born, child and adolescent health services will continue to be disrupted for many years to come. The COVID-19 epidemic has caused widespread economic damage, pushing more women and girls into extreme poverty. UN Women's research shows that in Sub-Saharan Africa, women make up 20% of task force members and only 15% of task force leaders. In addition, existing gender inequalities have increased significantly, with 85% of national COVID-19 task forces led by men, despite women comprising 70% of the health and social workforce. Mostly, disruptions in health services often leave women with limited access to health services.

Africa's economic structure is quite singular, manufacturing is lagging, and its poor taxation system and heavy internal and external debt pressure make it difficult to invest effectively in public health. Africa bears 25% of the global burden of disease, but health spending is less than 1% of global spending, and the total value of health products consumed is less than 2%. Poverty has prevented many African countries from establishing robust health systems. Before the outbreak of the pandemic, in 2019, the number of people living in extreme poverty in Africa was estimated to be 478 million, which is over 1/3 of the total population in Africa; however, 490 million people were estimated in Africa live under the poverty line of 1.90 PPP$/day in 2021, and this is 37 million people more than the situation that without the pandemic (UNCTAD, 2021).

African public health systems are under dual pressure from infectious diseases as well as non-communicable diseases (NCDs). Africa has long been hit by infectious diseases, with Ebola, Lassa fever, malaria, cholera, yellow fever, AIDS, schistosomiasis and other infectious diseases plaguing Africa all year round. Moreover, Africa has become one of the fastest-growing regions in terms of NCDs, such as cancer, cardiovascular disease, respiratory disease, and diabetes becoming the leading cause of mortality. Besides, unhealthy dietary habits, such as excessive intake of foods high in sugar, fat, oil and salt, are one of the imperative reasons for the rising rate of chronic diseases in Africa. According to the WHO, between 50 and 88% of deaths in seven African countries, mostly small island nations, are due to NCDs. The report also indicates that in seven other countries, the majority of them being Africa's most populous, the diseases claimed between 100,000 and 450,000 lives annually (WHO, 2022a). In particular, the situation of vulnerable older people is beset with difficulties as well. A study conducted by WHO's Africa Research Department in the first half of 2021 reports that the response has been hampered by the prevalence of NCDs associated with rapidly aging populations. In addition, 22% of older adults were economically active before the pandemic, as the COVID-19 pandemic took hold, older people suffered a significantly higher risk of complications from COVID-19, and the CFR increased obviously with age due to the reduced immunity (WHO, 2021). Although this has been mitigated in some countries through direct vaccination to the most vulnerable people, hesitancy among older adults due to inadequate vaccine promotion and budgetary and logistical constraints to vaccination have resulted in many older adults not being vaccinated. Thus, infectious diseases and NCDs under the COVID-19 pandemic, made Africa more stressed.

25.3 China–Africa Health Collaboration Under the BRI

According to the United Nations (UN), the 2030 Agenda is determined to end poverty and hunger in all their forms and dimensions and to ensure that all human beings can fulfil their potential in dignity and equality and a healthy environment. China and Africa have a long history since the 1950s. In terms of health aid to Africa, China sent the first medical team to Algeria in 1963. Since 2014, China has been actively involved in fighting against Ebola in Guinea, Liberia and Sierra Leone three West African countries. The outbreak of Ebola made public health become a new area of China's medical assistance to Africa. In response, China–Africa health sector collaboration will help Africa to build a modern healthcare system with the capability to cover the whole of its population.

25.3.1 China–Africa's Proposal for Global Health Collaboration

With the Health Silk Road (HSR) addressed under the Belt and Road Initiative (BRI), China and African countries have been able to increase their health cooperation through personnel, capacity, and technology more effectively. According to Organization for Economic Co-operation and Development (OECD), the BRI is a large project aimed at improving regional cooperation through better connectivity among countries lying on the ancient Silk Road and beyond (OECD, 2018). China's BRI strategy expands the scope of public engagement and aims to build collaboration and connectivity with other developing countries. The BRI 2014 proposed the global health perspective and the next year, the National Health and Family Planning Commission unveiled the Three-Year Plan for Belt and Road Health Exchange and Cooperation (2015–2017). During the plan, China decided to strengthen the high-level health visits among countries along the Road and Belt routes, and push for cooperation agreements with countries, especially neighbours. In June 2016, President Xi Jinping proposed for the first time to work together to build a "Health Silk Road" at the Legislative Chamber of the Uzbek Supreme Assembly in Tashkent. In 2017, the first Belt and Road Forum for International Cooperation was established in Beijing, and the HSR was announced and endorsed since then. HSR aims to promote healthy sustainable development in the countries along the routes. HSR stressed the importance of international health cooperation to mobilize the enthusiasm of civilian organizations, multinational corporations and social groups through the government to maintain people's health and well-being. Besides, HSR considers that sufficient equipment and funds for emergency response are a necessity in the public health emergency system. Meanwhile, medical personnel training, pharmaceutical research and development and investment in health industries all need sound cooperation among countries. China's efforts to integrate health into its development agenda were also appreciated by the World Health Organization (WHO), believing China will take responsibility for global health governance.

Africa is also making a great effort to work toward health issues. The Agenda 2063 is Africa's blueprint and master plan for transforming Africa into the global powerhouse of the future. According to AU, to create a qualified life condition and improve the standard of living, one of the key goals for Africa is to ensure that its citizens are healthy and well-nourished and that adequate levels of investment are made to expand access to quality healthcare services for all people. Aspiration 1 of Agenda 2063 notes the healthy and well-nourished citizens of African people. Thus, The AU's Africa Centres for Disease Control and Prevention (Africa CDC) was set up as the lead institution to support African countries in promoting health and preventing disease outbreaks by improving prevention, detection, and response to public health threats. But overall, the capacities for health policy implementation remain weak. Since the BRI is highly compatible with the African Union's (AU) Agenda 2063 and the development strategies of African countries, the desperate need for healthcare resources and infrastructure in Africa needs China's help urgently.

In 2016, Healthy China 2030 was issued mainly for promoting people's health in the coming 15 years. The development of a "Healthy China" is central to the Chinese Government's agenda for health and development, and has the potential to reap huge benefits for the rest of the world. China–Africa collaboration in the health sector offers a good example. China declares to help Africa to strengthen its health systems, trying to eliminate major infectious diseases and improve the health of African people. It is believed that China–Africa shared mutual benefits in safeguarding world peace and stability, and China has the power and the obligation to ease the fragility of Africa's health system. There is no denying that Africa is a vital place for global public health governance, and China needs to strengthen the African continent's health systems to cope with the public health threats and epidemic burdens.

25.3.2 China–Africa Health Collaboration in Post COVID-19 Era

The current COVID-19 health crisis broke out at the beginning of 2020, and the COVID-19 outbreak has been regarded as the worst humanitarian crisis since World War II. During the pandemic, health services and related infrastructure and logistics have topped the agenda worldwide.

As China endured the first pandemic wave, the AU and African leaders were the first to express their solidarity with China and believed China would overcome the difficulties since the effective national government is trustworthy. Africa stands with China and calls for stronger international cooperation against the pandemic. Several African countries, including Algeria, South Africa, Egypt, Morocco and other African countries have provided China with financial support, masks and other medical supplies during this time. Algeria was the first country that provided emergency medical supplies soon after the epidemic had surged in Wuhan, where the infection case was first discovered. The government supported 500,000 surgical medical masks, 300,000 pairs of medical gloves and 20,000 pairs of medical goggles to China in fighting against the virus (Lianhe, 2020). A manufacturer U-Mask from South Africa donated face masks to help China during the epidemic (Omondi, 2020). The Ministry of Health of Egypt announced to provide 10 tons of preventive medical items and equipment to China (Mena, 2020). Moroccan Foreign Trade Bank donated 150,000 surgical masks and 900,000 medical gloves to Hubei through the China Development Bank (CDB).[4] Besides, according to the Ministry of Commerce of the People's Republic of China, Equatorial Guinea and Djibouti, among the world's least developed countries, also donated $2 million and $1 million to the Chinese government, respectively.

Regarding the nongovernment aspect, employees of the Standard Bank in Africa joined the celebration of China's Lunar New Year under the theme "Wear Red for China" on 31 January 2020 (Opali, 2020). This is mainly to support Wuhan, China. Tens of thousands of employees participated in the event, all wearing red and holding banners and cards in support of Wuhan, China. The event presented the African people's solidarity with the Chinese when China is suffering from the huge shock of COVID-19. Craig Ebden who serves as the bank's head of China–Africa sales in South Africa said that the bank stands with China during tough times. Despite the difficulties brought about by the epidemic, the nongovernmental exchanges and cooperation not only enhanced the understanding and friendship between the people but also effectively promoted official exchanges and cooperation between China and Africa.

In the meantime, China carried out the most concentrated emergency humanitarian assistance on the most wide-reaching scale during the pandemic period. As of November 2021, China's foreign anti-epidemic assistance has benefited more than 150 countries and 13 international organizations. 37 batches of Chinese medical experts were sent to 34 countries to organize technical guidance and shared the anti-epidemic experience with others. Meanwhile, China donated $50 million to WHO and $50 million to UN agencies as well as other international organizations (The State Council Information Office, 2021). Since the first case was detected in Africa on 14 February 2020, China has taken immediate action by donating funds, hoods, protective clothing, testing reagents, respirators, goggles, infrared thermometers, forehead temperature guns, and other diagnostic instruments and equipment to African countries. Besides, China sent medical expert groups to the Africa CDC, Burkina Faso, Ethiopia, Zimbabwe and other African countries to assist them in fighting against the pandemic. Besides, throughout the videoconferencing many times, China shared anti-epidemic prevention and control measures with Africa and exchanged China's experiences with Africa.

China has made great efforts to help Africa and has become Africa's one of the most important donors and suppliers in terms of COVID-19 vaccines and related

---

[4] Data source: Bank of Africa.

materials. China was the first country to propose vaccines as a global public good, which should ensure accessibility and affordability for developing countries with a reasonable distribution of vaccines globally. By the end of 2021, according to the Ministry of Foreign Affairs of the People's Republic of China, China had provided more than 180 million doses of vaccines to Africa since African countries are heavily dependent on imports of drugs and seldom have a biotechnology sector.

In May 2020, China committed at the World Health Assembly (WHA) and provided $2 billion to assist the international community in fighting against the pandemic, especially in developing countries within two years.[5] To date, according to the China International Development Cooperation Agency (CIDCA), China has provided more than 120 batches of testing reagents and respirators and other anti-pandemic materials, 50 African countries and AU have received support from China's vaccine assistance, and medical experts from China are dispatched to 17 African countries.[6] What's more, China has sent 46 medical teams to Africa and has established partnerships; 15 Chinese traveling anti-epidemic medical teams and 43 China–Africa counterpart hospital cooperation mechanisms were urgently established, bringing epidemic prevention experience, programs and materials to Africa in the first place. In addition, China's Sinovac vaccines account for 6.4 percent of vaccines received by AU members including 17 percent of Sinopharm vaccines,[7] placing China second after the United States in the percentage of vaccines received by all AU member states.

China promised to help the AU build the Africa CDC and work with the African people to implement the "Health and Wellness" project within the framework of the Forum on China–Africa Cooperation (FOCAC). It will become the first disease prevention and control centre with modern office and laboratory conditions and complete facilities in Africa. The FOCAC was established in 2000 and has a platform for collective dialogue between China and African countries and a mechanism to promote practical cooperation. At the FOCAC on 29 November 2021, Chinese President Xi Jinping proposed that China–Africa build an even stronger community with a shared future. To fight against COVID-19, people and their lives should be the top priority. This has prompted the realization of the China–Africa Cooperation Vision 2035, and one of the aspirations is medical and health program cooperation. During the 8th FOCAC ministerial conference, China promised to provide another one billion doses of vaccines to African countries, including 600 million doses as a donation as well as 400 million doses to be provided through joint production by Chinese corporations. In addition, China will undertake 10 medical and health projects for Africa and send 1,500 medical personnel and public health experts to African countries. This is all to assist the AU in achieving its goal of vaccinating 60% of the total African population by 2022.

What's more, China advanced billions of dollars for African trade and infrastructure and wrote off interest-free loans to African countries to help African countries

---

[5] President Xi Jinping's talk on the 73rd session of the WHA.

[6] China's State Council Information Office's press conference in 2021.

[7] Data Resource: Africa CDC.

heal from the coronavirus pandemic. China will also work with G20 members to implement the G20 debt relief initiative and extend the debt relief period for African countries. It is necessary that the international community, especially developed countries and multilateral financial institutions, take tough action on the issue of debt relief for Africa.

Other civil society organizations also play an active role during the coronavirus pandemic; for example, the Jack Ma Foundation and Alibaba Public Welfare have provided a large amount of anti-epidemic materials to African countries three times. Such actions fully demonstrated the humanitarian help of Chinese enterprises.

The pandemic is spreading around the world, and both China and Africa are facing the arduous task of combating the epidemic and righting the economy. Admittedly, the practical problem of the African continent's sustainable development is a relatively long process that cannot be resolved easily due to its deep-rooted structural flaws. However, China–Africa health collaboration is a significant part of China's health strategy to participate in global health diplomacy and governance. In the meantime, China would like to take the responsibility to assist and access funding sources to establish Africa's health care system. As a result, it is believed that China–Africa health sector collaboration must protect people's lives, influence the socio-economic effect and minimize the negative impact of the epidemic.

25.4 Challenges for the China–Africa Collaboration in the Health Sector

Although certain achievements have been made, there are several inherent long-term societal and health cooperation challenges faced by the two countries in the era of the post-pandemic. Language barriers are the most obvious limitation that hinders Chinese medical teams to organize an effective workload management deployment. The language system of the African continent is quite complex, including English, French, Portuguese, Spanish and other official languages. Communication became a great deal of pressure on the medical teams between the two countries. Except bring rich experience in fighting against epidemics, Chinese medical teams also conduct medical exchanges with local staff through training programs after arriving in African countries. However, the difficulties in language communication seriously restricted the work of medical aid to Africa, leading to an ineffective communication and even misunderstanding in the process of analysing, diagnosing and formulating treatment plans. This affects the efficiency of medical aid to Africa and has a negative impact occasionally. In addition, due to the language and culture barriers, some African people do not understand China's aid policies to Africa, which is not conducive for the emotional connections between Chinese and African people.

The poor state of infrastructure has plagued the African region for years. Africa has poorly constructed health facilities, underdeveloped health systems, and inadequate medical infrastructure. According to the World Health Statistics 2022, unsafe drinking water, sanitation and lack of personal hygiene led to approximately 870,000 deaths globally in 2016. The mortality rate in the African region was 45.9 per 100,000 people, four times higher than the global average (11.7 per 100,000 people) and more than 150 times higher than that in the least affected European region (0.3 per 100,000 people). However, the density of doctors in the African region is as low as 3 per 10,000 people, and these conditions lead to many limitations in health cooperation between China and Africa. Besides, the inefficient delivery of COVID-19 vaccines in Africa faces intractable issues. The cold storage transportation technologies are severely lacking in Africa. It is believed that the transportation process of the COVID-19 vaccine is complicated, with high requirements on temperature and technology, as well as strict compliance with transportation standards. The COVID-19 vaccines have to keep in cold storage to prevent them from spoiling during transportation. Through vaccine transfers, it is quite important to reduce the loss of vaccines, to ensure the quality of vaccines and make them reach the designated vaccination centres safely. However, the cold storage transportation technology in Africa is not well developed, especially in poor and remote regions.

The disappointing health conditions in the African region led to urgent action from the international community, while the impact of China's assistance is limited to a certain degree. On the one hand, China's medical devices do not meet the usage habits of African medical staff. Although China has brought a large number of medical and health instruments to Africa, many of these medical devices use Chinese panels and enforce China's electrical standards, which makes the African staff hard to understand and cannot get used to the devices. More importantly, a lack of a systematic after-sales system makes it tough to obtain local maintenance when China's medical devices have problems. In addition, the recognition of Chinese pharmaceutical products in Africa is facing difficulties, which restricts the potential of trade and cooperation between China and Africa. Although the BRI is an opportunity for the Chinese pharmaceutical industry to broaden the African market, China is not the main character of imports into the African pharmaceutical market. Except for a few artemisinin-based antimalarial drugs, there are several barriers to the drugs made in China getting into the main African pharmaceutical market. Besides, Chinese enterprises are not familiar with the African medical and health product regulations. Precertification by the WHO is a prerequisite for access to the African market, but China's medical products cannot pass the WHO precertification easily, which hinders the entry of Chinese pharmaceutical products into the African market.

Western countries' medical and health assistance to Africa is mainly implemented through international organizations and Non-governmental Organizations (NGOs). The International Red Cross, the Global Fund (known as the Global Fund to Fight AIDS, Tuberculosis, and Malaria), the Global Alliance for Vaccine and Immunization (GAVI), the Coalition for Epidemic Preparedness Innovation (CEPI), and the Coalition for Epidemic Preparedness Innovation (CEPI) are among the most influential health organizations in Africa. While China–Africa medical and health cooperation is mainly carried out through government and bilateral channels, which limits China's cooperation with local African civil society organizations and multilateral international organizations. As a result, the influence of international organizations and NGOs on Africa's health sector is much more significant compared to China. More than that, the long-standing monopoly of Western governments and multinational enterprises in the African pharmaceutical industry threatens China's external health sector cooperation that cannot be denied.

Still China attaches greater importance to cooperating with African countries and is committed to safeguarding Africa's security and development in the new era. China has established a relatively comprehensive medical assistance system for Africa, including donating medicine and medical devices, improving local healthcare infrastructure, and actively sending trained medical personnel abroad, offering scholarships to local people. The cooperation between China and Africa in the field of health has been quite effective, but China–Africa health cooperation still faces many challenges due to the inadequacy of the existing healthcare system and the special nature of the epidemic itself. What's more, the status and influence of health cooperation in China's foreign policy toward Africa remain hard. This may be related to China's lack of a thorough national strategy for health diplomacy. The development of a national or global strategy for health could further enhance China's discourse in the field of health diplomacy. Usually, developed countries earn more domestic support and have a greater voice in international negotiations and global public health governance. Most African countries have poor public health infrastructure but huge pharmaceutical market demand, which has been a major competition ground for developed country governments, multinational pharmaceutical companies and others. As a result, China should actively combat the pandemic and participate in global public health governance to improve the capacity and level of public health governance in African countries.

25.5   Recommendations for the China–Africa Health Collaboration (Post-COVID-19)

Africa is the region that contains 70% of the least developed countries with the most rapidly expanding population in the world. Maintaining a stable development environment is necessary for African countries. While the frequent outbreak of crises in the field of public health has deeply affected Africa's security and development in particular. China–Africa medical and health cooperation has a history of nearly 60 years and has been a crucial part of China–Africa relations. Over the past 60 years, China's medical teams have grown gradually, with more than 20,000 medical workers from all over China supporting more than 50 countries in Africa. China treated patients and trained medical technicians, and the efforts have been appreciated by African local people as well as the international community. The outbreak of the pandemic has highlighted the importance and priority of China–Africa health cooperation. Consequently, strategic and timely cooperation in China–Africa health development is much needed for financial support, medical supplies, vaccine research and development, technical capacity and talent training.

In brief, the health sector collaboration between China and Africa has been fruitful on the whole but still needs to be further improved with great efforts. As mentioned before, language barriers hindered the collaboration between the two countries. As a result, instructions for drugs and devices from China should be translated into local official languages with different targets effectively. Overcoming the language barriers could minimize the misunderstanding, enhance mutual understanding and enhance the substantive China–Africa health cooperation.

What's more, the Chinese government should take strict control of the quality of drugs and medical devices and timely adjustments to the actual situation of the recipient countries as well as the variant development of diseases. China has to strengthen field research in the early stage of aid and introduce medical devices and medicines according to the local drug management standards to prevent the introduction of prohibited drugs and devices. Besides, the National Medical Products Administration (NMPA), the Ministry of Commerce, the Ministry of Foreign Affairs and other relevant departments should strengthen coordination to accelerate the process of internationalization of Chinese medicine. Meanwhile, China has to strengthen its coordination with international organizations and African governments and organizations. The coordination of relevant departments and governments as well as Chinese pharmaceutical enterprises should promote the recognition of Chinese pharmaceutical standards and products in Africa. Thus, establishing a good presale and after-sales service system to minimize waste and damage to medical devices is essential as well. By doing so, Chinese medicines can be pre-certified by the WHO as soon as possible.

To promote the cooperation between China and Africa in the field of health, China should strengthen the training of personnel involved in relevant positions in vaccine transportation, especially the training of transportation personnel in recipient countries. It is important to cooperate with African organizations and governments to help relevant practitioners to better operate the business and improve the efficiency of vaccine transportation. Also, it is better to consider encouraging third-party pharmaceutical cold chain logistics and production enterprises to participate in the action of aid to Africa and participate in the distribution to reduce the transportation of finished products and improve delivery quality.

In response to inadequate medical and health systems in Africa, China should actively support African countries in developing local medical education. Compared to developed countries such as the United States, Australia, Canada and so on, China lacks global health programs on a large scale. Thus, the Chinese government should strengthen aid to medical education in schools such as medical schools and nursing colleges. China–Africa joint medical vocational schools should be encouraged, especially in vast rural and remote areas. Meanwhile, accelerating the construction of the China–Africa Friendship Hospital is crucial for China–Africa health cooperation. China needs to support the development of key specialties in African hospitals, strengthen the training of health personnel, improve hospital facilities, and carry out all-around academic and technical exchanges and cooperation in medicine, education and research. Such measures can help African countries improve the level of medical and health services and hospital management capacity. To promote the development of medical and health systems and medical education systems in Africa, both infrastructure facilities and medical and health talent are needed. Thus, China needs to strengthen cooperation in medical education, and train more medical and health professionals locally.

In terms of public relations, China–Africa medical and health cooperation should be publicized from China to Africa through the media correctly. China–Africa cooperation includes various aspects. The China–Africa economic and trade cooperation has developed rapidly in recent years especially. According to the latest data released by the General Administration of Customs of China, the total bilateral trade between China and Africa in 2021 reached $254.3 billion, up 35.3% year on year.[8] The Annual Report on China–Africa Economic and Trade Relations 2021 reports China remains Africa's largest trading partner, even during the COVID-19 epidemic. Under the framework of BRI, China provides technology, capital resources and development experience to African markets to support the African region's development. However, Western countries speculated and even criticized China. Therefore, China needs to strengthen the information disclosure and publicity so that Chinese people can better understand the importance of China–Africa health cooperation including the results achieved. While in Africa, China–Africa health cooperation should be publicized in local languages and through local media, making African people better understand the Chinese culture and China–Africa friendship. It is better to strengthen China–Africa cooperation with local media and encourage people-to-people exchanges to reinforce mutual feelings between the two countries. Moreover, China needs to establish the power of responsibility, and share aid strategies with international organizations. Also, China has to strengthen its ties with international organizations as well as NGOs in the health sector and emphasize multilateral partnerships in aid efforts to Africa. Not only will it share the aid information and improve efficiency but will also increase China's voice and influence in multilateral health governance at the regional and global levels. Encouraging Chinese NGOs and civil society organizations to participate in aid projects in Africa can portray a respectable and change such negative perceptions of China in western countries. Moreover, in the post-pandemic era, China and Africa should reinforce their cooperation in international affairs. Strengthening international cooperation and firmly supporting the work and authority of the UN, and the WHO in international affairs and world public health is particularly important for both China and Africa. Only by doing the measures mentioned above, China can improve aid effectiveness and the transparency of information on China–Africa health cooperation in today's international community.

---

[8] Data source: General Administration of Customs of China.

25.6 Conclusion

It is believed that China and African countries are communities with a shared future. From the HSR to a "Global Community of Health for Mankind", China's cooperation on public health has always been a key component of the BRI. China–Africa collaboration on health is a crucial part of the global efforts against COVID-19. As the ongoing COVID-19 pandemic continues to disrupt livelihoods, China–Africa needs to work together productively for mutual benefits to collaborate in the health sector. Therefore, it is necessary to strengthen China–Africa health cooperation and jointly build a China–Africa health community. China should not only strengthen the construction of hardware health facilities but also need to contribute efforts in training professionals, carrying out public health information exchange and strengthening the technical cooperation to improve Africa's public health systems. China will continue to give full support to Africa in the fight against the epidemic and continue to provide medical assistance, dispatch medical experts, and supply medical services to African countries. China needs to help Africa improve its public health prevention and control system, strengthen Africa's capacity to prevent and respond to diseases, and better benefit the people of Africa. Although the pandemic's impact damaged China–Africa relations to some extent, China–Africa has created unprecedented opportunities for bilateral health cooperation. The post-pandemic poses a grave threat to the health and lives of millions of people in Africa, but China–Africa solidarity against the epidemic fully reflects the brotherhood between the two sides in overcoming the difficulties. Under the framework of the BRI as well as FOCAC, most African people feel healthcare became more affordable and accessible with the help of the Chinese government. The brotherly friendly cooperation in the health sector presents a deep connection between China and Africa. To fight against the epidemic, China strongly supports Africa to get back to work, develop the economy, as well as ensuring and improving its livelihood. Even though there are some difficulties and challenges during the epidemic, China and Africa always stood by each other and worked together. China is willing to work with the UN, WHO, and other international organizations to help Africa and enhance cooperation in the health sector. Both China and Africa would work to build an even stronger community of shared future and advance cooperation under the BRI, establishing a new milestone in China–Africa relations. Developing Africa's health sector should be a responsibility shared by the international community. By doing so, China–Africa health collaboration could benefit both countries and the world.

References

Business, O. E. C. D., & Outlook, F. (2018). *China's belt and road initiative in the global trade, investment and finance landscape.*
Lianhe, L. (2020). *Peaches and jade of friendship.* China Daily. https://global.chinadaily.com.cn/a/202006/01/WS5ed4584ca310a8b241159c65.html. Accessed: 10 Jan 2023.

Mena (2020). *Egypt sends 10 tons of medical equipment to China.* Ahram Online. https://eng lish.ahram.org.eg/NewsContent/1/64/362713/Egypt/Politics-/Egypt-sends--tons-of-medical-equipment-to-China.aspx. Accessed: 10 Jan 2023.

Omondi, J. (2020). *South African manufacturer donates masks to Wuhan.* GGTN. https://africa.cgtn.com/2020/02/03/south-african-manufacturer-donates-masks-to-wuhan/. Accessed: 10 Jan 2023.

Opali, O. (2020). *Standard bank staff in Africa join the celebration of China's Lunar New Year.* China Daily. https://global.chinadaily.com.cn/a/202002/01/WS5e345abfa3101282172 73e9f.html. Accessed: 10 Jan 2023.

Pandey, K. (2020). *Africa 'fighting COVID-19 in the dark', says global aid body.* Down-ToEarth. https://www.downtoearth.org.in/news/bite-size/africa-fighting-covid-19-in-the-dark-says-global-aid-body-72627. Accessed: 10 Jan 2023.

The State Council Information Office-The People's Republic of China. (2021). *SCIO press conference on China's COVID-19 assistance and international development cooperation.* http://eng lish.scio.gov.cn/m/pressroom/2021-11/02/content_77846343_6.htm. Accessed: 10 Jan 2023.

UNCTAD. (2021). *Facts and figures.* Press release. https://unctad.org/press-material/facts-and-fig ures-7. Accessed: 10 Jan 2023.

World Health Organization (2021). *Assessing the impact of COVID-19 on older people in the African region.* A study conducted by the World Health Organisation Regional Office for Africa.

World Health Organization. (2022a). *Deaths from noncommunicable diseases on the rise in Africa.* WHO Africa. https://www.afro.who.int/news/deaths-noncommunicable-diseases-rise-africa. Accessed 10 Jan 2023.

World Health Organization. (2022b). *Report on the strategic response to COVID-19 in the WHO African region -1 February 2021 to 31 January 2022.*

Chapter 26
A Critical Evaluation of Health Diplomacy as an Altruistic Underpinning of China–Africa Relations

Dylan Yanano Mangani⬤, Marcia Victoria Mutambara⬤, and Richardson Shambare

26.1 Introduction

26.1.1 Towards an Analytical Framework in Africa

China–Africa relations are often seen in a historical-ideological perspective that emphasises an anti-imperialist, anti-colonial and anti-capitalist trajectory dating back to the epoch of the liberation struggle for independence of the African continent in which China played a pivotal role (Mangani, 2019). In the broadest sense, Africa constitutes a low-priority strategic region of developing countries for China's foreign policy. This economic and political reality might explain why several African policies are procedural decisions made under existing China's foreign policy objectives. This means China's African policymaking is disseminated at the working, ministerial and departmental rather than crafted at the top, strategic levels of the Politburo Standing Committee (PSC) of the Chinese Communist Party (CPC) (Sun, 2014). Therefore, two ministries carry the primary mandate for executing China–Africa relations. First, the Ministry of Foreign Affairs (MFA) is primarily responsible for implementing existing China's political foreign policy objectives in Africa. Second, the Ministry of Commerce of the People's Republic of China (MOFCOM) dispenses China's

D. Y. Mangani (✉)
Department of History and Political Studies, School of Governmental and Social Sciences, Nelson Mandela University, Port Elizabeth, South Africa
e-mail: Dylan.Mangani@mandela.ac.za

M. V. Mutambara
Gender–Based Violence Cluster, Health Economics HIV/AIDS Research Division (HEARD), University of KwaZulu-Natal, Berea, Durban, South Africa

R. Shambare
Faculty of Management and Commerce, University of Fort Hare, Alice, South Africa

© The Author(s) 2024   489
M. Muchie et al. (eds.), *China-Africa Science, Technology and Innovation Collaboration*, https://doi.org/10.1007/978-981-97-4576-0_26

economic and foreign aid aspects outlined in its foreign policy objectives (Lin et al., 2016; Yoshikawa, 2022).

China has pursued an attractive foreign policy towards Africa that is guided by a rejection of interference in many African states' affairs. This includes protecting African states at international foras, granting financial loans at reasonable interest rates, public health diplomacy, such as donating vaccines during the coronavirus (COVID-19) pandemic and embarking on numerous developmental projects across the African continent. Such policies demonstrate Beijing's readiness to collaborate with Africa within the context of Global South-South cooperation. Scholars like Grepin et al. (2014) emphasise that health-related aid is a key priority with China, and it now ranks among the top 10 bilateral global health donors. However, limited literature focuses on its involvement and effects on public health (Tambo et al., 2016). China–Africa relations are encapsulated in the 1964 'Eight Principles for Economic Aid and Technical Assistance to Other Countries and the 2006 'China's African Policy' that promulgated the bedrock for the Forum on China–Africa Cooperation (FOCAC). Scholars like Lin et al. (2016) point out that the China–African policy document granted the forum on China–Africa cooperation a significant role in constructing the China–Africa relationship. Since 2000, at each FOCAC summit, the Chinese government has been committed to assisting African countries in critical areas, including health. In addition, the FOCAC programs promulgated a series of other health-related instruments, which include: the African Human Resources Development Fund in 2000; the first China–Africa Forum on Traditional Medicine established in 2002, a 5 billion USD China Africa Development Fund in 2006; and the first ministerial forum on China–Africa Health Development in 2013.

The epoch of the 1960s' was marked by a developmental trajectory in China's global affairs. Traore (2021) gives credence to multiple regions, such as Africa, as instrumental avenues through which China sought significant roles in world affairs. The themes of state sovereignty and mutual economic and human capital development became defining features of China's 1964 and 2006 economic and technical blueprints towards Africa. This was a departure from the traditional Western-sponsored aid customarily conditioned by political interference and socio-economic structural adjustment demands (Tshuma, 2022). Arguably, some commentators have described China's support of Africa as opportunistic. They have analysed China's novel approach within the confines of a 'soft power' strategy that utilises appealing economic levers such as health aid. A key feature of China's health aid that has often been discussed in the literature is what is dubbed as the 'no strings attached' approach, which "never imposes ideology, values and development models on other countries, especially African countries". Concerns have been raised regarding the latter claims that China's approach challenges the vision of a democratic Africa in which good governance, democracy and human rights are respected (Brookes & Shin, 2015; Shein & Fan, 2014). Regardless of its stated 'no strings attached' policy, critics emphasise that China seeks to harvest political gains from health aid-receiving countries (Weston, 2011; Wang and Sun 2014). Some commentators also claim that China's health diplomacy in Africa is a move to secure stable supplies of oil and other important natural resources, as China imports approximately one-third of its oil from Africa and significant amounts of minerals and raw materials mainly from Sudan, Angola, and the Democratic Republic of Congo (The Emory Global Health Institute, 2020). These repertoires of engagement tend to re-interpret and reshape the structure of multiple regions, such as Africa, through a substantial approach to their immediate socio-economic and political needs (Traore, 2021; Zhang, 2010).

26.1.2   An Analysis of China's Softpower Strategy

The concept of a soft power strategy belongs to the generic tenets of the Liberal theory of international relations. It is an antithesis of the classical Realist theory of states' conduct that is conflictual and competitive and therefore utilises military force, brinkmanship and coercive diplomacy. The absence of a security guarantee and the anarchic nature of the international system justify why states are opportunistic rather than cooperative (Duguri et al., 2021). The Liberal theory of international relations is based on the concepts of self-restraint and cooperation among states. The theory emphasises the role of international institutions and economic imperatives as levers to contain states' aggression in the international system. Therefore, this school of thought incorporates the component of 'positive sum cooperation into frames of reference that denote the 'soft power strategy. Consequently, Nye (2017) defines the 'soft power' strategy as the ability of an international actor to sway other international actors by means other than sanctions, military pressure or threats but with co-optive power. This ability to influence other actors stems from the attractiveness of a state's repertoires of engagements, such as culture, economic incentives and health aid (Irgengioro, 2021).

Nye (2017) conception of the soft power strategy is rooted in the component of culture. Accordingly, Nye (2004) argues that the more states share a commonly accepted culture, the easier their soft power towards each other will be. However, this argument is marked by flawed reasoning that utilises the Western culture of democracy as an exceptionalist representation of this theory. This chapter considers this example inaccurate, arguing that though the Chinese cultural model is generally conceived as authoritarian and, therefore, less attractive in Western societies. Elsewhere across the world, the Chinese culture shares comparable and acceptable tenets in multiple regions such as Africa. This means that although most Western societies have a negative perception of general abuse of human rights, undemocratic practices and an authoritarian Chinese state. The majority of African states have adopted what is now known as the 'Look East' policy explained by the then Zimbabwean leader Robert Mugabe. According to Mugabe:

> We must look to the East, where the sun rises. It is weakening, and China is entering the scene. The comfortable place enjoyed by the United States and Europe in the past is gradually being reduced by the language, values, ideas, history and products of China, which are rapidly being assimilated in Africa (cited in Traore, 2021: 619).

China's soft power strategy is instrumental in its projection as a rising global power and warding off negative public perceptions as a worldwide threat to the United States of America's dominance. Analysing health diplomacy within the prisms of the 'soft power strategy helps to understand China's policy towards global public health. In a sense, China's health diplomacy is seen as a repertoire of engagement that promotes its interests as well as alleviating diseases and pandemics in multiple regions, such as Africa. On the other hand, a forward-looking explanation of the 'soft power' strategy posits that shared mutual cultural norms and ideas make states pursue altruistic motives independent of immediate national interests. Optimists contend that historical and ideological considerations such as a common fight against capitalism, colonisation and imperialism shared between China and the African continent, and public health exists within the framework of the Global South-South cooperation, with a developmental trajectory seen as an end itself (Gray & Gills, 2016; Killeen et al., 2018).

26.2   Towards a Conceptual Understanding of Health Diplomacy

Health diplomacy is the interaction between international political relations and public international health initiatives (Brown et al., 2018). In the past, traditional diplomacy focused on political and socio-economic issues that concerned the security of states. In the post-Cold War era, factors such as the end of the bipolar conflict between the United States (US) and the Soviet Union; the rise of Russia, Brazil and China; and the emergence of transnational diseases of global significance such as HIV/AIDS, Ebola, the Severe Acute Respiratory Syndrome (SARS) and the latest COVID-19 demonstrated how non-military threats could severely curtail the global economy. These global threats to public health have reinforced the need for mutual agreements, treaties and public foras to improve public health while strengthening relations among states. Several states, such as China, have integrated health issues into their foreign policy agendas by increasing their willingness to pool resources to develop health sectors across regions such as Africa (Killeen et al., 2018).

There are three distinct approaches to global health diplomacy: core health diplomacy, multistakeholder diplomacy and informal health diplomacy. Core health diplomacy is first understood in the traditional Westphalian notion of interactions between state governments. This type of bilateral diplomacy is issue-based, implements policies, and negotiates health matters between states. Bilateral agreements primarily characterise this approach, involving ministerial representatives, health experts and officials. According to Rudolf (2021), the Chinese government has entered into fifty-six agreements categorised as health. Another subset of core health diplomacy includes multilateral agreements and treaties. These can be classified as diplomatic health agreements that fall under the scope of multilateral institutions such as the World Health Organisation (WHO) and other international organisations vested in global public health (Kartz et al., 2011). These are multi-actor negotiation activities that craft and police the global health regime.

The second approach is multistakeholder diplomacy. This approach is derived from the Liberal theory of international relations that emphasises the role of state and non-state actors in solving common transnational and global issues. Significantly, the multistakeholder approach departs from the traditional role played by state actors to the role of agencies in their countries as actors that enter into health agreements. For example, the Chinese Centre for Disease and Control Prevention may agree with a specific African health ministry. These agreements are in the form of memorandums of understanding that are not legally bound by international law (Chaban & Knodt, 2015). Multistakeholder diplomacy encompasses global initiatives and international institutions to broaden the conception of health diplomacy by increasing the role played by non-state actors. This is based on the position that transnational health challenges such as COVID-19, HIV/AIDS and Ebola pandemics severely impact states' resource mobilisation capacity. These constraints impact the state's role in global health affairs and would require non-state actors to play an appendage role.

Thirdly, the informal health diplomacy approach encompasses the role of individual private funders, NGOs and research in confronting global health challenges. Private funders such as the Gates Foundation, Doctors Without Borders and other non-state actors have contributed toward global public health in areas of HIV/AIDS, Malaria, Tuberculosis and combating global hunger. Their role is twofold—providing essential health services in line with the objectives of missions such as Feed the Children and realising the health foreign policy objectives of specific states. In addition, there has been a proliferation of research output in areas of health as an integral part of informal health diplomacy. Research initiatives have taken the form of collaborations among several actors, including academic institutions, private entities, government agencies and laboratories. Over the years, partnerships between academic institutes in countries such as the US and counterparts in developing regions have taken shape in the form of clinical trials or peer-to-peer scientific research partnerships. These partnerships appear beneficial in the context of providing data on the one hand and the transfer of technology or skills on the other hand to the host counterpart.

26.3 The Evolution of China–Africa Health Relations

The Chinese First National Health Congress in 1950 laid the basis of global health approach for China. At this conference, a health policy was formulated whose objectives sought to promote a 'peasantry' focused health care system. The Congress determined that the Chinese health policy would service the subaltern, proletariat and downtrodden masses, including labourers, peasants and general workers. This policy was extended to the African regions in the following years (Killeen et al., 2018). The 1960s earmarked the development of China's influence in global affairs. In 1964 'Eight Principles for Economic Aid and Technical Assistance to Other Countries formed the bedrock of a developmental trajectory in China's international relations underpinned by a novel and attractive health diplomacy in Africa. As the Cold War intensified between the US and the Soviet Union, China's foreign policy veered between wading off American imperialism and the Soviet Union's revisionism. Another crucial factor was China's multilateral health governance exclusion from institutions such as the United Nations and the World Health Organisation during this period. To gain a foothold in multiple regions, such as Africa, in 1964, the Chinese premier Zhou Enlai contended that "Our assistance to Asian and African countries is keenly important for our competition with the imperialists and revisionists for the middle strip. This is a critical link. It is material assistance. It will not work without material" (Wang & Sun, 2014:4).

In 1963, the Chinese premier Enlai inaugurated China's core health diplomatic endeavours in Africa. These culminated in first China's bilateral relations with Algeria. Underpinning these engagements was the immediate need to resuscitate Algeria's constrained health system. In response, the Chinese premier dispatched a team of 13 medical experts, supplies and equipment to provide an efficient health-care system (Youde, 2010). The medical team was deployed to the rural and under-serviced parts of Algeria. This reinforced the peasantry-focused objectives of China's health diplomacy in Africa. The instrumentalisation of health aid characterised the period between 1963 and 1978 to expand China's foreign policy. Some commentators have refuted the claims of political opportunism influenced by the misgivings of Cold War politics in suggesting that China's health diplomacy in Africa demonstrated altruism (Hutchinson as cited in Youde, 2010:154). A closer analysis of this period remark positively on Chinese health aid in which a peasantry-focused health system was introduced by 'barefoot doctors'. This implies how Chinese doctors integrated and worked among the poor constituencies across the African continent in Tanganyika, Zanzibar, Algeria, Mali, Guinea, Congo Brazaville and Somalia. Of importance is a sustainable and developmental health drive that was focused more on the application of preventative primary health care rather than emergency care. This helped in promoting the host country's health infrastructure. In complementing this, China expanded its conception of health diplomacy through aid projects that built more than 100 hospitals since 1970. The reputation of China's health diplomacy gradually spread across the African continent, leading to most African states supporting the Chinese bid to resume its seat in the UN (Killeen et al., 2018; Ashan, 2011).

The chapter seeks to proffer alternative perspectives of China's health aid towards Africa beyond the traditional conceptions of hegemonic power's interests and political factors. Following the end of the Cold War, China's health aid toward Africa decreased as it focused on a broader internalisation and internationalisation of its national agenda. There are many reasons for this shift in foreign policy. The tenure of Deng Xiaoping and his subsequent neo-liberal market reforms; the disintegration of the Soviet Union and the subsequent vacuum created; and internal political shifts in several African countries led to a realignment of China's health goals. One commentator observed, "It is hard to make a case that Africa matters very much to China ... they [Africans] count for little in the overall scheme of Chinese foreign policy objectives" (Segal, 1992:115). Another explanation is found in the shifts in political circumstances in Africa. The advent of civil wars in Somalia and the Democratic Republic of Congo, and the adoption of the International Monetary Fund's economic structural adjustment programs dampened China's commitment to health aid regime in Africa.

26.4 Contemporary China–Africa Health Relations

In contemporary China–Africa relations, the 2006 *China's African policy* underpins China's renewed African commitment towards realising a set of horizontal and developmental programs at bilateral and multilateral levels. Significantly, the policy was crafted under the auspices of the China–Africa Cooperation Forum (CACF) promulgated in Beijing in October 2000. The forum laid the basis for a renewed commitment by China as the leader of the developing region of the world. China pledged to double its aid in Africa in strategic areas such as health, offering a US$ 5 billion loan. China also reinforced its pledge to Africa by cancelling a debt of US$ 1.5 billion owed by African states. Furthermore, the Chinese government emphasised a commitment to expanding aid projects in the health sector by pledging to build 30 hospitals across the continent. The pledge was also accompanied by an allocated US$ 37 million in the form of grants for malarial drugs and the construction of demonstration centres for malaria treatment.

26.4.1 China's Horizontal Health Diplomacy Approach

The horizontal health approach seeks to confront health challenges on a comprehensive and long-term basis. This is achieved by creating functioning systems and lasting institutions that can address diseases sustainably. Therefore, the approach seeks to address the efficacy of primary health care. When a country's primary health care system is competent, the better its systems and cost-effectiveness and the higher the impact on health. The horizontal approach is community-focused, seeking to empower and capacitate communities to address their health problem, and is also focused on preventive health care (Mostafavi et al., 2020). In Africa, the Chinese health model appears forward-looking. Far from deploying thousands of physicians in at least 47 countries, the Chinese have capacitated the African health system by constructing health facilities and training African medical personnel.

Regarding health projects, China has left an indelible print in supply procurement and healthcare facilities construction. Between 2013 and 2016, China modernised the General Hospital of Gagnoa in the Ivory Coast, the Tanzanian Abdulla Mzee Hospital and other facilities such as the Zimbabwean Friendship Hospital. Such infrastructure development is integral to the FOCAC, an important trajectory of the Chinese foreign policy toward Africa (Killeen et al., 2018). China's health diplomacy also demonstrates a capacity development component. Since the beginning of the

Millenium, China has funded training courses for over 14 000 African students and, at the same time, offered bursaries for African medical students to study.

In its post-2006 *China's African policy* health diplomacy approach, China has invested in capacity building regime that has seen several training workshops from a multistakeholder diplomacy approach that incorporates non-state actors such as Guilin Pharmaceutical Company and Beijing Holley-Cotec, Chinese pharmaceutical companies in the anti-malarial campaign in Kenya. Malaria is regarded as the killer disease in Africa. According to Ashan (2011), the United Nations International Children's Emergency Fund (UNICEF) reported that at least a million children die from malaria-related deaths annually in Africa. An estimated three billion people are at risk of malaria globally, and eighty per cent of this number and malaria deaths occur in Africa (Gui Xia et al., 2014). Therefore, China–Africa health relations have broadened in the realm of an anti-malarial campaign focused on the disbursement of medical experts, training workshops, donations of medical equipment and medicines and the establishment of an anti-malaria centre.

The concept of 'barefoot' doctors is figurative for medical experts who work with communities and the masses, and it remains an integral part of China–Africa health relations. China has dispatched medical experts to malaria-infested regions in Africa. These medical experts are responsible for administering free medication such as Cotecxin, which has proven to be an alternative drug to quinine. In African states such as Mali, the continued use of quinine produces other health complications such as hemiplegia, partial paralysis or weakening of muscles in one part of the body (Drave et al., 2020). Chinese interventions from medical experts have included using acupuncture needles as the most effective way of combating complications resulting from side effects of medications such as quinine.

The Chinese malaria response toward Africa demonstrates multistakeholder and informal health diplomacy approaches. A consortium of non-state pharmaceutical Chinese companies has been engaged in the disbursement of anti-malarial drugs and sales registration of these entities to penetrate the African market. In addition, the Commerce Ministry of China designated the Jiangsu Centre for Verminosis Control and Prevention as the global human resources development hub. The centre has run programs that include training about 170 medical officials from 43 countries across Africa. These programs included commencing two anti-malarial programs for Cameroon, Kenya and Madagascar (Anshan, 2022). Another dimension to the multistakeholder health approach is demonstrated in Uganda's three-day anti-malarial campaign workshop. The workshop was part of China's Africa's health policy in the fight against malaria, leading to its commitment to financing training programs and medication in Uganda.

Another critical aspect of China's fight against malaria in Africa is located within the role of Chinese non-state actors in preventing and controlling malaria. In 2007, Guangzhou University of Traditional Chinese Medicine's Tropical Medicine Institute partnered with Moheli Island in Comoros Islands in a project called Mass Drug Administration (MDA). The project was threefold: to educate the locals on malaria and malarial prevention strategies, treat Malaria patients and establish a competent and effective anti-malaria system in Moheli. Using WHO-recommended methods such as artemisinin-based combination therapies (ACTs) produced significant results in Comoros Islands (Wang et al., 2018). To complement these, the Chinese government paid about US$ 320 000 for the disbursement and use of anti-malarial drugs in the Moheli Islands over five years. Using ACTs clears the human body of parasites that can cause malaria. Research suggests that the source of malaria is the human body, and therefore mosquitos can only acquire it from humans. Once ACTs are administered in the human body, parasites are cleared from the body, making it unlikely for mosquitoes to receive and transmit them to other human beings.

One of the most significant aspects of Chinese health diplomacy in Africa is the establishment of anti-malarial centres. Since the promulgation of the 2006 *China's African Policy*, China has deployed delegations of teams across the African continent in the form of experts and medical experts who set up centres in key countries such as South Africa, Liberia, Madagascar, Kenya and Zambia. The delegation comprised trainees from academic institutions, science labs and hospitals. This signified China's readiness to provide scientific, research and medical aid in the fight against malaria in Africa. Accordingly, the Chinese experts could transfer knowledge and skills to African experts.

26.4.2   The New Health Silk Road and the Chinese Belt and Road Strategy

The Belt and Road initiative is foregrounded in the global Chinese initiative of expanding and providing an alternative political, social and economic framework that seeks to build a global developmental framework. Although the strategy was conventional infrastructure projects oriented, its scope has been expanded to incorporate healthcare services, medical equipment, and telecommunications infrastructure. Another essential component is that instead of being led by state entities, there has been a proliferation of non-state and private entities in what is known as the new Health Silk Road (HSR) (Habibi & Yue Zhu, 2021). The HSR can be traced to the 2015 announcement by the Chinese National Health Commission on enhancing health matters under the Belt and Road Strategy. Chinese President Xi Jinping reiterated this framework in a legislative assembly speech in Uzbekistan that focused on increasing cooperation in medical rescue, epidemic control, traditional medicine and infectious disease notification. This culminated in the Chinese concept of the Health Silk Road.

The internationalisation of the concept became apparent in 2017 when Dr. Tedros, the Chief of the WHO, observed that:

> Health Silk Road [proposed by China], which strengthens and renews ancient links between cultures and people, with health at its core, is indeed visionary … [which] contains the fundamentals to achieve universal health coverage: infrastructure, access to medicines, human resources, and a platform to share experience and promote best practices (as cited in Habibi & Yue Zhu, 2021: 4).

The internationalisation of the HSR gave the initiative a new face couched in utilising international foras such as the WHO and technological interventions in combating health challenges in multiple regions of the world such as Africa. The thrust of the HSR initiative in Africa is demonstrated by the lesser role played by the Chinese government. This is based on the limits of state intervention in sustainability and transparency. The HSR has a two-thronged approach: expanding health infrastructure to include pharmaceutical products and increasing the multistakeholder participation of the private sector. In terms of its role in the HSR, the private sector has revolutionised the global Chinese pharmaceutical regime. The most efficient approach has been the production of low-priced and efficient drugs that have penetrated the African market. These drugs are within reach of many African consumers. In the fight against malaria, a private entity named New South Group from Guangdong, China, has produced drugs like artesunate that have become a priority in some countries such as the Comoros Islands. Another private entity, Sansheng Pharmaceutical Plc, is known for supplying antibiotics across the African continent. In August 2018, Sansheng Pharmaceutical Plc established a subsidiary in Ethiopia for the same purposes. This led to its growth in creating jobs and exporting drugs in the East African region (United Nations Economic Commission for Africa, 2020). By establishing a local subsidiary in Ethiopia, China adopted a forward-looking policy: job creation and skills-oriented transfer. Fosun Pharma, based in Shanghai, and POE Neusoft Medical Systems Company Ltd were also deployed and involved in the developmental health trajectory in manufacturing products that assist in primary health care.

26.4.3 China's Management of Covid-19 in Africa

In December 2019, COVID-19, a virus caused by SARS-CoV-2, emerged in the Wuhan region of China. Unabated, the virus quickly spread to the rest of the world and became a global pandemic. In response to the significant losses of life and sickness, countries across the globe reacted swiftly with the introduction of mitigating measures such as good hygiene, social distancing, and movement and travel restrictions (Osseni, 2020). The most significant of these measures became national lockdowns that limited economic activities, leading to job losses and less production. Such eventually put off balance most of the economies in the world. It was anticipated that the African region would be hit hard the most given its health systems challenges.

In light of the pandemic, global leadership response to the COVID-19 pandemic fell to China. In 2020 during a virtual WHO conference, the Chinese leader Xi Jinping outlined initiatives to assist Africa in fighting COVID-19. According to Habibi and Yue Zhu (2021), these initiatives included (i) establishing a working partnership between Chinese and African hospitals, (ii) fast-tracking the construction of the African centre for disease control and prevention headquarters, (iii) allocating US$ 2 billion to fight against the pandemic (iv) internationalising COVID-19 vaccines at an affordable rate, especially towards developing regions such as Africa (v) a working partnership with other G-20 member states to operationalise debt suspension programs for poor and developing countries. China's response to the COVID-19 pandemic in Africa followed the HSR trajectory that focused on the role played by non-state private actors. China also incorporated aspects such as the fusion of health assistance with technology and a broad-based vaccination initiative to effectively assume a global leadership role.

Informal diplomacy became apparent with the increasing role of individual public figures such as China's Jack Ma, who pledged to provide testing kits, protective masks and face shields to each African country. To further this, Ma pledged to support African medical facilities with online training on the prevention and management of COVID-19 through his Jack Ma and Alibaba foundations (Johnston, 2020). Furthermore, a host of private entities supported the Chinese government's call to provide leadership in the fight against COVID-19 in Africa. Zhende Medical Co.Ltd provided protective equipment, while Huajin Group provided 2 million yuan. Though informal diplomacy primarily played a decisive role in the fight against COVID-19 in Africa, the Chinese government intervened. In April 2020, the Chinese government dispatched a plane to Ghana with medical equipment, including medical gloves, surgical masks, protective clothing and testing thermometers.

The Chinese approach to COVID-19 in Africa has incorporated a technology-medicine component. This component seeks to strengthen African healthcare systems by incorporating aspects such as cloud computing, big data applications and sophisticated digital telecommunication technology for mitigating the pandemic. Telemedicine technology uses technology tools by medical professionals to treat patients remotely. The COVID-19 pandemic led to mitigating measures such as social distancing and national lockdowns. The Chinese government provided telemedicine assistance to Africa to overcome challenges associated with the restriction of travelling and movement. In countries such as Ghana and Mozambique, the Chinese government partnered with central hospitals in setting up telemedicine technologies.

Another aspect of China's health diplomacy is its vaccination program introduced during the pandemic. Lee (2021) propounds that vaccine diplomacy can be viewed as a natural extension of Chinese soft power, including prior engagement in health diplomacy. As an instrument of soft power, vaccines play a pivotal role in fostering a country's image that is favourable as "few areas of diplomatic goodwill connect more with the humanitarian nature of international citizenship than medical assistance" (Bier & Arceneaux, 2020, para. 9). Vaccine diplomacy appears to be an appealing instrument for projecting soft power which aligns with the ideas of Nye (2008: 94) who describes soft power as "the ability to affect others to obtain the outcomes one wants through attraction rather than coercion or payment". China's vaccination diplomacy towards Africa was promulgated during the 2020 China–Africa summit. The summit reiterated China's commitment to prioritising African health systems in its global health diplomacy. At the summit, Chinese leader Xi Jinping stated, 'We pledge that once the development and deployment of COVID-19 vaccine are completed in China, African countries will be among the first to benefit' (Van Staden & Shan Wu, 2021).

As of August 2021, it is estimated that China had donated about 40 million doses to Africa. Sinovac and Sinopharm are the Chinese vaccines, and these were donated to frontline workers, vulnerable populations and first responders. To enhance China's vaccine outreach on the continent, Egypt, Algeria and Morocco were building their local production capacities to ease the shortages of Chinese vaccines on the continent. This led to the approval of producing the Chinese vaccine and a partnership between Egypt and the Chinese Sinovac Biotech entity. Though China has demonstrated its leadership capabilities during the COVID-19 pandemic, some commentators have been sceptical about the figures for vaccines produced vis-à-vis the number of vaccines donated to Africa. Benabdallah (2021) opines that while China promised to donate a billion doses of vaccines to Africa, the figure of vaccines manufactured is over a billion and Africa received less than five per cent of those vaccines. This starkly contrasts China's neighbouring countries that have procured a more significant portion of China's vaccines. Another observation is that while China sought to capacitate Egypt, Algeria and Morocco to manufacture vaccines locally, there is a limitation in the number of vaccines they can produce owing to the intellectual property rights of these vaccines.

26.5 Conclusion

The chapter set out to interrogate three main questions pertinent to the scholarship of China–Africa relations in health diplomacy: *What is the amount of China's health aid? What are the criteria utilised for making political decisions for aid in Africa? And Which instruments determine the impact of aid?* Over the decades, China has demonstrated its willingness to revolutionise the African public health systems through various repertoires of engagements. These range from dispensing medical experts, the construction of health facilities, the donation of medicines and an increase in the involvement of state and non-state actors. Of importance is the role played by China in forward-looking anti-malarial and COVID-19 campaigns in Africa. Several factors can help explain this. China–Africa relations are encapsulated in the 1964 'Eight Principles for Economic Aid and Technical Assistance to Other Countries and the 2006 'China's African Policy' that promulgated the bedrock for the Forum on China–Africa Cooperation (FOCAC). These blueprints are shaped by historical and ideological affinities in China and Africa as they fought against international imperialism, colonialism and capitalism.

China has pursued a novel approach in its relations with Africa known as the 'soft power' strategy. It is an antithesis of the classical realist theory of states' conduct that is conflictual and competitive and therefore utilises military force, brinkmanship and coercive diplomacy. Soft power relies on non-military and non-coercive means that are attractive and persuasive. These methods follow a developmental and positive-sum approach that is mutually beneficial, such as health diplomacy. Given the underdeveloped and overwhelmed African health systems, China's health approach has been interpreted as developmental and friendly towards African states. Though the surface demonstrates a rising global power that regards Africa as a developmental and equal partner, the undercurrents suggest China has used health diplomacy to confront her global isolation challenges during the Cold War era. In some instances, China has invested health aid to strategic African countries endowed with natural resources, such as the Democratic Republic of Congo, Sudan and Angola protecting its geopolitical interests.

It is noteworthy that in the post-2006 China–Africa relations epoch, there has been an increase in horizontal health initiatives that have led to the combat of malaria and the recent COVID-19 pandemic by the Chinese state and non-state actors under the auspices of the New Health Silk Road and the Chinese's Belt and Road Strategy.

References

Anshan, L. (2022). *China and Africa in global context: Encounters, policy, cooperation and migration.* Oxon: Routledge. *Chinese MediCal Cooperation in Africa.* Uppsala: Nordiska Afrikainstitutet.

Benabdallah, L. (2021). *China–Africa public health cooperation and vaccine diplomacy.* LSE IDEAS.

Bier, L. M., & Arceneaux, P. C. (2020). Vietnam's "underdog" public diplomacy in the era of the COVID-19 pandemic. USC Center on Public Diplomacy. April 23. https://www.uscpublicdip lomacy.org/blog/vietnam%E2%80%99s-%E2%80%9Cunderdog%E2%80%9D-public-diplom acy-era-covid-19-pandemic. Accessed: 25 June 2022.

Brookes, P., & Shin. J. H. (2015). China's influence in Africa: implications for the United States, https://www.heritage.org/asia/report/chinas-influence-africa-implications-the-united-states. Accessed: 23 June 2022.

Brown, M. D., Bergmann, J. N., Novotny, T. E., & Mackey, T. K. (2018). Applied global health diplomacy: Profile of health diplomats accredited to the UNITED STATES and foreign governments. *Globalisation and Health, 14*(2), 1–11.

Chaban, N., & Knodt, M. (2015). Energy diplomacy in the context of multistakeholder diplomacy: The EU and BICS. *Cooperation and Conflict, 50*(4), 457–474.

Drave, A., Napon, C., Dabilgou, A., Ouedrago, S., & Kabore, J. (2020). Clinical and etiological characteristics of hemiplegia at the University regional hospital center Ouahigouya. *World Journal of Neuroscience, 10*, 22–28.

Duguri, U. S., Hassan, I., & Ibrahim, K. Y. (2021). International relations, realism, and liberalism: A theoretical review. *International Journal of Social and Humanities Extension, 2*(1), 1–6.

Gray, K., & Gills, B. K. (2016). South-South cooperation and the rise of the Global South. *Third World Quarterly, 37*(4), 557–574.

Grépin, K. A., Fan, V. Y., Shen, G. C., & Chen, L. (2014). China's role as a global health donor in Africa: What can we learn from studying under-reported resource flows? *Globalisation and Health, 10*(1), 1–11.

Habibi, N., & Yue Zhu, H. (2021). The health silk road as a new direction in China's belt and road strategy. In S. L. Africa (Ed.), *Center for global development + sustainability.* The Heller School, Brandeis University.

Irgengioro, J. (2021). Soft power instruments: An assessment of China's soft power and Sinophobia in Central Asia. In F. Bossuyt & B. Dessein (Eds.), *The European Union, China and Central Asia* (pp. 181–200). Routledge.

Johnston, L. A. (2020). World trade, E-commerce, and COVID-19. *China Review, 21*(2), 65–86.

Kartz, R., Kornblet, S., Anorld, G., Lief, E., & Fischer, J. E. (2011). Defining health diplomacy: Changing demands in the era of globalization. *The Milbank Quarterly, 89*(3), 503–523.

Killeen, O. J., Davis, A., Tucker, J. D., & Meier, B. M. (2018). Chinese global health diplomacy in Africa: Opportunities and challenges. *Global Health Governance: The Scholarly Journal for the New Health Security Paradigm, 12*(2), 4–29.

Lee, S. T. (2021). Vaccine diplomacy: nation branding and China's COVID-19 soft power play. *Place Branding and Public Diplomacy, 19*, 1–15.

Lin, S., Gao, L., Reyes, M., Cheng, F., Kaufman, J., & El-Sadr, W. M. (2016). China's health assistance to Africa: Opportunism or altruism? *Globalisation and Health, 12*(1), 1–5.

Mangani, D.Y. (2019). *Changes in the Conception of Nationalism in Zimbwabwe: A Comparative Analysis of ZAPU and ZANU Liberation Movements 1977–1990* (Doctoral dissertation).

Mostafavi, F., Piroozi, B., Mosquera, P., Majdzadeh, R., & Moradi, G. (2020). Assessing horizontal equity in health care utilisation in Iran: A decomposition analysis. *BMC Public Health, 20*(914), 1–9.

Nye, J. (2004). *Soft power: The means to success in world politics*. Public Affairs.

Nye, J. (2008). Public diplomacy and soft power. *The Annals of the American Academy of Political and Social Science, 616*(1), 94–109.

Nye, J. (2017). Soft power: The origins and political progress of a concept. *Palgrave Communications, 3*(1), 1–3.

Osseni, I. A. (2020). COVID-19 pandemic in sub-Saharan Africa: Preparedness, response, and hidden potentials. *Tropical Health and Medicine, 48*(48), 1–3.

Rudolf, M. (2021). *China's health diplomacy during Covid-19: The belt and road initiative (BRI) in action*. German Institute for International and Security Affairs.

Segal, G. (1992). China and Africa. *Annals of the American Academy of Political and Social Sciences, 519*(1), 115–126.

Shen, G. C., & Fan, V. Y. (2014). China's provincial diplomacy to Africa: Applications to health cooperation. *Contemporary Politics, 20*(2), 182–208.

Sun, Y. (2014). *Africa in China's foreign policy*. John L. Thornton China Center and the Africa Growth Initiative.

Tambo, E., Ugwu, C. E., Guan, Y., Wei, D., & Xiao-Nong, Z. (2016). China–Africa health development initiatives: Benefits and implications for shaping innovative and evidence-informed National health policies and programs in sub-Saharan African countries. *International Journal of MCH and AIDS, 5*(2), 119.

Traore, S. E. (2021). The strategy of China in Sino-African relations. *Open Journal of Political Science, 11*(4), 614–629.

Tshuma, D. (2022). *What if Africa stops receiving foreign aid? The risk of reversing development gains in a Covid-19 world*. European Union Institute for Security Studies.

The Emory Global Health Institute. (2020). *Can Global Sanitation 2020 Contribute to China's Prosperity?* https://repository.gheli.harvard.edu/repository/10678/. Accessed: 23 June 2022.

United Nations Economic Commission for Africa. (2020). *Review of policies and strategies for the pharmaceutical production sector in Africa: Policy coherence, best practices and future perspective*, Addis Ababa: United Nations Economic Commission for Africa.

Van Staden, C., & Shan Wu, Y. (2021). Vaccine diplomacy and beyond: New trends in Chinese image-building in Africa. *African Perspectives Global Insights*, June, (pp. 1–20).

Wang, Q., Yun, Y., Zhang, H., Guo, J., Wu, W., Deng, C., & Song, J. (2018). Experience and inspiration of the mass drug administration programme with artemisinin-piperaquine in Moheli Island of Comoros assisted by China. *Global Health Journal, 2*(3), 1–7.

Wang, X., & Sun, T. (2014). China's engagement in global health governance: A critical analysis of China's assistance to the health sector of Africa. *Journal of Global Health, 4*(1), 1–4.

Weston, J., Campbell, C., & Koleski, K. (2011). China's foreign assistance in review: Implications for the United States. US-China Economic and Security Review Commission. chrome-extension://efaidnbmnnnibpcajpcglclefindmkaj/https://www.uscc.gov/sites/default/files/Research/9_1_%202011_ChinasForeignAssistanceinReview.pdfAccessed 24 June 2022.

Xia, Z., Wang, R., Wang, D., Feng, J., Zheng, Q., Deng, C., Abdulla, S., Guan, Y., Ding, W., Yao, J., & Qian, Y. (2014). China–Africa cooperation initiatives in malaria control and elimination. *Advances in Parasitology, 86*, 319–335.

Yoshikawa, S. (2022). China's policy towards Myanmar: Yunnan's commitment to Sino-Myanmar oil and gas pipelines and border economic cooperation zone. *Journal of Contemporary East Asia Studies, 11*, 1–18.

Youde, J. (2010). China's health diplomacy in Africa. *China an International Journal, 8*(1), 151–163.

Zhang, B. (2010). Chinese foreign policy in transition: Trends and implications. *Journal of Current Chinese Affairs, 39*(2), 39–68.

Chapter 27
Public Perception of Traditional Chinese Medicine in Africa and the Impact of COVID-19: Based on a Study of Africans

Wei Liu and Xuanzhen Zhang

27.1 Introduction

Africa has long been an important partner of China. However, due to factors such as its persistent economic underdevelopment (Kutor, 2014), limited medical technology, unbalanced distribution of medical resources, flawed healthcare systems, the limited capacity for the production, research, and development in the pharmaceutical industry (Oleribe et al., 2019), as well as increasing population, Africa is not in a position to settle the medical problem and make healthcare security accessible to African people. As a result, African countries rely heavily on imported drugs and offer a potentially lucrative market for pharmaceutical products (Saied et al., 2022). Renowned as a treasure of Chinese culture, traditional Chinese medicine (TCM) boasts a time-honored history, unique theories, and technical methods. With the help of modern technology, significant progress has been made in the development of TCM. Consequently, affordable and effective Chinese herbal medicine has the potential to significantly address the demand for medical services and pharmaceutical products in African countries.

At the 8th Ministerial Conference of the Forum on China–Africa Cooperation held in Dakar, Senegal on 29–30 November 2021, China and its African partners released the *China–Africa Cooperation Vision 2035*, in which China pledged to support Africa's health policy, improve the prevention and control system for infectious disease, promote the medical research and development of traditional medicine,

W. Liu (✉)
Faculty of European Languages and Cultures, Center for International Migration Studies, Guangdong University of Foreign Studies, Guangzhou, China
e-mail: liuwei@scau.edu

X. Zhang
School of International Organizations, Beijing Foreign Studies University, Beijing, China

© The Author(s) 2024
M. Muchie et al. (eds.), *China-Africa Science, Technology and Innovation Collaboration*, https://doi.org/10.1007/978-981-97-4576-0_27

enhance the accessibility and affordability of medicines, and help to reduce the infection rates of diseases such as AIDS, tuberculosis, and malaria in Africa. With its many advantages, TCM products and services are well-suited to the medical needs of African people, leading to steady growth in Sino-African trade in TCM. Compared to the China-Africa bilateral trade in TCM in 2012, the total import and export of TCM between China and Africa tripled to nearly US $80 million in 2017, and the trade in TCM services has thus become a new growth driver for China-Africa cooperation (Li,2018). Despite the great increase in TCM trade between China and Africa, the actual situation is far from promising. According to the data presented at the Conference about the Conditions for Internationalization of the Pharmaceutical and Health Industry from 2018 to 2019, In 2018, the total import and export of TCM products in China went up 10.99% year-on-year to US $5.768 billion, which accounted for 5.02% of the total import and export of pharmaceutical products in China. The export of TCM products in the same year surged by 7.39% year-on-year to US $3.909 billion, while the import of TCM products rose by 19.38% year-on-year to US $1.859 billion (Luo, 2019). Although China–Africa TCM trade tripled to nearly US $80 million in 2017 compared to 2012, it still accounts for a small proportion of the overall volume of TCM trade, indicating significant potential for further growth.

Research indicates that TCM faces several unresolved challenges in the African market. First, TCM gains a relatively low market share in Africa, and there are still apparent market access barriers for TCM in African countries. Second, as TCM is still grabbing a limited market share for low-end pharmaceutical products with numerous market rivals, further exploring the pharmaceutical market in Africa would be challenging. Third, China lacks an appropriate strategy for promoting TCM to the rest of the world (Hu, 2019). There is still considerable pressure to promote TCM in the African market. Understanding the medical needs, perceptions, and imaginations of Africans about TCM is critical to expanding its market share in Africa. Without relevant knowledge about the public perception of TCM in Africa, developing countermeasures or suggestions would be nearly impossible.

As the COVID-19 pandemic continues to ravage the world recently, African countries are still grappling with an unabated health crisis of coronavirus. As of January 2, 2023, the Africa CDC Centers for Disease Control and Prevention reported that the cumulative number of confirmed COVID-19 cases in African countries had surpassed 12,216,748, with reported deaths amounting to 256,542. This trend is expected to intensify as the pandemic continues to spread globally, posing an unbearable burden on healthcare systems and presenting a grim picture overall. Dr. Matshidiso Moeti, the World Health Organization Regional Director for Africa, emphasized the need for effective containment measures, stating that the sustained and widespread transmission of the virus could overwhelm health systems, and curbing a large outbreak is costlier than preventive measures being undertaken by government. Against this backdrop, TCM has played a crucial role in the fight against the pandemic. For instance, Professor Liu Jingyuan from Beijing University of Chinese Medicine shared invaluable anti-pandemic experience about enhancing immunity to viruses with Radix Codonopsis (党参: Dang Shen) and Astragalus Root (黄芪: Huang Qi). Mr.

Shen Shaoping, deputy governor of Shennongjia Forestry District in Hubei Province, introduced a recipe for effectively blocking disease transmission with TCM. Wu Xiangjun, general manager of Yiling Pharmaceutical in Shijiazhuang, even pointed out the miraculous effect of Lianhua Qingwen (连花清瘟: Lian Hua Qing Wen) in treating COVID-19 infected patients (Wang, 2020). TCM can offer Africa new solutions to improve medical and healthcare services in the context of health crises, which can benefit personal security and social development in African countries.

27.2 Literature Review

27.2.1 Health Conditions in Africa

Many African countries continue to face significant health challenges, with a health-care system that includes both traditional herbalists and modern medical practitioners. In urban areas, medical treatment is typically provided by foreign physicians, while rule and remote regions often rely on traditional healers, such as witch doctors, who prescribe medicinal plants. Despite the popularity of local traditional herbalists in Africa, most of them lack systematic medical knowledge (Yang & Song 2013). Despite this, the use of medicinal plants remains a fundamental component of African traditional healthcare, and in many rural areas, traditional healers are the most accessible and affordable health resource available to the local community, and at times, the only therapy available (Mahomoodally, 2013).

27.2.2 Traditional Medicine in Africa

Traditional medicine refers to health practices, approaches, knowledge, and beliefs incorporating plant, animal, and mineral-based medicines, spiritual therapies, manual techniques, and exercises, applied singularly or in combination to treat, diagnose, and prevent illnesses or maintain well-being (Fokunang et al., 2011). The concept of traditional, complementary, and alternative medicine (TCAM) refers to a set of healthcare practices, either indigenous or imported, that are not part of the mainstream healthcare system (James et al., 2018). In the African context, this may include local herbal medicines or products, indigenous healthcare practices, such as traditional bone setting, and imported complementary and alternative medicine products and practices, such as acupuncture or chiropractic.

African traditional medicine is defined as one of the holistic healthcare systems comprised of three levels of specializations, namely divination, spiritualism, and herbalism. Traditional healers provide healing services based on prevalent culture, religious background, knowledge, attitudes, and beliefs in their community. This chapter focuses on different types of African healing systems, traditional healers, traditional practices, and modern herbalism, while also describing the phytochemical and pharmacological evidence of traditional African herbs (Chaitanya et al., 2021). Despite being poorly recorded, African traditional medicine (ATM) is considered one of the oldest and most diverse medical systems. It is interwoven with cultural practices and religious beliefs, making it a holistic approach that involves both the body and the mind (Mothibe & Sibanda, 2019). To address the prevailing distrust between modern and traditional medical practitioners and promote regulation, standardization, and collaboration, it is essential for both groups to acknowledge their respective strengths and weaknesses and to approach the challenging task of human health with genuine concern (Abdullahi, 2011). For more than 80 percent of Africans, traditional medicine is the first or only resort. Clinical evidence supports the effectiveness of traditional medicines in treating and managing health conditions, such as malaria, sickle cell anaemia, diabetes, and HIV/AIDS. World Health Organization continues to support further research into the safety, efficacy and quality of these medicines to facilitate access to standardized African traditional medicines.

In Ethiopian traditional medicine, the scope of healing extends beyond the cure of diseases and includes the protection and promotion of human physical, spiritual, social, mental, and material well-being. In Ghana, traditional healing encompasses both physical and spiritual aspects and is practiced by herbal spiritualist collectively known as "bokomowo", who possess occult knowledge of divination, exorcism, and spiritual herbalism. Traditional healers in Ghana are also known by various local names, such as "gbedela" (Ewe), "kpeima" (Dagomba), "odunsini" (Akan), and "isofatse" (Ga). Ghanian excellence in traditional and alternative medicine has led to the development of standardized herbal medicine, which is now an essential part of modern herbalism (Chaitanya et al., 2021). A study was conducted to assess the traditional knowledge and use of medicinal plants by Algerian traditional healers. Forty traditional healers were face-to-face interviewed in three different Algerian regions (West, Kabylia, and Sahara), and the collected data were analyzed using quantitative indices (Belhouala & Benarba, 2021). Traditional medicine is part of everyday life in South Africa, where a wide variety of concepts and beliefs exists within the South African traditional healing system. Herbalism plays an important role in traditional healing (Chaitanya et al., 2021).

27.2.3 Traditional Chinese Medicine in Africa

Chinese medical teams in Africa have played a significant role in laying a solid foundation for the development of TCM on the continent. TCM has gained popularity in the African market due to its affordability and effectiveness, with TCM clinics receiving recognition and acceptance in several African countries, such as Côte d'Ivoire and Burkina Faso (Xiong & Han, 2011). The "going global" strategy for TCM and the potential for increased medical cooperation between China and African countries like Ethiopia and Uganda suggest a promising investment environment for TCM in Africa (Ke & Chen, 2018).

The Chinese National Administration of Traditional Chinese Medicine has signed a memoranda of understanding on TCM cooperation with six African countries, including Malawi, Tanzania, Comoros, Ghana, Ethiopia, and Morocco, establishing a framework for intergovernmental cooperation. TCM has gained recognition in North African countries, where frequent TCM exchanges with China take place. Algeria, Tunisia, and Morocco are among the first African countries to welcome Chinese medical teams. In East African countries facing health challenges due to pandemics, TCM can play a significant role in pandemic preparedness and response. Chinese medicine played a leading role in the prevention and treatment of infectious diseases, such as AIDS and malaria, in African countries, particularly in Tanzania, Comoros, and Malawi. TCM was introduced to Western Africa at a later stage, whereas in South Africa, the economic foundation is more solid, the development of modern medicine is way ahead of the rest of Africa and traditional medicine gains greater recognition. For instance, South Africa has affirmed the legal status of acupuncturists, offering policy guarantees to Chinese medicine in its legislation (Pei, 2018). In Gauteng province, scholars conducted the first study on South African patients' views on TCM. Findings indicate that TCM is considered as an alternative medical healthcare option resonating with African herbal medicine. Most participants agree that TCM is cost-effective and has no side effects. As traditional medicine is often relied upon in rural areas where medical services are lacking, TCM may serve as a viable healthcare option for these populations (Hu & Venketsamy, 2022). The development of TCM in Africa encounters challenges related to the lack of objective standards to evaluate the quality and quantity of TCM, and difficulties in controlling the active ingredient content. The heavy metals in TCM sometimes even exceed the normal standard, and the TCM labels and instructions are yet to be better adapted to the African market (Xiong & Han, 2011). In addition, China still needs to educate foreign language talents with TCM expertise and establish a distinctive system for TCM diagnosis and treatment (Zheng & Guo, 2020). Furthermore, challenges persist regarding the investment environment for TCM in African countries, including political instability, inadequate government regulatory capacity, relatively poor infrastructure, and a scarcity of skilled labor, all of which pose obstacles to the development of TCM in Africa (Ke & Chen, 2018).

27.3   Research Methods

We conducted a field study among African populations residing in China and Africa, obtaining 144 valid questionnaires through both online and offline distribution. Through analyzing the sample data, we aimed to identify factors influencing the adoption of TCM among Africans and provide constructive recommendations for promoting TCM in Africa.

Our study focuses on a sample of 144 Africans, consisting of 66% males and 34% females. The age distribution is as follows: 3.5% are under 18 years, 39.6% are between 18 and 25 years, 29.9% are between 26 and 30 years, 20.1% are between

31 and 40 years, 4.2% are between 41 and 50 years, and 2.8% are between 51 and 60 years. The majority of the participants have received high school education or higher, and come from English and French-speaking countries including Togo, Algeria, the Democratic Republic of Congo, Senegal, Cameroon, and Burkina Faso. We employed general and scale questionnaires as research instruments and utilized SPSS 22.0 for data entry and correlation analysis. The following section presents some preliminary findings from our survey.

27.4   The Preference of African People for Types of TCM

Many Africans hold preconceptions about TCM, such as their preferences for TCM dosage forms, sources, and taste, despite their limited access to TCM and lack of life experience in China.

27.4.1   Solid Dosage Form is Most Favored by Africans

Table 27.1 shows that 60% of the respondents prefer solid dosage form, while 47% favor liquid dosage form, and less than one-fifth of these respondents chose semi-solid dosage form. Solid dosage forms are easy to quantify, convenient to carry, take, and preserve; liquid dosage forms can be easily absorbed, less irritating, and highly effective, and most TCMs are suitable to be decocted in liquid form.

Our study indicates that Africans tend to prefer tablets as the solid dosage form for TCM. Tablets are popular because of their portability, ease of use, and accurate dosage. Additionally, the standardization of tablets helps reduce arbitrary drug use, which can promote trust in medication among African populations. Pills and capsules rank only second to tablets, due to their ability to enhance drug stability and conceal their smell with film coatings. While capsules can enable precise drug release, powders and granules are found to be the least favored solid dosage forms among African respondents.

Our research shows a notable discrepancy in the preference for different types of liquid dosage forms among African respondents. Among tonics, medicinal wine, and oral liquid, oral liquid is favoured the most due to its small dosage, pleasant taste, and quick absorption. Additionally, oral liquid can be easily prepared using herbal tonics. Both tonics and medicinal wine are preferred by more than 40% of

Table 27.1 Which dosage form do you prefer?

Option	Number	Percentage (%)
Liquid	67	46.53
Semi-solid	26	18.06
Solid	86	59.72

African respondents. Tonics are valued for their flexibility in dosage adjustment based on symptoms, aligning with the traditional perception of TCM. However, tonics, which are bitter and inconvenient to carry, are influenced by factors such as the herbs used, decoction devices, time, and fire. Therefore, TCM practitioners in Africa are supposed to improve the stability of tonics to ensure their effectiveness. Medicinal wine, commonly used for internal consumption, offers high solubility for TCM ingredients and promotes efficient absorption. Moreover, it can be taken with meals and minimizes the liver's digestive burden (Duan et al., 2017), making it a popular choice among African respondents.

Injection and aerosol are not popular among Africans. These modern drug delivery methods, which integrate TCM and advanced technology, are primarily reserved for emergencies due to concerns such as the pain of injection, inconvenient administration, and the potential for medication misuse. African respondents do not exhibit a strong preference for either topical or internal creams, which are two types of semi-solid dosage forms. While both forms are generally acceptable, they are not favored over other dosage forms by the majority of respondents.

27.4.2 Drugs Made of Plants are Favoured by Africans

Our research indicates that 80% of respondents prefer botanical ingredients in medicine, as plants are viewed as natural remedies and have a long tradition of use in Africa. This shared cultural background between China and Africa, along with the similarities in the use of herbal medicine, can help to build trust in TCM among African populations.

Likewise, a majority of African respondents (51%) prefer natural minerals. Natural mineral-based drugs show rapid curative effects and contain essential substances for the human body, such as cinnabar sedative pill (朱砂安神: Zhu Sha An Shen). Additionally, mineral-based health products like vitamin and calcium tablets are widely accepted in Africa, making it easier to introduce TCM based on natural minerals to African people. The similarity in the use of mineral-based ingredients in Western medicine and TCM further contributes to building trust among Africans.

Animal ingredients in medicine are least preferred by African respondents (29%). With growing concerns about environmental and animal protection, there is increasing interest in promoting harmony between humans and animals. Moreover, the problem of animal poaching in Africa has led to an ecological imbalance and endangered many species. Given TCM's use of animal ingredients, there is a potential risk of popular hostility towards harming animals. Therefore, it may be more appropriate to avoid challenging mainstream animal protection groups when promoting TCM in Africa.

27.4.3   Africans Prefer a Sweet Taste to Bitter Taste

According to our research, the majority of African respondents (89%) prefer sweetness over bitterness, while 70% explicitly express a dislike for bitterness. The preference for sweetness is rooted in human instinct, as it is an essential taste for the body. However, TCM often has a bitter taste, which is not well-received by African consumers. To address this, sweet substances may need to be added to TCM products to improve their palatability. Our research on TCM dosage forms reveals that the representation of "medicine" is crucial for its acceptance in the African market. TCM enterprises can increase their chances of success in Africa by developing pharmaceutical products in the form of proprietary medicine that aligns with African preferences.

27.5   Analysis of the Attitude of African People Towards Traditional Chinese Medicine

To promote the use of TCM products among Africans, it is important to understand their willingness to use them for disease treatment. Our research shows that only 29.17% of African respondents occasionally purchase TCM products, while 25% of respondents never purchase them. This indicates a low frequency of TCM product purchases among African consumers and suggests a lack of established purchasing habits. To increase the trade volume of TCM between China and Africa, it is necessary to understand how TCM is perceived by Africans, as this directly influences their purchasing behavior.

27.5.1   The Safety of Traditional Chinese Medicine

African consumers prioritize safety (51.39%) and effectiveness (41.67%) as the most important factors when purchasing TCM products, followed by convenience (15.97%) and price (11.11%). This reflects their expectation that TCM should meet the basic requirements of medicine. However, their emphasis on safety and efficacy also suggests a level of scepticism or concern toward TCM.

27.5.2   African People's Attitude Towards TCM Treatment

The results of our survey indicate that Africans generally have a positive attitude toward TCM, with high satisfaction rates among those who have tried TCM services.

Specifically, 18.82% of respondents reported being very satisfied with the effectiveness of TCM, while 57.65% expressed satisfaction. Dissatisfaction rates were low, with less than 6% of respondents expressing dissatisfaction. This suggests that Africans have a favorable view of TCM treatment and are willing to purchase TCM services, indicating a promising future for TCM in the African market. In fact, 53% of respondents expressed willingness to try TCM services, while only 9% were unwilling, signaling strong competitiveness for TCM treatment in the African market.

A significant proportion of respondents claim to have tried massage (32%), acupuncture (22%), and medical food therapy (22%), and those who have received such TCM services are quite fond of them. Both massage and acupuncture help to improve blood circulation and alleviate muscle fatigue, offering effective remedies for insomnia, enteritis, and eczema. In contrast, medical food therapy, unlike conventional medications, can treat chronic conditions and enhance overall physical health. These TCM treatments are not unique to African cultures and are commonly used in other regions of the world, such as Europe and North America. Hence, they present a viable avenue for promoting TCM in the African market.

Our survey revealed that moxibustion (5%), fire cupping (11%), and skin scrapping (7%) are less commonly used TCM treatments among African people. These treatments may not be widely accepted or may cause discomfort and skin bruising during the treatment process. Therefore, moxibustion, fire cupping, and skin scraping appear to be less popular and accepted in the African market.

27.5.3 The Effectiveness of Traditional Chinese Medicine

Effectiveness (53.47%) and quality (49.31%) are the most important factors for TCM according to respondents. Dissatisfaction with the effectiveness of TCM is low, with a majority expressing satisfaction (48.15%) and great satisfaction (16.67%). This suggests that most respondents recognize the effectiveness of TCM, and poor efficacy is not a significant factor in their decision not to purchase TCM products or services. The majority of respondents (58.2%) agree that TCM is effective in curing diseases, indicating potential support for TCM products and services in the African pharmaceutical market. Only a very small percentage (9.16%) hold opposing views. TCM appears to be competitive with traditional African medicine, as only a minority (16.79%) believe that African traditional medicine is more scientifically based. Furthermore, nearly 80% of the population would be willing to try TCM if access barriers were overcome.

27.5.4 The Changes in Africans' Attitude Towards TCM

According to our survey, 45.8% of the respondents believe that "as the relationship between China and Africa improves, TCM will become more acceptable to them", while 46.57% of these respondents express support for "the promotion of TCM and the construction of a Chinese Medicine Culture Communication Centre in Africa". This suggests that TCM has gained a positive reputation and significant support among the African population. With the comprehensive and in-depth development of the China–Africa relationship, more Africans will likely become receptive to TCM culture, and support its promotion and the establishment of relevant cultural centres. From this perspective, it would be quite feasible for TCM to enter the African market.

The COVID-19 pandemic has significantly impacted African countries and influenced their perception of TCM. Among respondents, 35% agreed that TCM has received increased coverage in African media during the pandemic. However, only 24.42% believed that TCM treatment options were more commonly used in African hospitals during the pandemic. Approximately 33% of respondents agreed that the COVID-19 pandemic has increased their understanding of TCM, while 28% disagreed. Furthermore, 30.53% of respondents believed that TCM has become more attractive to them since the onset of the pandemic. The pandemic has facilitated a greater willingness among African people to learn about TCM, and nearly 30% of respondents find it more appealing. This global health crisis can thus be viewed as an opportunity for TCM to enhance its image and expand its presence in the African market.

27.6   The Promotion of Traditional Chinese Medicine

The promotion of TCM can be regarded as a critical prerequisite for enabling its successful penetration into the African market. Only when the public is well-informed of TCM, will they be able to choose and purchase relevant products and services.

27.6.1   Promotional Channels

Effective product promotion is essential for gaining market share, and the choice of promotional channel is critical in modern society. Our research indicates that the internet and television (63.19%) are the preferred channels for Africans to learn about medication and its effects, followed closely by doctors (57.64%) and drugstores (55.56%). Family and friends (40.28%) and other channels (7.64%) also play a role in shaping opinions. Pharmaceutical product advertising on the Internet and television can effectively promote products and educate African consumers. Doctors and drugstores can also contribute to informing the public about pharmaceutical products. Additionally, the influence of family and friends should not be overlooked, highlighting the importance of promoting TCM within the Afro-descendant community to encourage recommendations from repeat customers.

Our survey results indicate that 53% of respondents prefer leaflet distribution, 47% are interested in online advertisements, 35% favor offline advertisements, and only

31% show an inclination toward celebrity endorsement. This highlights the high acceptance rate of leaflet distribution, and the effectiveness of online advertising, which is dynamic, precise, and targeted. On the other hand, celebrity endorsement is not favoured by Africans, with 27.78% of respondents showing dislike or even hatred towards it. TCM enterprises are advised to avoid using this promotion method in the African market and make necessary changes to align with the preferences of African consumers.

27.6.2   The Effect of Promotion

Various advertising methods and promotional channels can have diverse effects on audiences.

According to our survey, approximately 60% of African respondents find the promotion of pharmaceutical products to be useful or very useful, while 11.81% find them to be slightly useful. Only 16.67% of respondents hold opposing views, indicating that drug promotion is an effective means of promoting pharmaceutical products with a high conversion rate.

Over 50% of the African respondents perceive pharmaceutical product promotion as useful or very useful. This suggests a high level of acceptance and effectiveness of this advertising method in the African market. Notably, the African respondents prefer advertising methods and similar promotional channels over marketing campaigns for pharmaceutical products.

27.6.3   Language Barrier for Promotion

A significant percentage (51.9%) of the respondents believe that language barriers hinder the understanding of TCM concepts in the African market. To overcome this obstacle and enhance the promotion of TCM products and services in Africa, TCM product descriptions and illustrations should be translated into local languages to improve their effectiveness. Likewise, there is a need to cultivate professionals who possess a strong foundation in TCM and proficiency in foreign languages to facilitate effective communication and cooperation between China and Africa in the field of medicine, otherwise, it would be difficult to meet the demand for relevant talents in this field for better exchanges and cooperation between China and Africa.

27.6.4 Health-Related Cooperation Between China and Africa

TCM presents great potential in the African market, and promoting TCM products and services in Africa holds great significance. While traditional Chinese and African medicines may differ greatly, they also share similarities and embody their respective cultures, both of which have yet to be fully introduced to the African market. The discovery of artemisinin by Tu Youyou, a researcher at the China Academy of Traditional Chinese Medicine, resulted in a significant breakthrough and earned her the 2015 Nobel Prize in Physiology or Medicine. This achievement demonstrates the valuable knowledge and original thinking contained in ancient TCM literature, which should be thoroughly explored and inherited (Zhu & Lin, 2016). The discovery of artemisinin has profound implications for the health of African people and presents an important opportunity for the development and advancement of TCM and African traditional medicine. During the opening ceremony of the Beijing Tong Ren Tang Africa Company in Johannesburg on November 16, 2016, Dr. Nkosazana Dlamini Zuma, the Chairperson of the African Union Commission, praised the philosophy of Chinese medicine and its four foundations of diet, exercise, rest and relaxation, and a positive mental attitude as potentially beneficial for African health systems and medical professionals. Through face-to-face interviews with African respondents, this study provides valuable insights into their perceptions of TCM, which is of great practical significance for promoting TCM in African countries.

As for the dosage form of medicine, most African respondents prefer solid dosage forms for pharmaceutical products, such as tablets and pills. They prefer plant-based ingredients rather than animal-based or mineral-based ingredients, making Chinese herbal medicine a more acceptable option. Additionally, as for the taste of medicine, Africans prefer sweetness to bitterness.

Regarding the acceptability of TCM in Africa, safety is the most important factor for African people when purchasing TCM products and services, with price considered the least significant. Thus, TCM enterprises should prioritize the safety of TCM products and services, and invest more resources in promoting their safety to address the concerns of African people. African people generally exhibit a favourable outlook toward TCM treatment services. Many individuals have already engaged in various forms of TCM treatment, such as massage, acupuncture, and food therapy. Such interventions have been perceived as effective, acceptable, and readily adoptable by African consumers. Effective pharmaceutical promotion in the African market requires the use of appropriate advertising methods and channels. Our survey found that leaflet distribution is the preferred method for advertising pharmaceutical products among African consumers. Online and television advertising is also likely to get recognized by African consumers. Recommendations from friends and family are highly valued, but celebrity endorsements are less popular. TCM enterprises should adapt their promotional strategies to the African market and improve the quality of their advertisements to better cater to the preferences of African consumers.

Overcoming language barriers is critical yet challenging for TCM products and services to penetrate the African market, necessitating precise translation of their descriptions into local African languages. Given the wide diversity of local languages across the continent, developing expertise in African languages and cultures is crucial for unlocking the consumption potential of TCM products and services in the African market. The perception of TCM among Africans is influenced by various factors, including the relationship between China and Africa and the emergence of major diseases and pandemics in Africa, such as malaria and coronavirus. Our research shows that the favourable China–Africa relationship has led to an increase in the recognition of TCM culture by more than half of the African respondents, who are satisfied with its therapeutic effects. Moreover, the COVID-19 pandemic has resulted in greater attention toward TCM treatment options, leading to an improved understanding of TCM products and services. This presents a significant opportunity for the internationalization of TCM.

The *Belt and Road Development Plan of Traditional Chinese Medicine (2016–2020)* presents a viable opportunity to establish a platform for qualified TCM enterprises to penetrate the African market. This will enable TCM trade and services to thrive, create a top-notch TCM brand to meet the demands of the African market, and fully unlock the potential for exporting TCM health services to African countries. For this purpose, TCM should be transformed into products, technology, and services that serve the "Belt and Road" initiative. The *Forum on China–Africa Cooperation-Dakar Action Plan* (2022–2024) also encourages Chinese enterprises to collaborate with African counterparts in the pharmaceutical industry, supports China–Africa cooperation in traditional medicine, and advocates for the use of traditional medicine in the prevention and treatment of major infectious diseases such as coronavirus and malaria.

With the advancement of the "Belt and Road" initiative in Africa, China and Africa have reached a common understanding of the development of traditional medicine, leading to increased acceptance of TCM among local Africans. However, TCM still faces significant challenges and obstacles in the region. Despite these challenges, TCM has the potential to grow and thrive in Africa alongside traditional African medicine, drawing on the time-honored history and cultural heritage of both regions.

Acknowledgements The project members also include Chen Rihui, Zhang Linrui, Xiao Ruilin and Luo Ying.

References

Abdullahi, A. A. (2011). Trends and challenges of traditional medicine in Africa. *African Journal of Traditional, Complementary and Alternative Medicines.* https://doi.org/10.4314/ajtcam.v8i5S.5

Belhouala, K., & Benarba, B. (2021). Medicinal plants used by traditional healers in Algeria: A multiregional ethnobotanical study. *Frontiers in Pharmacology.* https://doi.org/10.3389/fphar. 2021.760492

Chaitanya, MVNL., Baye, H. G., Ali, H. S, & Usamo, F. B. (2021). Traditional african medicine. In H. A. El-Shemy (Ed.), *Natural medicinal plants.* https://doi.org/10.5772/intechopen.96576.

Duan, D. X., Zhou, S. Y., Chen, Y., et al. (2017). Analysis of common dosage forms of traditional chinese medicine and their application. *Asia–Pacific Traditional Medicine, 13*(16), 66–69.

Fokunang, C. N., et al. (2011). Traditional medicine: Past, present and future research and development prospects and integration in the national health system of Cameroon. *African Journal of Traditional, Complementary and Alternative Medicines.* https://doi.org/10.4314/ajtcam.v8i3. 65276

Hu, M. (2019). The development of african pharmaceutical market, and the China–Africa cooperation in the pharmaceutical industry. *Journal of Zhejiang Normal University (social Sciences), 44*(1), 12–13.

Hu, Z., & Venketsamy, R. (2022). Traditional Chinese medicine to improve rural health in South Africa: A case study for Gauteng. *Health SA.* https://doi.org/10.4102/hsag.v27i0.1871

James, P. B., Wardle, J., Steel, A., et al. (2018). Traditional, complementary and alternative medicine use in Sub-Saharan Africa: A systematic review. *BMJ Global Health.* https://doi.org/10.1136/ bmjgh-2018-000895

Ke, X. T., & Chen, Y. (2018). Analysis of China's international investment in Chinese medicine in Africa. *Cooperative Economy and Technology.* https://doi.org/10.13665/j.cnki.hzjjykj.2018. 15.014

Kutor, S. K. (2014). Development and underdevelopment of the African continent: The blame game and the way forward. *Research on Humanities and Social Sciences, 4*(14), 14–20.

Li, Y. (2018). A new growth driver for China–Africa cooperation: The trade of traditional Chinese medicine between China and Africa reached $80 Million in 2017. *Finance.* China.com.cn. http:// finance.china.com.cn/news/20180820/4735267.shtml. Accessed: 20 Aug 2018.

Luo, N. Y. (2019). The import and export of traditional Chinese medicine increased by more than 10% in 2018. *China Net of Traditional Chinese Medicine.* http://www.cntcm.com.cn/2019-03/ 21/content_58300.htm. Accessed: 21 Mar 2019.

Mahomoodally, F. (2013). Traditional medicines in Africa: An appraisal of ten potent african medicinal plants. *Hindawi Publishing Corporation.* https://doi.org/10.1155/2013/617459.Accessed: 10October2013

Mothibe, E. M., & Sibanda, M. (2019). African traditional medicine: South African perspective. *Traditional and Complementary Medicine.* https://doi.org/10.5772/intechopen.83790

Oleribe, O. O., Momoh, J., Uzochukwu, B. S., Mbofana, F., Adebiyi, A., Barbera, T., Williams, R., & Taylor-Robinson, S. D. (2019). Identifying key challenges facing healthcare systems in Africa and potential solutions. *International Journal of General Medicine, 12*, 395–403. https:// doi.org/10.2147/IJGM.S223382

Pei, A. D. (2018). Bringing more benefits to African people with traditional Chinese medicine— Interview with Wang Xiaopin, Director General of Department of International Cooperation, National Administration of Traditional Chinese Medicine. *China Investment.* https://www.cnki. com.cn/Article/CJFDTotal-ZGTZ201818009.htm. Accessed: Sept 2018

Saied, A. A., Metwally, A. A., Dhawan, M., Choudhary, O. P., & Aiash, H. (2022). Strengthening vaccines and medicines manufacturing capabilities in Africa: Challenges and perspectives. *EMBO Molecular Medicine, 14*(8), e16287. https://doi.org/10.15252/emmm.202216287

Wang, R. J. (2020). *The efficacy of traditional Chinese medicine in the fight against the COVID-19 pandemic received much attention from the international community.* https://m.haiwainet.cn/ middle/3544627/2020/0425/content_31776604_1.html. Accessed: 25 Apr 2020

Xiong, J. X., & Han, L. (2011). Study on countermeasures for the development of Chinese medicine in Africa. *China Pharmaceuticals, 20*(22), 13–14.

Yang, J. H., & Song, Q. (2013). Overview of the development of Chinese medicine in Africa. *World Journal of Integrated Traditional and Western Medicine.* https://doi.org/10.13935/j.cnki.sjzx.2013.02.001

Zhu, J. P., & Lin, M. X. (2016). The inspiration of Tu Youyou's Nobel Prize: Innovation in Chinese medicine based on inheritance. *Journal of Dialectics of Nature, 238*(01), 38–45. https://doi.org/10.15994/j.1000-0763.2016.01.005

Zheng, X., & Guo, Y. D. (2020). Planting the seeds of Chinese medicine in Africa, *Health Industry,* June 2020. http://www.tzzzs.com/type_gwqk/43326.html

Part IX
China-Africa Collaboration in Manufacturing

Part IX
China–Africa Collaboration
in Manufacturing

Chapter 28
Riding the Dragon: A Framework for Productive Collaboration

Stephen E Little

28.1 Introduction

The research in this chapter reprises the evolving relationship between China and the African continent over the last half century in order to establish the context of Africa–China Science Technology Innovation (STI) collaboration. The title echoes Zheng and Williamson's (2007) book 'Dragons at Your Door: How Chinese Cost Innovation Is Disrupting Global Competition'. It demonstrates how Chinese cost innovation presents a distinctive approach that presents a strong challenge to Western models for the introduction of innovative technologies. This contrasting marketing and pricing policy achieves much faster market penetration for innovative products by pursuing sales volume and rapid diffusion, rather than setting a premium for novelty to create high profit margins.

By eliminating the initial premium pricing period and moving directly to volume sales technologies such as medical scanning were available immediately on an affordable basis. Consequently, Chinese companies have expanded both domestic and export markets simultaneously. In addition, this approach also delivers a mass-market income to finance the penetration of the niche markets to which some higher cost competitors have retreated.

Rodney (1976) points out that when fifteenth century Portuguese traders first visited West Africa, they encountered societies with a comparable level of technological development. In the same century Zheng He visited the Horn of Africa with an imperial Chinese fleet. China is therefore one of a long succession of traders drawn to African resources through history, but which now brings the insight of a nation that has had itself confronted the problems created by the incursions of outside interests.

Present Address:
S. E Little (✉)
Centre for Pan-African Studies, Hume Institute for Postgraduate Research, Lausanne, Switzerland
e-mail: stephen.little@humelausanne.ch

© The Author(s) 2024
M. Muchie et al. (eds.), *China-Africa Science, Technology and Innovation Collaboration*, https://doi.org/10.1007/978-981-97-4576-0_28

The innovations which enabled Chinese participation in a global manufacturing network include both improvements in information and communication technologies and the associated transformation of shipping through the relatively neglected innovation of the standard shipping container (Levinson, 2006). The development of secure and reliable global logistics in conjunction with the 'virtual adjacency' facilitated by modern telecommunications has contributed to the disruption of the long-established model of multinational development and technology transfer. Orderly flows of technology from centre to periphery (Hirsch, 1967) and the ability of the established centres of production to dictate terms have been overturned (Zheng & Williamson, 2007).

In 1960 Ghana had a higher per capita Gross Domestic Product than South Korea but by 1997 Korea had broken the US$10,000 average income barrier, and that country had embraced manufacturing successfully and moved into key fields of high technology and innovation. However, if the pathway followed by the Asian Tigers and by China is blocked for African countries by the established dominance of these newer players, what are the prospects for development? (Kaplinsky & Morris, 2007).

China's approaches to the challenges of rapid growth and development does have value for Africa, despite obvious and significant contextual differences. China's internal processes of development address problems familiar to African policy makers, despite the different tools and resources available to them. Chinese policy statements acknowledge the key issue of uneven development and the need to redress both the outcomes and the governance of the development process using the term 'Harmonious Society'. These and related issues were highlighted in the 11th five-year plan, 2006–2010 while subsequent plans have focused increasingly on the development and purchasing power of internal markets to rebalance an initial export dominated growth agenda. During the current 14th five-year plan, 2021–2025 this has been articulated as a specific 'Dual Circulation' policy aimed at the development of internal markets for manufacturers in parallel with export markets (Paterson, 2021) clearly indicating an awareness of the problem of the 'middle income trap' and the need to reduce dependence on exports.

The following sections examine the institutional and human capacity necessary both for effective engagement in STI collaboration and the incorporation of its benefits into economy and society.

28.2   Global China's Engagement with Africa

China's current engagement with Africa developed in both the Cold War and post-Cold War periods. As colonial administrations were replaced by independent governments across the African continent, European and Western anxieties were focused on the stand-off between East and West. The continent became a site of proxy conflicts and Western engagement was often predicated on suppression of perceived Communist influence. During the Cold War China's interest in Africa was restricted by distrust of the Soviet Union and states aligned with that country.

The evolution of Africa–China collaboration during the post-Cold War period reflects the realignment from a bipolar global divide which relegated significant regions to a subsidiary role in a great power narrative. While the Cold War lasted, African nations were courted by both sides of the ideological divide. With the dissolution of the Soviet Union and the associated 'Eastern Bloc' Europe and North America turned inwards at the point at which China began to open up to inward investment and engagement in a globalising world economy.

The immediate post-Cold War period saw significant regions of the planet relegated to a subsidiary role in evolving networks of multi-polar relationships (Dicken, 1998; Ohmae, 1995). Global economic integration led to an increase of outward flows of resources from the African continent, including human, intellectual and social capital. Africa, and in particular sub-Saharan Africa, were marginalised as part of what has become known as the 'Global South', in a system dominated by North America, Asia and Europe.

This relative neglect of Africa by the former Cold War antagonists from the 1990s onwards allowed China to build new relationships with African countries through the exchange of natural resources for infrastructure investment and development. By the early twenty first century China's interest in Africa had rekindled competition for African resources and attention, contributing to an end of post-Cold War neglect reflected in a number of studies of China's new relationships with Africa reviewed by Mohan (2008).

The removal of Cold War divisions also allowed previously suppressed connections to be re-established leading most notably to renewed regional economic and cultural synergies across a greatly expanded European Union (Delamaide, 1995). Delamaide's argument applies equally to the prospect of re-configuring post-colonial relationships across the African continent where externally imposed divisions and boundaries are hampering regional synergy and development.

While the differences between Africa and China in scale and economic activity, population and global influence are obvious, both share experience of the impact of external interventions and incursions, and the subsequent legacy of patterns of development influenced by external interests. Both have to different degrees dealt with the developmental consequences of infrastructure optimised around the outward movement of resources and the introduction of technologies deployed in support of external priorities rather than any internal development agenda (Headrick, 1981).

Understandably by the twenty-first century China's new relationship to the African continent was attracting considerable attention and speculation over its impact on both indigenous capacity for development and the motivation behind it (Kolodko, 2020; Li, 2016; Mohan, 2008). By the beginning of the second decade of this century China was recognised as an integral part of a global manufacturing system seeking energy and other natural resources from Africa to sustain continuing modernisation. As noted above, the emerging dominance of China, and to an increasing extent India, in manufacturing is potentially compromising conventional development pathways driven by industrialisation for African and other developing countries.

However, a further shift in the dominant techno-economic paradigm (Perez, 1985, 2002) has been made more probable by the pandemic and associated disruption.

The pandemic lockdown forced individuals and organisations along an accelerated learning curve on the use of virtual space and virtual synchronisation while revealing the fragility of the physical supply chain underlying a potentially brittle globalisation.

Such a shift is in any case necessary to address the challenge of uncontrolled man-made climate change which is already impacting the least developed countries most severely.

The prospect of a disruptive 'Fourth Industrial Revolution' provides an opportunity for actors previously marginalised to contribute meaningfully to the development of a new paradigm as governments, civil and commercial organisations realign their activities (Mazibuka-Makena & Kraemer-Mbula, 2021). STI collaboration is an essential resource for these actors.

Successive five-year plans have ensured that China is well placed to contribute to this turn since the needs of remoter regions are being addressed through alternative routes and appropriate technologies for development. The concept of 'Green GDP' has been adopted to measure domestic wealth against environmental losses and set out in a national accounting study report (Gore, 2006).

China's high technology agenda and the scale of application mean that the cost of relevant technologies to countries in Africa and elsewhere is changing dramatically. For example, extensive deployment of solar technologies in Tibet and other western regions has reduced unit costs for wider markets. Equally, the commitment to coal fired power generation for the medium-term future has led to significant advances in clean combustion technologies, including retro-fitting of carbon capture technologies which may allow African coal reserves to be considered in transition planning.

28.3 Engaging with the Challenges of Collaboration

For African countries contact with China provides the obvious economic benefit of trade and inward investment into resources, and also an alternative market to those offered by the West. However, by the first decade of the twenty-first century it was clear that in addition to these benefits, China's pattern of modernisation, and the problems it has engaged with offered lessons of potential value to African decision makers (Little, 2007).

China shows African policy-makers an alternative pathway to a high technology, high performance economy. The course followed from a predominantly agricultural base was different from that in the West and different from the model promoted by Western-dominated international institutions. The potential benefits and challenges of engagement and collaboration with China are therefore complex and in many ways path-dependent (Penrose, 1959).

The global circulation and transformation of technologies and techniques underline the value of cross-national STI collaboration in identifying more direct routes to the diffusion and adoption of innovations. The pathway from discovery to invention and innovation and on to the deployment of new technologies must be considered in detail. Science is a globalised activity which relies upon international publication and collaboration. While invention may take place in nationally funded public research institutions, equally it may also be carried out by the network of collaborators within one or more multinational corporations. However, innovation will be more influenced by regional and local conditions and in the case of Chinese medical scanners the driver was the need to improve the capabilities of second-tier regional hospitals at affordable cost. As the context in which these linked processes take place becomes more significant, local knowledge and capacities become more significant. As a consequence, the same technology deployed in different locations may be understood and utilised in different ways, even within the same transnational corporation (Kirlidog, 1997).

The vulnerability of global supply networks depending on widely dispersed centres of production has been exposed by the disruption of the Covid-19 pandemic and further undermined by the outbreak of armed conflict between Russia and Ukraine which threatens the integrity of global food supplies. This has prompted speculation of the 'end of globalisation' and new regionalisms. The aftermath of these shocks coupled with the prospect of an equally disruptive 'Fourth Industrial Revolution' provides an opportunity for previously marginalised actors to engage in the development of a more robust 'new normal' (Mazibuka-Makena & Kraemer-Mbula, 2021).

However, in order to exploit this window of opportunity it must be remembered that successful technology transfer is limited by absorptive capacity (Cohen & Levinthal, 1990), in both organizational and individual senses, which highlights the need for both technical learning and social learning. The latter is the key to both innovation and effective deployment of new technologies. While technical learning is relatively straightforward, social learning leads to the reformulation of practices and transformative gains in performance and productivity (Sproull & Kiesler, 1991). This learning inevitably draws upon local experience and context.

Kaplinsky (1994) identifies social learning around available technologies as a key to the development of alternative development pathways. Engagement with Chinese and other Asian companies have enabled African undertakings to improve their efficiency and productivity by adapting features and practices that do not require intensive capitalisation. This aspect of social learning has been evident in China's own modernisation. Tao (2011) shows that the early stage of Chinese E-business companies mimicked Western models without the Western infrastructure so that, for example, internet purchases were delivered by bicycle courier and paid for cash on delivery allowing Western business models to be refined and adapted to local conditions, while the indigenous technical infrastructure caught up. Ultimately this form of social learning allowed Chinese companies such as Alibaba, Haier, Weibo and Tiktok to equal or surpass their Western counterparts.

There is evidence of similar capacity to develop business models and practices appropriate to Africa. In Rwanda drones have been re-purposed for the delivery of medical supplies (Fortune, 2019) and the model developed by TESSA—Teacher Education in Sub-Saharan Africa—which began with peer-to-peer support for teachers in remote rural communities via cell-phone text messaging now offers a range of on-line support resources.[1] The local, grass roots evolution of these systems avoided the pitfalls inherent in numerous attempts to implement imported solutions highlighted by Mawere and van Stam (2019).

Sustainable skill and capacity development of indigenous individuals and organisations is essential to achieve this. It should be seen as an essential precondition of effective engagement with the opportunities offered by STI collaboration with China and other partners.

For STI collaboration to be successful and fully productive for African institutions, communities and individuals, there must be focus on capacity building to facilitate the management of this progression from generic knowledge creation to specific application.

28.4   Geography and Governance, an Institutional Foundation for Collaboration

An obvious difference between China and Africa is that between a territory governed by a single state and one with more than fifty states. Issues of scalability and cost will be crucial in the identification and selection of opportunities to adapt components of the Chinese STI strategies for African conditions. While external connections to the global economy are a dominant issue for the continent, the management of intra-continental relationships is equally important. The complex network of regional economic groupings and cross-border arrangements reflects the pre- and post-colonial origins of the nation states and their pathways to independence. Ohmae (1995) advocates the exploitation of regional synergies, across national boundaries in order to agglomerate the most advanced regions of neighbouring economies. However, this does little for the problem of uneven national development.

In addressing such issues China enjoys greater leverage with potential external collaborators than any individual African nation or regional association and can negotiate transfers of knowledge and technology to address specific problems of relative development. For example, pressure has been put on international companies around issues of both location of inward investment and of corporate social responsibility (Zheng & Chen, 2006). However, a continental system of collaboration should make it possible to develop an African agenda around equivalent demands for the reciprocal input of both financial capital and knowledge as part of any development proposal.

Reflecting China's geographic and economic diversity there is evidence of a toleration of local policy variation within the structure of government that can be regarded as a softening of central direction in favour of 'de facto federalism' (Zheng, 2006). However, dealing with issues at a scale which would be national or provincial for China requires cross-border and inter-regional collaboration in Africa.

As with every other continent African regional and continental collaboration has waxed and waned over time. A range of regional associations have been established

---

[1] https://www.tessafrica.net/.

across the continent, some with overlapping membership: Arab Maghreb Union (AMU/UMA), Economic Community of West African States (ECOWAS), East African Community (EAC), Intergovernmental Authority on Development (IGAD), Southern African Development Community (SADC), Common Market for Eastern and Southern Africa (COMESA), Economic Community of Central African States (ECCAS), and Community of Sahel-Saharan States (CEN-SAD).

The most significant initiative is the Continental Free Trade Area (CFTA) framework agreement now signed by fiftyfour African countries.[2] This has ambitious long-term goals to deepen integration among African Union members in support of the African Union's Agenda 2063.[3] Such an association offers the prospect of harmonization and better coordination of trade regimes as well as the elimination of challenges associated with multiple and overlapping trade agreements across the continent. A report by the World Bank anticipates that the CFTA could lift 30 million Africans out of extreme poverty, boost the incomes of nearly 70 million people, and generate $450 billion in income by 2035 (World Bank, 2022).

While these ambitious long-term goals are being pursued, more limited and focused forms of cooperation are delivering worthwhile benefits in a shorter timeframe. One significant and practical step towards the longer-term objectives of the CFTA is the establishment of the Pan-African Payments and Settlements System (PAPSS), allowing payments among companies operating in Africa to be done in any local currency (U.S. International Trade Administration, 2022).

Attempts to optimise African air routes for African needs have led to Routes Africa, a regular intra-Africa route development forum[4] which helps individual actors to contribute to regional synergies across the continent. The remnants of colonial frameworks have in the past required commercial flights between African cities to be made via the capitals of former colonisers in Europe.

A number of states have contributed to regional collaborations to achieve infrastructure development beyond anything they could achieve individually. A set of projects has delivered circum-continental undersea fibre optic trunk lines providing Africa with the communication infrastructure and connectivity enjoyed by competing regions. The SAT-3 cable along the west coast of Africa, and the East African Submarine Cable System (EASSy) have been delivered by multinational consortia from the countries where the cables come ashore and has overcome conflict over pricing and access to capacity for partners. Equally significantly, landlocked countries such as Rwanda have negotiated access to high-quality fibre-optic connections through negotiation and collaboration with neighbours.

Nevertheless, the rapid development of physical and electronic communication infrastructure in China demonstrates the advantage enjoyed by a single jurisdiction with a strategically managed transition from state monopoly control to a regulated market environment. Several Chinese companies, not least ZTE, and Huawei, both founded in the 1980s, have become global players, rivalling incumbents such as

---

[2] https://au-afcfta.org/

[3] https://au.int/en/agenda2063/

[4] https://www.routesonline.com/events/routes-africa/.

Ericsson and Samsung in telecommunications and mobile telephony. They have contributed to infrastructure development across Africa and in China their technologies complement the expansion of domestic air routes in conjunction with a nationwide high-speed rail system.

The development of other large-scale infrastructure has also been influenced by differences in jurisdiction and governance between Africa and China. While large scale power and irrigation projects such as the Kariba and Volta dams, once the trademark projects of external funders typified by the World Bank, have lost favour in Africa, China remains committed to mega-projects (Flyvbjerg et al., 2003), including some first proposed in the 1950s. Following the Three Gorges project, the South-North Water Transfer project to move water to relatively arid northern provinces began construction at the end of 2002 and is scheduled for completion in 2050. This involves three canals running 1300 kms across the eastern, middle and western parts of China, linking the country's four major rivers—the Yangtze, Yellow, Huaihe and Haihe (Li, 2016). This system is expected to take 50 years to complete and cost US$59 billion and reflects the ability of the Chinese government to commit substantial funding to very long-term projects.

Whether or not projects of this scale are appropriate to African needs, Gathanju (2006) demonstrates the problems they face by describing the jurisdiction of the Nile waters. Post-colonial problems have been created by adherence to a 1929 treaty drawn up by a colonial power, Britain. This privileges Egyptian use of Nile water over the claims of ten upstream countries including Ethiopia, Sudan and Kenya. Such legacies require sensitive negotiation around conflicting interests, and the construction of the Grand Ethiopian Renaissance Dam (GERD) on the Blue Nile has exacerbated tensions and disagreement, not least because Ethiopia was neither involved in the original negotiations nor included in the Treaty. Yet even where cross-border interests are not in conflict, working across separate national jurisdictions can cause difficulties.

28.5 Managing Resources and Building Capacity

Water is just one of a number of critical resources identified as a potential source of conflict in this century. Recent armed conflicts, both in the form of international interventions and separatist civil conflicts have been attributed to contestation over oil and other strategic mineral resources. A Chatham House report published in 2012 (Lee et al., 2012) analyses the wide range of potential resource conflicts and proposes forms of international collaboration which give some prospect of reducing or eliminating these in the future. Competition for resources is predicted to result from the pressure of climate change, both directly in terms of water shortages and less directly in terms of competition over the many relatively rare resources required by the technologies being developed for climate change mitigation.

In the context of climate change mitigation and reduced resource consumption a significant proportion of scientific and technical research is now directed at the more efficient and effective use of existing resources through increased energy efficiency and through new combinations of AI (artificial intelligence) and indigenous knowledge in improving agricultural production (Hirschmann, 2022). At the same time the dramatically increased use of distance learning and communications technology due to the pandemic will help to redress regional resource imbalances in human capacity.

Despite this, resource rich countries may not necessarily derive the most benefit from their assets. Instead development can be distorted, Humphreys et al. (2007) address the 'resource curse' which blights development and Shaxson (2007) argues strongly that, for African countries, oil resources can bring more problems than benefits.

Even in more stable periods the volatility of commodity prices is particularly problematic for any economy that is heavily dependent upon primary and extractive industries. In addition, policymakers must work in the current context of climate emergency, and competition for the limited resources required for decarbonisation and Green Transition together with the prospect of climate uncertainty impacting on global food production.

The need to develop pathways and roadmaps to low carbon, circular economies has become widely recognised. The European Commission adopted a new circular economy action plan (CEAP) in March 2020.[5] This is seen as a prerequisite to achieve the EU's 2050 climate neutrality target and to halt biodiversity loss. The transition to a circular economy is expected to reduce pressure on natural resources and create sustainable growth and jobs. The whole-life management of the environmental and resource impacts of the production and use of goods erodes the distinction between manufacture and services and offers opportunities for value creation and employment throughout the life cycle.

For countries currently relying upon primary industries this means that an emphasis on increasing output could shift to extracting greater value from existing production levels. Indonesia's current ban on the export of nickel ore to encourage onshore processing is one instance of such strategies and has resulted in significant investment from China, its principal customer (Reuters, 2022). The 'dual circulation' model adopted by China suggests that managing resources and adding value in-country and building domestic consumption in parallel with exports offers a possible resolution. However, it is a solution requiring capacity development for all stages of the STI process.

28.6   The Human Dimension of Collaboration

The disadvantages of marginality within the global technical system and a legacy of externally driven infrastructure can be addressed through access to the information and communication technologies which have facilitated globalisation. While these

---

[5] https://eur-lex.europa.eu/legal-content/EN/TXT/?qid=1583933814386&uri=COM:2020:98:FIN.

technologies require skills and capacities which are scarce, the technologies them-selves can be used to leverage existing resources so that the necessary skills can be developed. The global pandemic prompted a worldwide switch to virtual collabora-tion. However, if the current inequitable situation is not to be reproduced in a new global infrastructure, these changes need to take account of African priorities and requirements.

Where access to affordable and reliable technologies is available, African users have demonstrated a capacity to make effective use of them on their own terms. For example, there is distinctiveness to African usage of mobile telephony (Gough & Grezo, 2005). In this regard the Economist (2005) argues that cell phone technology has been highly effective in triggering grass-roots economic activity across Africa, with rapid and high levels of take-up and widespread access by the turn of the twenty first century. Use of current communication technology builds on the appropriation of earlier generations of technology, such as the development of social capital around indigenous capacity in radio broadcast (Mytton, 2000) and African forms of social use for a technology, the portable radio, a personal artefact in the West, was managed initially as a social resource (Spitulnik, 2000), as were cell phones in Indian villages under the auspices of the Self-Employed Women's Association and Grameenphone. (Little, 2005).

The appropriation of key technologies is essential to any knowledge-based self-development strategy (Okpaku, 2006). The development of mobile Fintech applica-tions shows that in many respects Africa matches China in accessing the potential benefits of mobile telephony, if not in terms of profitability and global presence for indigenous manufactures and providers, in impact on general economic activity.

Currently the STEM (science-technology-engineering-mathematics) subject set central to technical innovation is being extended to STEAM reflecting the need to incorporate contextually and culturally specific knowledge. The 'A' represents arts plus the range of creative and design thinking which contributes to the creation of products and services that are both effective and valued by the target users (Colucci-Gray, 2019). It also provides a framework for the incorporation of indigenous and traditional knowledge into the process, typified by the global profile of Chinese traditional medicine (Xiang et al, 2022) and the incorporation of traditional practices and knowledge into current agricultural practice in Africa (Ogumanan, 2021).

All of this suggests that Africa has the capacity to absorb and adapt technologies to indigenous requirements and to identify those elements of China's development policies that can be deployed in an African context. The 'information challenge' (Grieco et al., 2006) presented by the global economy has already been met by China while Africa is engaging with it with increasing effectiveness for both economic and human development.

In the modern era China has benefited from the circulation of human capital since the Indemnity Scholarships, set up by the United States in the first decade of the twentieth century with money taken from China as compensation for the Boxer Uprising. These skill transfers set a pattern for the return of educated individuals which continued after the Communist revolution in 1949.

At the start of the twentieth century the creation of the Chinese Republic was supported by overseas Chinese Associations and subsequently the Chinese diaspora played a key role in investment prior to the wider acceptance of China as an investment destination. Redding (1996, 1998) analyses the nature of capital accumulation and the characteristics of Chinese family business networks. In the past they had accomplished significant international mobilisation of capital. Indeed, some observers suggested that the transition to more formal and institutional financial arrangements would become an obstacle to further economic development.

More generally financial remittances from international migrants have become recognised as a significant resource for development in less advantaged regions, not least within and between the member states of the European Union and between the coastal and interior provinces of China. Both foreign exchange and household income are enhanced in the country of origin.

Newland (2004), in a review of the role of diasporas in poverty reduction in their countries of origin, argues that policy in support of such networks should take as much account of the social and political activities of these trans-national communities as the straightforward financial dimension of their interaction. For many countries trans-national communities of professionals are a major source of foreign direct investment, market development, technology transfer and more intangible flows of knowledge, new attitudes and cultural influence. This transfer of human capital and capacity has been termed 'social remittance'. Levitt defines social remittances as the "ideas, behaviours, identities and social capital that flow from receiving country to sending country communities" (Levitt, 1998, p.357). This social remittance is reflected by the emergence of various technological, scientific and social networks connecting migrants with each other and with home countries.

There are well established links between many African countries and their overseas population (e.g. Henry & Mohan, 2003) and such linkages are able to mobilise much more than financial capital. The social capital created by movement between cultures and locations can provide a critical contribution to development and resource mobilisation. Miller and Slater (2000) describe an early example of how information and communication technologies can assist the maintenance of connection to homeland and identity in diasporic communities. The use of the internet as a social space by Trinidadians away from their home is mirrored in the use of the internet in support of both cultural and business activities in the African diaspora and enforced physical separation during the pandemic has accelerated the development of such interactions.

Across Africa and beyond many independence leaders benefited from the experience of movement between their country and the imperial centres. Unsurprisingly, social linkages and return migration are a significant resource for development in both Africa and China, as 'brain drain' becomes 'brain circulation'. Saxenian (2002) suggests that Asian engineers working in 'Silicon Valley' retained and cultivated links with engineers and businesses back home through various social networks and aided development, particularly that of the Indian software industry by providing knowledge and market access. Kale et al., (2005) show the important role of overseas Indian scientists working in the development of R&D capabilities in Indian pharmaceutical firms. These intellectual capital flows contributed to key modernisations and technical development in science and technology in the 1950s and continue with specific encouragement from the present government.

However, for both China and Africa the most numerically significant flows are internal ones brought about by uneven development, resource shortages and, in Africa's case, related conflict. In the face of the need to develop human capital the ability of African nations to attract return migration and foster overseas linkages becomes critical. The attractiveness to emigrants of return, either on a permanent, temporary or even virtual basis becomes an important factor in harvesting social capital flows and bolstering capacity for engagement in STI collaboration.

Government encouragement of both overseas engagement and return migration can be seen as an aspect of 'soft power'. Nye (1990) defines soft power as the ability of a state or other political body, to indirectly influence the behaviour or interests of other political bodies through cultural or ideological means. Soft power distinguishes the subtle effects of culture, values and ideas from more direct means, such as military or economic incentives. Thompson (2004) argues that China has incorporated this approach in their economic strategy in. Africa too has an opportunity to utilise soft power in dealings with both its diaspora and its trading partners.

28.7   Tracking Knowledge and Capacity

The first step in knowledge management is to find out what you know (Little & Ray, 2005). While scientific and technical publications are now accessible globally through either open access or subscription journals, easily identifying the human capacity behind these publications is a different matter. The Brazilian government pioneered a national system for identifying researchers, their topics and their collaborators. The Lattes Platform[6] is an information system established in 1999 and maintained by the National Council for Scientific and Technological Development to manage information on science, technology, and innovation related to individual researchers and institutions working throughout Brazil (Mena-Chalco & Cesar, 2009).

Researchers and institutions are required to maintain their records up to date. While the platform can be used as an information repository for individual researchers it can also be used to conduct performance evaluations at the organization level and serve as a means of identifying and mapping the nation's resources in science and technology. It also offers the capability to track individuals from the level of high school intern to Nobel prize-winning scientist, the movement of human capital and the interaction and collaboration between individuals and research groups. In contrast to the cumbersome assessment systems favoured by many other countries, the system utilises data routinely reported and demonstrates the potential of so-called 'Big Data'.

---

[6] https://lattes.cnpq.br/.

At the other end of the scale is the development of Community Asset Mapping (CAM) which seeks to identify existing resources within communities and to find ways of using them to leverage limited government resources as opposed to simply measuring deprivation and absence. This approach, first formulated in the USA by McKnight and Kretzmann (1993), has been applied in both Western Europe, particularly in rural and urban areas of relative deprivation, and in sub-Saharan Africa where local authorities have minimal formal resources.

While these two frameworks address radically different sections of society, they share the objective of identifying resources and potential synergies. They both echo the role of knowledge mapping in knowledge management (Little & Ray, 2005) and provide the means to identify both space and resources for innovation and entrepreneurship. The potential benefits of combining these frameworks can already be seen in one significant STI collaboration.

In South Africa, the African-European Radio Astronomy Platform (AERAP),[7] a stakeholder forum convened to define priorities for radio astronomy cooperation between Africa and Europe, is demonstrating that it is possible to combine these two perspectives. While operating as a high-level North–South scientific collaboration it has also established an educational outreach to engage children from an early age in astronomy in order to cultivate an interest and appreciation of STEM disciplines. This is seen as an integral part of their science capacity building remit.

28.8 A Framework for STI Collaboration

China's emergence as a key contributor to global capacity in manufacturing, technology and science and the consequent disruption of established pathways of diffusion of innovation has opened new possibilities for both collaboration and adaptation of technologies. From an African perspective the need is to identify areas of STI collaboration which will deliver the most assistance to continental development.

The European Union has developed medium to longer term STI programmes through successive 'Horizons' frameworks and is now promoting a range of mission-based policies (Mazzucato, 2021). These are focused on discovery and innovation within the framework of the United Nations Sustainable Development Goals (SDGs) which in turn provide governments with the means of identifying and agreeing policy priorities, both domestically and internationally.

STI cooperation offers the prospect of developing an indigenous pathway of discovery, invention and innovation directed towards African needs and priorities through the STEM to STEAM reformulation. In turn this offers fresh insights to outside collaborators, just as the reworking of Western models by Japan and China for their requirements has delivered wider benefits. However, as argued above, successful technology transfer involves both technical learning and social learning. The latter is the key to absorptive capacity of new technologies. Appropriate and sustainable

---

[7] https://aerap.org/sitemap.php

skill and capacity development at all levels is essential and the AERAP initiatives demonstrate that this is achievable.

While the widespread acceptance of the SDG's provides a framework for the identification of areas critical to sustainable development, the human dimension of knowledge transfer and capacity building requires appropriate spaces whether physical, virtual or institutional. Vygotsky's "zone of proximal development" (see Cole, 1985) offers a paradigm of how adjustment to a new consensual knowledge sharing culture might be assisted. Actors within organisations and across collaborations may be supported mutually by a scaffolding provided by institutionalised practices and structures, increasingly embedded in information systems, while the underlying values and assumptions are assimilated.

As noted earlier, science is a globalised activity and the languages of technology and engineering are international. However, as innovations are moved to implementation specific locations in context, regional and national variation becomes relevant.

O'Hara-Devereaux and Johansen (1994) argue that differences between work cultures, both professional and corporate, and the primary cultures in which collaborators are embedded can be bridged in a "third space". For them the synergy between these levels is a potential resource, but the tendency towards convergence determined by a dominant culture is seen as an obstacle to cross-cultural working. Culture needs to be de-composed into issues related to the historical, geographical and institutional setting in which organisation and individual must operate. The business recipes and frameworks grounded in these differences offer a view of "culture" of direct and practical value to actors (see for example Marceau, 1992).

There are a number of clichés in wide circulation based on Chinese sayings. A favourite in management and business circles is that 'the ideogram for threat can also be read as opportunity'. It is also widely understood that to be 'living in interesting times' is not desirable and that' the longest journey starts with a single step'. All three of these are relevant to the situation in which we are seeking to extend science technology and innovation collaboration between Africa and China. China has enjoyed some advantages over Africa in terms of governance both sharing a legacy of external intervention and patterns of development influenced by external interests.

China has demonstrated an alternative development pathway, most recently addressing uneven development with a dual circulation model for internal development and external trade while Africa has demonstrated capacities to adapt and innovate technologies and processes for local priorities.

Growing economic, technical and scientific collaboration between Africa and China offers the opportunity to co-develop the skill base to move from discovery to invention, innovation and implementation to the benefit of all.

This is the context in which African actors should approach effective engagement with the opportunities offered by STI collaboration with China.

References

Cohen, W. M., & Levinthal, D. (1990). Absorptive capacity: A new perspective on learning and innovation. *Administrative Science Quarterly, 35*(1), 128–152.

Cole, M. (1985). The zone of proximal development: Where culture and cognition create each other. In J. V. Wertsch (Ed.), *Culture communication and cognition: Vygotskian perspectives.* Cambridge University Press.

Colucci-Gray, L., Burnard, P., Gray, D., & Cooke, C. (2019). *A critical review of STEAM (science, technology, engineering, arts, and mathematics).* Oxford University Press.

Delamaide, D. (1995). *The new superregions of Europe.* Plume/Penguin.

Dicken, P. (1998). *Global shift: Transforming the world economy* (3rd ed.). Paul Chapman.

Flyvbjerg, B., Bruzelius, N., & Rothengatter, W. (2003). *Megaprojects and risk: An anatomy of ambition.* Cambridge University Press.

Fortune. (2019). 'How Delivery Drones Are Saving Lives in Rwanda' http://fortune.com/2019/01/07/delivery-drones-rwanda/ accessed 23/0519

Gathanju, D. (2006). The Nile Basin: water water.... *The World Today, 62*(8–9), 30–31.

Gore, L. L. P. (2006). How Green GDP become fashionable in China. *EAI Background Brief no.273.* Singapore: East Asian Institute, National University of Singapore.

Gough, N., & Grezo, C. (2005). Introduction. In: *Africa: the impact of mobile phones.* Vodafone Policy Paper Series No. 2, March 2005.

Grieco, M., Colle, R., & Ndulo, M. (Eds.). (2006). *Meeting the information challenge: The experience of Africa.* Cambridge Scholars Press.

Headrick, D. R. (1981). *The tools of empire: Technology and European imperialism in the nineteenth century.* Oxford University Press.

Henry, L., & Mohan, G. (2003). Making homes: The Ghanaian diaspora, institutions and development. *Journal of International Development, 15*(5), 611–622.

Hirsch, S. (1967). *Location of industry and industrial competitiveness.* Clarendon.

Hirschmann, S. (2022). Upscaling the benefits of pull-technology for sustainable agricultural intensification in East Africa. Presented to online workshop: *Excellence Lab: Boosting Research and Innovation Cooperation between Africa and Bavaria on Environmental Protection and Circular Economy*, Bayerische Forschungsallianz (Bavarian Research Alliance), July 2022.

Humphreys, M., Sach, J. D., & Stiglitz, J. E. (2007). *Escaping the resource curse.* Columbia University Press.

Kale, D., Wield, D., Little, S., Chataway, J., & Quintas, P. (2005). Diffusion of knowledge through migration of scientific labour in India. *Paper presented at Colloquium on Researching Innovative Themes in Skilled Mobility*, Centre for the Study of Law and Policy in Europe (CSLPE), University of Leeds.

Kaplinsky, R. (1994). *Easternization: The spread of Japanese management techniques to developing countries.* Ilford, Frank Cass.

Kaplinsky, R., & Morris, M. (2007). Do Asian drivers undermine export-oriented industrialization in SSA? *World Development, 36*(2), 254–273.

Kirlidog, M. (1997). *Executive computing in a developing country: A case study evaluation of Turkish experience.* PhD Thesis, University of Wollongong NSW.

Kolodko, G. W. (2020). *China and the future of globalization: The political economy of China's rise.* I.B. Taurus.

Lee, B., Preston, F., Kooroshi, J., Bailey, R., & Lahn, G. (2012). *Resource futures.* Chatham House.

Levison, M. (2006). *The Box: How the shipping container made the world smaller and the world economy bigger.* Princeton University Press.

Levitt, P. (1998). Social remittances: Migration- driven, local level forms of cultural diffusion. *International Migration Review, 32*(4), 927–957.

Li, A. (2016). Technology transfer In China–Africa relation: Myth or reality. *Transnational Corporations Review, 8*(3), 183–195.

Little, S. E. (2005). Plugging the gap: information systems research and the 'unwired' world. UK academy for information systems. In *10th Annual Conference 'Information Systems Unplugged: developing relevant research*. University of Northumbria, March 2005.

Little, S. (2007). Models of development: Finding relevance for Africa in China's experience of development. In M. Kitissou (Ed.), *Africa in China's global strategy* (pp. 182–197). Adonis and Abbey.

Little, S. E., & Ray, T. E. (Eds.). (2005). *Managing knowledge: An essential reader* (2nd ed.). Sage Publications.

Marceau, J. (Ed.). (1992). *Re-working the world: Organisations, technologies and cultures in comparative perspective*. de Gruyter.

Mawere, M., & van Stam, G. (2019). eLearning in an African place: How 'alien' eLearning models are failing many. In *15th International Conference on Social Implications of Computers in Developing Countries (IFIP 2019)*, 1–3 May 2019, Dar es Salaam, Tanzania.

Mazibuka-Makena, Z., & Kraemer-Mbula, M. (Eds.). (2021). *Leap 4.0: African perspectives on the fourth industrial revolution*. Mapungubwe Institute for Strategic Reflections.

Mazzucato, M. (2021). *Mission economy: A moonshot guide to changing capitalism*. Penguin.

McKnight, J. L. & Kretzmann, J. P. (1993). *Building Communities from the Inside Out: A Path Toward Finding and Mobilizing A Community's Assets* ACTA Publishers

Mena-Chalco, J. P., & Cesar Junior, R. M. (2009). ScriptLattes: An open-source knowledge extraction system from the Lattes platform. *Journal of the Brazilian Computer Society, 15*(4), 31–39. https://doi.org/10.1007/BF03194511

Mohan, G. (2008). China in Africa: A review essay. *Review of African Political Economy, 35*(1), 155–173.

Miller, D., & Slater, D. (2000). *The internet: An ethnographic approach*. Berg.

Mytton, G. (2000). From Saucepan to Dish: Radio and TV in Africa. In R. Fardon & G. Furniss (Eds.), *African broadcast cultures: Radio in transition*. Oxford: James Currey.

Newland, K. (2004). *Beyond remittances: The role of Diaspora in poverty reduction in their countries of origin*. Migration Policy Unit.

Nye, J. S. Jr. (1990). 'Soft power'. *Foreign policy*, 80 Autumn. (pp. 153–171).

Ogumanan, C. (2021). From science, technology, and innovation to 4th Industrial Revolution strategies in Africa: the case for indigenous knowledge systems. In Z. Mazibuko-Makena & E. Kraemer-Mbula (Eds.), *Leap 4.0 African perspectives on the fourth industrial revolution*. Mapungubwe Institute for Strategic Reflections.

O'Hara-Devereaux, M., & Johansen, R. (1994). *Globalwork: Bridging distance, culture and time*. Jossey-Bass.

Ohmae, K. (1995). *The End of the Nation State: The rise of regional economics*.

Okpaku, J., Jr. (2006). Towards a knowledge-, science and technology-based African self-development strategy. In M. Grieco, R. Colle, & M. Ndulo (Eds.), *Meeting the information challenge: The experience of Africa* (pp. 15–24). Cambridge Scholars Press.

Paterson, S. (2021). *For, by and from the party; defining the parameters of dual circulation*. Hinrich Foundation.

Penrose, E. T. (1959). *The theory of the growth of the firm*. Wiley.

Perez, C. (1985). Microelectronics, Long Waves and World Structural change: New Perspectives for Developing Countries *World Development, 13*(3), 441–63

Perez, C. (2002). *Technological revolutions and financial capital: The dynamics of bubbles and golden ages*. Edward Elgar.

Redding, S. G. (1996). Weak organizations and strong linkages: Managerial linkages and Chinese family business networks. In G. Hamilton (Ed.), *Asian business networks*. deGruyter.

Redding, S. G. (1998). The changing business scene in Pacific Asia. In F. McDonald & R. Thorpe (Eds.), *Organizational strategy and technical adaptation to global change*. Macmillan.

Reuters. (2022). Indonesia's Q2 FDI jumps nearly 40% y/y, biggest rise in a decade. *Reuters*, Jakarta, July 20. https://www.reuters.com/markets/rates-bonds/indonesias-q2-fdi-jumps-nearly-40-yy-biggest-rise-decade-2022-07-20 (Accessed: 18 May 2023).

Rodney, W. (1976). *How Europe underdeveloped Africa.* Howard University Press.

Saxenian, N. (2002). Trans-national communities and the evolution of global production networks: The case of Taiwan China and India. *Industry and Innovation, 9*(3), 183–202.

Shaxson, N. (2007). *Poisoned Wells: The Dirty Politics of African Oil.* Palgrave Macmillan.

Spitulnik, D. (2000). Documenting radio culture as lived experience: Reception studies and the mobile machine in Zambia. In R. Fardon & G. Furniss (Eds.), *African broadcast cultures: Radio in transition.* James Currey.

Sproull, L., & Kiesler, S. (1991). *Connections: New ways of working in the networked organization.* MIT.

Tao, Y. (2011). *The application of E-business in residential development in the People's Republic of China.* PhD Thesis, Open University, UK.

The Economist. (2005). Mobile phones and development. *The Economist,* July 7.

Thompson, D. (2004). Economic growth and soft power: China's Africa strategy. *China Brief, 4*(24), 07.

U.S. International Trade Administration. (2022). *African Continental Free Trade Area.* 3 January. https://www.trade.gov/market-intelligence/african-continental-free-trade-area. Accessed: 18 May 2023.

World Bank. (2022). *The African Continental free trade area.* Washington DC: World Bank. https://www.worldbank.org/en/topic/trade/publication/the-african-continental-free-trade-area. Accessed: 18 May 2023.

Xiang, L., Chen, Z., Wei, S., & Zhou, H. (2022). Global trade pattern of traditional Chinese medicines and China's trade position. *Front Public Health.* https://doi.org/10.3389/fpubh.2022.865887

Zheng, Y. -N. (2006). De-facto federalism and dynamics of central-local relations in China. *Discussion Paper 8,* Nottingham: China Policy Institute.

Zheng, Y. -N., & Chen, M. -J. (2006). China moves to enhance corporate social responsibility in multinational companies. *Briefing Series Issue 11,* Nottingham: China Policy Institute.

Zheng, M., & Williamson, P. (2007). *Dragons at your door: How Chinese cost innovation is disrupting global competition.* Harvard Business School Publishing.

Chapter 29
The Role of China and Western Countries in the Economic Globalization of Africa: Who is Behind the Recent Economic Renaissance of Africa?

Addis Kassahun Mulat and Rahel Belete Balkew

29.1 Introduction

Economic globalisation denotes the widening and intensification of international linkages and interactions in trade, investments and economic policy orientation in the World (Adejumobi, 2002; World Bank, 1996; Rosenau, 1997). The degree to which countries have moved toward globalization and economic integration varies widely. For most Sub-Saharan African in the 1980s and 1990s, following the World Bank and IMF's Structural Adjustment Program (SAP), a series of reform programs that are aimed at transforming the economy from a quasi-communist command structure to a market economy have been initiated and implemented.

For the last two decades, China's interest in Africa has grown exponentially (Geda & Meskle, 2008). With a total trade of USD 200 billion in 2019, China is Africa's biggest bilateral trade partner. This has been perceived by the Western nations as a severe threat, leading to their renewed interest in Africa. Consequently, these major trading partners—Great Britain, France, and U.S.A—see themselves as competitors for the African markets (Sautman & Hairong, 2007). At the same time, as the African countries become the attractive destination for foreign investment, the western countries, especially, the United States and some members of EU have increased economic relations with the Sub-Saharan African countries since 2000 through Africa Growth and Opportunity Acts (AGOA), Everything but Arms (EBA) and other initiatives.

A. K. Mulat (✉)
Kilimanjaro Global Consulting Training and Innovation Hub (KIH), Addis Ababa, Ethiopia
e-mail: addiskassahun@gmail.com

R. B. Balkew
Tshwane University of Technology, Pretoria, South Africa
e-mail: etgrace@gmail.com

© The Author(s) 2024   541
M. Muchie et al. (eds.), *China-Africa Science, Technology and Innovation Collaboration*, https://doi.org/10.1007/978-981-97-4576-0_29

Thus, the main focus of this study is whether the Chinese and Western OECD countries' economic relations and collaborations among African countries contributed to their economic growth and integration into the global market.

29.2 Literature Review

There are diverse studies conducted in China-Africa relations versus western countries and roles in the economic growth in Africa. Dijk (2009) argued that there is a strong Chinese government support to Chinese enterprises in Africa. China trades with almost all 53 countries in Africa. However, that China's six biggest export destinations absorb over half of its total exports to the continent. For other researchers like Brautigam (2010), China was on track to become the African continent's largest trading partner, outpacing Britain and the United States. Others like Fourie (2012); Stein and Uddhammer (2023) examine the influence of China's development on developing country elites where China is viewed as only one source of potential 'lessons', and its elites often embed its experiences within a wider East Asian development trajectory.

It seems that both the West and China are actively competing for Africa's natural resources, for markets and for political influence, using several strategies and tactics to ensure and expand their interests. Anton (2012) analyses the main actions and reactions taken by the U.S. in Africa, regarding China's presence on the continent of Africa and concludes that the U.S. is using diplomatic, military and economic instruments to counter China's influence on the continent. The U.S. is employing a "smart power" policy characterized by cautiousness. Some researchers argue that AGOA is one of these policy responses.

29.3 Methodology

The researchers have employed a panel data design for the period covering 2000–2011 and focused on analysing the influence of Chinese and selected OECD countries economic globalization variables, such as Foreign Direct Investment (FDI), Foreign AID (Official Developmental Assistance), Africa's Openness (Export plus Import as a ratio of GDP), its domestic investment and human capital on the economic growth (in terms of real GDP growth) selected African countries that have economic relations with China and the selected traditional western OECD partner countries (USA, United Kingdom, France and Germany) and based on their historical ties.

Table 29.1 provides the list of selected oil & minerals versus non-oil and minerals exporting African countries.

The study has been conducted through International published and compiled data from UNCTAD, World Bank, Centre for Global Development/CGD Media based database, COMTRADE and African Development Indicators (ADI) and World

Table 29.1 List of Selected Oil & Minerals versus Non-Oil & Minerals Exporting African Countries

Oil/Petroleum and Mineral exporting African countries include:	Non oil and mineral (agricultural commodities) exporting countries include:
Angola, Botswana, Cameron, Egypt, Gabon, Libya, Niger, Nigeria, Senegal, Sierra Leon, South Africa, Sudan, and Zambia	Benin, Burkina Faso, Burundi, Cape Verde, Ethiopia, Ghana, Kenya, Mali, Rwanda, Seychelles, Swaziland, Tanzania, Togo, Uganda

Source: ADI 2013

Development Indicators (2012) of the World Bank. Given the objective of estimating a GDP growth function, the researchers have used econometrics model to look at the different impact channels of the variables. In panel data analysis, the transfer of testing for unit root and stationarity from univariate time series to large panel data contributed to do the job effectively and made a significant increase to the power of those tests (Hadri and Kurozumi, 2008). The use of panel data also provides a means of resolving of reducing the magnitude of a key econometric problem that often arises in empirical studies. By utilizing information on both inter temporal dynamics and the individuality of the entities being investigated, one is better able to control in more natural way missing or unobserved variables.

29.3.1   The Econometric Growth Model

This study adopted Romer (1989), Sachs and Warner (1997) among others, to identify the bare bone drivers of economic growth in African countries, but the same has been augmented for foreign official financial flows: AID and FDI, to capture the effect of the later on growth dynamics of African economies. For practical estimation of the models panel co-integration approach is used to estimate two separate growth equations for each category African countries. For both categories of countries, the estimated growth equation will take the following with RGDP as dependent variables with arguments as its explanatory variables:

Where, RGDP denotes Real GDP growth (RGDP), Openness (EXP + IMP as a ratio of GDP). The equation is further augmented with the region involved with and/ or without to make comparison and to identify effects on RGDP growth.

29.3.2   Unit Root Test Results

The order of integration of the variables in this study is determined using unit root tests. Phillips–Perron Fisher Panel Unit Root Test is applied, and the result of the test is given in Appendix A. The unit root tests conducted using Phillips–Perron

Fisher Panel Unit Root Test revealed that all variables have unit root in their level, thus had to be differenced to achieve stationarity. Before practical estimation of the model, test for unit root is carried out and all variables are found to be integrated of order one, I (1), hence co-integration analysis is justified to be undertaken. The variables have also passed a test for the co-integration, and the Error Correction Model (ECM) of relationship is estimated to see the short run and long run determinants of growth. All the variables used in estimation are in their natural log transformation to tackle for the problem of unforeseen heteroscedasticity, autocorrelation and non-normality of errors. The presence of long run relationship co-integration is tested for the two categories of countries growth models in this study using Kao Residual Co integration test. As can be seen from the reported Tables 29.2 and 29.3, the presence of co-integration is confirmed, as the null of no co-integration is rejected following a p-value of zero, in favour of the alternative of there is co-integration between the variables of the study.

Table 29.2 Result of co-integration test for oil and minerals exporter african countries growth model

Kao residual co-integration test		
Series: LAID_CHINA LAID_OECD LFDI__CHINA LFDI__OECD LINV LOPNESS LRGDP		
Observation periods: 2000 2011		
Included observations: 144		
Null Hypothesis: No co-integration		
Trend assumption: No deterministic trend		
Newey-West automatic bandwidth selection and Bartlett kernel		
	t-statistic	Prob.
Augmented Dickey–Fuller (ADF)	− 2.981899	0.0014

Table 29.3 Result of co-integration test for non oil and minerals exporting countries category growth model

Kao residual co-integration test		
Series: LRGDP LAID_CHINA LAID_OECD LINV LOPNESS LFDI_CHINA LFDI_OECD		
Observation periods: 2000- 2011		
Included observations: 210		
Null hypothesis: no co-integration		
Trend assumption: no deterministic trend		
Newey-West automatic bandwidth selection and Bartlett kernel		
	t-statistic	Prob.
Augmented Dickey–Fuller (ADF)	− 4.280290	0.0000

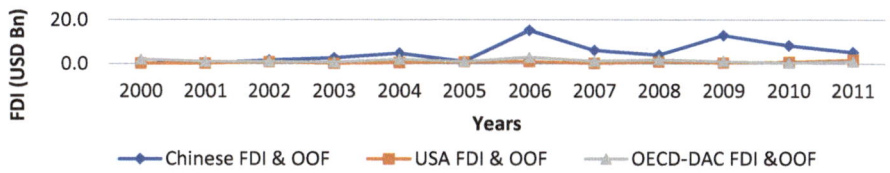

Fig. 29.1 Chinese, OECD and US Foreign Direct Investment (FDI) inflows to Africa from 2000 to 2011

29.4   Presentation, Analysis and Interpretation ff Data

29.4.1   Foreign Direct Investment (FDI)

The revival of FDI flows to Africa both from traditional and non-traditional partners in the last decades could be taken as an indicator of the improving investment climate, the less riskiness of investing in Africa as well as the high return to investment in the continent. However, the annual inflow FDI to Africa represents a mere 4% of the global flows and about 10% of flows to the developing countries in 2010–2011 (UNCTAD, 2012). Figure 29.1 shows that Chinese, OECD and US Foreign Direct Investment (FDI) in Africa from 2000 to 2011 in billion USD. This is the author calculation from Centre for Global Development (CGD) Database 2000–2011 which was available data during the time of the research at the Centre. DAC represents the Development Assistance Committee of OECD which manages most official and Other Official Financial flows.

Figure 29.1 shows that the Chinese FDI and Other Official Flows to Africa are by far more significant than both U.S. and other OECD countries originated FDIs. There is an increasing trend in the Chinese FDI to Africa since 2000 while frequent variations and up downs in both US and OECD inflows. Most researchers on China-Africa noted the resources seeking motive and market seeking behaviour from China's FDI in Africa because China's investments always go to Africa countries that have natural resources like minerals, cocoa beans, rubber, cashew nuts, hide, skin and oil (Cheng et al., 2010; Rotberg 2008). Others researchers like Fourie (2012) examine the influence of China's development on developing country elites where China is viewed as only one source of potential 'lessons', and its elites often embed its experiences within a wider East Asian development trajectory.

29.4.2   Foreign Aid

The graph below shows an increasing and stable flow of foreign aid (Official Development Assistance) from the US to Africa than from China in general (Fig. 29.2).

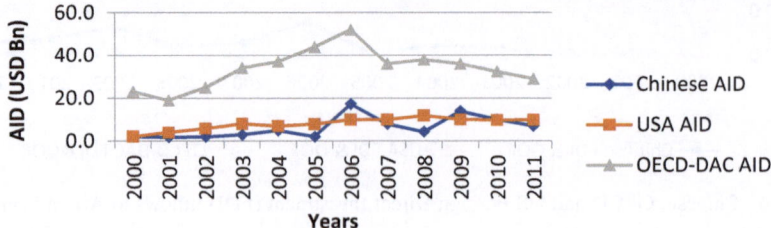

Fig. 29.2 Chinese, OECD and US Foreign Aid Inflows to Africa 2000–2011. Source: The Author calculation from Centre for Global Development (CGD) Database 2000–2011

The Chinese foreign aid witnesses more variations than the US and OECD Aid. The graph also indicates that Africa received significant Official Development ment Assistance (foreign aid) from OECD followed by the US than the Chinese counterpart.

29.4.3   Openness to Trade

Openness as the export and import as ration of GDP is one of the key indicators of integration of economies with the global market. More open economies believed to benefit from global economic integration. The study utilizes merchandise trade because it constitutes the lion's share of Africa's trade with the world.

Figure 29.3 above shows there is a positive trajectory of exports and imports in Africa and there is an indication of openness to trade in Africa with the outside markets. Greater trade openness arguably promotes economic growth in Africa through increasing competitiveness and providing access to international markets, as well as by enabling importation of raw materials and capital goods. However, the results vary across the country groups. The strongest long-run effect of trade openness on output is found in investment-driven economies, followed by the factor-driven category (Geda & Yimer, 2015; Yimer, 2022a, 2022b).

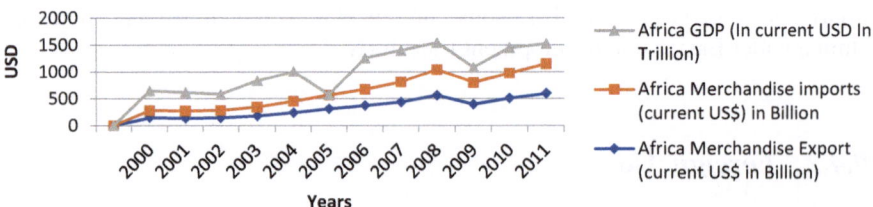

Fig. 29.3 Africa's merchandise export, import and nominal GDP

As can be seen from graph below that most of Africa's exports go to OECD category but the share of Africa's exports going to the OECD is on continuous decline.

Figure 29.4 further shows that USA has large significant contributions in Africa's exports compared to the other African traditional trade partners from the OECD countries. Though combined together the traditional partners of Africa are still more important than that of China as a main destination for Africa's exports, they are continuously being displaced by the emerging China, as an important Africa's export destination.

On the import side (Fig. 29.5), the share of Africa's imports from China is growing very rapidly. Like the case of exports, this also shows that, China is emerging to the international trade market with a growing importance as Africa's important source of imports. Figure below shows the importance of China and selected OECD members to Africa's imports.

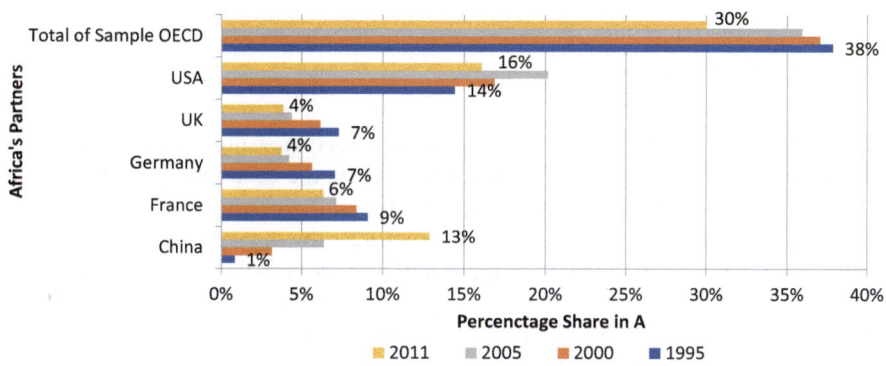

Fig. 29.4 Importance of Chinese and Selected OECD Members on Africa Export

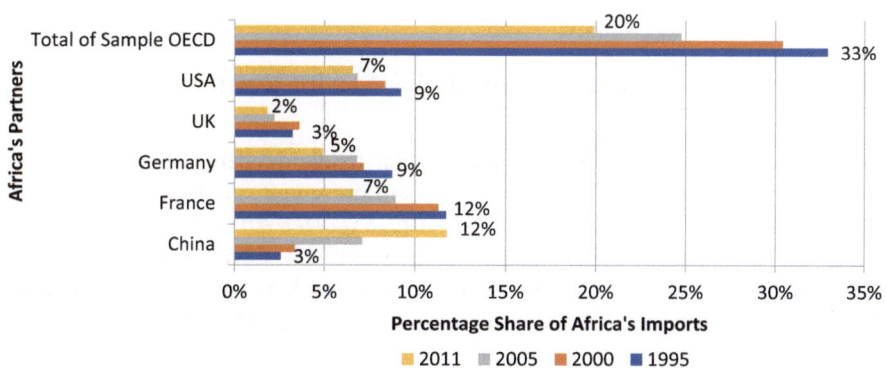

Fig. 29.5 Importance of China and Selected OECD members to Africa's Imports. Source: Author's computation using UNCTAD 2012 data

29.4.4 Domestic Investment and Human Capital

Africa is the world's most youthful continent, and its people are its greatest source of wealth. Human capital—the education, skills, and health of people—will play a pivotal role in the transformation of African economies. Developing human capital in Africa requires a massive and coordinated effort to strengthen the quantity, efficiency, and impact of investments in people (World Bank, 2019).[1] As noted by Kaplinsky and Farooki (2009), Africa's presence in the global economy closely reflects its economic weaknesses. One of the key underlying factors affecting human capital in Africa is the low level of empowerment of women and girls. Public spending in sectors that underpin the formation of human capital is lower than in all other regions of the world except for South Asia and, at its current rate, will be insufficient for countries in Africa to reach the targets set by the Sustainable Development Goals for 2030 (World Bank, 2019).

29.4.5 Economic Growth

Generally, the real GDP of both African economies is increasing. One can see further the implication and significance of this difference on the economic growth measure in the two African economies of non-oil & mineral exporting and oil & mineral exporting African countries from the econometrics growth model results.

Figure 29.6 shows Africa's total real GDP and its growth for the period 2000–2011. It shows the Real GDP and Percentage of Real GDP Growth from the author's computation using WDI 2013 data. The overall real GDP growth rate of African economies together increased from 3% in 2002 to 6% in 2007 and declined to 3% in 2009. The total real GDP of the continent has registered a steady increase for the period considered here.

29.5 Presentations of Econometrics Model Results

As a point of direction to smooth comparison and to avoid repetition, both the results of the two categories of African economies are presented side by side (*the results in the parenthesis are for the Oil & Mineral Exporter African countries*) in the analysis section. Since the short term econometrics results do not show the long term economic growth effects, the study mainly presents the long term economic analysis of the econometrics results in the remaining sections due to page limitation.

---

[1] The State of Human capital in Africa: Powering Africa's Potential Through Its People by the World Bank, 2019.

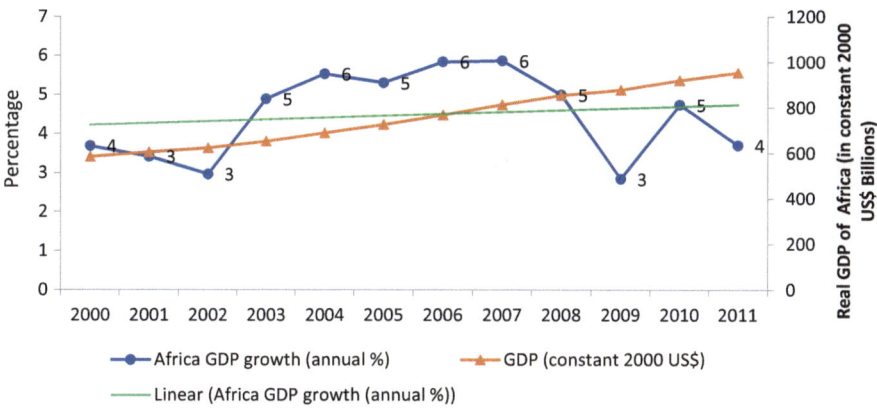

Fig. 29.6 Real GDP in constant 2000 USD and percentage of real GDP growth

29.5.1   Foreign Direct Investment

One can see how the Chinese and OECD economic globalization indices are statistically measured in terms of FDI, the results for Non-oil & mineral exporting versus Oil & Mineral exporting African countries as has been presented with the respective coefficient and p-values. From the Table 29.4 below which is the full model that includes both China and OECD, in the long term, China FDI has a positive coefficient of 0.09 (0.12) with 0.00 (0.00) significant p-value for Non-oil & mineral exporting versus Oil & mineral categories (in parenthesis) while OECD FDI has a positive 0.06 (0.09) coefficient with 0.44 (0.24) p-values for the two categories indicating that China's FDI has a more significant and stronger effect to African GDP than the OECD FDI for both categories of countries. The Interpretation of the results is that (Table 29.4):

(i).   A 1 unit increases in the Chinese FDI inflows will have on average 0.09 (0.12) unit increase in Real GDP growth in Non-Oil and Minerals versus the Oil &Mineral Exporters African countries, respectively.

(ii). Both Chinese and OECD FDIs are significant, but Chinese FDI is more significant in the long term than OECD.

29.5.2   Foreign Aid

It has been seen that foreign aid is the external official development assistance which used to fill gaps in the economic activities of the host countries. The results for non-oil & mineral exporting versus oil & mineral exporting countries can be presented in the same manner and show that in the long term, China Aid has a positive coefficient of 0.06 (0.04) with 0.00 (0.02) significant p-value while OECD aid has a positive

Table 29.4 EVIEWS 7.1 results outputs (full model with CHINA and OECD presence)

Non oil & mineral exporters African Countries (full model)				Oil & mineral exporters African Countries (full model)		
Dependent variable: D(LRGDP)				Dependent variable: D(LRGDP)		
Method: Panel EGLS (Cross-section weights)				Method: panel EGLS (Cross-section weights)		
Sample (adjusted): 2001 2011				Sample (adjusted): 2001 2011		
Cross-sections included: 14				Cross-sections included: 11		
Total panel (unbalanced) observations: 142				Total panel (unbalanced) observations: 115		
Variable	Coefficient	t-statistic	Prob.	Coefficient	t-statistic	Prob.
*Long-term impact analysis**						
Real GDP (− 1)	− 0.26	− 2.34	0.02	− 0.29	− 2.63	0.01
OPNESS (− 1)	0.02	0.2	0.84	0.04	0.35	0.73
DOMESTIC INV (− 1)	0.16	4.41	0.00	0.14	1.99	0.05
AID_CHINA (− 1)	0.06	2.89	0.00	0.04	2.28	0.02
AID_OECD (− 1)	0.5	2.57	0.01	0.60	3.32	0.00
FDI_CHINA (− 1)	0.09	2.84	0.00	0.12	3.76	0.00
FDI_OECD (− 1)	0.06	0.77	0.44	0.09	1.19	0.24
LABOR FORCE (− 1)	0.21	0.79	0.43	0.83	2.14	0.04
@TREND	0.04	2.33	0.02	0.06	5.37	0.00
	Effects specification			Effects specification		
	Weighted statistics			Weighted statistics		
R-squared	0.72	Mean dep. var.	0.09	0.73	Mean dep. var.	0.11

NB*: As explained previous section under the Econometrics Model approach, test for unit root is carried out and all variables are found to be integrated of order one, I (1) and no further differencing is necessary for order two I (2) or I (− 2), hence co-integration analysis is and all the variables have also passed a test for the co-integration, and the Error Correction Model (ECM) of relationship is estimated to see the short-run and long-run determinants of growth

0.50 (0.60) coefficient with 0.01 (0.00) *p*-value. The results clearly indicate that Chinese Aid is significant for both categories but is more significant for the non-oil while OECD-aid is also significant for both but is more significant the Oil & Mineral exporter's category. When we compare the Chinese and OECD aid we can see that OECD aid has more effect than Chinese aid (it has higher coefficients with strong significant *p*-value).

29.5.3 Openness (Export Plus Import as a Ratio of GDP)

The long term result of openness variable shows that it has a positive coefficient of 0.02 (0.04) with a probability 0.84 (0.73) indicating in-significant long term effect on Real GDP in Africa. Generally, the empirical study findings on openness and economic growth have mixed results. Some of the studies that investigate the robustness of the relationship between trade openness and long-run economic growth found no evidence that trade openness is directly, robustly and significantly correlated with economic growth in the long-run (Eris & Ulasan, 2013). Others (Baliamoune, 2002, 2008); Hodey et al., 2015) found that there is evidence for positive and significant relationship between openness and economic growth in high-income countries where there is export diversification stating that greater openness to trade may enhance growth in countries with relatively high income implying that threshold incomes effects may be crucial to the effectiveness of openness but tends to depress it in lower-income countries.

29.5.4 Domestic Investment

In the long term, domestic investment has a positive coefficient of 0.16 (0.14) with a high significant *p*-value of 0.00 (0.05). The finding shows that domestic investment has significant effect on Real GDP in both categories but has more significant effect on Real GDP growth of the Non-oil & mineral (Primary Agricultural Goods) exporting African countries than the Oil & mineral exporters.

29.5.5 Human Capital (as Measured by Labour Force)

With respect human capital indices, the long term measure indicates that labour force has a positive coefficient of 0.21 (0.83) with a probability 0.43 (0.04) indicating labour force has more significant effect on Real GDP for Oil & Mineral exporting countries than the Non-Oil & Minerals exporting category. The results might indicate a better human capital in the oil & mineral exporting economies than their non oil & mineral counterpart.

29.5.6   Relative Significance of Chinese and OECD Economic Globalization Indices on Real GDP Growth of Africa (Partial Models)

(i). There is a significant difference between Chinese and OECD economic glob-
      alization indices of FDI indicating that China-FDI is more significant than
      OECD-FDI (it is insignificant for both categories of African economies). The
      effect is stronger for Oil & Mineral exporters. That is, a 1 unit increase in
      Chinese FDI has 0.09 (0.12) unit increase in RGDP while OECD FDI has 0.06
      (0.09) unit increase given all other variables held constant.

(ii). There is a significant difference between Chinese and OECD Aid in that OCED
      aid has more significant effect to African GDP than the Chinese aid for both
      categories. That is, a unit increase in the Chinese aid has only a 0.05 (0.03)
      increase in RGDP while OECD aid has a 0.50 (0.60) unit increase with signifi-
      cant p-value. The result also shows that Chinese aid is more significant for Non-
      Oil & Mineral Exporters than the Oil & Mineral exporters while OECD-aid is
      more significant for oil exporters than non-oil exporters' category.

As can be seen from the long run adjustment coefficient, in the presence of China
and OECD aid and FDI inflows to both categories of African economies, 26%(29%)
of any deviation from the long run equilibrium of growth trajectory will be corrected
each year to attain long run equilibrium in the Non-oil & Mineral and Oil & Mineral
exporting African countries, respectively. This means that jointly China and OECD
has significantly impact on the long term growth of the Oil & Mineral exporter
countries of Africa. Comparatively, in the oil and mineral exporting countries, China
and OECD Aid and FDI flows to this category will help more, jointly with other
variables, to correct any deviation from the long run growth trajectory, by 29% each
year, while the same is around 25% for non-oil and mineral exporting countries.

29.5.7   Africa's Own Contribution on Its Own Real GDP Growth

In order to see the relative importance of OECD and China in African growth trajectory, it is essential to see the effect on the two categories of African economies by
including and excluding out both China and OECD Aid and FDI flows. Table 29.5
shows the real effect on the African Real GDP growth without both China and OECD
countries first. If there is no FDI and Aid inflow (without China and OECD economic
variables in the analysis) to both categories of African economies:

(i). The long run equilibrium adjustment coefficient drops to 10% (12%), which
      was 25% (29%) with the presence of China's and OECD's aid and FDI inflows
      in these economies. This means that without China and the OECD external
      financial inflows, only 10% (12%) of any deviation in the long run growth

Table 29.5 EVIEWS 7.1 Results outputs (without both China and OECD)

Non oil & mineral exporters without both China and OECD				Oil and mineral exporter without both China and OECD			
Dependent variable: D(LRGDP)				Dependent variable: D(LRGDP)			
Method: Panel EGLS (Cross-section weights)				Method: Panel EGLS (Cross-section weights)			
Sample (adjusted): 1998 2011				Sample (adjusted): 2001 2011			
Total panel (unbalanced) observations: 184				Total panel (unbalanced) observations: 115			
Variable	Coefficient	t-statistic	Prob.	Coefficient	t-statistic	Prob.	
Long term impact							
REAL RGDP (− 1)	− 0.10	− 3.32	0.00	− 0.12	− 3.09	0.00	
TRADE OPEN (− 1)	0.01	0.60	0.55	0.07	1.54	0.13	
DOMESTIC INV (− 1)	0.09	3.65	0.00	0.1	2.72	0.00	
LABOR FORCE (− 1)	0.01	0.79	0.43	0.04	3.02	0.00	
@TREND	0.01	2.79	0.00	0.00	0.05	0.96	
	Weighted statistics				Weighted statistics		
R-squared	0.44	Mean dep var.	0.08	0.62	Mean dep. Var.	0.11	

trajectory of non-oil and oil & mineral exporting countries will be corrected each year to revert to equilibrium. Hence, we can say that for both categories both China and OECD FDI and Aid inflows are important for a better long run growth.

(ii). There is still long-term economic growth with the positive contribution from the contribution of its own domestic investment and labour force including openness index. However, the openness to trade index is not significant. This is in line with some of the empirical studies which argued that there might be thresholds of incomes necessary for openness to trade to become significantly determine economic growth.

29.5.8   Chinese Economic Globalization Effect on Africa's Real GDP Growth

Table 29.6 shows, without the OECD Aid and FDI inflows but with China to both categories of African economies, the long-term growth adjustment coefficient has increased to 29% (23%)from when both are absent which was 10% (12%) for both categories. These indicate that China has an important say in the growth trajectory of African countries.

Table 29.6 EVIEWS 7.1 Results outputs (without OECD but with China- Partial Model)

Non oil and mineral exporting African Countries (without OECD but With China)				Oil and mineral exporter African (without OECD but with China)		
Dependent variable: D(LRGDP)				Dependent variable: D(LRGDP)		
Method: panel least squares				Method: panel least squares		
Sample (adjusted): 2001 2011				Sample (adjusted): 2001 2011		
Total panel (unbalanced) observations: 142				Total panel (unbalanced) observations: 115		
Variable	Coefficient	t-statistic	Prob.	Coefficient	t-statistic	Prob.
Long term impact						
REAL GDP (− 1)	− 0.29	− 4.31	0.00	− 0.23	− 2.27	0.02
OPENNESS OF TRADE (− 1)	0.05	0.68	0.50	0.02	0.17	0.86
DOMESTIC INVST (− 1)	0.15	3.46	0.0007	0.13	1.84	0.06
AID_CHINA (− 1)	0.015	1.30	0.19	0.03	1.56	0.12
FDI_CHINA (− 1)	0.03	1.91	0.05	0.03	1.11	0.27
LABOR FORCE (− 1)	0.28	0.84	0.40	0.11	0.33	0.73
Effects specification				Effects specification		
Cross-section fixed (dummy variables)				Cross-section fixed (dummy variables)		
R-squared	0.61	Mean dependent var.	0.08	0.59	Mean dependent var.	0.104

(a). The long run growth equilibrium adjustment coefficient increased from 10% (12%) from the models where both OECD and China are absent to a higher RGDP equilibrium adjustment coefficients 29% (23%) for both categories of African countries with the presence of China alone.

(b). OECD Economic globalization effect on Africa's Real GDP growth: Table 29.7 below shows without China's FDI and AID inflows but with OECD economic variables presence to both categories of African economies, the long- run growth- adjustment coefficients have increased to 22% (22%) from when both China and OECED were absent, which were 10% (12%). Thus, the results clearly indicate that China has an important say in the growth trajectory of African countries.

(c). Comparing the impact of OECD with the impact of China's inflows on the oil & mineral exporting African category, it seems that China is slightly more significant than the OECD in the oil and mineral category too.

Table 29.7 EVIEWS 7.1 results outputs (without CHINA but with OECD)

Non oil & mineral exporter African countries (without China but OECD)				Oil & mineral exporter African countries (without China but OECD)			
Dependent variable: D(LRGDP)				Dependent variable: D(LRGDP)			
Variable	Coeff.	t-Statistic	Prob.	Coeff.	t-Statistic	Prob.
REAL GDP (− 1)	− 0.22	− 3.1	0.00	− 0.22	− 3.31	0.00
TRADE OPENNESS (− 1)	0.01	0.16	0.87	0.03	0.34	0.73
DOMESTIC INVEST (− 1)	0.16	4.08	0.00	0.13	2.61	0.01
AID_OECD (− 1)	0.117	1.95	0.05	0.02	0.29	0.77
FDI_OECD (− 1)	0.08	2.25	0.02	0.13	2.97	0.00
LABOR FORCE (− 1)	0.33	0.73	0.46	0.88	1.89	0.06
@TREND	0.02	1.34	0.18	0.04	2.55	0.01
	Effects specification					
Cross-section fixed (dummy variables)						
R-squared	0.67	Mean dep. var.	0.08	0.66	Mean dep. var.	0.10

29.6   Key Findings, Conclusion and Recommendations

29.6.1   Findings

Overall, the key findings from the above regressions results for Non-oil exporting African countries are as follows:

(i).   Both China and OECD aid and FDI inflows to non-oil and mineral exporting countries are very crucial for growth in such African countries.

(ii).   Without both of them, there would be growth in such African economies, but it is much slower than the case of their existence.

The key findings from the above regression results for oil African exporting countries are:

(i).   For the oil and mineral exporting countries, both China's and OECD's Aid and FDI inflows to these countries are very crucial to enhance growth.

(ii).   Together, both China's and OECD's importance is greater for the oil and mineral exporting category than other wise.

(iii). China's importance is greater for the non-oil and mineral exporting category (28%) than oil category (23%) whereas OECD's importance is relatively the same for both categories (22%).

(iv). If there is no FDI and aid inflows to the oil and mineral exporting Countries from both OECD and that of China, still openness, domestic investment, and labour force affects growth positively both in the short run and that of the longer run.

29.6.2 Conclusions

(i). Both OECD and China are important for any category of African countries to fasten growth.

(ii). There is an indication that the traditional West (OECD) partners of Africa are losing their place rapidly to the emerging China in the last decade especially in the areas of export and import trade measures as shown graphically in the previous section.

(iii). In terms of real GDP growth, the impact of China and OECD together is slightly more important in the oil and mineral exporting countries (29%) than the non-oil (25%) categories. That is, in the oil and mineral exporting countries, China and OECD aid and FDI flows to this category will help, jointly with other variables, to correct any deviation from the long run growth trajectory, by 29% each year, while the same is around 25% for non-oil and mineral exporting countries.

29.6.3 Recommendations

The researchers offer the following recommendations on the basis of the findings revealed and the conclusions arrived at in the study:

1.1. African governments need to promote and further enhance the enabling environment to attract more FDIs into key sectors contributing to economic and social development in Africa. Africa will be benefited by diversifying its renewed partnership with the emerging global power in the Eastern Hemisphere, i.e. China in addition to its traditional partners like US and other OECD countries.

1.2. Since foreign aid (official development assistance) is contributing significantly to its economic growth, we recommend Africa to diversify its partners to get the most advantage from developmental aid to sustain its long term growth but they need to incline more on trade and investment than seeking for foreign aids for all their problems.

1.3. We also recommend the Chinese government to diversify and increase the provision of their foreign aid (developmental aid) to the other sectors (like education, health and environmental protection) as well. On the other hand, the OECD countries need to give attention to the African infrastructural development through "Marshal Plan" like support than the existing fragmented aid allocation that could not make transformational changes on the ground.

2.1. As the destination market for the African products becomes more open, it is important for the African countries to open their markets not only for the traditional partners of the west (OECD) and emerging trade partners like China but also for other markets as well. The study shows the share of the selected OECD countries is declining continuously in Africa's trade (export and import), it is important for OECD to open their market for Africa's products through different effective preferential trade schemes like the USA led initiative of Africa Growth and Opportunity Act (AGOA) before they are being displaced by China completely.

2.2. Since the role of domestic private investment is positive and significant, African government need to enhance the enabling environment for enhancing domestic saving to have a sustained internal generation of resources for their economic development.

2.3. The labour force in Africa is growing in number. We recommend that Africa needs to use its own and external resources to improve the quality of its human capital by allocating more in education, health and skill development. With respect to the external partners, Chinese government and investors in Africa need to give due attention for human capital development by providing skill developments and technology transfer in real sense.

3. The positive and significant real GDP growth (economic growth) of Africa shows that the need to sustain the economic partnership it has with China and OECD countries and expand its economic activities to attain a higher real GDP growth rate by exploiting their untapped potential in natural resources and make a sustained economic growth by enhancing both the Extra and Intra African trade and investment.

4. Africa itself has positively contributed for its own economic growth. The results showed without China and OECD its domestic economic indices have significant effect on its real GDP growth. The domestic private investment and its labour force have undeniable contributions to its growth. Boosting investment in key infrastructures such as transportation, communication and power is crucial for enhancing growth in Africa.

5. Focusing on few partners might not be a good long term strategy for Africa. We recommend Africa to diversify its partners to get the most advantage from sustained trade, FDI, and developmental aid.

6. Africa needs to have a strategy which accommodates both the Eastern and Western countries without remaining stuck in one ideology as before.

7. Africa needs to have also a clear strategy with China to equally contribute to the social sector development and domestic investors' growth which is instrumental for sustained economic development in Africa.

Appendix A: Phillip–Perron Fisher panel unit root test results

Variable	Test equation	At level critical values at 5% level of significance (*p*-value in parenthesis)	At first difference critical values at 5% level of significance (*p*-value in parenthesis)	Conclusion on the order of integration
LRGDP	Intercept	1.53 (1.00)	39.01 (0.03)	I (1)
LAID_ CHINA	None	7.486 (1.00)	150 (0.00)	I (1)
LAID_ OECD	Intercept	19.299 (0.74)	50.39 (0.00)	I (1)
LFDI_ CHINA	None	1.87 (1.00)	182.32 (0.00)	I (1)
LFDI_OECD	None	16.89 (0.85)	257.31 (0.00)	I (1)
LINV	Intercept	2.83 (1.00)	53.69 (0.00)	I (1)
LOPENESS	None	29.86 (0.19)	180.73 (0.00)	I (1)

References

Adejumobi, S. (2002). Economic globalisation, market reforms and the delivery of social welfare services in Africa. In T. Aina & C. S. L. Chachage (Eds.), *Globalisation and social policy in Africa*. CODESRIA.

Anton, M. (2012). *The U.S. economic, political and geostrategic response to China's presence in Africa*. Instituto Superior de Economia e Gestao, Universidad Technica De Lisboa.

Baliamoune-Lutz, M. (2002). Assessing the impact of one aspect of globalization on economic growth in Africa. *WIDER Working Paper Series DP2002-91*, World Institute for Development Economic Research (UNU-WIDER).

Baliamoune-Lutz, M., & Mavrotas, G. (2008). Aid effectiveness: Looking at the aid-social capital-growth nexus. *RePEc*. https://doi.org/10.1111/j.1467-9361.2009.00504.x

Brautigam, D. (2010). *The Dragon's gift: The real story of China in Africa*. Oxford University Press.

Eris, M. N., & Ulasan, B. (2013). Trade openness and economic growth: Bayesian model averaging estimate of cross-country growth regressions. *Economic Modelling, 33*(C), 867–883.

Geda, A., & Meskel, A. G. (2008). China and India's growth surge: Is it a curse or blessing for Africa? The case of manufactured export. *African Development Review, 20*(2), 247–272.

Geda, A., & Yimer. A. (2015). Determinants of Foreign direct investment inflows to Africa: A panel co-integration evidence using new analytical country classification. *AAU Department of Economics Working Paper No. 4*, Addis Ababa University.

Geda, A., & Yimer, A. (2022a). An applied dynamic structural macro-econometric model for Rwanda. *Studies in Economics and Econometrics*. https://doi.org/10.1080/03796205.2022.2135587

Geda, A., & Yimer, A. (2022b). The trade effects of the African continental free trade area: an empirical analysis. *The World Economy*. https://doi.org/10.1111/twec.13362

Hadri, K., & Kurozumi, E. (2008). Panel Stationarity Test with Structural Breaks. *Oxford Bulletin of Economics and Statistics*. https://doi.org/10.1111/j.1468-0084.2008.00502.x

Hadri, K., & Kurozumi, E. (2010). *Simple Panel Stationarity Test in the Presence of Cross-Sectional Dependence*

Hodey, L. S., Oduro, A. D., & Senadza, B. (2015). Export diversification and economic growth in Sub-Saharan Africa. *Journal of African Development, 17*(2), 67–81. https://doi.org/10.5325/jaf rideve.17.2.0067

Kaplinsky, R., & Farooki, M. (2009). *Africa's cooperation with new and emerging development partners: Options for Africa's development*, UN-OSSA.

Romer, P. M. (1989). Human Capital And Growth: Theory and Evidence. https://doi.org/10.3386/w3173

Rosenau, J. (1997). The complexities and contradictions of globalization. *Current History, 96*, 613.

Rotberg, R. I. (Ed.). 2008. *China into Africa: Trade, Aid, and Influence* Brookings Institution Press. www.brookings.edu.

Stein, P., & Uddhammer, E. (2023). China in Africa: The role of trade, investments, and loans amidst shifting geo-political ambitions. *Occasional Papers*, Observer Research Foundation.

Sachs, J. D., & Warner, A. M. (1997). Sources of slow growth in African Economies. *Journal of African Economies, 6*(3), 335–376.

Sautman, B., & Hairong, Y. (2007). Friends and interests: China's distinctive links with Africa. *African Studies Review, 50*(3), 75–114.

The World Bank (2004–2011). *African development indicators (IDI) database*.

The World Bank. (2019). *The State of Human capital in Africa: Powering Africa's potential through its people*, Washington D.C.

United Nation Conference on Trade and Development. (2012). [Online] Available at: www.unctad.org.World Development Indicators/WDI (2004–2012). Different series.

Van Dijk, M. P. (2009). *The new presence of China in Africa*. Amsterdam University Press. https://doi.org/10.5117/9789089641366

World Bank. (1996). *Global economic prospects and developing countries*.

World Investment Report (2001–2012). Different series. New York and Geneva: United Nations.

Yimer, A. (2022a). The effects of FDI on economic growth in Africa. *The Journal of International Trade & Economic Development*. https://doi.org/10.1080/09638199.2022.2079709

Yimer, A. (2022). When does FDI make a difference for growth? A comparative analysis of resource-rich and resource-scarce african economies. *International Finance*. https://doi.org/10.1111/infi.12423

Chapter 30
Does China–Africa Economic Partnership Following the Right Trajectory?

Mulatu F. Zerihun

30.1 Introduction

There are clear assertions that declare Africa can claim twenty-first century if aided by its development partners to overcome the development traps that kept it confined to a vicious cycle of underdevelopment, conflict, and untold human suffering for most of the twentieth century (Almeida, 2013; World Bank, 2000). It is believed that the new century provides unique opportunities for Africa to leapfrog its stages of development. These opportunities include increasing political participation in the continent, the end of the Cold War, and globalization along with information and communications technology (World Bank, 2000, p. 1).

Africa has long aged overall partnership with the West given the geographic proximity and colonial legacy. Early this century Africa has embarked on new relationships like China–Africa, India-Africa, Brazil-Africa, South–South relationships and G77 countries based on what the continent had set for itself in the African Union (AU) and the New Partnership for Africa's Development (NEPAD). The role of these 'new' and 'old' development partners in assisting Africa to meet its short-run and long-run development goals in this century has always been questionable given the hidden agendas associated with the so called "partnership". Broadly defined, China–African relations refer to the historical, political, economic, military, social and cultural connections between China and the African continent. The focus of this study will be on the economic analysis of China–Africa partnership in trade, aid, and foreign direct investment (FDI). The major question here is whether this partnership can indeed significantly help Africa in realizing its objectives in this century.

M. F. Zerihun (✉)
Department of Economics, Faculty of Economics and Finance, Tshwane University of Technology, Pretoria, South Africa
e-mail: Zerihunmf@tut.ac.za

© The Author(s) 2024
M. Muchie et al. (eds.), *China-Africa Science, Technology and Innovation Collaboration*, https://doi.org/10.1007/978-981-97-4576-0_30

After decades of trade engagement with China, many African countries have been claiming that there is economic growth momentum induced in their respective economies. Despite the progress achieved in the past decades, cooperation between China and Africa also faces numerous challenges ahead. Some frictions and disputes have emerged in bilateral cooperation in ideological, diplomatic, and economic affairs in recent years. As more and more of the world's attention shifts to the continent's rich natural resources and its enormous market potential, China is expected to encounter increasingly harsh competition from other countries. There is little evidence whether China's renewed, and most probably lasting, involvement in Africa will serve the continent better than the decades of multifaceted relationship with Western governments, which have scarcely delivered on their promises. There are also parties who are predominantly suspicious of China–Africa trade partnership. Such optimism and suspicions about China's foreign policy towards Africa give rise to a new area of research and international meetings around the world. However, researches so far in this area are shallow and largely journalistic commentaries.

The motivation for this study was a need for a solid empirical investigation to base policy considerations and analyses of various aspects of China–African trade, aid, FDI of this emerging field of research. So far, many literatures in this are largely descriptive, lacking in depth analysis, limited scope and covering too few countries, mainly speculative journalistic works. This study doesn't attempt to fill these gaps, rather this study assesses the overall trajectory of China–African economic partnership since the early 1990s using descriptive trend analysis. Therefore, the central research question of this study is to establish whether the China–Africa partnership is economically justifiable and following the right trajectory. The general objective of this study is to assess the potential trajectory of China–African economic partnership. The more urgent task for Africa is to clearly define its interests with China and to pursue these interests with consistency and vigour. This study finds that China–Africa economic partnership has revived the economic growth momentum among many African nations. In addition, given the promising productive economic partnership and the neutral[1] position of China in terms of political affairs, its economic cooperation with Africa can be adapted to follow the right trajectory for the benefit of both parties (Oqubay & Lin, 2019; Regissahui, 2019). Africa's exports to China have generally exceeded its imports, excepting during the 2015/16 commodity crisis, and the outbreak of the pandemic in 2020 when the region's imports exceeded exports. The stronger increase in Africa's exports to China (41.14%) than its imports (25.46%) enabled Africa to enjoy a trade surplus of US$10.8 billion in 2021 from the sluggish deficit of US$0.19 billion in 2020 (African Export–Import Bank, 2022). However, this study alerts the worrisome trade balance condition of the top 10 deficit African

---

[1] We respect Africa, love Africa and support Africa. We follow a *'five-no' approach* in our relations with Africa: no interference in African countries' pursuit of *development paths* that fit their national conditions; no interferences in African countries' *internal affairs*; no imposition of *our will* on African countries; no attachment of *political strings* to assistance to Africa; and no seeking of *selfish political gains* in investment and financing cooperation with Africa. We hope this 'five-no' could apply to other countries as they deal with Africa...Ultimately, it is for the peoples of China and Africa to judge the performance of China–Africa cooperation. (Xi, 2018).

countries with China, (Ethiopia, Kenya, Togo, Malawi, Morocco, Benin, Algeria, Liberia, and Egypt and Nigeria) and calls for careful structural adjustments.

30.2   Central Problem of the Research

The modern world is notable for its high levels of international trade. Economists broadly agree that international trade offers an opportunity to eradicate poverty and promote development. However, it is agreed that international trade must be carefully regulated to avoid social injustice, economic inequality, political instability, environmental degradation, and cultural dispossession (Trade Law Centre for Southern Africa, 2009, p. 1). It is important to investigate the case of China–African partnership from the points mentioned above. However, there are no in-depth empirical studies on the impacts and prospective impacts of China's engagement in specific sectors, countries, and regions in Africa. Also, there are no comprehensive compilation of the theoretical approaches to China–Africa partnership in various economic engagements and how these relate to each other. Overall analysis in this study will be located within the body of literature concerning International Political Economy Particularly International Trade Theories and contemporary literatures on China in Africa. Key problems associated with China–Africa partnership are neither fully addressed nor satisfactorily contextualized in the current debate regarding China and Africa. This study will be in these raging debates, investigating and explaining whether China–Africa trade partnership has been following the right trajectory or not.

The post-1980 period brought several changes that enhanced the economic and political power of the North in the world economy at the expense of the South: the narrowing 'policy space' prevented the South from pursuing optimal trade policies that could have counteracted the deterioration in its terms of trade; the neoliberal actors in policy-making, both domestic and international, eliminated the effectiveness of the state as a developmental force in pursuing industrial development. The Northern retaliation in the form of increased protection of its own markets against Southern exports turned the terms of trade further against the South. In addition, international institutions changed the rules of the game that sustained imbalanced growth in the South by keeping its innovation gap with the North wide open. With the recent rise of large-sized developing countries including Brazil, India, China, and Russia in international trade, the Southern intra-regional game can also be thought of as 'a prisoner's dilemma' from a strategic viewpoint. Thus, even when the large country assumption fails to hold for many small developing countries, their best response strategies to cope with the growing competitive pressures in the world market lead them to outcomes where they are worse-off. Moreover, it is much harder for many small countries in Africa to reach cooperative outcomes with bigger economies like that of China.

30.3   Literature Review

This section presents a literature review on China–Africa relations in trade, aid, and foreign direct investment. North–South economic partnerships are often not successful due to politicization and donor-driven type of engagements. Thus, the South–South economic partnerships are the way forward as they attempt towards monetary integration among developing and emerging market economies, collectively known as South–south monetary cooperation. Africa has embarked on new relationships like; China–Africa, India-Africa, Brazil-Africa, South–South relationship and G77 countries based on what the continent had set for itself in the African Union (AU) and the New Partnership for Africa's Development (NEPAD). China–Africa relations refer to the historical, political, economic, military, social and cultural connections between China and the African continent. Many African countries have been claiming that there is economic growth momentum induced in their respective economies. However, there is little evidence whether China's renewed, and most probably lasting, involvement in Africa will serve the continent better than the decades of multifaceted relationships from Western governments. There are also parties who are predominantly suspicious of China–Africa trade partnership. The following three subsections (30.3.1–30.3.3) present the details of South–South integration, the economic history of China–Africa partnership, and the economics of China–Africa partnership, respectively.

30.3.1   Brief Overview of South–South Integration

As the dominant economies South–South economic integration, China and India have been the main drivers of the growing trade ties between Africa and Asia (African Export–Import Bank, 2022). South–south economic partnerships such as China–Africa and India-Africa are not well explored. When we discuss about south–south economic partnership, we need to consider the *original sin hypothesis*. The original sin hypothesis put simply is *'the inability of a country to borrow abroad in its own currency and measured as the ratio of foreign currency–denominated gross debt to foreigners as a share of total gross debt to foreigners'* (Eichengreen & Hausmann, 2005). This is the dominant case among developing countries.

The original sin concept (Eichengreen, 2008, Fritz & Metzger, 2006) evidences the particular importance of the denomination and composition of domestic and external debt for economic growth and partnership. By definition, SSI is pursued by countries which that accumulate debt in foreign currency, thereby most often suffering from a restricted lender of last resort function, balance sheet effects in the event of a currency devaluation and original sin and, as a result, small and undiversified financial markets. While levels and composition of internal and external debt may vary among the participating countries, SSI needs to deal with the specific monetary constraints of countries with economic partnerships.

Intraregional hierarchies in terms of original sin and net creditor/net debtor rela-tions play a crucial role in the success of an SSI project (Fritz & Mühlich, 2010). The authors further argue for a clear hierarchy in terms of indebtedness in foreign currency seems to provide favourable conditions for a successful SSI and may provide further perspectives for regional monetary integration and financial market development. In this sense, both stronger and weaker countries could benefit from regional monetary integration, with the larger economies establishing potentially stabilising leading roles. Fritz and Mühlich (2010) conclude that *'intra-regional hierarchies, involving differing levels of original sin and indebtedness in foreign currency, constitute a major success factor for intra-regional exchange rate stabilisation and enhanced regional monetary SSI'*.

A regional monetary arrangement potentially generates economies of scale in regional financial markets. Thus, the potential stabilization gains of SSI need to be understood as a monetary strategy, including a specific exchange rate regime choice of the integrating countries. Emerging market economies are excluded from economic blocs based around the international key currencies. Given the international trend toward building economic blocs, it seems fruitful to understand the exchange rate regime options for developing and emerging market economies, from the perspec-tive of their relation to the latter—*instead of the usually applied corner solutions perspective* (Priewe, 2006).

30.3.2 An Overview of the Economic History of China–Africa Partnership

Snow (1988) was one of the first historians who tried to analyse China's engagement with Africa. Snow essentially narrates China's engagement from its beginnings in 1955 and throughout the Cold War era. Given prevailing ancient ties between Africa and China, Rodrigues (2008) argues that there is a significant normative framework behind China's engagement of Africa and experience of shared history. Although China was never a colonising power, its economic relations with Africa can be traced to 200 B.C. (Lyakurwa, 2008, p. 1). The West (academics, practitioners, governments, corporations, international financial institutions, and nongovernmental organisations) are alarmed at the pervasiveness of Chinese re-engagement with Africa (Gaye, 2008). African academics have been preoccupied with analysing China's current engagement in terms of implications for the continent. Undoubtedly, China has built an informal economic influence that spans all sectors of Africa to foster economic relations that cater to the needs of its booming economy. Marks (2007, p. 2) put it succinctly: "Almost every African country today bears examples of China's emerging presence, from oil fields in the east, to farms in the south and mines in the centre of the continent". The big question remains: Is the Chinese scramble for Africa a manifestation of new imperialism?

China has established an informal economic influence, and in the process acquired indirect political influence on the continent, in the post-independence era. The nature of the relationship is voluntary, and thus, devoid of coercion. It relies on diplomatic manoeuvres, with Africa willingly embracing Chinese economic expansion on the continent at the expense of competitors, the Western powers or former colonial powers. This is what is called the "soft power diplomacy" (Gaye, 2008; Naidu & Davies, 2006). To argue that China has a beneficial impact on Africa is in line with Warren's view, in Baylis, Smith and Patricia (2020), which attributes positive elements to economic imperialism. The positive effects of China are manifested as short- or medium-term benefits accruing to Africa. One would be tempted to view this as exceptional phenomenon associated with Chinese investment and trade. However, China reaps more advantages than Africa given the fact that it is a more developed market economy than any African country it relates with. Economic asymmetries point to overwhelming benefits accruing to China. Typical of imperialistic relations is the trade that perpetuates Africa's mono-economies and dependence on the export of raw materials. This imbalance can thwart Africa's effort to diversify economies from the production of raw materials to that of processed goods, as well as condemns Africa to underdevelopment and marginalisation in the globalised economy (Naidu & Davies, 2006). This is particularly true regarding unequal trade between China and its major trading African countries, namely, South Africa, Nigeria, and Sudan. For example, South Africa's trade deficit with China increased from US$24 million in 1992 to US$400 million in 2001, and it has been growing since. In 2005, South Africa exported US$1.4 billion to China and imported US$5.2 billion worth of goods, a deficit of US$3.8 billion (Alden, 2005: 161).

The growing engagement of China on the African continent, which is particularly visible in the natural resource sector, is a very heated and controversial topic. China's appetite for Africa's minerals and oil has rung alarm bells in the West and raised many concerns among Western academics, NGOs, and politicians. There is a tendency to describe the current China–African relations under the rubric of neo-colonialism, and China is frequently pictured as a menace (danger) to the continent's long-term development. This might seem like a curse due to the likely incidence of the "Dutch disease" phenomenon and the unsustainable exploitation of natural resources, but these motives again need to be examined objectively.

There is, of course, another side of the coin. The Chinese (re)emergence in Africa has brought high hopes for the continent's economic revival. China–Africa trade and foreign investment has surged at an unprecedented rate. Overall, China's rise vis-a-vis Africa's manufacturing sector could be a positive catalyst for change, enlarging the market for exports from Africa to China, stimulating competition and, in turn, innovation within Africa. The IMF estimates that Africa's growth overall is close to 6%, the highest in 30 years, due in large part to China's growing investments. According to the World Bank report (2004, 2008) China's aid to Africa is focused on two aspects: Africa's infrastructure sector (also called hardware sector), where China is playing an enormous role and, through in-depth analysis of the sector, reaching the objective and positive conclusion that it is a fact that China has been serving Africa's economic growth as a proactive driving force (Gu & Carey, 2019; McKinsey, 2020; OECD/ACET, 2020).

However, Africa's leadership needs to rise to China's challenge and cooperatively develop realistic and achievable objectives for common development and prosperity. As a starting point, a closer alignment of Forum on China–Africa Cooperation (FOCAC) with NEPAD priorities would go a long way to structuring a common development agenda. Africa's bargaining power is limited by its lack of industrialisation and dependence on primary products as its main source of export, but effective management of its competitive advantages can translate into benefits for China's economy and the citizens of Africa simultaneously. In effect, this requires that Africa transform the 'minerals curse' into a 'vector for socio-economic development'. Africa's leadership must avoid switching from a dependency on the West to a dependency on China and focus on building an authentic partnership with China's through bilateral interaction as well as the FOCAC process. Africa should develop a joint venture with China to address common development difficulties, while ensuring mutual benefit, rather than allowing a new asymmetrical relationship with China to replace old neo-colonial links with the West.

30.3.3   The Economics of China–Africa Partnership

Though it is hard to find in-depth economic analysis of China–African economic engagements, the existing literature often has mixed results and obscure conclusions. There are extreme positions prevailing and biases based on ideological orientations rather than factual analysis. Rodrigues (2008, p. 12), following Alden (2007, p. 5), in his attempt to classify the existing literature on China as Africa' uses three typologies: China as a development partner, as an economic competitor, and as a coloniser. The same typology is adapted here to understand the central messages of different authors in the area. When we see China as a development partner, we mean that China is committed to transmitting its development to the African continent (Rodrigues, 2008). At the same time Chinese capitalism seems to be a "better fit" for Africa than Western liberalism has been (Rodrigues, 2008). Rodrigues (2008) constructs an argument that is more supportive of the view that China's engagement tends to be increasingly underpinned by severe economic competition. It is understood that because Africa opened its doors to China's competition that exploitation sometimes occurs, but in these views, this exploitation is more of an economic than a political nature. This Chinese model of development can hold, according to those who see China as a development partner, another important lesson. It can allow Africa to learn how to organise trade policy; how to move from low- to middle-income status; and how to educate for quick payoff (Chan-Fishel, 2007, p. 139).

There are many authors arguing that China is an economic competitor in some specific sectors with other international actors in Africa while in other sectors it arrives as a dynamic stimulant (Burke et al., 2007; Wang & Elliot, 2014). The argument is that China is first and foremost a development partner with great potential while its behaviour is framed within what the authors call "coalition engagements"—a collaborative state business approach to foreign policy. Under this approach private resources from China's MNCs and SOEs are put together with the political and diplomatic clout of China, resulting in an active and innovative model of engagement, which is flexible yet decisive in its nature and constitutes the central model of China's strategic engagement with the continent (Rodrigues, 2008). In the academic world, China is sometimes portrayed as an imperial power that should be unwelcome in Africa. This is the case with the views of Marks (2007) and Lee (2006) that express great concern over China's colonising tendencies in Africa and blame them on aspects such as the arrival of neoliberal tendencies at home. In fact, the views of two big critics of understanding of China as a colonial power and of the new scramble for Africa, Sautman and Hairong (2012), support the argument that such unfounded affirmations should be avoided for the sake of constructive debate. The fact is that the development lessons China holds for Africa do hold paradoxes. Investigating these paradoxes is the gist of this study.

30.4   Methodology, Data Analysis and Discussions

This section presents the methodology, data analysis and the study's findings.

30.4.1   Data and Methodology

Data used in this are obtained from UN Comrade (April 2023), Ministry of Finance of China, the International Trade Centre (ITC), the Statistical Bulletin of China, the U.S. Bureau of Economic Analysis and Yearly Bulletin of China's Foreign Direct Investment (2022). A Microsoft Excel spreadsheet was used to capture all the collected secondary data. Furthermore, the latest version of Statistical Package for the Social Sciences (SPSS-version 28) was used to analyse data. Additionally, two methods of statistical procedures which include the descriptive and inferential analysis were utilised to analyse and present data accordingly. Descriptive analysis was presented using the frequencies, with mean and standard deviation being used as measures of central tendency and dispersion respectively (Yang & Lee, 2019, p. 54). The determination of the Spearman's correlation coefficient (ρ) can be obtained from Eq. (30.1) for observation without ties.

$$\rho = 1 - \frac{6 \sum d_i^2}{n^3 - n} \tag{30.1}$$

For observations with ties (which is our case), Eq. (30.2) holds thus:

$$\rho = \frac{\sum^{(}X_i - X')(Y_i - Y')}{\sqrt{\sum^{(}X_i - X')^2 \cdot \sum^{(}Y_i - Y')^2}} \tag{30.2}$$

30.4.2 Data Analysis and Discussions

In general, the correlation analysis between Chinese FDI flows to African Countries and their annual GDP growth rate shows a weak trend. As shown in Fig. 30.1, Chinese FDI annual flows to Africa, also known as OFDI ("Overseas Foreign Direct Investment") in Chinese official reports, have been increasing steadily since 2003. Flows surged from US\$ 75 million in 2003 to US\$ 5 billion in 2021. They peaked in 2008 at US\$ 5.5 billion because of the purchase of 20% of the shares in Standard Bank of South Africa by the Industrial and Commercial Bank of China (ICBC). Table 30.1 presents descriptive statistics showing that measures of central tendencies and measures of dispersion show that GDP growth and Chinese FDI data are not normally distributed. Jarque–Bera's test of normality also confirms that there is evidence to reject the null hypothesis of normal distribution for the variables under discussion since the p-values are lower than usual significance levels.

China Africa Trade As depicted in Figs. 30.1 and 30.2, China–Africa bilateral trade has been steadily increasing for the past two decades. However, weak commodity prices since 2014 have greatly impacted the value of African exports to China, even while Chinese exports to Africa remained steady (China Africa Research Initiative, 2023) due to the supply chain disruption by COVID-19, the value of China–Africa trade in 2020 was US\$176 billion, down from US\$192 billion in the previous year. But the value bounced back in 2021 to US\$251 billion. In 2021, the largest

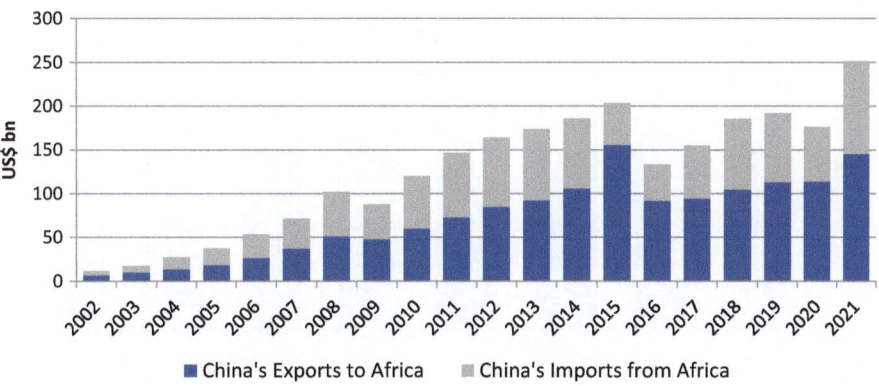

Fig. 30.1 China–Africa Trade between 2002 and 2021 (Unit: USD billion). *Source*: UN Comrade (April 2023)

Table 30.1 Descriptive statistics

	GDP Growth (Annual)	Chinese FDI flow to Africa ($billion)
Mean	4.029130	53.32272
Median	4.372019	6.350000
Maximum	86.82675	4807.860
Minimum	− 50.33852	− 814.9100
Std. Dev	6.252441	205.3300
Skewness	1.582886	15.28016
Kurtosis	51.04134	346.7818
Jarque–Bera	80,839.98	4,154,306
Probability	**0.000000**	**0.000000**
Observations	837	837

Source: Author form Yearly Bulletin of China's Foreign Direct Investment (2023)

exporter to China from Africa was South Africa, followed by Angola and the Democratic Republic of Congo. In 2021, Nigeria remained the largest buyer of Chinese goods, followed by South Africa and Egypt" (China Africa Research Initiative, 2023). Figure 30.1 shows China–Africa Trade between 2002 and 2021 (Unit: USD billion).

As illustrated in Table 30.2, it is imperative to note that the trade balance condition of the top 10 deficit African countries with China is worrying. Namely, Ethiopia,

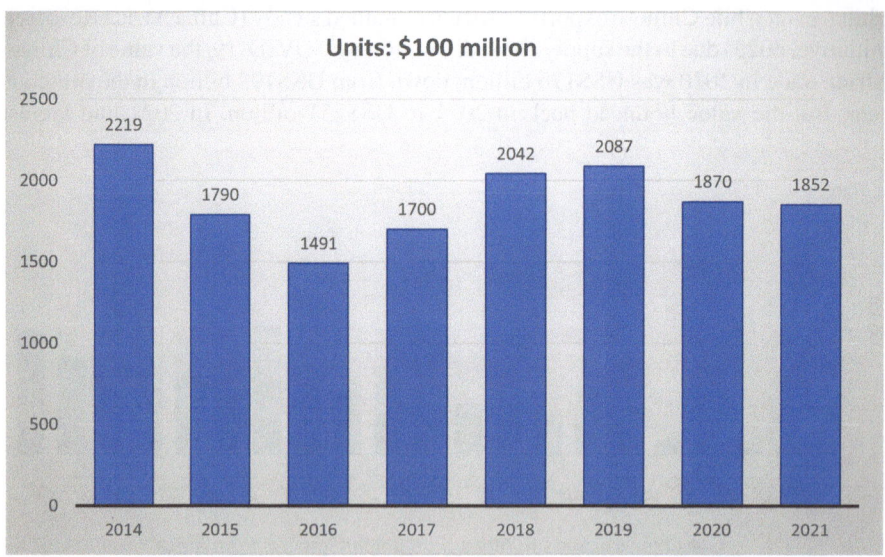

Fig. 30.2 China Africa Trade Volume from 2014 to 2021. *Source*: Author's computation of data collected from the International Trade Centre (ITC) Database, 2022

Kenya, Togo, Malawi, Morocco, Benin, Algeria, Liberia, and Egypt and Nigeria in China–Africa trading and calls for careful structural adjustments.

Chinese FDI flows to Africa As shown in Figs. 30.3, 30.4 and 30.5 Chinese FDI flows to Africa have exceeded those from the U.S. since 2013, as U.S. FDI flows have generally been declining since 2010. Chinese FDI flows to Africa have certainly grown substantially since 2004 and continue to increase. However, the opportunities to reach the size of European or US FDI are limited because these countries have already carved up opportunities leaving relatively little space for newcomers. The flow of Chinese FDI to Africa is small in comparison to flows to South Asia, East Asia, and Southeast Asia and to flows to Latin America.

China–Africa Cooperation Fig. 30.6 shows the top 20 destination countries for African investment by number of projects. According to Sun (2021), the 2035 Vision

Table 30.2 China–Africa balance of trade (10 leading surpluses and 10 leading deficits) in 2022

Country	Surplus (Unit: millions of Dollars) Between China and African countries	Country	Deficit (Unit: millions of Dollars) between China and African countries
Angola	20,812	Ethiopia	− 936
Sudan	4725	Kenya	− 1271
South Africa	4103	Togo	− 1747
Congo	2769	Malawi	− 1810
Libya	2454	Morocco	− 2032
Zambia	2036	Benin	− 2148
Congo (DRC)	764	Algeria	− 2823
Mauritania	687	Liberia	− 4374
Namibia	252	Egypt	− 5123
Gabon	168	Nigeria	− 5625

Source: UN Comtrade (April 2023)

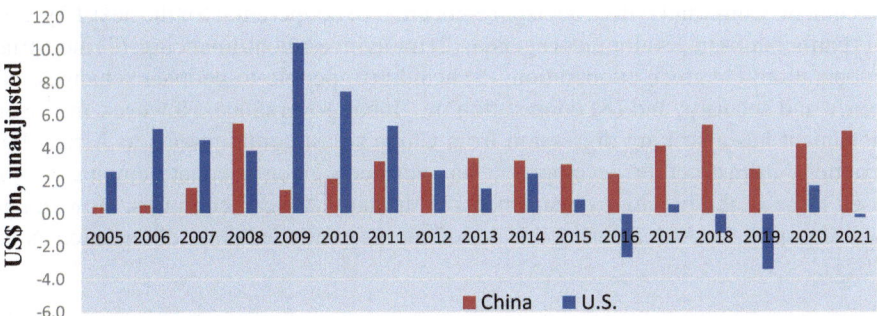

Fig. 30.3 Flow of Chinese FDI vs. US FDI to Africa. *Source*: UN Comtrade (April 2023)

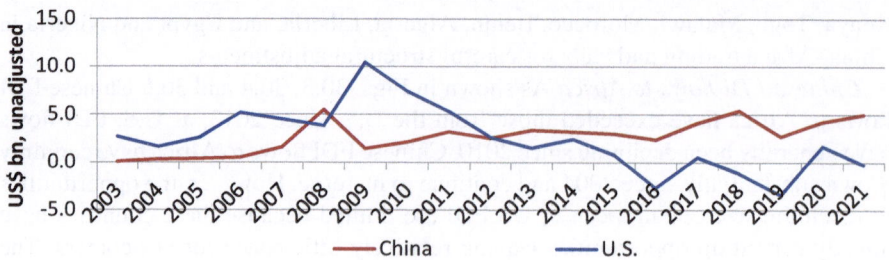

Fig. 30.4 Trend Analysis of Flow of Chinese FDI vs. US FDI to Africa. *Source*: UN Comtrade (April 2023)

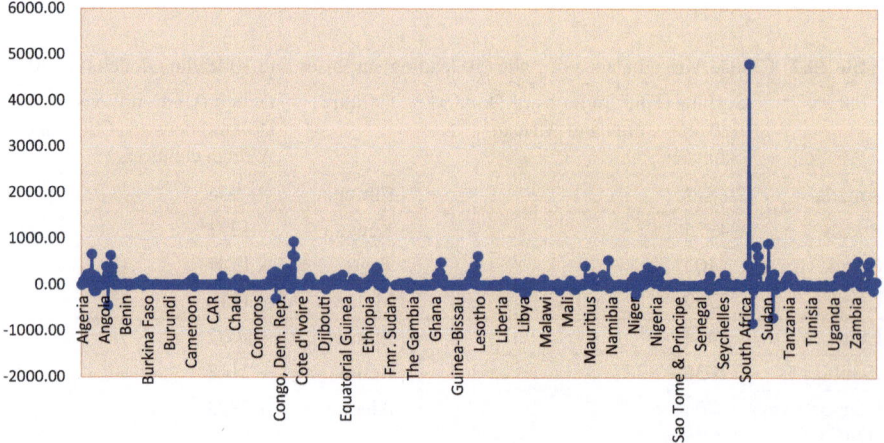

Fig. 30.5 Chinese FDI Flow to African Countries. *Source*: Yearly Bulletin of China's Foreign Direct Investment (2022)

defines the overall framework of China–Africa cooperation for more than a decade to come. It aligns with China's own 2035 Vision. According to the cooperation document, China and Africa envision eight areas of cooperation for the next 12 years: (1) partnerships in development agenda; (2) trade/investment/financing; (3) industrial cooperation; (4) green cooperation; (5) health; (6) people-to-people exchanges; (7) peace and security; and (8) cooperation on global governance. However, the vison document has a striking digression from China's past commitments to Africa. For example, infrastructure development has not been given adequate emphasis as it used to be in the previous engagements with many African countries. How things will transpire for Africa during this period merits continued close observation (Sun, 2021).

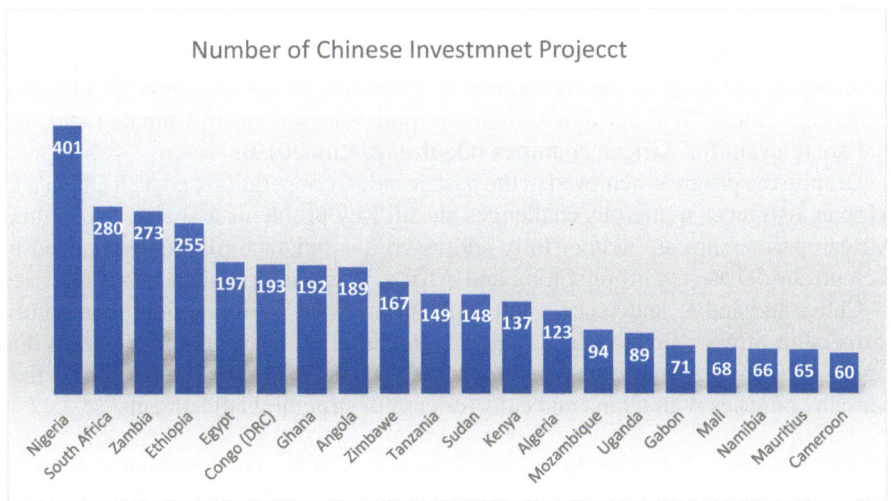

Fig. 30.6 Top 20 destination countries for African Investment by Number of Projects. *Source*: Yearly Bulletin of China's Foreign Direct Investment (2022)

30.5   Concluding Remarks

The motivation for this study was a need for a solid empirical investigation to base policy considerations and analyses of various aspects of China–African trade, aid, and FDI of this emerging field of research. Though it is late, relatively this is the right time for Africa to learn from past pitfalls and formulate a comprehensive trade partnership policy framework for the bright future and destiny of its people. For Africa to transform its economy the role of its partners matters a lot. Centuries of linkage to western markets have not brought about prosperity to Africa. Likewise, it remains less clear to Africa that China offers a unique opportunity to transform its economy and break away from structural dependence. However, the Chinese footprint in Africa gradually becomes more visible than any other western counterparts. One thing that is comprehensible is that China's own experience in transforming agriculture and increasing production can be a model for Africa. Thus, exploring this very important and current area of research can shed light on China–Africa economic partnership. The study also helps to understand both the threats as well as opportunities posed by China's engagement with Africa; it also helps to look at how Chinese economic partnership and natural resource exploitation compare with those of traditional western powers.

China's new internationalization strategy, the Belt and Road Initiative, is a strategic initiative that covers not only Africa, but other continents, and represents China's 'Going Global' strategy and the potential to attract FDI in light manufacturing (Oqubay & Lin, 2019). In addition, reaping the opportunities offered by China's economic rebalancing and the 'Going Out' of Chinese enterprises will primarily depend on the capacity of African actors to make the most of these opportunities. Considering China's global ambitions, African policymakers need to understand the contributions as well as the limitations of economic ties with China in a rapidly changing context. Without doubt, there are policy lessons from China's rapid rise that are relevant for African countries (Oqubay & Lin, 2019).

Despite the progress achieved in the past decades, cooperation between China and African also faces numerous challenges ahead. Key problems associated to China–African partnership are neither fully addressed nor satisfactorily contextualized in the current debate regarding China and Africa. There is optimism about the rises of China and India, and other south-south partners of Africa hoping that such a partnership offers a great chance to Africa in claiming twenty-first century. In this regard, this study draws attention to the trade balance condition of the top 10 deficit African countries with China and calls for careful structural adjustments.

Acknowledgements This book chapter was presented at the international conference on China–Africa Science, Technology, and Innovation Collaboration held at Future Africa Conference Centre from 9 to 10 October 2022 in Chengdu, China. The author is grateful for the comments and insights from the conference participants, the anonymous reviewers and the book editors who helped to improve the contents of this book chapter.

Declarations

Conflict of Interest The author declares no conflict of interest regarding the publication of this chapter.

References

African Export–Import Bank. (2022). *African Trade Report 2022: Leveraging the power of culture and creative industries for accelerated structural transformation in the AfCFTA era*, Heliopolis, Cairo, Egypt: African Export-Import Bank.

Alden, C. (2005), China in Africa, *Survival*, *47*(3): 147–164

Alden, C. (2007). China in Africa. London: *Zed Books*.

Almeida, E. C. (2013). Africa in the 21st century: prospects and causes: Effects on African States. Thoughts and Considerations, *JANUS.NET e-journal of International Relations*, 4(2).

Baylis, J., Steve, S., & Patricia, O. (2020). (eds.), The Globalisation of World Politics: An Introduction to International Relations, 8th edn., Oxford: *Oxford University Press*, ISBN 13: 9780192559586

Burke, C., Corkin, L. & Tay, N. (2007). China's Engagement in Africa: A Preliminary Scoping Study, *Centre for Chinese Studies*, Stellenbosch University. (pp 1-224)

Chan-Fishel, M. (2007). Time to Go Green: Environmental Responsibility in the Chinese Banking Sector, *Friends of the Earth – US and Bank Track, May*.

China Africa Research Initiative. (2023). The Paul H. Nitze School of Advanced International Studies Johns Hopkins University, Washington D.C.

Dullien, S., Fritz, B. & Mühlich, L. (2013). Regional Monetary Cooperation: Lessons from Euro Crisis for Developing Areas?, *World Economic Review 2*, (pp 1-23).

Eichengreen, B. (2008). *The Real Exchange Rate and Economic Growth. Commission on Growth & Development Working Paper*, No. 4, the World Bank.

Eichengreen, B., & Ricardo, H. (Ed.). (2005). *Other People's Money: Debt Denomination and Financial Instability in Emerging Market Economies.* University of Chicago Press. ISBN: 0-226-19455-8.

Fritz, B., & Metzger, M. (2006). Monetary coordination involving developing countries: The need for a new conceptual framework. In *New Issues in Regional Monetary Coordination* (pp. 3–25). Palgrave Macmillan Books, Palgrave Macmillan.

Fritz, B & Mühlich, L. (2010). South-South Monetary Integration: The Case for a Research Framework beyond the Theory of Optimum Currency Area, *International Journal of Public Policy,* 6(12):118–135.

Gaye, A. (2008). China's policy in Africa: an overview. In: Jobarteh, N., & Drammeh, O. *China in Africa: Reflection on China's growing economic influence in Africa.* A Publication of the African Center for Information and Development (ACID).

Gu, J., & Carey, R. (2019). China's development finance and African infrastructure development. *Oxford University Press.* https://doi.org/10.1093/oso/9780198830504.003.0008

International Trade Centre (ITC). (2022). Database (2022).

Lee, M. (2006). The 21st Century Scramble for Africa. *Journal of Contemporary African Studies,* 24(3), 303.

Lyakurwa, W. (2008). *American and Chinese activities in Africa—and African priorities for the future.* BRENTHURST DISCUSSION PAPER 6/2008, The Brenthurst Foundation.

Marks, S. (2007). Introduction. In: F. Manji, (Ed), *African Perspectives on China in Africa.* Cape Town: Fahamu - Networks of Social Justice.

McKinsey (2020). *Solving Africa's infrastructure paradox,* McKinsey & Company, https://www.mckinsey.com/industries/capital-projects-and-infrastructure/our-insights/solving africas-infrastructure-paradox.

Naidu, S., & Davies, M. (2006). China fuels its future with Africa's riches. *South African Journal of International Affairs, 13*(2), 69–83. https://doi.org/10.1080/10220460609556803

OECD/ACET (2020). *Quality infrastructure in 21st Century Africa: Prioritising, accelerating and scaling up in the context of Pida (2021–30).*

Oqubay, A., & Lin, J. Y. (2019). The future of China–Africa economic ties: New trajectory and possibilities. In A. Oqubay & J. Y. Lin (Eds.), *China–Africa and an economic transformation.* Oxford University Press.

Priewe, J. (2006). Economic divergence in the Euro area: why we should be concerned. In: E.Hein, J.Priewe, A.Truger (eds.): European integration in crisis. Marburg: Metropolis-Verl., ISBN978-3-89518-610-3. 2007, (pp. 103-130)

Regissahui, M. (2019). Overview on the China–Africa trade relationship. *Open Journal of Social Sciences, 7,* 381–403. https://doi.org/10.4236/jss.2019.77032

Rodrigues, D. G. A. (2008). *China's economic involvement in Mozambique and prospects for development-an analysis of the processes and impacts of major recent investments.* Unpublished MA Dissertation, University of Stellenbosch, South Africa.

Sautman, B. and Hairong, Y. (2012). The Chinese are the Worst? Human Rights and Labor Practices in Zambian Mining, *Maryland Series in Contemporary Asian Studies, 1*(3), Available at: https://digitalcommons.law.umaryland.edu/mscas/vol2012/iss3/1

Snow, P. (1988). *The Star Raft: China's encounter with Africa.* Weidenfeld and Nicolson.

Sun, Y. (2021). *FOCAC 2021: China's retrenchment from Africa? Commentary,* Global Economy and Development, Foreign Policy, John L. Thornton China Center, Africa Growth Initiative

Tralac (2009). The *world trade organization an African perspective, more than a decade later.* The Trade Law Center for Southern Africa (TRALAC). Available at: http: www.tralac.org

Wang, F. L., & Elliot, E. A. (2014). China in Africa: presence, perceptions and prospects. *Journal of Contemporary China, 23*(90), 1012–1032. https://doi.org/10.1080/10670564.2014.898888

World Bank (2000). *Can Africa Claim 21st Century?* Washington, D.C.

World Bank. (2004). *Patterns of Africa–Asia Trade and Investment: Potential for Ownership and Partnership.* Washington, D.C.: Africa Region, Private Sector Unit.

World Bank (2008). *Building bridges: China's growing role as infrastructure financier for Sub-Saharan Africa.* Washington, D.C.

Xi, J. (2018). *Speech at the Opening of FOCAC Beijing Summit (FOCAC VII).* 3 September.

Yang, J., & Lee, J. (2019). Application of sensory descriptive analysis and consumer studies to investigate traditional and authentic foods: A review. *Foods, 8,* 54.

Chapter 31
The Impact of Chinese Foreign Direct Investment on the Productivity Growth in the Ethiopian Manufacturing Sector

Aragaw Mulu Muhaba

31.1 Introduction

Foreign direct investment (FDI) is defined as a future investment in which an investor establishes foreign business operations or acquires foreign assets including initiating ownership or controlling interest in a multinational firm. In recent years, policy-makers, especially in developing countries, have come to the assumption that foreign direct investment (FDI) is needed to boost the growth of their economy. It is claimed that FDI can create employment, increase technological development in the host country and improve the economic condition of the country in general. Investment is treated as one of the most important factors determining economic growth in nearly all growth models (Solow, 1957; Romer, 1986; Lucas, 1988).

During the last 20 years, the recognition of the importance of foreign direct investment (FDI) within the developing policy and strategies of the emerging economy was growing. Due to this, a significant range of developing countries in transition became a lot of receptive to FDI and is exploring ways in which for increasing inflows (UNCTAD, 2009). The FDI inflows improve their economic growth, in particular through the development of domestic investment, work position creation, improvement in the balance of payments, the contribution in creating the direct added value by the foreign company production, facilitating knowledge and technology transfer, and the improvement of competitiveness of the local economy.

In the recent past, Ethiopia has attracted large projects and brand manufacturers in labor-intensive manufacturing in textile and garment, leather, and leather goods production. Hence, FDI capital flow to Ethiopia shows a lift up from 1.3 billion USD in 2013 to 3.6 billion USD in 2017 (UNCTAD, 2018). According to EIC, Ethiopia accounts for 18.5% of all jobs created through FDI in Africa. China's National Bureau

A. M. Muhaba (✉)
Faculty of Mechanical and Industrial Engineering, Bahir Dar Institute of Technology, Bahir Dar University, Bahir Dar P.O. Box 26, Ethiopia
e-mail: Aragaw.Mulu@bdu.edu.et

© The Author(s) 2024   577
M. Muchie et al. (eds.), *China-Africa Science, Technology and Innovation Collaboration*, https://doi.org/10.1007/978-981-97-4576-0_31

of Statistics (NBS) reports that the turnover on economic cooperation projects in Africa reached \$29 billion in 2011 compared to \$1.2 billion in 2000. So, China has become a major economic partner of sub-Saharan African countries.

FDI inflow from China to Ethiopia has increased considerably over the past period with more than 70% of it going into the manufacturing sector. Gebrehiwot et al. (2020) indicated that the top five sectors that captured the Chinese FDI between 1992 and 2018 are manufacturing, real estate, hotel, construction, agriculture, social service, and mining. The results indicate that of the total 5300 investments more than 1350 are from Chinese FDI to Ethiopia which accounts for more than 25% of the total FDI of Ethiopia between the periods 1992–2018.

In most developing countries, FDI is seen as a major source of getting the required funds for investments hence they offer incentives to encourage FDI (United Nations, 2005). In addition, FDI inflow produces externalities through technology transfer and spill-over effect, which have a long-run effect on the economy. Because of the need for a better understanding of FDI determinants, impacts, and implications, most policymakers are encouraging FDI research (Bijit, 2002).

Since the nature of FDI investments, their technological content, and the manner and conditions of the related technology transfer to domestic firms are not necessarily the same and have a trend that differs from one country to another; researchers should consider the impact of the origin of investment on the mechanism of technology transfer. In view of this, the current study attempts to integrate the System Generalized Method of Moments (SYS-GMM) estimator and STATA with a view to modeling more FDI impact indicators interaction for Total factor productivity (TFP) to fulfill the goal of productivity growth. Hence, the hypothesis set is that investment coming from Chinese and other countries does not impact in the same manner and degree TFP of Ethiopian Manufacturing firms.

31.2 Literature Review

In the empirical studies, the impact of FDI will clearly recognize through the economic growth of the sector of the host country. So, the authors study this growth, as a resource of productivity gain, a defined type of fund. The growth model simulates endogenous assets, in which technical progress is possessed, in terms of endogenous outputs from the evolution of human capital (Benhabib, 1994; Grossman, 1991; Lucas, 1988). The Model expresses the contribution of Human capital assets to the creation of new technological and organizational knowledge. Through the development of new knowledge by FDI, the economy maintains a good level of long-term growth, under conditions related to knowledge externalization.

The finding of empirical studies has categorized into two groups. The first category of empirical results considers FDI as a very important tool for the productivity of host countries, especially in less developed countries (LDCs). Thus, studies proved the existence of a productivity spillover effect from foreign firms to domestic firms. For example, Hoffman and Tan (1980) explore those foreign firms of Malaysia in the 1960s, which contributed up to 23% in investment and up to 17.7% in the Gross Domestic Product (GDP) increments. Similarly, the economic growth of Taiwan, China and Indonesia had made a positive impact on FDI (Chuang & Lin, 1999). Das (2007) indicated that Foreign direct investment (FDI) has been a vital element of China's reform and growth strategy, and foreign-invested enterprises have played a crucial role in China's growth and globalization endeavours. Likewise, Liu and Wang (2003) set out positive benefits on total factor productivity for the industrial sectors of China in 1995. Despite the positive externalities of FDI, there are controversies in the former findings.

The second category of empirical studies argues that FDI has no positive contribution to the productivity performance of domestic firms (Konings, 2001; Bruhn et al., 2014; Lemma & Kitaw, 2014). For researchers such as Kokko (1996), Aitken and Harrison (1999), the impact of FDI on the productivity of local firms is negative. It is confirmed by Djankov and Hoekman (2000) showing that a 10% increase in foreign investments led to a 1.7% decrease in the productivity of local firms.

In view of the importance of FDI studies for policymakers, it is believed that a study on the effect of FDI has attracted the attention of the scientific community in several aspects. However, with the gradual increment of FDI inflow in Ethiopia (among the top in sub-Saharan countries), the level of studies on the topic of the impact of FDI has been relatively low in literature, little has been done on TFP and efficiency change in the manufacturing sector.

In the case of Ethiopia, empirical research on FDI and TFP in the manufacturing sector is limited and not so conclusive. For example, Soderbom (2012); and Bigsten and Gebreeyesus (2007) analysed the CSA panel data on large and medium manufacturing industries to study issues such as performance, growth, and productivity of firms. Admit (1998) studied the technical progress of the Ethiopian manufacturing sector from 1976 to 1995 and found zero or negative TFP growth. He also found a variation in the trend of TFP growth across sectors. TFP increased in sectors such as tobacco, paper, plastic, and leather while it was stagnant or decreasing in other sectors. Similarly, Gebreeyesus (2008) investigated the productivity growth of the Ethiopian manufacturing sector using the annual CSA census of medium and large manufacturing industries. He found an annual average productivity growth of about 9.3% between 1996 and 2003, with the entry and exit of firms being the major source of productivity growth.

To sum up the earlier empirical studies of the Ethiopian manufacturing sector, the Authors focus on the main FDI impact determinant which is the absorptive capability to make the endogenous growth model. The other main determinant which is the origin of FDI was ignored in their studies. Due to the research gap in the literature, it is vital to perform the examination of the origin of FDI in the manufacturing sector of Ethiopia that highlights the predominance of those coming from the Chinese and other countries, particularly the US, UK, Spain, etc.

31.3 Materials and Methods

31.3.1 Indicators

In the analysis of the FDI's origin, the Ethiopian manufacturing sector considers a large number of Chinese investments in the form of equity participation in the sector. Chinese investments represent an annual average of 70%, respectively, out of the foreign investments in the sector over the period 2011–2016. Other weight indicators such as value-added, production, exports, body investment flows, turnover, and employees related to firms of China and other countries show their role in the Ethiopian industrial firms. Malmquist Total Factor Productivity indices and the level of the average wage are also considered indicators that characterize the sector performance of Ethiopian, Chinese, and other countries' firms.

31.3.2 Data and Methods

Data relied on the annual survey of low, large, and medium-scale manufacturing industries conducted by the Ethiopian Central Statistical Agency from 2011 to 2016. CSA has used a Stratified sampling method in each region. The nature of the data set was unbalanced panel data that covers 260 domestic and 52 foreign firms under 24 manufacturing industries with a total number of 1511 domestic and 213 foreign observations at the firm level. By aggregating the firms under the 24 industries in a balanced panel, a data set with 144 observations at the industry level was also used to conduct FDI spillover effect analysis by assuming that spillover to firms categorized under the same industry is the same. The data was cleaned, coded, and entered in the STATA version 14.

For the analysis of the data, both descriptive and inferential methods have been used. From the descriptive statistics, the mean, standard deviation, minimum, and maximum of the variables are used. The FDI effect estimation on TFP growth is usually biased due to endogeneity problems. Heckman's (1979) selection bias correction model is also used to test whether selection bias is present or not. The main biases that could affect our results are the simultaneity bias and the problem of heterogeneity of the estimated coefficients. The simultaneity bias results from a possibility of reverse causality effects (Impact of TFP on the explanatory variables) presence. As for the problem of individual heterogeneity of the estimated coefficients, it comes from the fact that the effect of foreign participation differs from one industrial sector to another. Since the data are a dynamic panel set, Generalized Instrumental Variables Estimation (GMM) has been employed to control the endogeneity problem and provides better estimation results in efficiency and robustness through the manipulation of endogenous variables by their respective differences and delays. The validity of the selected material can be confirmed or rejected through the Hansen test and

Sargan test as well as the autocorrelation tests proposed by Arellano and Bond AR (1) and AR (2).

The estimation was facilitated by using STATA version 14, specifically, using STATA's XTABOND2 user-written command by Rodman (2003) for estimation.

31.3.3 Econometric Model

Moreover, the choice of econometric model variables is based: (i) firstly, on the theoretical and empirical arguments justifying the relationship between FDI, human capital, trade openness, and TFP and, (ii) secondly, on the availability of data per branch in the case of the Ethiopian industry. Then Ethiopian manufacturing firms' TFP should be determined before estimating the econometric model. The total factor productivity (TFP) of the Ethiopian manufacturing sector represents the dependent variable. Its analysis is essential to assess the sector's performance in technology. Value addition of industry and firms is calculated using the following method:

$$y_{it} = F(L_{it}, K_{it}) = A_{it}L_{it}^{\alpha}LK_{it}^{1-\alpha} \tag{31.1}$$

where y_{it} is the value-added of the manufacturing industry a function of two inputs capital and labour; represents the level of productivity, which is assumed to vary across firms within each sector i and across time t; L and K are Labor and physical capital, respectively. The coefficients on the growth of labour and capital are simply their share in value-added.

Here the production model is adjusted to TFP using the log-linearization, so TFP is measured as follows:

$$\log Y_{it} = \log(TFP_{it}) + \alpha \log(K_{it}) + (1 - \alpha) \log(L_{it}) \tag{31.2}$$

$$\log(TFP_{it}) = \log Y_{it} - \alpha \log(K_{it}) - (1 - \alpha) \log(L_{it}) \tag{31.3}$$

Therefore, the relationship between FDI and TFP of the sector i in a given date t and explanation factors are described as follows:

$$TFP_{it} = \alpha_1 + \beta_1 FDIC_{it} + \beta_2 FDIOC_{it} + \beta_3 H_{it} + \beta_4 TO_{it} + \beta_5 TG_{it} + \beta_6 TFPFF_{it} + \varepsilon_{it} \tag{31.4}$$

where: - *TFP* is the total factor productivity of Ethiopian firms (independent variable). *FDI* (Foreign direct investment): Share of equity capital held by Chinese and other world countries' firms in the industrial sector of domestic firms within each sector i and across time t. FDIC (FDI Chinese); FDIOC (FDI Other Countries). *H* (human capital): in order to evaluate skilled Labour in the Ethiopian manufacturing industry, its proxy factor is the measure of the average of paid salaries gap. *TO* (Trade openness): Total exports of firm divided by total added value, X (exports), AV (added value).

$$TO_{it} = \frac{X_{it}}{AV_{it}} \times 100; \tag{31.5}$$

where X is export and AV is added value. *TG* (Technological Gap): the technological gap as defined by Wang and Blömstrom (1992) i.e., the ratio of TFP between the foreign firms and their Ethiopian counterparts. It is applied as a proxy of absorptive capability. *TFPFF* (total factor productivity of foreign firms): This variable is introduced to test the impact of competition generated by the presence of foreign firms in the total productivity of domestic firms. *FDI*H* (interactive variable of FDI and human capital): like Borensztein et al. (1998), the variable used to highlight their effect on the TFP growth.

31.4   Results and Discussion

The analysis of the FDI's origin in the Ethiopian manufacturing sector shows a high proportion of Chinese and other countries' investments in the form of equity participation in the sector. These investments represent an annual average of 70.44% and 41.94%, respectively, out of the foreign investments in the sector over the period 1992–2016. Of all foreign investment in Ethiopia, 49% is in the manufacturing sector, followed by investment in real estate and hotels (27%). Agriculture accounts for 11% of total foreign direct investment in Ethiopia. A better understanding of the sectorial distribution of FDI in Ethiopia can be obtained when we disaggregate the data between China and other countries. The vast majority of Chinese investments are engaged in the manufacturing sector in Ethiopia. At least 70% of Chinese enterprises invest in the sector.

Other weight indicators related to firms of these two countries show their role in the Ethiopian industrial fabric. Thus, the share of firms with Chinese and other countries funding in the value added is 32.12% and 19.12%, respectively. This share was 28% and 6% in production, 27% and 8% in exports, 33% and 7% in body investment flows, and 30% and 7% in turnover. Moreover, their firms employ successively 39% and 9% of the total workforce in the sector. The weight presence of these two countries in the industrial sector can have direct and indirect effects on the performance of local businesses.

As for other FDI impact indicators that characterize the sector performance of Ethiopian firms, Chinese and other countries, Malmquist Total Factor Productivity indices result is shown in Table 31.1.

The comparison of Chinese and other countries' affiliated firms' performance and domestic firms' one, highlights the significant differences in TFP, in particular over the sub-period 2011–2016. Thus, the Chinese and other countries-owned firms are successively 2.41 and 1.03 times more productive than domestic firms. This gap is the result of performance made by technological progress that is 2.56 and

Table 31.1 Performance comparison of Ethiopian firms in terms of TFP (2011–2016)

Chinese-affiliated/domestic firms			Other countries-affiliated/Domestic firms		
Technical efficiency change	Technological change	TFP	Technical efficiency change	Technological change	TFP
0.95	2.56	2.41	0.86	1.22	1.03

Source: Author computation from CSA survey Data

1.22 times better in Chinese and Other countries' firms, respectively, with respect to domestic ones. Over the whole period 2011–2016, Chinese affiliated firms are by far better than domestic firms in terms of annual average TFP (2.41 times) while these are at the same level of productivity as firms in other countries-funding (1.03 times). It is important to note that while the level of productivity is almost equivalent between the other countries and Ethiopian firms, there are significant differences in the contribution of technical efficiency and technological change in TFP. Thus, the contribution of technological progress to TFP is greater than 1.22 times in the case of other countries while the contribution of technical efficiency to TFP is 0.86 times more in the case of Ethiopia.

Tables 31.2 and 31.3 show the results of estimations using the GMM method to explain TFP by FDI by country of origin and by technology classification. Taking account of the origin of FDI, namely those from the Chinese and other countries, the TFP of the Ethiopian manufacturing sector is impacted in a different way. With respect to this, the impact of FDI coming from Chinese is positive and statistically significant However, the impact of FDI coming from other countries is negative and statistically significant. Therefore, the hypothesis that states that FDI can have different effects on TFP by country of origin is verified in our case study. These results can be interpreted as follows: in the case of other countries, their participations are concentrated, mostly in medium and high technology sectors where technological spillovers require an absorptive capability and high assimilation through qualified and skilled workers. Indeed, the other countries participating in medium and high technology industries exceeded 30% of overall foreign ownership in that sector.

However, Chinese participation does not exceed 10% in medium and high-technology industries. Rather, their presence is displayed in the low-technology branches where penetration rates sometimes reach over 50% as in the case of the woodworking industry and manufacturing of wooden artifacts textile industry and the furniture manufacturing industry, and various industries.

These findings mean that the transfer of technology through a foreign presence in the manufacturing sector applies only to low-technology or medium–low industries where the technology gap between foreign and local firms is negligible. Moreover, the impact of the technological gap (TG) on TFP, in our estimation, is negative and statistically significant, meeting the arguments behind the hypothesis that the technological gap between foreign and domestic firms reduces the technology transfer

Table 31.2 Impact of FDI on TFP growth of Ethiopian Manufacturing Industry per origin country: (2011–2016)

Independent variable	Generalized method moments		
	Regression number (1)	Regression number (2)	Regression number (3)
Constant	2.339239 (0.014)*	4.076809 (0.000)*	4.426724 (0.000)*
FDI Chinese	0.0001079 (0.059)*	0.1302012 (0.000)*	0.0022075 (0.000)*
FDI other countries	− 0.055531 (0.028)*	− 0.1302012 (0.628)***	− 0.0100533 (0.613)***
Human capital	0.1398812 (0.049)*		0.0342068 (0.019)*
Trade openness	0.108728 (0.011)*	0.1452835 (0.001)*	0.0837156 (0.005)*
FDI Chinese*Human capital		0.0171849 (0.534)***	
Total factor productivity of foreign firms			0.0059633 (0.000)*
Technological Gap			− 0.3424356 (0.000)*
Number of observations	484	484	484
Arellano–Bond test AR (1)	0.008	0.009	0.044
Arellano–Bond test AR (2)	0.157	0.376	0.068
Hansen test	0.13	0.208	0.488

Note: (*) significant at 5%; (**) significant at 10%; (***) no significant
Source: Author computation using Sys-GMM

Table 31.3 The impact (in %) of FDI on TFP growth in the Ethiopian manufacturing industry per origin country and using technology-based classification: panel of six years (2011–2016)

Independent variable	Low-technology manufacturing	Medium and high-technology manufacturing
Constant	1.752124 (0.000)*	1.78739 (0.002)*
FDI Chinese	0.0563169 (0.000)*	0.0877583 (0.000)*
FDI other countries	0.1243233 (0.005)*	− 0.1867934 (0.892)*
Human capital	0.14561 (0.000)*	0.2369586 (0.000)*
Trade openness	0.1061235 (0.001)*	0.1047378 (0.001)*
Number of observations	420	224
Arellano–Bond test AR (1)	0.034	0.018
Arellano–Bond test AR (2)	0.146	0.93
Hansen test	0.243	0.172

Note: (*) significant at 5%; (**) significant at 10%; (***) no significant

from the former to the latter firms, especially in a context characterized by a low level of skilled and required human capital (Haddad & Harison, 1993).

In addition, technological spillovers transferred through competition pressure (PTFE) brought by foreign firms, are proven to be statistically significant and positively affect the TFP of local businesses even with weak effects.

The human capital variable seems significant and has a positive effect on TFP. Therefore, human capital contributes to the improvement of productivity even if the current level of qualification of the Ethiopian workforce is not able to absorb and assimilate technologies from foreign firms including those other countries ones that are concentrated in medium and high technology industries. Moreover, the interaction between FDI coming from Chinese and other countries, considering the human capital variable was positive but not significant.

As for trade liberalization, it is statistically significant and positive, despite the trade deficit of Ethiopia with other countries, the Chinese, and other countries of the world. However, the import of equipment goods is the kind of investment to improve the TFP of the sector.

From Table 31.3, the impact of FDI on the TFP of Low-technology manufacturing is positive regardless of the country of origin (China and other countries). This confirms previous arguments arguing that industrial branches with low technology witness faster technology transfer since the Ethiopian firms have the absorptive capability and assimilation of such technologies.

On the other side, the effect of FDI on TFP in industrial branches with medium technology and high technology is positive in the case of other countries but negative in the case of the Chinese. This result illustrates that technology transfer can also be reached in medium and high-level technology industries per country of origin.

Regarding this point, it is appropriate to note that the level of productivity and the importance of control through the participation rate brought by foreign firms are likely to influence TFP differently among the branches with medium-technology and high technology. Thus, the comparison of performance in terms of TFP and its two components, namely technological change and technical efficiency change, between Chinese, other countries, and Ethiopian firms over the period 2011–2016, shows a significant gap, particularly in a technological change in. Consequently, Chinese firms are 2.56 times better than Ethiopian ones, considering the annual average, while the latter is quite similar to firms with other countries' participation (1.22 times). This differential in productivity, particularly in the case of the Chinese, generates vicious competition for domestic firms and negatively impacts productivity, especially in medium-technology and high-technology manufacturing where the absorptive capability of local firms is already low.

Similarly, the important participation rate of other countries' enterprises in medium-technology and high-technology branches reduces the diffusion of technological spillovers and limits the spreading and transmission of knowledge and managerial expertise to local firms. This is due, firstly, to the supremacy of their power of control and, secondly, to the differential in salaries which is 1.43 times higher than the salaries obtained in domestic firms.

Moreover, in the case of Chinese, where the power of control in medium and high technology sectors is low and the gap in terms of wages paid is negligible compared to domestic firms, the effect of their holdings on TFP is positive and is statistically significant.

Therefore, the robustness of the above results (Tables 31.2 and 31.3) seems to be confirmed. Thus, the Hansen test shows that the instruments used in the regressions are valid, as indicated by the (p-value) associated with this test that exceeds 10%. In addition, Arellano and Bond autocorrelation tests show that the second order autocorrelation hypothesis is rejected, as confirmed by the (p-value) related to test AR (2) which goes above 5% or 10%.

31.5   Conclusion

This study has evaluated the impacts of Chinese and other countries' FDI on TFP growth in the Ethiopian manufacturing sector. A Generalized Instrumental Variables Estimation is used in dynamic panel systems considering a set of 24 groups of the manufacturing sector in the period 2011–2016. A deterministic production frontier using linear programming technique is constructed to evaluate technical efficiency and measure TFP growth and decompose it into its components. Trans log production function with labour, capital inputs in the production function is used to determine the production frontier and time is included in the function to trace any movements in the production frontier through time. The analysis to compare the share of Total factor productivity has also been done to give further explanation about the growth of productivity.

Results of the study show that the consideration of the origin of FDI namely those from Chinese and other countries has a diverse way on TFP in this sector. This result is similar to the findings based on the frontier analysis. The frontier analysis has shown a negative and significant, particularly in the case of medium and high technology sectors, TFP growth in most groups in the period when the effect of the capital of Chinese origin is significant and positive what the technology classification restraint. Estimation results of the frontier model have also shown the existence of large inefficiencies in the manufacturing sector and the existence of capital using technical progress. The decomposition of TFP growth indicated that a large share of the growth comes mostly from technical change. Factors such as FDI, technological advances, and the import of technologies may have contributed largely to the progressive technical change. Therefore, the transfer of technology including that embodied in medium and high technology industries determined by the absorptive capability of the host country, namely Ethiopia. Thus, it is necessary to manage human skills in science and management and to provoke research, innovation, and development at the state level and at the enterprise level to progress the transfer capability of technology to local firms. With regard to the determinants, the difference in pay between other countries and local firms is likely to prevent the movement of skilled human skills and therefore limit the transfer of knowledge and managerial know-how. Finally, the differential in terms of productive performance (due to competition) between other countries' firms and their local counterparts acts negatively on the productivity of the domestic firms. With regard to the other components, it is found that the effect of scale efficiency and technical efficiency are negative or slightly positive. Constant returns to scale and/or decreasing returns to scale and technical inefficiencies make the two components have a limited role in improving TFP.

References

Aitken, B., & Harrison, A. (1999). Do Domestic firms benefit from direct foreign investment? Evidence from Venezuela. *The American Economic Review, 89*, 605–618.

Benhabib, J. S. (1994). The role of human capital in economic development. *Journal of Monetary Economics, 34*, 143–173.

Bijit, B. (2002). *Foreign direct investment research issues*. Rutledge.

Bigsten, A., & Gebreeyesus, M. (2007). The small, the young, and the productive: determinants of manufacturing firm growth in Ethiopia. *Economic Development and Cultural Change, 55*, 813–840.

Borensztein, E., de Gregorio, J., & Lee, J. W. (1998). How does foreign direct investment affect economic growth. *Journal of International Economics, 45*, 115–135.

Bruhn, N. C. P., & Calegario, C. L. L. (2014). Productivity spillovers from foreign direct investment in the Brazilian processing industry. *Brazilian Administration Review, 11*, 22–46.

Chuang, Y., & Lin, C. (1999). Foreign direct investment, R&D and spillover efficiency: Evidence from Taiwan's manufacturing firms. *Journal of Development Studies, 35*, 117–137.

Das, D. K. (2007). Foreign direct investment in china: Its impact on the neighbouring asian economies. *Asian Business & Management, 6*, 285–301.

Djankov, S., & Hoekman, B. (2000). Foreign investment and productivity growth in Czech enterprises. *World Bank Economic Review, 14*, 49–64.

Gebrehiwot, B. A. (2020). *Chinese investment in Ethiopia: Contribution, challenges, opportunities, and policy recommendations*. Policy Studies Institute. https://psi.gov.et/research-reports-2/fil e/124. Accessed: 15 Feb 2023.

Gebreeyesus, M. (2008). Firm turnover and productivity differentials in the ethiopian manufacturing. *Journal of Productivity Analysis, 29*, 113–129.

Grossman, G. H. (1991). *Innovation and growth in the global economy*. MIT Press.

Haddad, M., & Harrison, A. (1993). Are there positive spillovers from direct foreign investment? *Journal of Development Economics, 42*(1), 51–74. https://doi.org/10.1016/0304-3878(93)900 72-U

Heckman, J. (1979). Sample selection bias as a specification error. *Econometrical, 47*, 153–618.

Hoffmann, L., & EE, T. S. (1980). *Industrial growth, employment and foreign investment in Malaysia*. Oxford University Press.

Kokko, A. (1996). Productivity spillovers from competition between local firms and foreign affiliates. *Journal of International Development, 8*, 517–530.

Konings, J. (2001). The effects of FDI on domestic firms: Evidence from firm-level panel data in emerging economies. *Economics of Transition, 9*, 619–633.

Lemma, Y., & Kitaw, D. (2014). The impact of foreign direct investment on technology transfer in the Ethiopian metal and engineering industries. *International Journal of Scientific & Technology, 3*, 242–249.

Liu, X., & Wang, C. (2003). Does foreign direct investment facilitate technological progress? Evidence from Chinese industries. *Research Policy, 32*, 945–953.

Lucas, R. (1988). On the mechanics of economic development. *Journal of Monetary, 22*, 342–367.

Romer, P. (1986). Increasing returns and long-run growth. *Journal of Political Economy, 94*, 1002–1037.

Roodman, D. (2003). *XTABOND2 STATA module for estimating dynamic panel data models.* Center for Global Development.

Soderbom, M. (2012). Firm size and structural change: A case study of Ethiopia. *Journal of African Economies, 21*, 126–151.

Solow, R. (1957). Technical change and the aggregate production function. *Review of Economics and Statistics, 39*, 312–320.

UNCTAD. (2005). *World investment report: Transnational Corporations and the Internationalization of R&D.* New York and Geneva: United Nations.

UNCTAD. (2009). *World investment report: Transnational corporation and infrastructure challenges.* New York and Geneva: United Nations.

UNCTAD. (2018). *World investment report: Investment and new industrial policy.* New York and Geneva: United Nations.

Wang, J., & Blomstrom, M. (1992). Foreign investment and technology transfer. *European Economic Review, 36*, 137–155.

Chapter 32
Post Scriptum

Mammo Muchie, Angathevar Baskaran, and Mingfeng Tang

The idea behind this book on China–Africa Science, Technology and Innovation Collaboration came out of the discussion from the Peoples Republic of China's Embassy in South Africa with global scholars that have been working for a long time on how the China–Africa relationship can create a mutually beneficial and sustainable development for all to gain and no one to lose with a solid bedrock to make all generations to achieve full success in all spheres of life by learning from the inspiration of the past to make the present and future secure, stable, peaceful by encircling the universe with very innovative and significant rich knowledge imagination.

The development experience of China can provide a rich lesson for Africa and the rest of the world. The development pathway of China to transform and reform the social, economic, health, education and science, technology and innovation domains can generate very tangible lessons to Africa and all the developing world. This edited volume of the book has focused on the on-going science, technology and innovation collaboration part between China and Africa. This is done to discover in a specific domain the existing China–Africa relationship by exploring and undertaking evidence and data analytic knowledge based on specific selected cases. What is remarkable is that the tangible knowledge generated by all the contributors demonstrates that the China–Africa relations on the science, technology and innovation area

M. Muchie (✉)
DSI/NRF SARChI Research Chair On Science, Technology and Innovation Studies, Tshwane University of Technology, Pretoria, South Africa
e-mail: muchiem@tut.ac.za

A. Baskaran
Department of Political Science, Public Administration and Development Studies, Faculty of Business and Economics, UM North-South Research Centre (UMNSRC), University of Malaya, Malaysia & Senior Research Associate, SARChI (Science, Technology and Innovation Studies), Tshwane University of Technology, Pretoria, South Africa

M. Tang
Sino-French Innovation Research Center (SFIRC), Faculty of Business Administration, Southwestern University of Finance and Economics, Chengdu, China

© The Author(s) 2024
M. Muchie et al. (eds.), *China-Africa Science, Technology and Innovation Collaboration*, https://doi.org/10.1007/978-981-97-4576-0_32

is mutually beneficial without any loss by one and gain by another. The many different chapters clearly demonstrate the China–Africa science, technology and innovation relation is going in the right direction.

The research has covered comprehensively the following areas, and the output is very inspiring: Evolution of China–Africa collaborations in science, technology and innovation, China–Africa collaboration in higher education, China–Africa research collaboration and training, China–Africa collaboration in agriculture and food security, China–Africa collaboration in environmental management and climate change, China–Africa collaboration in telecommunications, China–Africa collaboration in digital technologies, China–Africa collaboration in renewable energy, China–Africa collaboration in space technology applications, China–Africa collaboration in manufacturing, and China–Africa collaboration in health sector. The research output demonstrates productive and hard collaborative work in healthcare, poverty reduction and agriculture, trade promotion, investment, digital innovation, green development, capacity building, people-to-people exchanges, and peace and security between China and African states.

As China is said to have abolished absolute poverty in 2021 and achieved the SDGs no. 1 zero poverty, there is a lot Africa can learn from the real success of China. How can the on-gong science, technology and innovation collaboration between China and Africa promote what China is achieving from the 17 SDG Goals before the deadline of 2030? How can Africa draw lessons to benefit from the remarkable achievement by China? Africa can draw lessons from the Chinese sustainable development efforts to fulfil all the 17 SDG goals by using the science, technology and innovation collaboration without excluding all other spheres. The Chinese and African current development pathways must be critically evaluated to discover the appropriate inclusive ways to eradicate poverty, unemployment, conflicts, and inequality. The China–Africa mutually agreed partnership can generate the appropriate action and policy to achieve sustainable development for both China and Africa.

The research done by scholars from all parts of the world presented in the numerous chapters explain effectively how significant it is to employ clear indicators to measure success and failure and to find the necessary ways to achieve on time what is pursued by different African states by drawing lessons from the experience of China. China has explicitly articulated how to build a China–African Partnership to promote sustainable development and build a China–Africa Community with a Shared Future. The actions from China are numerous: There is the Dakar Action Plan that has produced a comprehensive China–Africa cooperation in all spheres: the medical and health, the poverty reduction and agricultural development, the trade promotion, the investment promotion, the digital innovation, the green development, the capacity building, the cultural and people-to-people exchange, and peace and security.[1]

The Science, Technology and innovation cooperation between China and Africa are numerous. Some of the examples covered in this edited book are the Belt and Road Science, Technology and Innovation Cooperation Action Plan and the China–Africa

---

[1] Source: https://www.focac.org/eng/zywx_1/zywj/202201/t20220124_10632444.htm.

Science and Technology Partnership Program. They are currently promoting cooperation on technology-supported poverty reduction, and fully leverage scientific and technological innovation to guide the sustainable economic and social development of Africa. The China National Space Administration is working with the African Union on the China–Africa Space Cooperation to use space technology to enhance cooperation in areas such as scientific and technological development, poverty and hunger elimination, promotion of social security, disaster prevention and relief, climate change, and eco-environmental protection. There is also remarkable focus on the youth with China working to make the African youth as change makers and game changers. There are several programs such as the "International Outstanding Young Scientists Exchange Program", the "Innovative Talent Exchange Project" and the "Africa Young Scientists in China Program" to provide young Africans training in applicable technology and science management. There is also the CubeSat Middle School Student Science Project to provide space exchanges between Chinese and African secondary school students.

Africa will benefit also by the China founded Global Initiative on Data Security to build communities with a shared future in cyber security. There are in Africa a number of the Confucius Institutes teaching Chinese language; and African languages are also being taught in China. There are also many research centres such as the Sino-Africa Joint Research Centre that are helping to build STI capabilities across Africa. There is also the Africa Centres for Disease Control and Prevention built as a Continent-wide Public Health Agency to safeguard Africa's health.

The book has done extraordinary contribution on the China–Africa relation by focusing principally on the exploration of the science, technology and innovation collaboration. The research demonstrates that a truly exemplary model relation is evolving between China and Africa that can generate win-win benefits based on highly valued principles. This demonstrates how all nations in the world must unite to create a sustainable "Great Green Wall" development in all spheres of life to remove the self-interest driven competition and promote shared pursuit to save humanity and the world with humane justice, moral intelligence and solidarity.

Index

© The Editor(s) (if applicable) and The Author(s) 2024

M. Muchie et al. (eds.), *China-Africa Science, Technology and Innovation Collaboration*, https://doi.org/10.1007/978-981-97-4576-0